Nanotechnology and Microelectronics:
Global Diffusion, Economics and Policy

Ndubuisi Ekekwe
Johns Hopkins University, USA

INFORMATION SCIENCE REFERENCE

Hershey · New York

Director of Editorial Content:	Kristin Klinger
Director of Book Publications:	Julia Mosemann
Acquisitions Editor:	Lindsay Johnston
Development Editor:	David DeRicco
Publishing Assistant:	Tom Foley
Typesetter:	Deanna Jo Zombro
Production Editor:	Jamie Snavely
Cover Design:	Lisa Tosheff
Printed at:	Yurchak Printing Inc.

Published in the United States of America by
Information Science Reference (an imprint of IGI Global)
701 E. Chocolate Avenue
Hershey PA 17033
Tel: 717-533-8845
Fax: 717-533-8661
E-mail: cust@igi-global.com
Web site: http://www.igi-global.com

Library of Congress Cataloging-in-Publication Data

Nanotechnology and microelectronics : global diffusion, economics and policy / Ndubuisi Ekekwe, editor.
    p. cm.
  Includes bibliographical references and index.
  Summary: "This book assesses the state of nanotechnology and microelectronics, and examines many issues, such as climate change, trade, innovation, diffusion, etc, with a theme focused on facilitating the structures for the adoption and penetration of the technologies into developing nations"--Provided by publisher.
  ISBN 978-1-61692-006-7 (hbk.) -- ISBN 978-1-61692-007-4 (ebook) 1.
Nanotechnology. 2. Microelectronics. 3. Technological innovations--Economic aspects--Developing countries. 4. Technology transfer--Developing countries. 5. High technology industries--Developing countries. I. Ekekwe, Ndubuisi, 1973-
  T174.7.N37325 2010
  620'.5--dc22
                                   2010010146

British Cataloguing in Publication Data
A Cataloguing in Publication record for this book is available from the British Library.

All work contributed to this book is new, previously-unpublished material. The views expressed in this book are those of the authors, but not necessarily of the publisher.

Table of Contents

Section 1
Foundations and Science

Section 2
Technology Transfer and Innovation

Section 5
Lessons from Agricultural Technology

Section 6
Regional Developments

Detailed Table of Contents

Section 1
Foundations and Science

This section discusses the general theme of knowledge economy, innovation, technology transfer and diffusion. It offers insights on the progress which has been made in the fields of nanotechnology and microelectronics over the years and discusses the broad implications of safety, intellectual property rights, and funding- both government and private. It also examines the trends, the transfer trajectories and the penetration of the technologies across the globe. Besides, it explains the science and technology, molecular manufacturing, upon which the future of commercial nanotechnology is largely hinged upon.

Chapter 1

Ndubuisi Ekekwe, Johns Hopkins University, USA & African Institution of Technology, USA

For many centuries, the gross world product was flat. But as technology penetrated many economies, over time, the world economy has expanded. Technology will continue to shape the future of commerce, industry and culture with likes of nanotechnology and microelectronics directly or indirectly playing major roles in redesigning the global economic structures. These technologies will drive other industries and will be central to a new international economy where technology capability will determine national competitiveness. Technology-intensive firms will emerge and new innovations will evolve a new dawn in wealth creation. Nations that create or adopt and then diffuse these technologies will profit. Those that fail to use technology as a means to compete internationally will find it difficult to progress economically. This chapter provides insights on global technology diffusion, the drivers and impacts with specific focus on nanotechnology and microelectronics. It also discusses the science of these technologies along with the trends, realities and possibilities, and the barriers which must be overcome for higher global penetration rates.

Chapter 2
Chris Phoenix, Center for Responsible Nanotechnology, USA

A wide variety of nanotechnology programs, both pedagogical and research-oriented, can incorporate some aspect of molecular manufacturing. Nanotechnology is developing from large tools that give us access to the nanoscale, to tools constructed at or near the nanoscale. To date, these nanoscale tools are not capable of accomplishing much of commercial interest; however, this will be changing with increasing rapidity over the next decade or two. Eventually, nanoscale tools will be capable of broad classes of nanoscale construction; this will enable the fabrication of increasingly complex and useful structures. Fabrication of structures with molecular precision is especially relevant. When nanoscale tools are developed to the point of being capable of building duplicate tools, a manufacturing revolution may occur; even before that point, there are both scientific and likely commercial benefits to developing capabilities in this area.

Section 2
Technology Transfer and Innovation

The fields of nanotechnology and microelectronics have evolved out of many years of scientific and technological innovations. In this section, technology innovation, transfer and diffusion are discussed within the contexts of general technology and specificity of nanotechnology. The section also examines the factors that accelerate open innovation, technology creation and why in most cases, some nations exhibit fatigue in technology adoption, lacking resilience in the diffusion process. Kondratieff Cycle, technology trends and knowledge dissemination instruments are discussed.

Chapter 3
Nazrul Islam, Cardiff University, UK

This chapter explores trends in nanotechnology knowledge creation across scientific disciplines and technology domains, and helps to understand the dissemination of nanotechnology knowledge. In relation to intense global competition in nanotechnology, this study exhibits a forward-looking approach in characterizing nanotechnology research and development trajectories. This research adopts hybrid research methodology, including both quantitative and qualitative methods. The findings imply that nanotechnology knowledge creation and dissemination trends have appeared to bridge divergent disciplines, emphasizing the importance of collaborative research networks among scientists to co-create, share and disseminate nano-knowledge across groups, institutions and borders.

Chapter 4
Annamária Inzelt, Financial Research Ltd, Hungary

Although the impact of open innovation on a global scale on the collaboration between universities and foreign industry is clearly important, empirical evidence from the field is lacking. This paper investigates the collaboration between Hungarian universities and foreign companies in research and development. The paper attempts to provide a relevant picture of the research-related linkages of Hungarian universities and foreign companies by employing secondary data processed from various data-banks. The analysis suggests that foreign direct investment and foreign companies play major roles in the internationalisation of research during this second decade of the transition process.

Chapter 5

Nanotechnology is currently seen as a paradigm shift towards scientific revolution or 'nano revolution'. This chapter discusses the nano revolution within the global context. It is interesting to see that the governments around the world have formulated policies to manage the research and development (R&D) efforts and exploit the potential of nanotechnology to increase industry's ability in the global economy. The chapter analyses the successive waves of technological change based on Kuhn's model of scientific change and Schumpeter's model of Kondratieff cycles. As nanotechnology would have significant impacts on virtually every commercial sector, many countries commit to foster nanotechnology developments. This chapter will focus on nanotechnology framework policy recommendations. The policies and research activities of the most preeminent nations discussed in this chapter represent global research trend towards nano revolution in the next decades.

Economic zones are facing difficulties to control their endogenous process of technology creation. Technology poverty does exist. How to ensure that fragile economic zones become more resilient in capturing technology? How to achieve a competitive advantage in non-agile economies and get a larger category of their populations benefit from it? These challenges have taken all serious developmentalists by storm. Unfortunately, the so-called 'technology transfer' does not yield to the outcomes it promotes resulting to mitigated results in non-agile countries. This paper argues that in an unpredictable and changing environment, technology diffusion is part of a complex change process which needs to be dealt with in a comprehensive manner. It cannot take place without revisiting the capability formation dilemma in non-agile economies. The technology-driven capability should therefore be supported by strategic policies and public private institutions. Technology diffusion may appear as the systemic link between shared wealth creation and collective efficiency.

Chapter 7

This paper examines the adopter fatigue phenomenon in the diffusion of nanotechnology and microelectronics innovations. It is hypothesized that innovations spread through a social system in an s-curve and that the speed of technology adoption is determined by two variables p and q where p represents the speed at which adoption takes off and q the speed at which later growth occurs. However, this two-variable model has been criticized as an over-simplification of a complex reality hence the need to examine adopter fatigue phenomenon defined as the hesitation, delay, or refusal by an individual to adopt an innovation on account of prevailing circumstances including the rapid evolution of new technologies. This phenomenon is particularly relevant to nanotechnology and microelectronics products and processes which are characteristically continuously refined and upgraded.

Section 3
Industry, Policy and Experiences

In this section, industry, policy and case studies for both nanotechnology and microelectronics, and technology in general are discussed. It looks at how nanotechnology and microelectronics are shaping and transforming industries and how policies continue to influence their penetrations across the globe. The issues surrounding technology licensing and R&D, barriers stalling the global diffusion of emerging technology to developing nations along with the mechanism for assessing maturity and performance of both nanotechnology and microelectronics are examined. Case studies on national policies as they pertain to developing knowledge economies and a study evaluating university-industry network on nanotechnology are also presented.

Chapter 8

Nanotechnology is becoming a transformative element for the manufacturing sector into the knowledge economy. Access to relevant knowledge is a critical factor in this transformation as manufacturing firms cluster in peripheral suburbs away from the knowledge intensive ring of central business districts. Results from a project conducted in South-West Sydney shows that informal university-industry networks raise the awareness of firms to the potential of nanotechnology applications, their willingness to invest in nanotechnology R&D and the number of university-industry cooperation initiatives and business-to-business partnerships. Results from the project also suggest that, despite the importance for firms of being involved in global networks, access to local knowledge and local networks is significant for the innovation process of small and medium enterprises (SMEs).

Chapter 9

The literature on cooperative R&D did not pay much attention to knowledge sharing ex-post innovation through technology licensing, which is a common phenomenon in many industries. We show how licensing ex-post R&D affects the incentive for cooperative R&D and social welfare by affecting R&D investment and the probability of success in R&D. Licensing increases both the possibility of non-cooperative R&D and social welfare.

Chapter 10

Neslihan Aydoğan-Duda, İzmir University of Economics, Turkey
İrge Şener, Çankaya University, Turkey

Nanotechnology is the science that focuses on the control of matter at the atomic scale. It has the potential to create many new materials and devices with wide-ranging applications, such as in medicine, electronics and energy production. There are many entry barriers which can affect nanotechnology penetration in developing and emerging nations. This chapter discusses such barriers for Turkey. Despite about 10 universities having nanotechnology programs, the number of nanotechnology firms in the country is still low. Using combinations of interviews, surveys and literature, these issues that continue to stall the commercialization of discoveries in Turkey are examined.

Chapter 11

Nazrul Islam, Cardiff University, UK

This chapter aims to provide a new readiness matrix called 'innovative manufacturing readiness levels (IMRLs)' to evaluate and assess the areas of micro and nanotechnology maturity including their performance. The study employs a case study approach through which the practicability and applicability of the IMRLs conceptual matrix were verified and confirmed. A case study with laser-based manufacturing technologies explores the stages of micro and nano technologies (MNTs)' maturity, including the key issues and performances that contributed to the development of a new assessment tool. Concerning intense global R&D competition in MNTs, this study exhibits a forward-looking approach in assessing MNTs maturity and performance. A generic conclusion is reached by which product designers and technology managers position themselves and take into account risk reduction exercises related to MNTs. The novelty of the research could be that organizations, which develop and use MNTs, have an opportunity in applying such a specific assessment matrix to quantify the technology readiness of unreleased MNTs.

Section 4
Ethics, Regulation and Environment

The fields of nanotechnology and microelectronics are involved in broad ethical complications owing to their diversity and relative infancy. With these burdens come the need for regulations towards ensuring safety in their products, processes and tools as well as protecting the environment and human. In this section, ethics, regulation and the climate control are examined. The climate economics and finance, the intellectual property rights challenges of microelectronics and nanotechnology along with information and communication technologies are discussed.

Chapter 12

Shaikh M. Rahman, Texas Tech University, USA
Ariel Dinar, University of California, Riverside, USA
Donald F. Larson, World Bank, USA

The Clean Development Mechanism (CDM) of the Kyoto Protocol is an innovation that combines greenhouse gas abatement targets with sustainable development objectives. This chapter provides an estimate of the overall growth pattern of the CDM and makes projections about CDM activity during and beyond the first commitment period of the Kyoto Protocol commitments under current rules. The results imply that if the emission reduction targets remain unchanged beyond the first commitment period, further expansion of the CDM pipeline is unlikely.

Chapter 13

Ahmed Driouchi, Al Akhawayn University, Morocco
Molk Kadiri, Al Akhawayn University, Morocco

Information and communication technologies, nanotechnologies and microelectronics are progressively challenging the current state of intellectual property rights. This is related to the economic features underlying these technologies. The directions of changes in intellectual property rights are found to require further coping with the overall chain of innovation and with the uncertainty that can be embedded in the new trends of technological development.

Chapter 14

Adrian Muller, University of Zürich, Switzerland

In this chapter, we argue that the countries in the Global South can gain from stringent own climate policies. This is so, as in the current situation, the south tends to be dominated by the climate policies of northern countries and climate finance largely supports single projects and technology transfer that are not embedded in a broader policy framework in southern countries. Adopting own stringent policies could help to counteract this and to channel these financial means to their most beneficial use. This could help southern countries to follow an agenda that is different from the fossil fuel based development path of the north. Such a "green new deal" could be a promising economic and technological development strategy. Stringent climate policies would strengthen the southern countries in the international climate negotiations and southern countries could take the lead in the climate change mitigation debate.

Chapter 15

Nicola Cantore, Overseas Development Institute, UK and Università Cattolica del Sacro Cuore, Italy
Emilio Padilla, Univ. Autónoma de Barcelona, Spain

An abundant scientific literature about climate change economics points out that the future participation of developing countries in international environmental policies will depend on their amount of pay offs inside and outside specific agreements. Also, scholars recently outline that a perceived fairness in the distribution of emissions would facilitate a wide spread participation in international agreements. In this paper we overview the literature about distributional aspects of emissions by focusing on those contributions investigating past trends of emissions distribution through empirical data and future trajectories through simulations obtained by integrated assessment models.

Chapter 16

Chi Anyansi- Archibong, North Carolina A&T State University, USA
Silvanus J. Udoka, North Carolina A&T State University, USA

Nanotechnology is science at the size of individual atoms and molecules. At that size scale, materials have different chemical and physical properties than those of the same materials in bulk. Research has shown that nanotechnology offers opportunities to create revolutionary advances in product development. It also has the potential to improve assessment, management, and prevention of environmental risks. There are however, unanswered questions about the impacts of nanomaterials and nanoproducts on human health and the environment. This chapter describes state-of the-science review, exposure assessment and mitigation, and potential macro ethical issues that must be considered to mitigate risk implications of emerging technologies such as nanotechnology.

Section 5
Lessons from Agricultural Technology

This section uses agriculture and agricultural technology to help to understand technology adoption, diffusion and penetration trajectory as they pertain to developing nations. Agriculture being one of the most common industries offers lots of insights on the struggles many developing nations have faced in adopting new technologies. Where these nations have failed and succeeded in agricultural technologies could provide indicators on how well they can do on nanotechnology and microelectronics adoption. The lessons could provide the needed directions in developing plans for global transfer of emerging technologies like nanotechnology and microelectronics.

Chapter 17

Alejandro Nin-Pratt, International Food Policy Research Institute, USA

This chapter discusses the economic impact of science-based research in agriculture. Global agriculture was transformed in the 20th century by the Green Revolution that resulted from applying Mendelian genetics to crop and animal breeding. Developments of biotechnology in the last 20 years marked the dawn of a gene revolution that is thought to replace Mendelian genetics as the driver of technical change in agriculture. In recent years and still far from reaching the full potential impact of biotechnology in agriculture, developments in nanotechnology promise to further push the research and innovation frontier in agriculture. In this new environment, the private sector emerges as the main actor in agricultural R&D displacing the public sector, which played a central role during the Green Revolution period. However, more public investment in R&D rather than less and new institutions will be needed in developing countries if they are to benefit from the new technologies.

Chapter 18

Taiwo E. Mafimisebi, Federal University of Technology, Nigeria

Africa's economic development will result from conscious efforts directed towards diversification and increased productivity in its low-performing agricultural sector. Technology development, transfer and uptake, which are low for now, are indispensible necessities in this respect. This chapter reviews the characteristics, importance, constraints and technology adoption process of African agriculture to identify factors that enhance or hinder technology uptake. Case studies of the welfare-enhancing impacts of adopted agricultural technologies were examined under use of fertilizers, improved varieties and biotechnology. Useful lessons for development and transfer of nanotechnology and micro-electronics to Africa were highlighted.

Chapter 19

Saikou E. Sanyang, National Pingtung University of Science and Technology, Taiwan

Agricultural technology development and transfer is a driving force for national development. The perspective is to reduce poverty and hunger facing people by adopting new measures to raise income and attain household food security. This is attainable through the establishment of research institutions, extension services, farmer organizations, and public private participation. The importance of technology development and transfer approaches in developing countries has been recognized as a tool for economic development. Technology development and transfer economics create employment opportunities, reduce poverty, and enhance economic growth. This chapter examines the components of agricultural technology development and transfer and offers implied lessons for emerging technologies like nanotechnology and microelectronics transfer and diffusion.

Chapter 20

Edwin M. Igbokwe, University of Nigeria, Nigeria
Nicholas Ozor, University of Nigeria, Nigeria

The early years of the green revolution heralded a new era of technology adoption and increasing productivity in agriculture. This momentum has not been sustained, giving rise to food shortages and widespread poverty in developing countries. The paper reviews processes and models of technology transfer in agriculture in developing economies and concludes that previous efforts were not demand driven and therefore lacked the ingredients for diffusion. The drivers of technology transfer are discussed. A number of factors responsible for the low rate of technology transfer especially the absence of public policies on technology transfer are identified and linked to the transfer of emerging technologies, mainly biotechnology and nanotechnology. The paper recommends the development of public policies, development of the private sector, establishment of partnerships between the two sectors and development of infrastructures especially in rural areas.

Section 6
Regional Developments

In this section, progress and regional developments on nanotechnology and other emerging technologies are discussed. It covers nanoscience and nanotechnology in Chile, Mexico, Brazil and Argentina. It also discusses innovations, trade policies and management challenges as they affect technology transfer in Africa. An empirical study on technology penetration pattern in a developing nation for a new technology is provided.

Chapter 21

  Adolfo Nemirovsky, LatIPnet Inc., USA
  Fernando Audebert, University of Buenos Aires, Argentina
  Osvaldo N. Oliveira Jr., USP, Brazil
  Carlos J. L. Constantino, UNESP, Brazil
  Lorena Barrientos, Universidad Metropolitana de Ciencias de la Educación, Chile and
    Universidad de Chile, Chile
  Guillermo González, Universidad de Chile and CEDENNA, Chile
  Elder de la Rosa, Centro de Investigaciones en Óptica, México

Latin America (LA) can count some strong research centers with a tradition of research excellence in certain disciplines such as medicine and biology, nuclear technology, metallurgy and materials, among others. Latin American countries have generated networks of researchers across disciplines, centers, etc. within a country, and linking two or more countries in the region (e.g., Argentina-Brazil Bi-National Center for Nanoscience & Nanotechnology, CABN). Additionally, collaborations have extended beyond LA, mainly to the EU and the USA. In general, these programs have been quite successful in the generation of interdisciplinary nanoscience and nanotechnology (N & N) research. The relation between academia and industry has been improving in the last few years, but it is still weak. In particular, funding incentives for N&N efforts have encouraged joint efforts and contributed to new dimensions in collaborations. This chapter reviews the state of nanoscience and nanotechnology in Chile, Brazil, Argentina and Mexico.

Chapter 22

Technology is generally seen as a significant tool for development while technological innovations connote better ways of achieving results. This paper assesses different areas countries can experience technological innovations and notes that most African countries are lagging below expectations in this regards using secondary data sourced from International Telecommunication Union (ITU), United Nations Statistical Divisions (UNSTAT), among others. From the analytical perspective, the paper established that the low levels of technological innovations in Africa is one of the major reasons why the continent remains in the low developmental echelon compared to other regions of the world. Thus, this paper submits that adequate efforts should be placed on functional education, health system and technology related innovation programs. Besides, Africa and indeed all developing world must revamp their infrastructures, especially transportation, power and communication towards development in the 21st century.

Chapter 23

No nation can succeed economically without a strong and solid scientific educational base particularly in this era of knowledge economy. In many developing nations, the resources to develop both the human capital and infrastructure for education are inadequate. Specifically, in Africa, the intellectual capabilities on nanotechnology and microelectronics research and education are still evolving and some foundation technologies like electricity and ICT needed to drive and support them are not available. Lack of management efficiency and good governance continue to stall progress in the continent. In this matrixed four sub-chapters, these issues are discussed including a new model, Generic and Incremental Value (GIV), proposed for African development.

Chapter 24

Trade between nations is very crucial in the process of economic and technological growth. Directly or indirectly, trade facilitates the process of technology innovation, transfer and diffusion. It offers the trajectory to evaluate and understand how technology penetrates economies and remains a good indicator to measure national progress on technology creation and assimilation. The growth link between international trade and economic development could be traced to the classical trade theory of Adam Smith, and David Ricardo and the modern neoclassical trade model of Heckscher-Ohlin (H-O). While there is no single model that captures the route to economic development, this chapter explores how African countries working closely can harness and utilize technological advancements to improve their share of global trade so as to accelerate their overall economic growth and development.

Chapter 25

Olalekan A. Jesuleye, National Centre for Technology Management, Nigeria
and Obafemi Awolowo University, Nigeria
Williams O. Siyanbola, National Centre for Technology Management, Nigeria
and Obafemi Awolowo University, Nigeria
Mathew O. Ilori, Obafemi Awolowo University, Nigeria

Considering the huge wastage associated with the present energy production and consumption pattern in Nigeria, solar electricity is acclaimed to be of great potentials as a viable alternative to fossil fuels and is being considered by policy makers to contribute to improving energy efficiency, security and environmental protection. The veracity of such claim is being ascertained in this study through analysis of solar electricity utilization for lighting, refrigeration, ventilation, water pumping and others by just 5% of about 100 million Nigerian rural dwellers who lack access to national grid. The study deduced that increase in rural access to solar electricity will yield tremendous carbon credits for Nigeria under the clean development mechanism and that generating more energy at cheaper cost will enhance policy support for green energy. This connotes a great future for microelectronics and nanotechnology in processing high efficiency multi-junction solar cells and nanosolar utility panel being optimized for utility-scale solar electricity systems.

Foreword

Nanotechnology and Microelectronics: Global Diffusion, Economics and Policy tackles some of the main challenges – technological, economic and cultural – that we are facing and will face during the coming decades.

Rapid scientific discovery and new progress show the great potential of a large number of applications and commercial opportunities, also enabled by the development of appropriate tools to look at and form matter at the small scale with atomic resolution. These tools include design, modelling, computer simulations, engineering, manufacturing and characterisation at the appropriate scale. Materials embed, contain and bring knowledge forward down the value chain. By realising advanced materials via the assembly and manipulation of atoms and molecules, we are able to develop and -where needed- maybe even re-invent industrial sectors. Innovation acme is probably nowadays in the information and communication technologies, first of all in microelectronics.

Electronics has been a key driver of the scientific and technological progress that made a major contribution to social and economic growth worldwide starting since the middle of the 20th century. Notable success stories such as mobile telephones and multimedia applications, digital media, computing and networking (especially the internet), cleaner production processes, safer vehicles, more performing medical systems... they all illustrate the crucial innovative role that microelectronics played.

With the internet the concept of a centre and a periphery of the world has begun to fade, bringing closer the so-called developed, emerging and developing nations. There is a dynamics between on the one hand the efforts to harvesting the benefits promised with nanotechnology and on the other hand to avoiding a possible substantial "nano divide" between countries, particularly from the so-called North and South of the World.

It seems that a new race has started. Nanotechnology and microelectronics re-opened a major competition in terms of industrial output, commerce and services that seemed somehow stagnated 20 years ago. Yesterday's winners will not necessarily be those of tomorrow. Evidently, it is not really a matter of "sharing the pie" of financial gains, but to realise an even "larger pie", from which all economic actors can benefit. It will be crucial to the vitality and attractiveness of economic systems (at a regional, national or supra-national levels, such as the European or African Union); they will have to transform the outcome of research into jobs, economic growth and a better quality of life. The optimum points of equilibrium between, quality and quantity, efficiency and flexibility, processes and products, mass production and product individualisation have to be constantly verified if not re-conceived.

A global system approach needs to be ever prominent and synergies should be sought, where appropriate. With its 7th Framework Programme for Research (FP7 - the scheme for funding scientific research and technological development up to 2013), Europe opened the doors to broad-ranging international cooperation. FP7 is a unique case world-wide in terms of its scope and dimension.

The world is one; there are challenges that we tackle more efficiently and effectively together. For this, we need to create a common and level playing field and share fundamental knowledge and best practices.

This book will surely help.

*This Foreword expresses the opinions of the Author and not necessarily those of the EU and it does not engage the Commission in any manner

Renzo Tomellini,
European Commission, Belgium

Renzo Tomellini *was born in 1960 and graduated in chemistry "cum laude" in Rome, in 1986. He worked in Italy as a researcher at the Centro Sviluppo Materiali1, a research centre corporate within steel industry. He was also a visiting researcher in Germany and France. His further education included management and business administration, leadership, European law and regulations. His career within the European Commission started in 1991, when he was scientific/technical responsible for ECSC² steel research projects. Between 1995 and 1999 he was managing the ECSC-Steel research and technological development programme. In 1999 he became the assistant to the director of "Industrial Technologies" in the Research Directorate-general. Amongst others, he prepared for the provisions to bring to its end the ECSC Treaty and to launch the new research fund for coal and steel (see the Nice Treaty). Meanwhile, since 1999 he promoted initiatives in nanotechnology. Until summer 2008 he has been Head of the Unit "Nano- and Converging Sciences and Technologies" and chaired the European Commission's interservice group "Nanosciences and Nanotechnologies", which he initiated in February 2000. He is currently Head of the Unit "Value-added Materials". He deposited 4 patent applications (a new source for atomic spectroscopy and some innovative sensors), published some 50 articles, drafted 4 standards on analysis and measurements, edited 12 books and 2 as co-author, created 2 newsletters and 3 webpages, and realised 4 films on science and research issues. He gives university courses on knowledge management to post-graduates.*

Foreword

It is a great honor and pleasure, serving as a Professor of Environmental Studies, an environmental specialist, a director of Clarion University Sustainability Institute, and as the Provost and Vice President for Academic Affairs, to contribute this foreword for *Nanotechnology and Microelectronics: Global Diffusion, Economics and Policy*. This comprehensive collection, which brings together experts from different fields to address issues germane to nanotechnology, is timely and relevant in the world that we live in today. Our divided world of the industrialized and the non-industrialized is on the verge of a transformation that is unequivocal. The authors of the following chapters cover topics that are collapsed into six sections: (1) Foundations and Science, (2) Technology Transfer and Innovation, (3) Industry, Policy, and Experiences, (4) Ethics, Regulation, and Environment, (5) Lessons from Agricultural Technology, and (6) Regional Developments. It is certainly because of the breadth and depth of the book and its multidisciplinary approach to the subject of nanotechnology that this book fills an important gap in the literature. It is my expectation that it would encourage strong voices from all sectors of the world, particularly from the developing countries that might be left out in the development of nanotechnology and suffer the negative consequences of the technology. The good news is that if managed properly, nanotechnology could be beneficial to both the developed and the developing countries.

I must emphasize the point that involving authors from multidisciplines has made this work rich in approach and rigorous in analysis. In writing this foreword, my comments are focused on the benefits, cost, and risk issues of investing in nanotechnology. As an environmentalist, I am always interested in the environmental impact of new technologies on ecosystems and humanity. The fundamental concepts surrounding nanotechnology deals with the deformation of materials by an atom or by impacting the material by a molecule. Such deformation brings into the environment an uncertainty which is unpredictable with regards to the possible consequences. Nanotechnology also involves the processing of materials which brings both positive and negative consequences. The creation of new materials through processing usually results in many useful new materials which enhance advancements in genetics, information technology, biotechnology, and robotics. More importantly, processing of materials results in exponential economic growth. The separation and consolidation of material leads to new materials that benefit humanity in the fields of engineering and medicine. Cures for cancer and other illnesses are quite possible. But it must be stressed, that it is quite possible that the new materials can cause new diseases and exacerbate existing health problems. The nano-particles must be studied and health and environmental impact analyses are imperative as the world moves forward with this great technology which rivals the industrial technology age. The invention and advancement of cluster science and scanning tunneling microscopes have resulted from investment in nanotechnology. The production of devices in parallel could enable the cost of devices to be cheaper and as such, provide consumers with cheap material. Although further applications of nanotechnology require the arrangement or manipulations of nanoscale component they have not been fully researched; tangible results have been received economically through the sale of nanotubes and nanowires.

Before providing a brief summary of the significance of the chapters by the authors in this volume, I would like to reiterate the sentiments offered by many scientists around the world who claim that despite the economic and technological gains of the nanotechnology, caution must be exercised on how fast the world is moving with the technology and what investments are being made in the technology. Rearranging the fundamental buildup blocks of nature inexpensively does pay tremendous dividends, but caution must be exercised in the proliferation of technology when it is used to produce weapons of mass destruction that might fall in the hands of terrorists and unstable governments. The use of nanotechnology in food, clothing, cosmetics, tooth fillings, paints, and other products have been brought to the world's attention with regard to the possible impacts of the nano-particles that are in the products. The side affects must be understood in order to inform the public. There is no doubt that through nanotechnology progress has been made in the early detection of prostate cancer and the positive results of nanotechnology are leading to investment by national and international laboratories to invest in the technology. Universities around the world are establishing new programs to expand research and teach new technology related courses. Governments, such as Indonesia, are investing in this "technology of the future." Understanding nanotechnology is leading to the re-examination of the natural phenomena that happens in biological science. Biological processes offer us the opportunity to observe nature's manufacturing process which can be translated into our attempts to build materials that can be useful for our civilization and life on earth in general.

The chapters in this book provide voices from the North and South of the world. They provide discussion of the significance of technological innovation in developing countries and the connection to economic development. There are significant trends emerging in the North; the South must position itself in such a way as to take advantage of such development and growth. Nanotechnology must be approached with a thorough knowledge of the ethics that must accompany expansion and adaption on a global basis. Technology transfer is an important aspect of this book and it is here that it must be stressed that building a true partnership between the North and South is necessary to have a successful transfer of technology and knowledge in order to have a sustained nanotechnology in developing countries.

The proliferation of information technology in Africa and other developing countries has led to some improvements in the economic vitality of developing countries. The implication of policy implementation must be understood and steps must be taken.

I thank Dr. Ndubuisi Ekekwe, the editor of this book, for asking me to write this foreword. I sincerely compliment the authors on an excellent job and wish them continued success in their scholarly research efforts. They have worked in a true collaborative spirit to put together a very important and holistic approach to understanding and thinking about a global perspective on nanotechnology particularly, as it relates to developing countries and the transfer of technology from the North.

Valentine U. James
Clarion University of Pennsylvania, USA

Valentine U. James is the Provost and Vice President for Academic Affairs at Clarion University of Pennsylvania, USA. He was the Dean of Graduate Studies at Fayetteville State University, USA from February of 2008 to June of 2008. He served as the Dean of the College of Humanities and Social Sciences at Fayetteville State University from August of 2004 – February 2008. He is also a full professor who specializes in Sustainable Development and Environmental Sciences with particular emphasis on the environmental issues of developing countries. He has held tenure track/tenured positions at University of Louisiana, Lafayette; University of Virginia in Charlottesville; Southern University in Baton Rouge, where he served as the Director of the Ph.D. Program in Public Policy and Interim Chair of the Master of Public Administration Department. Dr. James has authored and edited fourteen books. He is the recipient of scholarly awards such as The American Association for the Advancement of Science (AAAS) Diplomacy Fellowship, Fulbright –Hays Faculty Research Abroad Fellowship and The University of California Presidential Fellowship. He has worked for the United States Agency for International Development. Dr. James holds a Ph.D. in Urban and Regional Science from Texas A&M University in College Station, Texas; an M.A. in Environmental Science from Governors State University, University Park, Illinois and a B.S. in Biology from Tusculum College, Greeneville, Tennessee. He studied liberal arts at the American College of Rome in Rome, Italy.

Preface

Within the last two centuries, technology has emerged as a key driver of global economic growth. It has redesigned international competition in all major industrial sectors by enabling speed, efficiency and capacity in business processes and operations. It has become the most important enabler of national wealth creation and productivity. As the world moves towards knowledge-based economic structures and data-driven societies, made up of networks of citizens, organizations and countries, mutually and interdependently linked globally, the impacts of technology will remain central in commerce, industry and culture. Both in the short and long terms, this global technological progress—improvements in the techniques by which goods and services are produced, marketed, and brought to market—will remain at the heart of human progress and development. And the pace of technological innovation will continue to accelerate, disrupting markets and industries, along the way.

Increasingly, the world is experiencing major new dimensions in knowledge acquisition, creation, and dissemination. The trend has become a virtuous circle where new ideas facilitate new processes and tools which in turn drive new concepts. This progress has advanced to the point where researchers are able to work at the levels of atoms and molecule, evolving a new field, called nanotechnology. Nanotechnology is the science of minuscule molecule or a wide range of technologies that measure, manipulate, or incorporate materials or features with at least one dimension between approximately 1 and 100 nanometers (a nanometer is one billionth of a meter; the width of an average human hair is about 100,000 nm). At this scale, the laws of quantum physics supersede those of traditional and classical Newtonian physics, and materials change yielding to unique characteristics in chemical reactions, electrical, and magnetic properties. Nanotechnology offers the closest means to manipulate matter and life whose building blocks are at nanoscale.

Nanotechnology is a transformative technology and has the ability to bring about changes that can rival the Industrial Revolution of the late 18th and early 19th centuries where mechanization of industry, changes in transportation and introduction of steam engine had a profound effect on the socioeconomic and cultural conditions in the world. Heralded to underpin a new global turning point in human society, nanotechnology "has the potential to fundamentally alter the way people live". But it is not completely (scientifically) proven, still growing with only few nanostructures at commercial productions. In most cases, precision is lacking and controls are difficult with many of the concepts not economically viable with the present body of knowledge. It poses environmental and health challenges, though it can also be used in combating pollution and other environmental hazards by enabling advanced water purification and clean energy technologies. Its impact will be profound in medicine where it is leading many innovations; for instance, *in situ* nano engineered robots (as small as pills) offer the prospects for better medical diagnosis. The technology is broad with convoluted ethical and safety issues.

Nanotechnology is estimated to grow in excess of $1 trillion global market by 2015 with energy, textiles, and life sciences the leading sectors transitioning from labs to markets. This technology will drive a new global economy, nanomics or nanotechnology-driven economy and usher in a revolution that will advance genetics, information technology, biotechnology and robotics through low cost, high utility and high demand of its products.

While nanotechnology is an evolving technology, microelectronics has relatively matured. Microelectronics is a group of technologies that integrate multiple devices into a small physical area. The dimension is about 1000 larger than nanotechnology dimension; micrometer vs. nanometer. Usually, microelectronics devices are made from semiconductors such as silicon and germanium using lithography, a process that involves the transfer of design patterns unto a wafer. Products are called ICs, chips, microchips or integrated circuits. They are found in computers, mobile phones, medical devices, toys and automobiles. Contemporary, the world lives in the era of microelectronics as everything is enabled by microchips. Its impacts, arguably, are unrivalled in the human history. As engineers make the transistor sizes smaller to improve performance and reduce cost, microelectronics begins to converge with nanotechnology. This advancement comes at a huge price as power dissipation and noise in chips increase- potential limiting factors that could stall further progress in the industry unless novel architectures, materials and processes are developed. Possibly, nanotechnology could address many of these challenges as microelectronics transmutes into nanotechnology. Indeed, the ETC Group notes that "*with applications spanning all industry sectors, technological convergence at the nanoscale is poised to become the strategic platform for global control of manufacturing, food, agriculture and health in the immediate years ahead.*"

Together, nanotechnology and microelectronics are the engines of modern commerce, and are directly or indirectly enabling many revolutionary global changes. Whenever there is advancement in their performances, a dawn emerges in global economy bringing improvements in all areas of human endeavors. Yet, despite these pervasive impacts of these innovations on daily lives and businesses, the technologies have not diffused globally. Patents, academic journals and other metrics for ascertaining technology creation and innovation indicate that advanced nations dominate the creative sectors of these technologies and the global diffusion trajectory will flow from them to other parts of the world. This implies that the prospects of transferring these technologies around the world will involve an adoption and diffusion strategy from developing nations which lack inventive capability to create technology. Records show that in many previous efforts, these nations have failed to absorb new technologies effectively. However, owing to the expected impacts of nanotechnology, the abilities of developing nations to adopt and drive penetration in their economies will affect their economic viabilities in the long-run.

This book is written to assess the state of nanotechnology and microelectronics, and emerging technology in general. While some aspects focus on nanotechnology and microelectronics, others discuss technology transfer and diffusion within the generic technology context with no specific distinction. It examines many issues, climate change, trade, innovation, diffusion, etc, with a theme focused on facilitating the structures for the adoption and penetration of the technologies into developing nations. The problems which continue to undermine technology progress in developing nations along with suggestions that can accelerate progress are examined. The strategic importance of moving from dependence on minerals, commodities and hydrocarbons to nations that thrive on knowledge anchored on technology is emphasized. It is almost certain that nanotechnology will exacerbate the economic divide between the advanced and poor nations unless the latter develop new pragmatic technology policies. This book shares some insights from various experts on what these policies could be for a reliable, sustainable and profitable nanotechnology era.

The technologies are capital intensive and the returns are not immediate. In short, there exists a level of uncertainty in nanotechnology as many of the discoveries cannot be economically commercialized, at least with present technology. This calls for tripod partnerships among governments, firms and academic communities in structuring policies and mapping the technology roadmaps. Around the world, even in developed nations, governments have played and continue to play major roles in accelerating innovations in nanotechnology and microelectronics. The developing nations must not be on the illusion that markets forces alone can drive development in these areas. They lag well behind in both the technology creation and dissemination and spirited efforts must be made to facilitate adoption and improvements in the business environments.

Consequently, government interventions on infrastructure, education and business climate for these ultra knowledge-driven technologies must be paramount in national developmental plans. Critics argue that developing nations should focus on spending their limited resources on mundane activities like food production and water supply instead of investments in these emerging technologies. The problem with that argument is that food production, water supply and others are driven directly or indirectly by these technologies. Microchips continue to improve crop yields by enabling better sensors while water purification has a future anchored on nanotechnology. In this century, it makes no sense to separate activities from technology because technology leads the world and only those that invest and develop it will prosper. Investments in technology will bring progress and presence of technology clusters will continue to influence global technology diffusion trajectory. It is a continuum, where the presence of one technology enables another. Nanotechnology investment today could lead to breakthroughs in energy and food security tomorrow.

Nonetheless, nanotechnology must not be viewed as a fix to all the technology problems in the developing world; in other words, it must not be adopted without examining alternatives or immediate needs which may be more appropriate to the particular nation. Cautious and systematic approach is needed as these nations develop plans for the adoption of any aspect of nanotechnology or microelectronics. Without this strategy, the technologies may not be sustainable as previous technology adoption efforts have shown. For many developing countries, provision of power supply to their industries will be the beginning of wisdom as inadequate electricity remains a major reason for de-industrialization, especially in sub-Sahara Africa. By focusing on the basics and improving industrialization climate, conditions for high-tech economy will be nurtured.

This book explains how technology and technological progress are central to economic and social well-being, and why the creation and diffusion of goods and services are critical drivers of economic growth, rising incomes, social progress, and medical progress. It notes that political climate, corruption, stifling business environment, poor infrastructures, lack of innovation culture, poor economy regime, along with low technology literacy are major challenges which must be overcome. While the world discussed digital-divide in the information technology era, the future will potentially will be nano-divide. The reasoning is that nano will continue to enable economic concentration in developed nations (holders of core patents with economic rights) and developing ones will find it increasingly difficult to transition from their present states. It is up to developing nations to observe that global powers and respects are not won by gun powers anymore, rather by economic prosperity driven by technology creation.

Besides, with lack of innovation in developing nations, the disruption of global economic systems by nanotechnology can harm the developing nations since they lack the resilience and fluidity to react to market and industrial changes. The prospect of nano-weapons could be a concern in the hands of these unstable developing countries as they can self-destruct or destroy neighbors. Terrorism could escalate to a level not imagined, not just in the developed world, but globally as nanotechnology will make it easy to terrorize with devastating global impacts. The world could be visited with arms race and nuclear anti-proliferation could be relegated to the background with anti-nano (weapon)-proliferation upfront. If nanotechnology products could affect trade patterns with replacements of raw materials, the developing world would be the most affected as poverty could increase. Displacing their exports will increase global unemployment and that can pose global insecurity. The world within the last few centuries have depended on the raw materials of developing nations to sustain civilization, if nanotechnology can replace the needs of those materials, monumental upheavals could result in these countries with (soon) worthless cotton, copper, and rubber. Simply, the prospects of nanomaterials pose a huge security implication in the developing world.

Across the globe, many nations have developed initiatives towards transitioning discoveries to markets. Just like in the Industrial Revolution, which took half a century to come to fruition, nanotechnology is expected to advance overcoming many of the technical challenges that presently stall commercialization of many of the discoveries. As its standardization and safety improve along with ethical regulations, the global 'innovation economy' with be revamped. The new economy will witness new breakthroughs in computing where performance can be increased exponentially even at decreasing cost. Early detection of tumors, efficient and cheap solar cells delivering vast amounts of energy, effective HIV/AIDS prevention control, and hosts of other applications will be made possible. These impacts will be ubiquitous and most likely will be gradual and evolutionary, rather than very sudden. A look into the future of nanotechnology and microelectronics shows that any nation that fails to develop programs aimed at tapping their enormous benefits will compete internationally at disadvantaged positions. It will be catastrophic to misunderstand that Technology leads the world and mastering the process of creating, enabling and commercializing technology is one of the most important duties of any modern parliament or congress.

One major goal of this book is to highlight multifaceted issues surrounding nanotechnology and microelectronics and technology in general on the basis of economics, innovation, policy, transfer, and global penetration through comprehensive research, case studies, academic and theoretical papers. More than forty five experts spread in about twenty countries with its respective understanding, perspectives and resources provide a very broad audience to accomplish that. This book will be a useful reference for academics, students, policy-makers and professionals in the field of technology economics.

This book is organized into six matrixed sections. Section 1 is focused on the foundations and the science of nanotechnology and microelectronics. The first chapter discusses the science, trends and global diffusion of nanotechnology and microelectronics, highlighting some of the historical advancements in the technologies. The manufacturing process, molecular manufacturing, which is structured for building nanosystems, is explained in Chapter 2.

Section 2 focuses on technology transfer, diffusion and innovation in the contexts of both nations and organizations. Chapter 3 explains the latest trends in nanotechnology knowledge creation and dissemination, and Chapter 4 shares insights on collaborations in the age of open innovation. Chapter 5 discusses Kondratieff cycle of nano revolution with Chapter 6 explaining how economic agility of nations could affect capacity building for technology resilience and diffusion. Then Chapter 7 points out that fatigue could occur in diffusion of innovations especially in adopter nations.

Section 3 examines the industry, policy and experiences from nations and institutions. Chapter 8 highlights the case of a university in Sydney on firm innovation and university-industry networks. Chapter 9 discusses licensing and R&D, and Chapter 10 outlines nanotechnology industry entry barriers in Turkey. In Chapter 11, micro and nanotechnology maturity and performance assessment are discussed.

Section 4 considers the ethics, regulation, environment, and climate control challenges. It begins with Chapter 13 which examines the diffusion of the clean development mechanism. Then Chapter 14 looks at the intellectual property rights challenges under information and communication technologies, nanotechnologies and microelectronics. Chapter 15 discusses how the global south could benefit from climate finance, technology transfer and effective climate policies. It is followed by Chapter 16 that highlights emission distributions in post-Kyoto international negotiations, and Chapter 17 that outlines the ethical concerns in nanotechnology.

Section 5 examines some lessons within agriculture and agricultural technology which could be helpful for many developing nations adopting technology. Agriculture being their mainstay, it is natural they can relate to this industry. Chapter 18 discusses how the industry has moved from biotechnology to gene revolution and asks if nano revolution is the next for agriculture. Chapter 19 sees the patterns within the industry and connects them with adoption and development. In Chapter 20, the author gives lessons on technology development and transfer drawing from agriculture, and finally Chapter 21 discusses technology transfer and diffusion in developing economies from the perspectives of agricultural technology.

In the final Section 6, regional developments are highlighted. Its first chapter, Chapter 22 shares very comprehensive insights about nanoscience and nanotechnology on Latin America, covering Chile, Argentina, Mexico, and Brazil. Subsequent chapters are devoted to Africa. They are technological innovation and the continent's development in the 21st century (Chapter 23), and emerging technology transfer and policy (Chapter 24) which has four sub-chapters: thoughts on nanotechnology transfer, sustainability and management challenges, factors affecting nanotechnology and microelectronics transfer, and recent polices on science and technology. Others are trade policies and technology development (Chapter 25) and finally in Chapter 26, a technology penetration national case study.

In conclusion, it is important to note that penetration of nanotechnology and microelectronics into developing nations will not just benefit them alone; it will help to accelerate market growth for advanced nations that drive the industries. Technologies will remain major catalysts for wealth creation to nations that create and commercialize them. For developing nations that merely consume, lacking invective capability and depending on minerals, commodities and hydrocarbons, it is very imperative they change strategies because if nanotechnology era goes as heralded, economically, these nations could be imperiled. Just as R. Wright noted, *"Society becomes increasingly non-zero-sum as it becomes more complex, specialized, and interdependent,"* the whole concept of globalization is not win-win by default because knowledge and technology disparities exist. It is still early for any nation to get into the nanotechnology business by building its capacity- one that will be used to access national competitiveness in the near future.

Ndubuisi Ekekwe
Johns Hopkins University, USA

Acknowledgment

I would like to thank and acknowledge the reviewers, colleagues, family, and friends that have helped to make this book a possibility. Innumerable appreciations go to all the authors who contributed. The commitments and generosity of these experts, drawn from both the academic and business communities, lessened the challenge of producing this volume. To Elizabeth Ardner and Dave DeRicco of IGI Global, thank you for your guidance and direction throughout the course of this project.

The resources to develop this book were facilitated by multiple fellowships I received during my doctoral program in the Johns Hopkins University, Baltimore: The Whiting School of Engineering fellowship, United States National Science Foundation Engineering Research Center - CISST fellowship, Jay D. Samstag fellowship. The Jay D. Samstag fellowship partly provided the seed resources that facilitated the establishment of my nongovernmental, US-based, African Institution of Technology (AFRIT). Through AFRIT, I have developed relationships to more than twelve African universities, giving workshops and seminars on microelectronics and nanotechnology. My chapter in this book was funded from resources from this fellowship. I also appreciate the supports from the Electrical Manufacturing & Coil Winding Association and the United Kingdom Congress on Computer Assisted Orthopaedic Surgery-DePuy fellowships during the course of my program.

I also express my appreciation to the World Economic Forum and African Union Congress for their invitations. The Congress of African Economists provided lots of insights on the plights of Africa and developing world in general. I thank the African Union for providing the resources that enabled my participation.

Many contributors in this book have used data drawn from many sources: World Bank, International Monetary Fund, International Telecommunication Union, World Economic Forum, United Nations, African Development Bank, Asian Development Bank, European Union, Banco Interamericano de Desarrollo, Ford Foundation, among others. They have also interviewed experts and used primary data from universities and professional websites. To all these organizations and individuals, I want to express my appreciation for allowing them to use these data.

Having authored one of the earliest works on the influence of information technology in the Nigeria's banking and finance sectors while in Diamond Bank, Lagos, it becomes natural to thank my colleagues and advisors that enabled that effort: Dr. C. K. Ayo, Dr. O. Oyebanji, Prof. D. Iornem, and Prof. A. Aju. Those works were very instrumental in understanding how technology influences industrial dynamics and paved the way for this work. It also shaped my interests in technology transfer, diffusion and competitiveness despite a pursuit of joining the league that makes these innovations practically possible.

Let me express my gratitude to Prof. Valentine U. James and Renzo Tomellini for writing the forewords to this book. I also appreciate all the contributors, whose works could not be included in this book due to editorial constraints.

Finally, I am indebted to the United States National Science Foundation ERC-CISST Education & Diversity committee in the Johns Hopkins University where I served for more than four years. Through workshops and seminars, I picked vital ideas on the role of technology in any modern economy and how technology progress could be sustained. The Institute of Neuromorphic Engineering (INE) and University of Maryland, Baltimore provided funds that took me to the prestigious Telluride Neuromorphic Workshop where the future of neural microelectronics is crafted. All these broadened my universe of technology policy, economics, and practice in no small measure.

To my Ph.D. advisor, Prof. R. Etienne-Cummings, I am indeed very thankful. He nurtured the confidence in me to not just write about microelectronics and nanotechnology economics, but also to stay in the forefront of advancing the fields. He accommodated my continuous interests on technology economics and policy despite engaging in advancing the field of robotics, alternative energy and biomedical engineering.

Finally, to my wife and best friend, Christiana (Ifeoma), thank you for your support, love and helping to review sections of this book.

Ndubuisi Ekekwe
Johns Hopkins University, USA

Section 1
Foundations and Science

This section discusses the general theme of knowledge economy, innovation, technology transfer and diffusion. It offers insights on the progress which has been made in the fields of nanotechnology and microelectronics over the years and discusses the broad implications of safety, intellectual property rights, and funding- both government and private. It also examines the trends, the transfer trajectories and the penetration of the technologies across the globe. Besides, it explains the science and technology, molecular manufacturing, upon which the future of commercial nanotechnology is largely hinged upon.

Chapter 1
Nanotechnology and Microelectronics:
The Science, Trends and Global Diffusion

Ndubuisi Ekekwe
Johns Hopkins University, USA & African Institution of Technology, USA

ABSTRACT

For many centuries, the gross world product was flat. But as technology penetrated many economies, over time, the world economy has expanded. Technology will continue to shape the future of commerce, industry and culture with likes of nanotechnology and microelectronics directly or indirectly playing major roles in redesigning the global economic structures. These technologies will drive other industries and will be central to a new international economy where technology capability will determine national competitiveness. Technology-intensive firms will emerge and new innovations will evolve a new dawn in wealth creation. Nations that create or adopt and then diffuse these technologies will profit. Those that fail to use technology as a means to compete internationally will find it difficult to progress economically. This chapter provides insights on global technology diffusion, the drivers and impacts with specific focus on nanotechnology and microelectronics. It also discusses the science of these technologies along with the trends, realities and possibilities, and the barriers which must be overcome for higher global penetration rates.

INTRODUCTION: THE GLOBAL TECHNOLOGY DIFFUSION

Within the last two centuries, technology has emerged as a key determinant of sustainable growth and poverty reduction. It has become central to many modern developments across the globe and the most important competitive factor in the international economy. Before technology began to drive business operations and processes, global economic growth was flat for centuries and the world did not experience substantial progress in productivity. In other words, generations that lived more than three centuries apart might not have experienced substantial changes in their per capita incomes. But with the evolution of

DOI: 10.4018/978-1-61692-006-7.ch001

technological advancements shaping global commerce and industry, the world is experiencing new dimensions in wealth creation and productivity. Technology drives the modern world and national competitiveness is anchored on technological strength and innovation which encompasses the social and economic fabric of any economy (Chinn, 2006). It is the major factor that separates the rate and level of incomes between developing and developed nations.

Furthermore, the classification of nations into different categories of developments, advanced, emerging, and developing nations, indirectly translates into their different stages of technology capabilities. The state of global technology diffusion shows that developed nations continue to create the bulk of the new knowledge while developing countries depend on adoption and adaptation for technological progress as the latter lack inventive capacity (World Bank, 2008a). Nations have different abilities to process technological inputs, even as they have many ways of developing technological competence. For the developing nations, trade and importation of foreign technology goods creates local awareness and brings exposure to new technology (IMF, 2006). Most especially, their skilled diasporas contribute immensely in technology adoption and diffusion. Also, when multinational corporations (MNCs) invest locally through FDI, they bring knowledge of vital technologies and international markets. According to World Bank (2008a), the diasporas population is an important resource for their home country—a "brain bank"—that contribute to technology transfers by strengthening trade and investment links with advanced economies, providing access to technology and capital which contributes to domestic entrepreneurship and investment. They also provide technology and marketing know-how, facilitate FDI, and expand banking and other financial services in their home economies.

Historically, not many successful technologies have been transferred across the globe, provided technology transfer is not said to have occurred when an adopting nation imports technology products from innovating ones. Such a narrow context may erroneously imply that many developing nations have adopted steam engine by merely importing trains from developed economies. Technology transfer involves imparting knowledge, skills, capabilities and techniques which are involved in the whole production cycle. Where technology has been effectively transferred, changes in the production system and its compatibility with system needs, institutional framework, skills, financial capacity, and support of endogenous capacity with appreciation of the natural environment of the recipient country are visible (Dabic, 2008). For the adopting nations of technology, the prospects for progress will involve innovation system through their institutions, citizens, universities and research institutions (World Bank, 2009). Adapting existing technologies to meet local needs will be important and technology penetration rate can accelerate globally if low-income nations modernize their educational and trainings structures for efficiency and accountability (Hassan, 2007). This modernization is vital as sustainability depends on the development of knowledge citizens to lead the efforts for acceleration of adoption and adaptation stages. Efficient and new models of education designed collaborations between schools and firms are urgently needed while also allowing market to be the driver of technological improvements.

From the World Bank Knowledge Economy Index (KEI), there is a positive correlation between education, technology, innovation and GDP per capita of nations (World Bank, 2008b). Nations with high KEI show higher competencies in technological advancements while those with very low KEI are mainly non-innovating nations. These latter nations must depend on adoption and adaptation for technological growth and advancement since they lack the capacity to create new knowledge owing to poor facilities, small economies (lacking large scale advantage for funding) and human capital. These problems point out the fact that radical steps must be taken by developing nations if they expect a convergence in technology advancement between them and the developed economies. Their rates of technological advancements could be faster than high-income nations since they have lots of rooms to grow but convergence of technology penetration rates with advanced economies will require major policy changes from them. From the World Economic Forum (WEF, 2009a & 2009b) reports, technology innovation correlate positively with income levels; the more nations advance in technology creation and penetration, the more the incomes levels in those nations. This can also explain why emerging technologies like nanotechnology and genetics are usually associated with high-income nations. Capacity to sustain the diffusion of these emerging technologies is always positively correlated with quality tertiary technical education in the respective areas (Golding, 2006). Globalization with offering of more market access and FDI will continue to help developing nations; however, the most important indicator will be the human capital development. Retaining the best and brightest technical manpower in those nations will boost the prospects of catching up with the developed countries (Ekekwe, 2009a).

Unfortunately, the manpower is not the only problem that has stalled the transfer of emerging technologies to developing economies. Besides technological illiteracy and low skill level, the regulatory and political environments, lack of economic transparency, low intellectual property rights (IPRs), and infrastructural development hinder innovation and entrepreneurship. Others factors include corruption, civil strife, macroeconomic turmoil and state monopoly of industrial sectors (World Bank, 2008a). These explain why old and matured technologies like telephone (landline), water supplies, and electricity have not properly diffused in these nations despite decades of their existence. In many developing nations, software piracy is unchecked and remains the

major barrier to the innovation and growth of the local software industry. A sustainable technology transfer will not take place if there is no holistic absorption capacity in the transfer cycle. Mapping technology shows that penetration rate positively correlates with strong national IPRs. This is why governments and other institutions must work as partners to stimulate and nurture appropriate transfer environment. Opportunities are emerging from globalization and market-forces based economic models are being adopted by developing nations. These changes can facilitate the penetration of new technologies that enable better business processes, tools and services. However, according to World Bank (2008a), the reality is that these new technologies require the infrastructure of the matured ones to grow and be sustainable. In addition, weak relationship between R&D and the business communities, and absence of strong links to diffuse even local technologies (products of local universities and industrial R&D) undermine technology advancement and progress. It is an effect centered on poor foundation showing that while exposure is important, what matters most is the absorptive capability of these nations (World Bank, 2008a).

Specifically, for nanotechnology and microelectronics, the prospect of deep penetration in developing nation is very low. The key factors being the poor state of the basic amenities, human capital and huge capital investment needed for these technologies. The challenge is getting these technologies into the nation first since naturally technology spread takes long time to even appear in developing nations. The next problem will be how to effect the penetration when it has been brought into the country, including penetration within firms since they adopt at varying rates. Electricity and clean water technologies remain major challenges in developing nations despite many years of their existence. While few of the cities enjoy these technologies sparingly, majority of the citizens are yet to use them. But technology process is correlated with its penetration in each economy. For technologies discovered between 1950 and 1975, only a quarter of the developing nations (upper-middle-income nations) that have reached at least 5% penetration level have gotten up to 25%; the low-income ones fair very badly (World Bank, 2008a). Nonetheless, nanotechnology and microelectronics are not just products, but technologies whose products are penetrating global markets. Mobile phones, Internet, water systems, and electricity use microelectronics products, at least, since nanotechnology is still evolving. Estimate of the penetration rate of microelectronics will be better evaluated at the creative side of the technology; in other words, the capacities of the developing nations to develop and mass produce microelectronics products, instead of just consuming the products. This will involve having capacity in microelectronics design and fabrication. So while mobile phones could be penetrating rapidly in a developing nation like Nigeria, the microelectronic technology is not as the nation does not have a single modern operational cleanroom (Ekekwe and Ekenedu, 2007). These interrelationships show why the developed nations where technology is usually imported or acquired have a duty to facilitate global diffusion by structuring their licensing fees and technical know-how to enable easier acquisition by low-income nations. The developed nations will benefit in the long-run if more diffusion translates to more needs for their products; a symbiotic partnership in a globalized world.

KNOWLEDGE AND INNOVATION IN INTERNATIONAL ECONOMY

Knowledge rules the world. This is evident as many new firms operate on the capacity of knowledge without the luxury of massive natural resources. While natural resources are still very important for survival and growth of some business models, the world is experiencing a shift where knowledge is a major component of organizational factors of production. Today, at both national and organizational levels, progress in knowledge creation, acquisition and processing is an indication of the state of global competitiveness. Figure 1 shows the World Bank (2008b) Knowledge Economy Index (KEI) for three nations: Ghana, USA and Brazil representing a developing nation (Ghana), a developed nation (USA) and an emerging nation (Brazil). Also included are Western Europe and Africa. The figure shows a relationship between the level of innovation and education and the state of national KEI. To move from a low KEI, a nation must have the capacity to advance its educational program and innovation culture. Education not only helps in developing new knowledge, it also helps in diffusing technologies and established knowledge. That is why innovation and excellent education are closely related.

Further analyses using Figure 2 show specific indicators that contribute to the factors presented in Figure 1. In these indicators that include publications, patents, and university-company research collaboration, USA leads Brazil and Brazil leads Ghana in average. These data show that education plays major roles in advancing national KEI and development. It is a very vital component for innovation systems in any economy. It is also a source for creating and assimilating new knowledge. For nations that want to successfully adopt a new technology, education must play important role in this international economy. It offers organic succession pathways that will sustain any national progress in this area. For microelectronics and nanotechnology, without sound education, the sustainability of any of the transfer cycles will not be possible.

From the KEI, it is obvious that the developing nations have uphill tasks to equilibrate the global technology know-how disparity. The divide is huge and it will require substantial efforts to narrow. In nanotechnology and microelectronics, these nations do not have policies or instruments that can stimulate domestic innovation capability. So even when efforts are made to adopt emerging tech-

Figure 1. KEI of Ghana, Brazil, Western Europe, Africa and USA (most recent data compared with 1995: Source, World Bank)

Knowledge Economy Index
Comparison Group: All
United States
1995
Western Europe
1995
Brazil
1995
Africa
1995
Ghana
1995
0.0
2.0
4.0
6.0
8.0
10.0
Economic Regime
Innovation
Education
ICT
KAM 2008

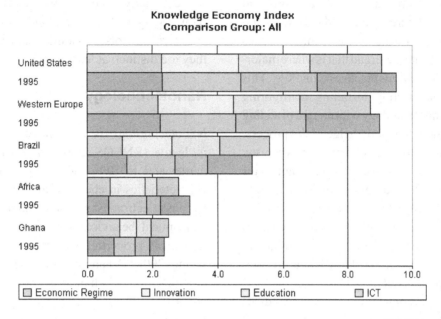

nologies through various means, lack of requisite skills will always affect adaptation process. That education content in KEI that improves national skill level is still lacking. Also, without innovation, improvements will stall and the values derived over time will be suboptimal. Infrastructural problems and out-modeled educational system means that developing nations will continue to lack the capacity to drive innovations in pushing the frontiers of science and technology. There is a tendency they will remain outside the horizon of creative nanotechnology and microelectronics for a long

Figure 2. Indicators of KEI for three nations: USA, Ghana and Brazil (Source: World Bank)

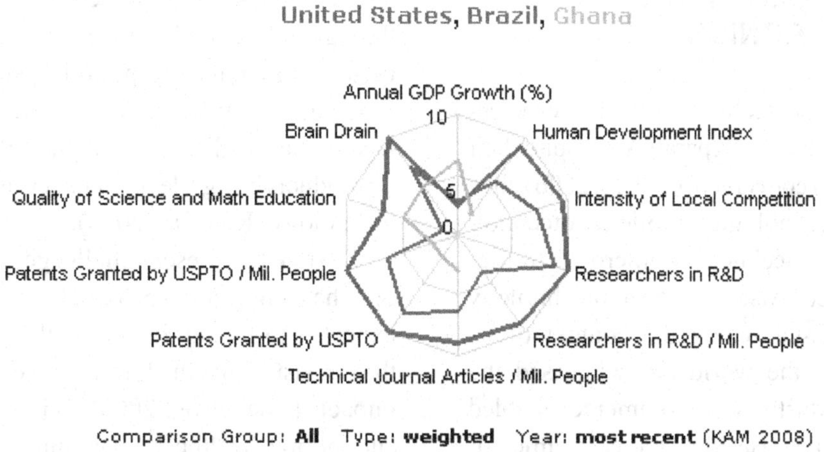

United States, Brazil, Ghana
Annual GDP Growth (%)
Human Development Index
Intensity of Local Competition
Researchers in R&D
Researchers in R&D / Mil. People
Technical Journal Articles / Mil. People
Patents Granted by USPTO
Patents Granted by USPTO / Mil. People
Quality of Science and Math Education
Brain Drain
10
5
0
Comparison Group: All Type: weighted Year: most recent (KAM 2008)

time. By not contributing in the innovation chain and creating knowledge, they will miss the wealth that technology enables at the upstream level, and remain perpetual participants at the downstream level. Unfortunately, the downstream level does not generate lots of wealth and that is where majority of developing nations practice technology. This sector means supporting, installing, configuring and maintaining technology with no sign of adding value or creating new knowledge in the process. This is analogous to the petroleum sector where the multinational firms practice in the upstream sector while the local firms dominate the downstream sector, basically, distribution and sales of crude oil products. However, the petroleum business wealth is localized at the upstream sector that requires continuous innovation and knowledge creation which these poor economies lack.

Table 1 shows comparative economic indices and factors for certain regions, categories and selected nations. For each of these regions and nations, there is a positive correlation between indices like Technology Readiness Index, Human Development Index, Networked Readiness Index, and Global Competiveness to Global Innovation Index. In other words, as a nation focuses on improving one of the indices, it will suddenly improve other ones too.

THE SCIENCE OF NANOTECHNOLOGY AND MICROELECTRONICS

In this section, nanotechnology and microelectronics will be discussed separately. Despite their future prospects of convergence (Roco, 2006), they are still distinct technologies. While nanotechnology is an evolving technology, microelectronics is largely matured. Also, while nanotechnology is usually discussed within the context of its future potentials, the world has witnessed the disruption in industry and commerce enabled by microelectronics. Besides, the economics of

microelectronics, micromics, is well understood, the one of nanotechnology, nanomics, is yet to shape properly. In general, they are two very close (converging) technologies, but in policy, science, ethics, risk factors, economics, and growth cycle, they are distinct, at least, now.

Nanotechnology

Nanotechnology is the science of minuscule molecule (RNCOS, 2006) or manipulation of matter at 1 to 100 nanometers yielding unique characteristics in chemical and biochemical reactions, electronics, physical, magnetic, thermal and optical behavior, mechanical strength and biological properties (Pourrezaei, 2007; Armstrong, 2008). A nanometer (nm) is one billionth of a meter; the width of an average human hair is about 100,000 nm. The technology is advancing with potentials to radically affect key aspects of human existence. It is an evolving (and potentially) disruptive technology that is transforming industries like electronics, materials and medicine. It has capabilities for low cost, high efficiency and high capacity in tools, industrial processes and products. The technology is not completely proven, still growing and only few nanostructures are at commercial productions. Largely, precision is lacking and controls are difficult and in most cases, some of the concepts are not economically viable with present body of knowledge. It poses environmental and health challenges as many of the materials are toxic, though nanotechnology can be used in combating pollution and other environmental hazards. Besides technical challenges, lack of standardization and public perception of its products bring a level of uncertainly about this technology (RNCOS, 2006).

Despite the present challenges, nanotechnology has enormous prospects. Many scholars expect it to be significant as the steam engine, the transistor, and the Internet in terms of societal impacts (Michelson, 2006). It is already used in automobile, healthcare, computers, and genetics,

Table 1. Comparative economic indices and factors for selected regions and nations

	GDP $bn, 2007	GDP, % annual growth 2002-07	Purchasing Power GDP,% of total, 2007	Purchasing Power $ per head, 2007	Int'l Trade Export of goods & services, % of tot, 2007	Int'l Trade Current acct bal, $bn, 2007
World	54,312	4.6	100.0	9,730		
Advanced Economies	39,131	2.7	56.4	35,780	66.4	-463
G7	30,419	2.4	43.5	37,380	38.4	-544
Euro area(15)	12,158	2.0	16.1	32,940	29.5	-30
Asia	5,724	9.0	20.1	3,840	13.2	384
Latin America	3,450	4.8	8.3	9,760	5.1	16
Eastern Europe	3,527	6.9	8.6	11,700	8.0	-45
Middle East	1,387	6.0	3.8	10,350	4.7	275
Africa	1,092	5.9	3.0	2,420	2.5	2
	GDP per head, $PPP, 2006	Human Dev. Index	Global Competitive-ness, 2009	Global Innova-tion Index	Technological Readiness Index	Networked Readiness Index
USA	43,970	95.1	#2	#8	#9	#3
Brazil	8,950	80.7	#56	#72	Above #24	#59
Ghana	1,250	53.3	#114	Above #110	Above #24	#103
World	9,250	74.7				

Sources: The Economist, 2008; UNDP, 2008; World Economic Forum, 2008; Boston Consulting Group and National Association of Manufacturers, March 2009

though not at a significant scale. It many cases, it offers only marginal innovation to existing products and processes. United States, Europe and Asia are the world's largest markets, dominating trade and collectively accounting for more than 80% of R&D; the BRIC nations (Brazil, Russia, India and China) have recently increased their nano-investments as they design their future developments around technology capabilities. The potential global wealth this technology will create besides possibilities that it can help provide cures for decade-old diseases makes it exciting. A nano-economy driven by nanotechnology could potentially exceed the micro-economy presently anchored on the powers of microelectronics. It is estimated to become a $1 trillion global market in 2015 (RNCOS, 2006). Energy, textiles and life sciences are the leading sectors transitioning from labs to markets. Low cost, high utility and demand for nanotechnology products will drive the nano-revolution which will help advance genetics, information technology, biotechnology and robotics.

The Science

Nanotechnology is not new; research has been done at nanoscale for many decades. What has enabled the sudden transformation in the scale and mass of nanotechnology focused research has emerged from the following factors (Maclurcan, 2005):

- Availability of modern tools like scanning probe microscopy, quantum mechanical computer simulation, soft x-ray lithography and synthesis technique for experimentation at nanoscale level

- Sudden recognition of nanotechnology as an emerging field which creates new levels of multidisciplinary collaboration and cross-fertilization amongst the sciences
- Desire to manufacture with ultimate precision on the atomic scale in a 'bottom-up' manner

The technology refers to a wide range of technologies that measure, manipulate, or incorporate materials and/or features with at least one dimension between approximately 1 and 100 nanometers (nm). From American Society for Testing and Materials (ASTM, 2006), such applications exploit the properties, distinct from individual atoms or bulk/macroscopic systems, of nanoscale components resulting from quantum effects and high surface areas; the laws of quantum physics supersede those of traditional and classical Newtonian physics. At nanoscale, quantum mechanical effects dominate material properties and with increased surface areas, electrical, chemical, conducting, optical and others properties of materials change. The technology offers the closest means to manipulate matter and life whose building blocks are at nanoscale. There are five phases of nanotechnology developments that have emerged (Saxton, 2007; Michelson et al, 2008);

- First generation (2000-2005) of "passive nanostructures" that incorporate nanoscale materials into coatings, aerosols, and colloids
- Second generation (2005-2010) of "active nanostructures" that are biologically or electronically dynamic
- Third generation (2010-2015) of "systems of nanosystems" that more fully integrate these materials into more complex organizational and manufacturing systems
- Fourth generation (2015-Beyond) of "molecular nanosystems" that lead to atomic

and molecular-level assembly. This stage captures the intelligent design of molecular and atomic devices, leading to "unprecedented understanding and control over the basic building blocks of all natural and man-made things."(Saxton, 2007).

- Fifth generation is the stage of singularity where the growth rate will seem to be infinite with production of products which today will seem like 'science-fiction'.

Present research works concentrate at the 1st and 2nd generations, though some minor works are taking place at the 3rd and 4th generation systems. This industry is multidisciplinary involving physics, chemistry, biology, computer science and engineering. Some of the major products/components of nanotechnology include nanomaterials, polymer nanocomposites, nanoparticles, and nanoclays. Others are nanotubes, inorganic nanoporous and microporous adsorbents, nanomagnetic materials and devices, nanocatalysts, nanofilms, nanoscale devices and molecular modeling, nanophotonic devices, advanced ceramic powders and nano ceramic powders, quantum dots, nanoelectronics, and nanosensors (RNCOS, 2006). It presently or will in future find applications in areas like automotive and transportation, life sciences, medicine and healthcare, instrumentation and tools, consumer products, photonics, energy, computers and communication, food and beverage packaging, aerospace and defense, environment and water, construction and structural materials, security and textiles, industrial process control and electronics (RNCOS, 2006; Pourrezaei, 2007). Many decade-old problems could be solved by nanotechnology if the enormous potentials are effectively harnessed and commercialized. In developing and emerging nations, it offers prospects to provide low-cost energy sources, control of HIV/AIDS transmission, water purification, food security and cheap housing.

For nanotechnology to shape global economic structures, the science must be translated into price-competitive and reliable products with market demand. Nanomanufacturing, the bridge between nanoscience and nanotechnology products must develop and advance. This may require the development of new technologies, tools, instruments, measurement science, and standards to enable safe, effective, and affordable commercial-scale production of nanotechnology products (Sargent Jr., 2009). A well controlled, low-cost, knowledge-intensive and minimal-labor component is required for mass production of the nanosystems.

The Trends

In the United States, many states are anchoring their technology-based economic development on nanotechnology. China is also investing heavily in this technology. The trend is anchored on the tripod of government, universities and industry partnerships towards developing IP (intellectual property) with commercial values (Armstrong, 2008; USDO, 2006).

Nanotechnology is maturing with many organizations focusing on areas with potential market impacts. There are more than 800 nano-related products in the market with $50 billion revenue (globally) in 2006 and there is potential for the industry to become a multi-trillion business in few years (Michelson, 2006). The next decade will usher the period for harvesting some of the basic works as application concepts are developed. Market focus and cost efficiency will dominate the next phase and nanoproducts and nano-driven industrial processes must demonstrate market viability. This is expected as investors begin to plan exit strategies and the global economy recovers from recession. As the technology diffuses globally, there is need for effective oversight mechanisms for environment and public health issues, internationally coordinated risk research strategies, expanded public awareness to change some negative perceptions (Michelson, 2008). Because public attitude to nanoproducts could affect

their market acceptability, R&D environment, and regulation (Sargent Jr., 2009), public education is very important. A number of potential challenges to the continuous progress in nanotechnology have been identified and they include Armstrong (2008):

- The industry is capital intensive and not many universities and firms can participate in the R&D
- The venture funds is still low due to the uncertainty of the technology
- Many nanoproducts are sub-standards with poor quality and replicability. The industry standardization process is poor and effective techniques to evaluate products safety, environmental and health issues are yet to be developed.
- Public perception and lack of understanding of the technology continue to undermine efforts for allocating investment and R&D funds
- The technology requires highly skilled workforce and support services which are not readily available, especially in the developing and emerging nations
- There are many legal uncertainties with un-standardized regulatory guidelines. Besides, the overlapping IPR poses a challenge along with increasing advocacies for customer protection (ASECO, 2008)

The diverse applications of nanotechnology pose difficult challenges and hence it is implicated in broad ethical issues that range from medicine to information management. This broadness calls for case-by-case ethical evaluation in the areas of research, development and dissemination (Michelson, 2008). This is necessary since there is the possibility of using the same nanotechnology innovation for both good and bad causes; so, application-regulation is more important than science-regulation. An engineer that manipulates molecule for HIV/AIDS transmission prevention could use the same raw materials to build bio-weapon. How these two activities are regulated must be different as regulating the science itself will not be sufficient. While the former is just, humane, sustainable, and necessary the latter is not.

Going forward, nanotechnology could be a displacing technology; however, majority of it will be a complementary one. From UNESCO-sponsored study in 1996, "nanotechnology will provide the foundation of all technologies in the new century" (Mooney, 1999; Maclurcan, 2005) and it will have huge societal impacts. The expected magnitude of the impacts is the major reason why the regulation of the technology and its applications must ensure safety, reliability and responsibility while enabling market success. These regulations must be supported by scientific studies, instead of being based on apocalyptic imaginations that derail the science and harm efforts to develop ethical guidelines, health, safety or environmental risks assessments (Nanotechnology Now, 2009).

Microelectronics

The remarkable success of information and telecommunication technology within the last few decades has been facilitated by the phenomenal growth of the microelectronics technology (Ekekwe, 2007). While nanotechnology has future prospects, microelectronics has already transformed global competition and commerce. It offers strategic advantages to firms, institutions and nations through its capacity to develop products and services cheaply and efficiently. It is the engine that drives present global commerce and industry.

The world has experienced many new dimensions in knowledge acquisition, creation, dissemination and usage (Radwan, 2009) courtesy of this technology. The advancement of Internet and digital photography could all be linked to better performance from microchips. When microelectronics technology advances, a dawn emerges in global economy in speed, efficiency and capacity (Ekekwe, 2007).

Microelectronics is considered a very revolutionary technology considering the disruptions it has brought to the dynamics of the global economy via its different applications since its invention by Jack Kilby in the late 1950s (see Table 2). Of the gross world product (GWP), estimated (2007) at about $55 trillion (currency) (The Economist, 2008), microelectronics contributes more than 10%. Microelectronics is very pivotal to many emerging industries in the 21st century with a central position in the global economy (Sicard, 2006). Because Internet, medicine, entertainment and many other industries cannot substantially advance without this technology, it has a vantage position in engineering education in many developed nations. These nations invest heavily in microelectronics education as in the United States, Canada and Western Europe where the MOSIS, CMC and Europractice programs respectively enable students to fabricate and test their integrated circuits for full cycle design and learning experience on integrated circuits. On the other hand, developing nations increasingly lag behind in adopting and diffusing this technology in their economies owing to many factors, which include human capital and infrastructure. The same problems that are hindering the global penetration of nanotechnology affect microelectronics. Absence of quality technical education has contributed to stall the transfer, diffusion and development of microelectronics in both the emerging and developing economies. This is why the technology despite a long history of success has not penetrated globally. Bottom-up creative technology diffusion model anchored on developing nation's tertiary institutions, and small and medium-scale enterprises (SMEs) is required for creative and sustainable microelectronic programs in these nations.

Just as there are public anxieties in some quarters on potential job disruption by nanotechnology, many expressed similar feelings (though, the scale is different) on microelectronics. As the technology was evolving, there was fear that it could cause major unemployment crises by enabling industrial automation. Fortunately, microelectronics has actually contributed to global economic growth and enabled a generation of new class of workers, knowledge workers. But it is not nanotechnology. Nanotechnology has the capacity to transform the traditional global raw materials supply industry which benefits many developing nations. Materials like copper, rubber and cotton that are shipped from low-income nations as major foreign earners to advanced ones could be substituted by alternatives created in the labs. This potential massive disruption in GDPs of these poor nations and resultant job displacements could be catastrophic, both politically and economically. Microelectronics offers an example to understand what the potential nano-driven economy could look like; but in many cases, it will not give an accurate picture if the potentials of nanotechnology are fully harnessed. A model for the diffusion of this technology in developing nations will involve a sound technical education program, multi-chip project, microelectronics academic network, etc (Ekekwe, 2009a).

The Science

While nanotechnology is an evolving technology, microelectronics has relatively matured. Micro-

electronics is a group of technologies that integrate multiple devices into a small physical area (FMNT, 2009). The dimension is about 1000 larger than nanotechnology dimension; micrometer vs. nanometer. Usually, these devices are made from semiconductors like silicon and germanium using lithography, a process that involves the transfer of design patterns unto a silicon wafer (Ekekwe, 2009b). There are accompanying processes which include etching, oxidation, diffusion, etc. Several components are available in microelectronic scale such as transistors, capacitors, inductors, resistors, diodes, insulators and conductors. The microelectronics can be divided to its subfields which in turn are connected to other micro related fields. These subfields are micro electromechanical systems (MEMS), nanoelectronics, optoelectronics and single electron devices (FMNT, 2009). Integrated circuits or microchips are typical microelectronic devices, which can be found in computers, mobile phones, medical devices, toys and automobiles. There is a high level of convergence between nanotechnology and microelectronics. The major difference lies in the size of the materials; nonetheless, the techniques are very different. Complementary metal oxide semiconductor (CMOS) transistor is the most common transistor used in the industry owing to its ease of integration and low static power dissipation (Ekekwe and Etienne-Cummings, 2006). Bipolar junction transistor is

Table 2. The Invention of the Transistor and Integrated Circuit

The Invention of the Transistor and Integrated Circuit
Throughout the first half of the twentieth century, radio valves played an important role within electronic products. In 1947, however, scientists at the AT&T Bell Laboratories developed a device that would revolutionize the whole economy: the transistor.
The first demonstration of the transistor was carried by William Shockley, John Bardeen and Walter Brattain, and the three would later receive the Physics Nobel Prize.
One of the most important discoveries related to the transistor was the fact that some materials were neither electrical conductors nor electrical resistors, they were in fact semiconductors. Silicon, for instance, is a semiconductor and William Shockley figured that he could change the properties of semiconductors by "doping" it with certain substances.
The interesting fact about the invention of the transistor is that AT&T failed to transform it into innovation. The invention was obviously patented, but the organization was not able to find promptly an application for the new device. They did an outstanding job with the invention, but failed to commercialize it. Precisely for that reason in 1952 AT&T decided to license out the transistor. For $ 25,000 companies like Sony and IBM acquired a technology that would produce billions of revenues in the coming years.
With the successful integration of circuit elements such as transistor, resistors and capacitors onto a die by Jack Kilby and Robert Noyce, the world of microelectronics was born.

Adapted from D. Scocco, 2006

another popular version. With the sizes of CMOS transistor in the nanometer range, the behaviors of the transistors are radically affected by parasitic noise and power dissipation. These problems pose potential challenges to the continuous progress of CMOS technology and microelectronics industry in general. The survivability of Moore's Law, (after Gordon Moore, co-founder of Intel Corp) which states that the numbers of transistors in a semiconductor die double every 18 to 24 months, is presently challenged if engineers cannot downscale the transistor size any further efficiently. This scaling has been the driver that has enabled microelectronics products to improve in speed, capacity and cost-efficiency. Many efforts have been geared to overcome the problems faced in the industry as transistors scale into the deep nanometer. They include improving the structure of the metals and polysilicon materials used in making the devices, more enhanced doping profile, new materials to keep the industry alive and well into the future (Ekekwe, 2006).

Trends

The Moore's law has been the gauge on the advancement of microelectronics (Ekekwe, 2009c). The ability to sustain the law for more than five decades shows the level of innovation in the industry. IBM and the Common Platform (a collaboration between IBM, Chartered Semiconductor Manufacturing, and Samsung Electronics developed to implement a common process technology across all three companies' semiconductor manufacturing facilities) already has a 32 nanometer (32 nm) process available. The 32 nm process is the next step after the 45 nm process in CMOS manufacturing and fabrication. Firms like Intel and AMD, major microprocessor vendors, are already working on the 32 nm process for logic. However, as transistors are scaled into deep nanometer scale, they will eventually reach the limits of miniaturization at atomic levels. Around that time, microprocessors

will contain about tens of billions of transistors. Intel Corp expects the end to come before 2020 with 16 nm CMOS manufacturing process and 5 nm gates due to quantum tunneling (Ekekwe, 2006). Lawrence Krauss (2004) expects an ultimate limit at about 600 years based on rigorous estimation of universal limits of computation (Krauss, 2004).

Despite all these predictions and discussions, it is important to note the level of resilience in microelectronics and technology in general with a deep history of overcoming obstacles. The extremely futurists expect the law to ultimately lead to a technological singularity, a period where progress in technology occurs almost instantly (Kurzweil, 2005)), however, *International Technology Roadmap for Semiconductors,* with objective to ensure advancements in the performance of integrated circuits and remove roadblocks to the continuation of Moore's Law is not that optimistic. Figure 3 shows the industry trend, measured by the size of lithography, over the years. The size of lithography is expected to reach 22 nm in 2013 and that is deep nanometer scale with all the associated quantum mechanical effects.

It is evident that the truism that the 'only constant is change' applies to innovation because innovation is change with higher value. Predicting the future of science based on the understanding of the present science makes no sense since inventions of tomorrow will unlock the realities of after-tomorrows. The key aspect of innovation is meeting not just the needs and expectations of the customers or the marketplace, but exceeding their perfections. Nanotechnology and microelectronics will remain central in creating innovative products and services and possibly herald the next technology that will come after them. Developing nations have opportunities to learn and it is imperative they do so quickly to avoid devastating impacts technologies and innovations will bring. The Irish story on innovation is a good case study (Table 3).

Figure 3. Technology scaling with year (source: Sicard, 2009)

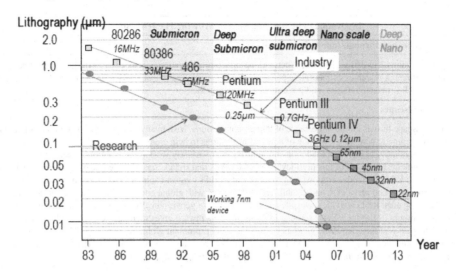

NANOTECHNOLOGY AND MICROELECTRONICS GLOBAL DIFFUSION TRAJECTORY

The U.S. National Nanotechnology Initiative defines nanotechnology as "the science, engineering, and technology related to the understanding and control of matter at the length scale of approximately 1 to 100 nanometers". However, "nanotechnology is not merely working with matter at the nanoscale, but also research and development of materials, devices, and systems that have novel properties and functions due to their nanoscale dimensions or components". A joint report by the British Royal Society and the Royal Academy of Engineering similarly defined nanotechnology as "the design, characterization, production, and application of structures, devices and systems by controlling shape and size at nanometer scale" (Sargent Jr, 2009). Microelectronics, on the other hand, is related to the study and manufacture, or microfabrication, of electronic components which are very small (usually micrometer-scale or smaller, but not always). Both technologies are converging as microelectronics transitions into nanometer regime. Patents, academic journals and other metrics for ascertaining technology innova-

tion indicate that advanced nations dominate these technologies and the global diffusion trajectory will flow from them to other parts of the world.

A global technological progress—improvements in the techniques (including firm organization) by which goods and services are produced, marketed, and brought to market—has been at the heart of human progress and development (World Bank, 2008a). This is pivotal to stimulating income growth and poverty reduction and shaping the social and economic structures of nations for many decades. Over the last few decades, many technologies have evolved and disrupted the global business and economic structures. Especially, the Internet has affected many traditional industries and introduced new dimensions to commerce. Firms that have knowledge as its factor of production continue to challenge established ones. In the midst of all these revolutions, microelectronics has been at the center, enabling these disruptions through its efficiency, speed or capacity. Nanotechnology is new, but promises enormous potentials that will transform all fields of human endeavors. But these technologies are skill-intensive and their global diffusion will follow a peculiar pattern, from developed nations that invent to developing ones that adopt. Owing to

Table 3. Innovation: the key to business growth: the Irish story

Innovation: The Key to Business Growth: The Irish Story
Collaboration and co-operation through innovation networks
Corporations today are pursuing a globally-distributed, network approach to innovation. Current university programs and company R&D activities reach across borders in search of collaborative partnerships. Companies can most easily reap the rewards of innovation through a global ecosystem in which firms, universities, and governments work together.
Ireland's innovation landscape
Ireland's innovation landscape thrives on the importance of human connections. Irish business policy brings together - in a unique, no-nonsense and highly pragmatic way - a wide range of national institutions to help create leading edge research programs. Government, funding agencies, regulatory authorities, academia and industry are constantly working as a national team, creating a fast-growing, dynamic research environment. The result of this high-level connectivity is that Ireland has become one of the new global centers for science- and innovation-based R&D. Ireland is empowering some of the world's biggest companies to research, develop and commercialize world-class products, processes and services. Long-established partnerships with global corporations have been at the core of Ireland's success in attracting leading edge R&D activities. Despite Ireland's small size geographically, its energetic, knowledge-based economy wins a disproportionate amount of Europe's R&D centers. In 2006 Ireland's inward investment agency, IDA Ireland, supported 54 R&D investment projects. The past year has seen R&D announcements by many prominent global corporations. The names speak for themselves: CISCO, GlaxoSmithKline, PepsiCo, Intel, IBM, Bristol-Myers Squibb. These corporations are actively supported by renowned global research organizations located in Ireland, such as Georgia Tech Research Institute and Bell Labs.
An integrated, collaborative strategy
The Irish Government pursues a carefully planned, integrated R&D strategy encompassing all of the key elements necessary to achieve world-class R&D. Its US$5 billion 'Strategy for Science, Technology and Innovation' will double the number of Ph.D. graduates and attract future generations of well-educated young people into research careers in knowledge-driven companies. It will substantially extend the physical infrastructure to support them. And, for the first time ever, eight government departments will co-ordinate all activity in relation to science, technology and innovation. IDA Ireland is one of the main players behind the new wave of national, collaborative R&D activity. It works closely with Science Foundation Ireland (SFI), the agency which consolidates links between industrial and academic research and funds such research. IDA Ireland and SFI have developed a range of new initiatives to encourage pooled projects and attract world-class scientists to carry out research in Ireland. This inclusive way of bringing together industry and academia has led to a boom in research projects. More than 10,000 researchers are working on cutting edge R&D projects in Ireland. Many of them have relocated from the US, Canada, Japan, the UK, Switzerland and Belgium. Ireland's Centers for Science, Engineering & Technology ('CSETs') link scientists and engineers in partnerships across academia and industry. One such CSET is CRANN, the Centre for Research on Adaptive Nanostructures & Nanodevices. CRANN's mission is to advance the frontiers of nanoscience. It provides the physical and intellectual environment for world-class fundamental research, and has partners in Irish and overseas universities.
Tax and intellectual property
Ireland's intellectual property laws provide companies with generous incentives to innovate. The Irish tax system offers huge support to turn brilliant ideas into the finished article. A highly competitive corporate tax rate of 12.5% is a major incentive. No tax is paid on earnings from intellectual property where the underlying R&D work was carried out in Ireland. Ireland recently introduced a new R&D Tax Credit, designed to encourage companies to undertake new and/or additional R&D activity in Ireland. It covers wages, related overheads, plant/machinery, and buildings. Stamp duty on intellectual property rights has been abolished.
People skills
The IMD World Competitiveness Yearbook 2006 rates Ireland's education system as one of the world's best in meeting the needs of a competitive economy. It also ranks the Irish workforce as one of the most flexible, adaptable and motivated. Ireland's young workforce has shown a particular flair for collecting, interpreting and disseminating research information. Major investment in education has provided a skilled, well-educated workforce; Ireland has more than twice the US/European per capita average in science and engineering graduates.
A track record of success

continues on following page

the high cost in these technologies, government will have major roles. From facilitating economic environments that will enable the technologies if adopted to thrive to inducing foreign firms with opportunities to exploit the technologies with huge profits, it does not seem that firms can accelerate their penetrations without efforts of governments. Especially in Africa where instability is common, it will be very challenging for firms to invest in the scale associated with these technologies (WEF, 2009a). This calls for government participation in the process of technology diffusion, especially

Table 3. continued

Ireland's success in innovation spans a wide range of businesses and sectors. For example, some of the most exciting Irish-based product development has been in medical technologies. Over half of all the medical technologies companies based in Ireland have dedicated R&D centers. Boston Scientific researched and developed the world's first ever drug-coated stent using researchers in Ireland. Bristol-Myers Squibb's Swords Laboratories is the launch site for several new healthcare treatments used to treat hypertension, cancer and HIV/AIDS. GlaxoSmithKline's latest Irish R&D project involves groundbreaking research into gastrointestinal diseases, in collaboration with the Alimentary Pharmabiotic Centre in University College Cork. Recently Microsoft marked its 20th Irish anniversary by opening a new R&D center, creating 100 new jobs. The centre is working on a wide range of projects, including Digital Video Broadcasting (DVB) and SmartCard security technology. Intel, a significant supporter of education and training in Ireland, is engaged in several research collaborations with leading Irish universities, including Trinity College Dublin, University College Cork and Dublin City University. Intel's Irish operation is the global headquarters for the company's Innovation Centres. Analog Devices' long established R&D operation is heavily integrated into its Irish operation. Its 335-strong team has sole responsibility for the global design, manufacture and supply of value added high voltage, mixed signal CMOS products.

An exciting future of world-class innovation

Lucent Technologies' Bell Labs, one of the world's most eminent research institutions, has established its Center for Telecommunications Value-Chain-Driven Research in partnership with Trinity College Dublin. It will undertake research aimed at realizing the next generation of telecommunications networks. Georgia Tech Research Institute's new Irish operation will be a critical component of Ireland's innovation infrastructure. It plans to build up a portfolio of research programs and collaborations with industry which at full operation will employ 50 highly qualified researchers. Wyeth is establishing a bio-therapeutic drug discovery and development research facility at University College Dublin. It will utilize new technologies to discover the next generation of therapeutic biopharmaceuticals for the treatment of a wide variety of diseases. At an academic level, just one illustration of the integration in R&D activity in Ireland is Dublin City University's Biomedical Diagnostics Institute. It is carrying out cutting-edge research programs focused on the development of next generation biomedical diagnostic devices. Ireland's success is based on a culture of co-operation and collaboration to win complex, high value, sophisticated investments. The country's strong business philosophy of inclusiveness, informality and teamwork are the foundations on which Ireland is fast becoming an important player in the development of global innovation networks.

Source: Radwan (2009), Business Week and IDA Ireland

in nanotechnology where R&D is very high with uncertainties in return. Possible government subsidies of nanotechnology economic development will lower research costs, increase industry participation, stimulate technology transfer, and assist firms that commercialize nanotechnology discoveries (Armstrong, 2008).

For nanotechnology and microelectronics to be adopted into developing and emerging economies, these nations must develop national level strategies. The focus will be to assist firms and universities acquire tools and resources needed for R&D owing to the capital-intensive nature of the technology. This will be followed in parallel with massive investments in human resources and skills to facilitate the adopting process. At both secondary and tertiary education levels, developing nations have a challenge to upgrade the quality of their science and technology programs. A strategy that coordinates both educational levels is important as the secondary students must remain organic feeders to the tertiary institutions. Also, there

should be more levels of collaborations among the various institutions. Efforts must be geared to attract multi-national corporations (MNCs) through tax holidays and other means to invest and assist in knowledge transfer to the local industry.

A strategy similar to the Pennsylvanian state (USA) Ben Franklin Technology Development Authority University Research Funding designed to promote stronger synergy between university-based research and development and the transfer of technology as it relates to economic and work force development (newPA, 2009) is needed in these nations. The aims which can be customized to individual nations are as follows:

- Developing and increasing new technologies, escalating technology transfer, and enhancing university-based resources and skills
- Increasing commercialization of applications and processes through university, industry and government collaboration

- Forming new spin-off companies that are deriving a significant portion of its commercial activities from the use of technology and/or know-how developed at a tertiary institution
- Leveraging of funding by the federal, state and local government, philanthropic foundations, strategic investors, and industry sponsored research
- Creating consortia-driven, educational and workforce development programs
- Developing strategies for financial sustainability

Similar policies developed and adopted by the EU could be used for developing and emerging nations (EU, 2004) in mapping their nanotechnology diffusion. They include the following steps:

- Increase investment and coordination of R&D to reinforce the industrial exploitation of nanotechnologies whilst maintaining scientific excellence and competition
- Develop world-class competitive R&D infrastructure ("poles of excellence") that take into account the needs of both industry and research organizations;
- Promote the interdisciplinary education and training of research personnel together with a stronger entrepreneurial mindset;
- Ensure favorable conditions for technology transfer and innovation to ensure that R&D excellence is translated into wealth-generating products and processes;
- Integrate societal considerations into the R&D process at an early stage
- Address any potential public health, safety, environmental and consumer risks upfront by generating the data needed for risk assessment, integrating risk assessment into every step of the life cycle of nanotechnology-based products, and adapting existing methodologies and, as necessary, developing novel ones

- Complement the above actions with appropriate cooperation and initiatives at international level (EU, 2004).

Also, initiative similar to the United States National Science Foundation will be very useful in developing nations. Efforts must gear towards funding technology centers and establishing technology clusters across national regions and developing technological entrepreneurship besides eliminating the barriers to technology penetration (See Table 4). Major assessment is necessary in developing nations because these are high investment technologies. With limited resources, regional cooperation between universities, firms, national laboratories for resources and information may prove vital. The same calls for provision of international access to tools, manpower and infrastructure to poor nations at this stage of the technology cycle.

Many factors will continue to affect the trend of these technologies as well as their location and localization. Presence of centers of excellence, good universities, technology clusters, government labs, infrastructure, and skilled workers will shape the trajectories of diffusions. The more nations or regions have these capabilities, the easier for them to transfer the technologies. Advancements bring complexities, especially in microelectronics, and firms require pool of highly trained skill workers to remain competitive. Also, the design stage, manufacturing systems and validation stage have become complex and expensive. A modern semiconductor plant exceeds $3 billion in investments and not every firm can afford that during this time of global credit meltdown. So, presence of technology clusters and access to the right technology will continue to influence the diffusion trajectory.

Evolving economic dynamics and the intense competition in the ultra cost-intensive and –sensitive industries, has changed the ways many semiconductor firms operate. While few are still integrated device manufactures (IDMs), the business model of total in-house design and manufacturing for a typical product, have resorted to new models, focusing on distinct technological markets (Scott, 2007). Some of these firms operate under the following categories (Scott, 2007):

- Companies devoted solely to developing Semiconductor Intellectual Property (SIP), known as the 'Chipless' business model
- Companies that concentrate only on product design, referred to as the 'Fabless' business model
- Companies that offer contract semiconductor manufacturing to other companies, known as the 'Foundry' business model
- Companies that perform testing and packaging for other companies on a contract basis.

These new models offer the optimal paradigm for diffusion as barriers to entry is radically reduced when compared to when firms have to be IDMs. As firms focus on specific markets/areas where they have competence, the industry is emerging with higher level of coopetition, collaboration and cooperation. This evolving paradigm will favor developing and emerging nations that can deploy their scarce resources to develop and nurture a segment of the industry. Their universities and firms, who can share tools and equipment, could concentrate on the testing and packing, 'chipless', and 'fabless' models that seem to require lower cost-investment when compared with 'foundry' and IDM models. As they progress, they will incorporate other areas and possibly this makes it easier for them to connect into the network of semiconductor or nanotechnology knowledge creation. Unfortunately, the realities are that poor financial structures, unstable economic regime, unskilled human resources, social instability, inadequate and basic amenities may not enable them to participate in any of the models. So, the divide continues as they remain importers and consumers of these technologies, instead of

partners in developing them. Without breaking into the value chain, especially at the upstream sector of the technologies where knowledge is created, developing nations will have difficulties to advance economically. The future of global economy is rooted on knowledge with microelectronics and nanotechnology central to exploiting that knowledge.

REALTIES AND POSSIBILITIES WITH NANOTECHNOLOGY AND MICROELECTRONICS

Since technology and technological progress are central to economic and social well-being, the creation and diffusion of goods and services are critical drivers of economic growth, rising incomes, social progress, and medical progress (World Bank, 2008a). Developing nations lack behind in both the technology creation and dissemination. Their pace of technology adoption is low and the technology landscape remains poor. Though over the years, FDI, trade, and exposure to international technology has improved the penetration rate, the gap compared with high-income nations is still very large. The political climate, corruption, stifling business environment, poor infrastructures, lack of innovation culture, poor economy regime, along with low technology literacy are major challenges which must be overcome. It is well established that one of the key factors to technology adoption and transfer is knowledge barriers. While the world discussed digital-divide in the information technology era, the future will potentially will be nano-divide. The challenge is that nano will continue to enable economic concentration in developed nations (holders of core patents with economic rights) and developing ones will find it increasingly difficult to transition from their present states. Besides, with lack of innovation in developing nations, the disruption of global economic structures can harm the developing nation since they lack the resilience and fluidity to react to changes. The prospect of nano-weapons could be a concern in the hands of these unstable developing countries as they can self-destruct or destroy neighbors. Terrorism could escalate to a level not imagined, not just in the developed world, but globally as nanotechnology will make it easy to terrorize with devastating global impacts. The world could be visited with arms race and nuclear anti-proliferation could be relegated with anti- nano (weapon)-proliferation. If nanotechnology products could affect trade patterns with replacements of raw materials, the developing world would be the most affected as poverty could increase. Replacing the exports will increase global unemployment and that can pose global insecurity. The world within the last few centuries have depended on the raw materials of developing nations to sustain civilization, if nanotechnology can replace the needs of those materials, monumental upheavals could result in these countries with worthless cotton, copper, rubber, among others.

Nanotechnology must not be viewed as a fix to all the technology problems in the developing world; in other words, it must not be diffused without examining alternatives which may be more appropriate to the particular nation. Cautious and systematic approach is needed as these nations develop plans for the adoption of any aspect of nanotechnology. Without this strategy, the technology may not be sustainable as previous efforts have shown. An assessment of national activity by HDI (Human Development Index) groupings shows that the strength of developing country engagement with nanotechnology correlates positively with HDI rank (Maclurcan, 2005). The same would be expected in the Knowledge Economy Index; innovation either in process or products correlates with economic regime. As developing nations improve their KEI, there is expectation that technology transfer capacity will improve.

While many policies have been centered at national level, some developing nations may require protection from predatory firms which

may take advantage of them. It may be necessary to implement international policy that will guide nanotechnology transfer so as to protect poor vulnerable nations without expertise to understand the risk aspects of the technology. It is almost certain that nanotechnology will exacerbate the economic divide between the advanced and poor nations; nonetheless, this policy will protect them from nano-waste and –dump. EU and United States dominate major policies in nanotechnology and those policies will likely evolve to become international standards. Broader discussion will arguably result to standards that will be universal, equitable and accommodative of the views of other nations. Now is the time to implement universal policies towards a reliable, sustainable and profitable nanotechnology era for all stakeholders.

Funding

Governments and investors (equity and capital funding) will be very important in the growth of nanotechnology (Harper, 2009). But the uncertainties in many areas of nanotechnology mean that funding could be a challenge during a time of global economic meltdown where investors have lost money and government lost revenues. Harper (2009) reports that in the US 19,300 firms received venture funding while only 351 were venture capital-backed IPOs since 2002, and only about 13% of the VC exited. As the 2009 global liquidity crises lessens, funding in 2010 for emerging technologies will be expected to improve and many nanotechnology firms will be ready for IPOs. The US government stimulus money will also offer source of funding that will help the technology sector in general. Lux Research, a consulting firm, estimates that global nanotechnology venture capital investment in 2007 was $702 million, of which 90% went to U.S. based firms (Sargent Jr., 2009).

United States has appropriated about $9.9 billion for nanotechnology R&D via National Nanotechnology Initiative since its inception in

Table 4. Potential barriers to implementing emerging technologies in developing countries

Barrier	Comments
Absorptive capacity	Inadequate ability to recognize, place value upon, internalize and apply new knowledge
Attitudes and perception	Acceptability, perceived needs based on a needs analysis, attitudes towards technology, concepts of development and aid, and focus on the problems to be solved (i.e. being people driven and problem-oriented not kit-driven)
Cultural and community issues	Language, cultural views towards technology, sharing of resources within the community, appropriateness of a specific technology within a given culture or community, literacy requirements, gender issues and access issues
Legal and ethical Issues	Privacy, confidentiality, security, malpractice potential, insurance, jurisdiction, copyright, patents for new technologies and treatments, other intellectual property issues
Technical issues	Access to electricity grid and alternative power supplies, power schedules and reliability, UPS back-ups, ongoing maintenance of computers. Inappropriate access devices and inappropriate Internet technologies including low bandwidth. Insufficient language and cultural adaptation of content and the digital divide
Environmental issues	Effects of weather, temperature, humidity and dust on equipment. Security and accessibility of equipment. Isolation, transport issues
Sustainability issues	Ongoing upgrades of technology, ongoing costs, issues cost-effectiveness
Practical issues of working internationally	Corruption, borders and customs in equipment transport, nationally-imposed barriers to information access or dissemination or to information privacy, donor-imposed barriers, time zones and communication issues of working in remote geographical areas
Health care infrastructures	In health, insufficient means to implement health care and take full advantage of leapfrog ICT technologies, e.g. lack of treatment facilities, drug delivery systems, inadequate cold chain facilities for vaccines

(Source: McConnell, 2008)

2000; other 60 nations have similar programs (Sargent Jr., 2009). About $12.4 billion was spent in 2006 from both the private and public at roughly equal percentages (Sargent Jr., 2009); the estimated R&D is about $12 billion annually (Michelson, 2008). Two BRIC nations, China and India, have made nanotechnology central to their future developments (Parker, 2008) and other nations are implementing similar polices. South Africa remains the only Sub-Sahara African nation with a sound nanotechnology strategy.

Intellectual Property

Technological progress drives national economic development. Increasingly, sustainable national wealth depends on knowledge creation, acquisition and diffusion. The global economic growth rate was insignificant before the industrial revolution and ever since the world has experienced new technology changes. These technologies have brought enormous growths to global wealth. United Nations Conference on Trade and Development (UNCTD, 1986) has noted that as technology diffuses across national boundaries and firms, intellectual property rights (IPRs) affect the dynamics of technology innovation- the invention and commercialization. Ineffective IPRs will slow the diffusion process as that contributed to lack of technology innovation for many centuries. The drive to innovation is usually attributed to potential economic benefits; without IPRs, this drive could be slowed. This explains why developing nations with low IPRs experience low technology achievement. With no legal structures to protect property, firms may not like to transfer technology since it can be stolen easily by a competitor. Given the dearth of avenues to protect new ideas and economically benefit, developing nations will continue to have problems in technology adoption and diffusion (Coe, 1997). The ability for developing nations to adopt and acquire the tacit knowledge (organizational and managerial-process innovations) of nanotechnology and microelectronics will depend on their capacities to protect property rights in these fields. While it is important for these nations to attract FDI, license, collaborate and trade for the purpose of technology transfer, it is imperative to understand that only a strong IPRs legal system can encourage MNCs to efficiently develop relationships for transfer. There is a positive correlation between IPRs and technology transfer. The advocacy for open IP structure for nanotechnology (Thakur, 2008) especially for the benefits of developing nations may not materialize. The technology is very expensive that open source model will hinder the progress; only the motivation for wealth creation will ensure innovation and sustainability.

Environmental, Health and Safety Implications

Just like nanotechnology is very broad, the environmental, health and safety issues are varied. The broad spectrum of these issues concern all aspects of nanotechnology activities: laboratory, workplace, consumers and the environment. The rapid rate of advancement and drive to make commercial gain could result to sacrificing safety for commercial gains. Many stakeholders believe that concerns about potential detrimental effects of nanoscale materials and products on health, safety, and the environment—both real and perceived—must be addressed for a variety of reasons, including (Sargent, 2009):

- Protecting and improving human health, safety, and the environment
- Enabling accurate and efficient risk assessments, risk management, and cost-benefit trade-offs

- Creating a predictable, stable, and efficient regulatory environment that fosters investment in nanotechnology-related innovation
- Ensuring public confidence in the safety of nanotechnology research engineering, manufacturing, and use
- Preventing the negative consequences of a problem in one application area of nanotechnology from harming the use of nanotechnology in other applications due to public fears, political interventions, or an overly-broad regulatory response
- Ensuring that society can enjoy the widespread economic and societal benefits that nanotechnology may offer.

Many nations have developed structures and guidelines for monitoring and regulating many aspects of nanotechnology business (from research to products) to implement acceptable public health and safety levels. The EU has grouped the relevant legislation under four categories - chemicals, worker protection, products and environmental protection (EU, 2004).

Global Market Disruption

Nanotechnology upon maturity will radically change the structure of the global markets. As the technology advances, developing nations, if unprepared, could witness dramatic consequences in terms of trade and employment opportunities. One area of these concerns were captured by Friends of the Earth (see Table 5).

DRIVER FOR GLOBAL INNOVATION ECONOMY

The cyclical boom and bust of leading national economies demonstrate the levels of complexity that confront regulators, investors, markets and industries. While technology has positively increased both individual and national wealth, which for centuries was static, it has also brought major management challenges. The world economy can accelerate very fast, but can also fall as well. As the present global recession shows, the world has not decoupled the capacity to use technology to mitigate this vicious cycle. While all nations are hurt, some are severely hurt since they lack the technology that can help them accelerate progress during recovery. Over time, the real economic progress rate is the difference between the rates of boom and bust averaged over the time period. A 21st century economy, dubbed the innovation economy, is expected to be an economy with advanced manufacturing and knowledge-intensive jobs through emerging technologies like nanotechnology and green technology via ultra competitive global industries and workers. Medicine, finance, capital markets, entertainment, and indeed all industries will become technology industries. Their survival and growth will depend on the level of technology innovations used. That brings the challenge of how the world can use technology to actually 'create the future' and hence effectively 'predict it' with certainty. The world needs a technology solution that will provide powers that can mitigate factors that contribute to global economic downtown. That technology must have the capacity to drive all global industries seamlessly by enabling the right type of tools that will help regulators stay ahead of capital markets, ahead of mortgage crises, ahead of disease outbreaks, etc, and offer corrective measures before the economy reaches that decelerating turning point. An innovation economy must not be an economy of boom and bust. The major questions are these: are there algorithms which the powers of nanotechnology could help to unlock? Can nanotechnology prevent this cycle by offering more powers to computational systems to extract information and knowledge, requisite to development, and thereby manage the complex global economic variables, preventing busts? Can nanotechnology evolve an era of continuous economic growth and capitalism? Certainly, time will tell.

CONCLUSION

Technology will continue to drive global economic and social progress bringing improvements in people, processes and tools while shaping institutions and nations. Nanotechnology will be very central to new economic disruptions in the next few years. Just as microelectronics has brought phenomenal progress in industrial efficiency, capacity and speed, nanotechnology will drive a new international economy whose competition will be technology. From health to entertainment

Table 5. Global market disruption

Global Market Disruption
In the short-medium term, novel nanomaterials could replace markets for existing commodities, disrupt trade and eliminate jobs in nearly every industry. Industry analysts Lux Research Inc. have warned that nanotechnology will result in large-scale disruption to commodity markets and to all supply and value chains: 'Just as the British industrial revolution knocked hand spinners and hand weavers out of business, nanotechnology will disrupt a slew of multibillion dollar companies and industries'.
Technological change and the social disruption it brings have been with us for millennia. What will be different this time is that we are confronting the potentially near simultaneous demise of a number of key commodity markets where raw resources (eg cotton, rubber, copper, platinum) may be replaced by nanomaterials, with subsequent structural change to many industry sectors.
The displacement of existing commodities by new nanomaterials would have profound impacts for economies everywhere. However it would have the most devastating impact on people in the Global South whose countries are dependent on trade in raw resources - 95 out of 141 developing countries depend on commodities for at least 50% of their export earnings.
Cotton is an example of an important commodity that could be displaced by the introduction of novel nanomaterials. There are currently an estimated 350 million people in the world directly involved in the production of cotton. Countries in the Global South such as Burkina Faso, Benin, Uzbekistan, Mali, Tajikistan, Cote D'Ivoire, and Kazakhstan rely on cotton as a major source of revenue.

Source: AZoNano (2009)

and to neuromorphics, these two technologies will be pivotal to solving many global problems like environmental and climate issues, HIV/AIDS cures, cancer, Parkinson disease, etc. Nations that develop and commercialize nanotechnology will reap enormous economic benefits and this technology could possibly restructure the dynamics of global competitiveness in this new millennium. Unfortunately, it has the possibility of evoking global crises if many raw materials usually imported from developing nations are engineered through nanotechnology in developed nations thereby depriving the poor economics their sources of foreign earnings. But the developing economies have opportunities to invent and structure themselves to depend on knowledge, instead of minerals, to avoid obsolescence in the emerging innovation economy. Planning for nanotechnology economy, nanomics will be very important as convergence of nanotechnology and microelectronics will redesign all global economic, political and social structures. A Nano World Order of nano-ethics, anti-nanoweapon-proliferation, nano-civilizations, and nanomics. New embodiments of knowledge, driven by issues, holistic in approach, strategic and proactive in plans, and designed for intergenerational commitment, and sustainability will be needed to manage the convoluted and complex factors that range from ethics to global safety.

ACKNOWLEDGMENT

This chapter was partly funded by Samstag Fellowship-awarded by Whiting School of Engineering, Johns Hopkins University (JHU), USA. * This chapter was written while in JHU.

REFERENCES

Alliance of Social and Ecological Consumer Organisations (ASECO). (2008). ASECO opinion on nanotechnology, Copenhagen, Denmark, 2008

Armstrong, T. O. (2008). Nanomics: The Economics Of Nanotechnology And The Pennsylvania Initiative For Nanotechnology. *Pennsylvania Economic Review, 16*(1).

ASTM. (2006). *American Society for Testing and Materials Publications*. Retrieved from http://www.astm.org/

AZoNano. (2009). *Nanotechnology Led Changes To Manufacturing, Defence, Farming, Human Development and the Possibility of Large Scale Social Disruption As Predicted By The Friends of The Earth*. Retrieved from http://www.azonano.com/Details.asp?ArticleID=1876

Boston Consulting Group and National Association of Manufacturers. (2009 March). *The Innovation Imperative in Manufacturing: How the United States Can Restore Its Edge*. Boston, USA.

Chinn, M. D., & Fairlie, R. W. (2006 December). The determinants of the global digital divide: a cross-country analysis of computer and internet penetration. *Oxford Economic Papers.*

Coe, D. T. (1997). North-South R&D spillovers. The Economic Journal, 107(134).

Dabic, M. (2007). *Gaining from Partnership: Transfer Technology - Issues and Challenges in Transitional Economies*. Retrieved from http://inderscience.blogspot.com/2007/10/call-for-papers-gaining-from.html

Ekekwe, N. (2002). *Telematics and Internet Banking: Implications, SWOT analyses and Strategies-A case study of Diamond Bank Limited*. MBA Dissertation, University of Calabar, Nigeria.

Ekekwe, N. (2009a). Towards competitiveness and global outsourcing: practical model for microelectronics diffusion in Africa. In *Int'l Conference on Industry Growth, Investment and Competitiveness in Africa (IGICA)*, Abuja, Nigeria.

Ekekwe, N. (2009b). *Reconfigurable Application-Specific Instrumentation and Control Integrated Systems*. PhD Dissertation, Electrical & Computer Engineering, Johns Hopkins University, Baltimore.

Ekekwe, N. (2009c). Adaptive Application-Specific Instrumentation and Control Integrated Microsystems. Koln, Germany: LAP Academic Publishing.

Ekekwe, N., & Ekenedu, C. (2007), Challenges and Innovations in Microelectronics Education in Developing Nations. In *IEEE International Conference on Microelectronic Systems Education*, San Diego, CA, USA.

Ekekwe, N., & Etienne-Cummings, R. (2006). Power dissipation sources and possible control techniques in ultra deep submicron CMOS technologies. Microelectronics Journal, 37(9), 851–860. doi:10.1016/j.mejo.2006.03.008doi:10.1016/j.mejo.2006.03.008

European Union (EU). (2004). Towards a European strategy for nanotechnology. In EU Policy for Nanosciences and Nanotechnologies, Brussels, Belgium.

FMNT. (2009). *What is microelectronics*. Retrieved from http://www.fmnt.fi/index.pl?id=2408

Golding, D. (2006). *United Kingdom Technology Strategy Board Annual Report*.

Harper, T. (2009). Nanotechnologies In 2009:Creative Destruction or Credit Crunch? Cientifica.

Hassan, M. H. A. (2007). Reforming universities is key to technology transfer. Trieste, Italy: Academy of Sciences for the Developing World.

International Monetary Fund (IMF). (2006). Integrating Poor Countries into the World Trading System. *Economic Issues, 37.*

Krauss, L. M., & Starkman, G. D. (2004 May). Universal Limits on Computation. *Physical Review Letters.*

Kurzweil, R. (2005). The Singularity is Near. New York: Penguin Books.

Maclurcan, D. C. (2005 October). Nanotechnology and Developing Countries Part 2: What Realities? *Journal of Nanotechnology Online.*

Maclurcan, D. C. (2005 September). Nanotechnology and Developing Countries Part 1: What Possibilities? *Journal of Nanotechnology Online.*

McConnell, H. (2008). Leapfrog technologies for health and development: Technological innovations. *Global Forum Update on Research for Health, 5.*

Michelson, E. S. (2006 October). *Nanotechnology Policy: An Analysis of Transnational Governance Issues Facing the United States and China.* Project on Emerging Nanotechnologies, Woodrow Wilson International Center for Scholars, Washington, DC.

Michelson, E. S., Sandler, R., & Rejeski, D. (2008). Nanotechnology. In M. Crowley (Ed.), From Birth to Death and Bench to Clinic: The Hastings Center Bioethics Briefing Book for Journalists, policymakers, and Campaigns (pp. 111–116). Garrison, NY: The Hastings Center.

Mooney, P. (1999). The ETC century erosion, technological transformation and corporate concentration in the 21st century. Development Dialogue, (1-2): 1–128.

Nanotechnology Now. (2009). *Nanotechnology Introduction.* Retrieved from http://www.nanotech-now.com/introduction.htm

New, P. A. (2009). *Pennsylvania Initiative for Nanotechnology*. Retrieved from http://www.newPA.com

Oshikoya, T. W., & Hussain, M. N. (1998). Information Technology and the Challenge of Economic Development in Africa. *Economic Research Papers*, 36.

Parker, R. (2008). Leapfrogging Development through Nanotechnology Investment: Chinese and Indian Science and Technology Policy Strategies. In *China-India-US Workshop on Science, Technology and Innovation Policy.*

Pourrezaei, K., Carpick, R., & Anthony, P. G. (2007). The Nanotechnology Institute: A Comprehensive Model for Nano-Based Development (pp. 1–148). Proposal to the Ben Franklin Technology Development Authority, Drexel University, University of Pennsylvania, Ben Franklin Technology Partners of Southeastern Pennsylvania.

Radwan, I., & Pellegrini, G. (2009 February). *The Knowledge Economy Gateway to Nigeria's Future*. Abuja, Nigeria: World Bank.

RNCOS. (2006 August). *The World Nanotechnology Market*. Noida, India.

Roco, M. C. (2003). Nanotechnology: convergence with modern biology and medicine. Current Opinion in Biotechnology, 14, 337–346. PubMeddoi:10.1016/S0958-1669(03)00068-5doi:10.1016/S0958-1669(03)00068-5

Sargent Jr., J. F. (2009 February). *Nanotechnology: a policy primer*. Washington, DC: Congressional Research Service.

Saxton, J. (2007 March). *Nanotechnology: The Future is Coming Sooner Than You Think*. Washington, DC: Joint Economic Committee.

Scott, I. (2007 December). *Revitalizing Ontario's Microelectronics Industry*. Information Technology Association of Canada.

Sicard, E. (2009). Electromagnetic Compatibility of IC's. Toulouse, France: Advances and Issues, Past, Present and Future.

Sicard, E., & Bendhia, S. (2006). Basics of CMOS Design. New York: McGraw Hill.

Thakur, D. (2008, November), The Implications of Nano-technologies for Developing Countries - Lessons from Open Source Software, workshop on Nanotechnology, Equity, and Equality Center for Nanotechnology in Society, Arizona State University The Economist. (2008). *Pocket World in Figures* (2009 Ed.). New York: Profile Books Ltd.

United Nations Conference on Trade and Development (UNCTD). (1986). *Periodic Report 1986: Policies, Laws, and Regulations on Transfer, Application, and Development of Technology* (TD/B/C.6/133). New York: United Nations.

United Nations Development Programme (UNDP). (2008). Human Development Indices: A statistical update 2008. New York: HDI.

U.S. Department of Commerce (USDO). (2006 August). *Resource Guide for Technology-Based Economic Development: Positioning Universities as Drivers, Fostering Entrepreneurship, Increasing Access to Capital*. Washington, DC: Economic Development Administration, U.S. Department of Commerce.

World Bank. (2008a). Global Economic Prospects-Technology Diffusion in the Developing World. Washington, DC.

World Bank. (2008b). *Knowledge Assessment Methodology (KAM)*. Retrieved from http://www.worldbank.org/kam

World Bank. (2009). *World Development Report 2009: Reshaping Economic Geography*.

World Economic Forum. (WEF). (2008). *The Competitiveness Report 2008-2009*. Geneva, Switzerland.

World Economic Forum. (WEF). (2009a). The Africa Competitiveness Report 2009. Geneva, Switzerland.

World Economic Forum. (WEF). (2009b). The Global Enabling Trade Report 2009. Geneva, Switzerland.

Chapter 2
Molecular Manufacturing:
Nano Building Nano

Chris Phoenix
Center for Responsible Nanotechnology, USA

ABSTRACT

A wide variety of nanotechnology programs, both pedagogical and research-oriented, can incorporate some aspect of molecular manufacturing. Nanotechnology is developing from large tools that give us access to the nanoscale, to tools constructed at or near the nanoscale. To date, these nanoscale tools are not capable of accomplishing much of commercial interest; however, this will be changing with increasing rapidity over the next decade or two. Eventually, nanoscale tools will be capable of broad classes of nanoscale construction; this will enable the fabrication of increasingly complex and useful structures. Fabrication of structures with molecular precision is especially relevant. When nanoscale tools are developed to the point of being capable of building duplicate tools, a manufacturing revolution may occur; even before that point, there are both scientific and likely commercial benefits to developing capabilities in this area.

NANOTECHNOLOGY AND MOLECULAR MANUFACTURING

The field of nanotechnology owes much of its popularity to molecular manufacturing (also called molecular nanotechnology or MNT), yet many nanotechnologists are only vaguely familiar with it. In general, nanotechnology involves creating or studying nanoscale structures–less than 100 nm in size. This includes a very broad range of fields, from computer chips to medical diagnostics.

Making nanoscale structures is difficult. Although some tools, such as scanning probe microscopes, can interact with the nanoscale directly, such tools are expensive and difficult to use. Indirect methods of creating nanoscale structures, including various techniques of chemistry, can be used, but the results can be difficult to characterize and control, and of limited complexity.

DOI: 10.4018/978-1-61692-006-7.ch002

As long ago as 1959, Richard Feynman pointed out that nanoscale tools could be built, which could perform advanced manufacturing operations with atomic precision (Feynman, 1959). Since the early 1980's, Eric Drexler has been describing various approaches to building the tools, calculating the potential performance of the resulting manufactured products, and promoting awareness of the implications of advanced molecular manufacturing (Drexler, 2009). Drexler's 1986 book *Engines of Creation* (Drexler, 1986) introduced the public to the word "nanotechnology," which he coined to refer to molecular manufacturing; the book generated substantial interest. When molecular manufacturing was first described in detail by Drexler in the late 1980's and early 1990's, it was seen as a transformative technology. There was a large gap between what was technically possible at the time, and the possibilities that could be projected from the well-understood laws of physics and chemistry. At the same time, the payback of a successful molecular manufacturing program was seen as so extreme that several theorists, including Drexler, expected a crash program to develop it, even at a cost of multiple billions of dollars. Accordingly, the first view of molecular manufacturing was of an advanced manufacturing system being developed with few precursors, and creating revolutionary change.

Early publications, including several fictional treatments and non-fiction books written for the general public, led to a backlash of skepticism. Almost from the start, some scientists were willing to publicly state that molecular manufacturing would be impossible, or else centuries in the future. Skepticism persisted despite the increasing level of detail in analysis of (still theoretical) capabilities and mechanisms, notably including Drexler's 1992 book *Nanosystems* (Drexler, 1992).

An early concern about the safety of molecular manufacturing appeared in the form of warnings about the potential for small self-contained self-replicating systems to duplicate out of control and consume large amounts of biomass. Although

these warnings originated from a mid-1980's picture of molecular manufacturing systems as being bacteria-like, and did not apply to later designs published as early as 1992, they played a starring role in an article written by Bill Joy, then chief scientist of Sun Microsystems, in 2000 titled "Why the future doesn't need us" (Joy, 1992). Joy claimed that a single laboratory accident with nanotechnology could destroy the world. The article appeared in Wired magazine and gained broad attention.

The U. S. National Nanotechnology Initiative (NNI) gained funding in 2001, to the tune of almost half a billion dollars per year (almost $1.5 billion in 2009). The NNI continued the process of broadening the meaning of "nanotechnology" far beyond molecular manufacturing, to include most research involving nanoscale features. With major funding at stake, many nanotechnology researchers felt a strong incentive to explain to the public why their work could not lead to runaway nanomachines eating the biosphere. This incentive dovetailed well with previous skepticism and misunderstanding about the science behind molecular manufacturing, and a number of scientists published poorly-considered opinions claiming that molecular manufacturing was theoretically impossible.

In recent years, it has become more broadly accepted that at least some forms of molecular manufacturing are possible and may be worth developing, though discussion continues about the likely levels of performance that can be obtained from nanoscale machinery. As nanotechnology has advanced, it seems likely that molecular manufacturing will be developed incrementally, with early systems having substantially less performance than the possibilities calculated by Drexler.

A decade ago, as a set of theoretical projections of extraordinary future potential but no currently feasible prototype, molecular manufacturing would not have been of interest to most nanotechnology programs. However, as an incrementally developing technology with some capabilities already emerging, molecular manufacturing may be of interest to many programs. Since it will require integrating multiple functionalities–structure, actuation, control, characterization, design, and so on–it provides a good stimulus for interdisciplinary and inter-institutional work. Since so many different areas of nanotechnology are potentially useful as one or another component of the system, and the design space is largely unexplored, there is room for even nascent nanotechnology programs to do useful theoretical work or laboratory research.

Much, perhaps most, of the theoretical work on molecular manufacturing to date has focused on high-performance integrated manufacturing systems or their immediate precursors. With the steady advance of nanoscale technologies, it will soon become feasible, and then straightforward, to design and build primitive molecular manufacturing systems using existing nanoscale technologies. There is plenty of room for innovation, creativity, and theoretical research aimed at developing architectures for such systems. At this writing, molecular manufacturing systems based on biopolymer derivatives represent a vast design space that is very poorly understood and that likely contains many opportunities for commercially relevant research.

SELF-COPYING MANUFACTURING SYSTEMS

Almost any object humans use derives most of its value from being manufactured. Faster, less expensive, more flexible manufacturing of more functional and higher quality products has been a key trend in the world's economy. A key contributing factor to the cost of manufactured products is the cost of their manufacture, including the cost of the manufacturing equipment (which must itself be manufactured) and the cost of the labor required to run it. A fully automated manufacturing system that could build copies of itself efficiently–in effect, becoming capable of self replication (but

only when supplied with feedstock and instructions)–might drive down the cost of manufactured products substantially.

A blacksmith's shop contains most of what is needed to build a complete copy of the shop. Blacksmithing requires only a few operations–heat, cut, bend, weld, and so on. These operations, applied in the correct sequence and with the correct variations, can turn simple metal bars into a shop full of tools. Until modern machining was developed, blacksmithing was a key manufacturing technology. Unfortunately, blacksmithing, for all its theoretical simplicity, requires a highly skilled blacksmith. Its operations are not precisely repeatable, so each handmade product requires adaptation and error correction. By today's standards, even the most finely worked product of blacksmithing is crude and imprecise. And although it might take only a few weeks to manufacture the tools for a duplicate blacksmith's shop, it takes years to train a blacksmith.

Modern manufacturing depends on a planet's worth of infrastructure. To build a factory requires wire, sheet metal, cast metal in many shapes, plastic in many forms, concrete, a vast array of fasteners, light bulbs, roofing material, computer chips, and too many other things to list. Most of these things need to be prepared at other locations by specialized techniques. Even the raw materials must come from a wide range of wells, mines, and fields, and each factory can make only a tiny fraction of the objects that went into its construction. Although the world's manufacturing infrastructure can and does double itself, it takes many years to do so. Manufacturing capacity remains scarce and expensive.

There is a kind of manufacturing that does not rely on human-operated machines; instead, it uses fully automated nanoscale systems to produce a wide range of sophisticated chemicals and structures from extremely simple inputs. Its products are quite low-cost. This is, of course, agriculture. Unfortunately, the raw products of agriculture can only be modified by plant breeding (or, very recently and with substantial limitations, by genetic engineering). Microbial fermentation illustrates the low cost of products of self-copying manufacturing systems; a few pennies worth of yeast can make many bottles of wine or loaves of bread, replicating as many times as necessary to process the feedstock in just a few hours. However, making anything but food out of plants requires either advanced technology, or additional manufacturing processes, or both.

What is needed, for manufacturing to reach its full potential, is a manufacturing system that is fully automated, uses simple raw materials, and can build advanced products including duplicate manufacturing systems. Molecular manufacturing promises to accomplish this.

MOLECULAR MANUFACTURING TECHNOLOGY

The core idea of molecular manufacturing is to make products by massively parallel, computer controlled, mechanically guided chemical synthesis. Instead of traditional manufacturing processes that deal with bulk materials, molecular manufacturing deals with individual molecules. Biological cells use a form of molecular manufacturing to make the proteins and other components of their structure and machinery. However, cells are programmed by DNA rather than a more traditional and easily-controlled computer, and their products are mostly limited to biopolymers such as DNA and protein.

Molecules can be combined by a variety of methods. At one extreme is self-assembly, as is used in DNA-based manufacturing, where built-in properties of the molecules cause them to find the correct orientation in the product. Because the process of self-assembly relies on repeated trial and error, the bonds between components tend to be fairly weak, and the assembly time increases rapidly as products get more complex. Human-engineered self-assembled DNA constructs have

become quite intricate and large recently, such as a 40-nm box with a lid that opens and closes (Douglas S.M, et al, 2009). However, the self-assembly process of the largest structures can require a week.

Fortunately, self-assembly is not the only option available for making precise products out of molecules. Molecules can be mechanically guided into their desired locations, and induced to bond once they are positioned correctly. This is called *mechanosynthesis*. Scanning probe microscopes have been used to do mechanosynthesis. Since the component molecules do not need to encode their own positional information, they can be smaller and simpler than self-assembled molecules, and the process of guiding them to the correct location can be quite rapid–at least in theory.

A hybrid technique, in between chemistry and positional mechanosynthesis, would mechanically control the sequence of molecules that are added to a polymer. Some existing techniques for DNA synthesis use a similar approach, controlling whether a new DNA base is added to a chain using optical illumination, but this has a fairly high error rate. Mechanical control of the sequence of chemistry–sequential mechanosynthesis–may be faster and less error-prone than even advanced DNA-fabrication techniques. (Preventing unwanted reactions by mechanical mechanisms, while allowing desired reactions to happen via conventional chemistry, is a gray area between chemistry and mechanosynthesis.)

In a manufacturing system based on mechanosynthesis, the complexity of the product does not depend on the complexity of the input molecules, but on the complexity of the operations of the machines that guide them. If those machines can be externally controlled–robotic, in a sense–then the product complexity is not even limited by the physical complexity of the machines, but rather by the aggregate complexity of the sequential control program. This means that there is no fundamental barrier to a manufacturing system building a product as complex as itself, or even a copy of itself.

In order to build large quantities of stuff out of molecules, the manufacturing system must be massively parallel and each machine must be quite fast. In order to be fast, the system should be small, because smaller systems tend to have a higher operating frequency. A machine on the scale of a bacterium might place on the order of 100,000 molecules per second, enabling it to process its own mass of feedstock (and produce its own mass of product) in a few hours.

To date, most fields of nanotechnology have used techniques, such as self-assembly or bulk chemistry, that do not require machinery capable of manipulating individual molecules. A few examples of such machinery exist, but the capabilities are far from mature, whereas self-assembly and chemistry have been under development for decades and centuries, respectively.

Using mechanosynthesis to build large products, or even large quantities of small products, will require a massively parallel manufacturing system made of highly functional nanomachines. It is unclear whether such a system can be constructed by self-assembly. An alternative is to build a very small molecular manufacturing system by slow techniques, and then use that system to build bigger systems. Once an initial nanoscale module has been built, by whatever means, then only a few dozen doublings are required to reach kilogram or decimeter scale. Depending on the design of the system, a kilogram-scale system may or may not be able to make kilogram-scale integrated products. However, a system that can be scaled up via self-manufacture will necessarily be able to process its own mass of feedstock in a few days or less, into products of complexity equivalent or better than the fabrication system; this compares very favorably with traditional mechanical manufacturing processes.

Controlling a kilogram of individual nanomachines may be less difficult than one might first assume. The information delivered by the control system need only be as complex as the product, and the information delivery mechanism can be even simpler than that. For example, if each machine was to make an identical nanoscale product, then kilograms of that product could be made by broadcasting identical instructions to a suitable number of modules. Even for large heterogeneous products, the communication system need only route appropriate parts of the blueprint to appropriate manufacturing modules, using routing information contained in the blueprint. Vastly intricate products can thus be made by a manufacturing system having a relatively simple communication and control system.

Potential of Molecular Manufacturing

Several technical factors indicate that molecular manufacturing, once developed to a mature technology, has the potential to be revolutionary. According to scaling laws, smaller devices have a higher operating frequency and power density, implying that motors and computers could be vastly more powerful than today's versions. The precision of molecular bonding and the repeatability of the resulting structures indicates the potential for extremely low error rates in the finished products; this in turn implies that materials could approach their theoretical (defect-free) strength, and that many manufacturing operations may be done without error correction. Precise molecular surfaces (in contrast to today's MEMS) can be engineered to have very low friction and near-zero wear (Dienwiebel M, 2004).

Mechanically guided chemistry may enable building classes of molecules that would be difficult or impossible with conventional chemistry. Today's chemistry requires that each potential reaction site must be chemically distinct from sites where reactions are not desired. Increasingly sensitive ways of distinguishing between similar reaction sites has made organic chemistry a rich and exciting field, but the problem remains difficult. In contrast, mechanically guiding the reactants can cause the reaction to happen at sites distinguished only by position, such as a certain spatial coordinate on a crystal surface. This means that highly crosslinked molecules should be buildable with atomic precision–strong and machine-like.

The book *Nanosystems* by Eric Drexler is currently the most detailed study of the capabilities of products that a mature molecular manufacturing system should be able to build. The Productive Nanosystems Roadmap, produced by the Battelle Memorial Institute and Foresight Nanotech Institute (Alivisatos P, 2007), reaffirms these calculations: for example, nanoscale motors may have a power density of 1 megawatt per cubic centimeter; CPUs may fit into a cubic micron, run at 100 GHz, and use less than a microwatt.

Milestones Toward Molecular Manufacturing

As nanotechnology develops toward molecular manufacturing, there are several capabilities that must converge to produce a useful manufacturing system.

Until molecular manufacturing is developed, the kinds of nanoscale machines that can be built will be sharply limited. Self-assembly can only build systems of limited complexity. Lithography is limited in the kinds of shapes it can produce–in particular, most lithographic systems cannot build three-dimensional shapes, and cannot create atomically smooth surfaces on most faces of complex shapes. Chemistry is mostly limited to molecules that are too small to build machines with easily. Direct manipulation (e.g. manipulation or deposition via scanning probe microscope) is slow and exceedingly difficult as a manufacturing technique, though progress is being made.

A significant milestone is the ability of a molecular manufacturing system to duplicate itself in a sufficiently short time period. Without this capability, the scale and complexity (and thus the functionality) of systems that can be built will be limited by the nanoscale technologies used to build them. By contrast, a molecular manufacturing system that is capable of duplicating its own structure would be able, with about fifty doublings, to create a kilogram-scale manufacturing system. Roughly, every ten doublings increases the mass of the system about one thousand (10^3) times, and a bacteria-sized system is about 10^{-15} kilogram.

A system that could perform all the operations required to duplicate itself, but required a month to do so, would require fifty months to scale up. Before this was accomplished, a faster system would surely have made that effort obsolete. By contrast, a system that could duplicate itself in a day would require less than two months to scale up–and every week could see the manufacturing capacity increase more than one hundred times, until some other factor limited the desire for more manufacturing capacity.

Thus, a difference in fabrication speed of about thirty-fold makes the difference between a system that cannot be scaled up effectively, and a system that can make as much nanoscale machinery as desired in a matter of weeks. A molecular manufacturing system capable of processing its mass in less than a week is thus a significant threshold.

The impact of a molecular manufacturing system will be closely tied to the materials it produces. A system that works by self-assembly, in addition to being slow, may produce only weak materials, comparable to soft plastic. A system capable of making molecular bonds in polymer chains may produce materials comparable to advanced polymers or metals. A system that can form highly cross-linked structures–which may require positional control of chemistry, not just sequential control–may produce materials comparable in strength to diamond or carbon nanotubes. In addition, highly cross-linked structures may in general be stiffer, which is especially important at the nanoscale since stiffness scales adversely; stiff structures may also be used as mechanical springs, capable of faster relaxation than entropic springs. (Natural rubber is an entropic spring–entropy, not the springiness of chemical bonds, is responsible for its contraction. Proteins likewise depend on entropy for at least some of their shape. In contrast, a highly crosslinked molecule may derive its shape, stiffness, and strength largely from bond stress. Stiff molecules may be significantly easier to design.)

Another milestone relates to whether the molecular machines must be immersed in water, or whether they can be run dry. For example, many proteins require water to retain their shape, although some enzymes are used industrially in dry processes. Immersion in water would impose a substantial fluid drag penalty on the machines, slowing down their operation, so a dry system might be able to work up to two or three orders of magnitude faster than a wet system. Natural protein machines can have high efficiency as long as they work at a sufficiently slow speed; dry machines may have equally high efficiency at substantially higher speeds.

The cost of feedstock molecules will be a significant factor in the scope of impact of a molecular manufacturing system. The cost of synthesized molecules can vary by many orders of magnitude, depending on the difficulty of synthesis. If a few hundred molecules cost $100, then sensors may be about the only thing that can be built with them cost-effectively. If a billion molecules costs $100, as may be the case for DNA synthesis, then computer circuits (with one or a few molecules per transistor) may be commercially viable. If a milligram of product costs $100, then medicines may become cost-effective, although the lengthy time required to verify the effectiveness of any medicine may reduce the value of a molecular manufacturing system that could in theory make a wide range of new medicines. If a kilogram of product costs $100 or less, then a much wider range of products becomes cost-effective; at that point, product cost becomes less important than the range of capabilities that can be built into the system's products.

Molecular Manufacturing and the Environment

During development of molecular manufacturing, environmental impacts are likely to be low. The quantities used of any material will be small, and it seems likely that more waste material will be generated by the general operation of the laboratory than by the actual construction of molecular devices.

Once molecular manufacturing systems develop to the point of creating economically viable products, the environmental cost becomes more difficult to estimate. Depending on the materials used and the reactions employed, the waste stream may have a dry mass several times that of the product. As long as the feedstock is expensive on a per-gram basis, that will serve as an incentive to minimize the volume of waste. Given the range of possible chemistries that could be used in a molecular manufacturing system, it is impossible to say what elements or chemicals may be in the waste stream. However, since molecular manufacturing basically uses a few operations repeated many times, it is likely that the waste stream will contain a limited number of different chemicals, and it is possible that they may be cost-effectively recycled into feedstock.

The waste produced by molecular manufactured products must be compared with the waste produced by the products they would replace. Many high-tech products, such as computer chips, produce quite a lot of waste during manufacture. Of course, if and when molecular manufacturing reduces the price of a product sufficiently to increase demand for it, then the waste stream will increase proportionally. At the point where the technology has advanced to the point of building entire integrated products out of fairly inexpensive feedstock, it is likely that the waste stream can be engineered to be either innocuous or recyclable.

Sufficiently advanced molecular manufacturing should be able to make sufficiently advanced products in sufficient quantity to play a part in mitigation of environmental problems. Once the availability of manufacturing systems is no longer a gating factor in how much product can be created, rapid production of infrastructure such as solar collectors may become possible. On the other hand, a substantial increase in the general rate of production is likely to create environmental problems of its own, whether by using more material, energy, or land.

Controlling the Nanoscale

Initially, it will be difficult to bridge the gap between the large-scale source of control information (computers with conventional wires and actuators) and the small scale of self-assembled nanomachines. Some means of delivering signals to the nanoscale will have to be found. Several methods of signal transmission have been used with success in various problem domains. A closely related problem is using the signals to control nanomachines. It is not enough to create motion at the nanoscale; for programmed control of machines, multiple actuators must be accessible in arbitrary sequence.

A nanostructure on a surface can be reached electrically by lithographic techniques. Either the structure can be deposited randomly, its position determined by microscopy, and conductors deposited to the structure, or the conductors can be deposited first and the structure subsequently added. Future advances may include the lithographic creation of template structures to guide the self-assembly of nanostructures onto the surface where they can contact the conductor. The scales of lithographically patterned conductors and self-assembled molecular structures have recently overlapped–DNA structures 100 nm in size have been built. Thus, DNA-templated nanomachines can, at least in principle, be contacted directly by multiple electrical conductors. DNA can be attached to surfaces patterned by e-beam in selected positions and orientations (Ryan J. Kershner et al,

2009). Electrical signals might also be delivered by probes with nanoscale tips but positioned by micro- or macro-scale machines. This technology is under continuous development for integrated circuit diagnosis. Mechanical motion might also be delivered with such probes.

Computers can control the fabrication sequence of polymers, as in DNA synthesis. A DNA molecule can encode a lot of information, and deliver it to a nanomachine by binding (self-assembling) to a targeted part of it; thus, a synthesized molecule can convey information to the nanoscale. DNA molecules have been used to zip and unzip portions of DNA nanostructures; the resulting changes have been used, for example, to open and close DNA boxes and move "walkers" along tracks (Omabegho T, et al, 2009). Since DNA binds via self-assembly and is distributed by diffusion, DNA control of nanomachines is likely to be slow. In theory, a future system might be controlled by reading the sequence of bases directly from a DNA strand (or other sequenced polymer), as ribosomes do, rather than simply undertaking one action based on a sequence match with the entire strand. This would allow far faster operation under control of fabricated molecules.

Nanoscale actuators have been operated by changing the chemical composition of the solution bathing them. Some actuators change state depending on ion concentrations. However, this may not allow many different actuators to be addressed individually, and modifying ion concentrations may also affect the stability of some self-assembled structures including DNA. Small molecules may be used instead of ions, and ATP has been used to control bio-derived motors outside of cells. It may be possible to develop a family of motors that each use a related but distinct molecule as fuel, so that flushing through small molecules in sequence can power a programmed sequence of actions. Microfluidics may increase the speed with which molecules can be introduced for control and/or power.

Light has been used to control molecular actuators, as well as to do chemistry; patterns of light projected onto a surface have been used to control the synthesis of many DNA sequences in parallel (Gao X et al, 2001), and might similarly control reversible actuators. Some actuators already exist that can be switched back and forth by two different light frequencies. It may be possible to create a family of actuators that respond to multiple frequencies, perhaps by using frequency-specific optically active nanoparticles to capture the light energy and transfer it to the actuator molecule.

Acoustic energy has been proposed as a method of control of nanomachines, albeit rather advanced types with rigid cylinders and pistons to be displaced by pressure (Merkle R C, 1996). It is also possible that acoustic energy (perhaps in the form of surface acoustic waves) or radio waves might be used to activate resonators; a radio receiver has already been built that uses resonant mechanical vibration of a carbon nanotube to tune a radio signal (Jensen K et al, 2007).

Control of Fabrication Chemistry

The core of molecular manufacturing is the use of mechanically guided chemistry to fabricate useful components and eventually products. For the purposes of manufacturing, chemistry can be divided into several basic types, albeit with gray areas and overlaps between the types. First, there is conventional reactive chemistry, in which molecules are mixed together and reactions happen on their own. Which reactions happen is controlled by the properties of the reactants themselves. Second, there is sequence-controlled chemistry, and finally there is positionally controlled chemistry.

Conventional chemistry offers few opportunities to control the product (except by selecting the reactants, of course). Reaction conditions such as temperature and pressure can be adjusted to shift equilibrium between different products. Concentrations of reactants and products can be adjusted. But all in all, these are very indirect and limited means of control. Conventional chemistry does well to achieve yields of 95% desired product and 5% contaminant. Different reactant molecules can sometimes be chosen by performing the reaction in several steps and thus modifying the reactant in each step, but this tends to reduce the yield of the product.

Enzymes can, in some cases, be used to improve the results of conventional chemistry. Enzymes can make a reaction both faster, acting as catalysts, and more specific, improving yield. Enzymes are complicated and complex molecules that "recognize" certain features of the reactant molecules, binding reversibly to them and bringing them together in the right orientation to enhance the desired reaction. In some cases, enzymes can be quite specific, with yields well over 99%. Some enzyme-mediated processes, such as DNA copying, can have error rates better than one in a million. But enzymes are difficult to design, and being protein, they can tolerate only a limited range of conditions. Enzymes are thus of only limited utility in industrial chemistry. Designing hundreds of enzymes to make a molecular product containing thousands of atoms would be impractical, and anything larger is probably not worth considering.

There are some molecules, notably polymers, that can undergo a long sequence of reactions, modifying the molecule each time, with the reactive site having the same basic character each time so that only a few basic types of reaction need to be designed. Proteins being fabricated by ribosomes have this character: the end of the protein is chemically very similar regardless of which amino acid has just been added, and so the choice of which amino acid to add next is arbitrary. DNA being constructed by a DNA synthesizer also has this character. The growing end of the DNA chain is the same regardless of how many nucleotides have been added, or which ones. By controlling which nucleotide is available to be added next, and designing the reactions so that only one nucleotide can be added at a time, any desired DNA sequence can be constructed.

Once a suitable molecule is selected–one that can grow without changing the character of its reaction site–control of the sequence can be achieved in any of several ways. Ribosomes control the sequence by using additional molecules that encode the desired sequence: an mRNA strand contains the information, and matches with tRNA molecules that carry the appropriate amino acid. Thus, a single mRNA strand can, with the help of a ribosome, guide the manufacture of a protein of specific sequence from a floating mix of tRNA/ amino acid feedstock. In DNA synthesis, the reaction conditions are adjusted while the desired nucleotides are flushed past the growing strand. Control thus comes from non-chemical means. This has the advantage that control can come directly from a computer; it has the disadvantage that repeated flushing of the nucleotides is slow and wasteful.

In a molecular manufacturing system, the goal is to make useful products–indeed, the range of products should include copies of key (difficult to manufacture) components of the manufacturing system itself. The manufactured molecules will presumably need to take on a mechanical function–perhaps structural, perhaps enzyme-like. Thus, once a sequentially sequenced molecule is produced, it must fold or otherwise take on a desired shape. Protein folding is difficult to predict, though steps have been made in that direction (Prentiss M.C et al, 2006), and novel proteins have been designed and found to work as predicted (Kuhlman B et al, 2003). DNA folding is fairly easy to predict, and design of desired structures is rapidly becoming easier. Christian Schafmeister has designed a protein-derived polymer that forms a rigid structure due to rigid bonds between the monomers; it is synthesized in a floppy state, and then the chain is modified to become rigid with easily predictable angles between the various monomers (Gupta et al, 2006).

Not many polymers are currently known that have substitutable monomers, can be fabricated in controlled sequence, and can form useful structures once synthesized, and they are typically based on biopolymers–protein or DNA. An artificial polymer with these characteristics might be extremely interesting and useful in a variety of nanotechnology fields, including molecular manufacturing. This is one of many suitable areas of research for a new molecular manufacturing-focused nanotechnology program.

There are several possible ways control the sequence of fabrication of a polymer. The sequence of chemicals can be manipulated. The end of the polymer may be chemically deprotected either by bulk chemistry, or by an external signal (the signal is optical in some of today's DNA synthesis systems). The latter allows multiple different sequences to be built on different regions of the same substrate, but still requires the appropriate bulk chemical to be flushed past after the deprotection step. A mechanical molecular protection device, blocking or unblocking the end of the polymer without being bonded to it, would be a first step toward a system that could select the appropriate next monomer from a mix of monomers, eliminating the time-consuming and inefficient flushing step. For long polymer fabrication, the mechanism would have to step along the polymer to stay at the end.

It is tempting to think that a mechanical molecular device that was used to select monomers for polymer growth would be a step toward a mechanical molecular device that could protect or deprotect, and/or select molecules for addition to, a particular spot on a crystal surface. There are enough differences between polymer growth and crystal growth that this is likely over-optimistic. It might be better to start from scratch, designing a machine more like a scanning probe microscope, working in a system where the surface would remain unchanged until interacted with by the probe. Such an approach is already being used, although with a conventional large-scale scanning probe (Abeln G.C, 1998).

Early Molecular Manufacturing Systems

Several lines of technological development are converging on molecular manufacturing systems. DNA machines are being designed and built, including machines that template the construction of programmed sequences of DNA (Liao S. et al, 2004). Site-specific chemistry is being done with scanning probe microscopes. Even synthetic biology is raising the possibility of engineered synthetic cells that could be considered molecular manufacturing systems–though they are many years in the future.

Initial molecular manufacturing systems need not be capable of crafting duplicates of themselves. Instead, they can be self-assembled from molecules and nanoparticles. Since molecules and nanoparticles can be made in quantity, and self-assembly can proceed in parallel, a large number of manufacturing "workstations" can be built. Each workstation could then be controlled to produce its own copy of a desired product.

DNA is a very versatile molecule, capable of folding by self-assembly into a wide range of shapes. Adding additional strands of DNA can make DNA objects re-fold, join, or come apart, providing a sort of actuation–although this process is quite slow, requiring many minutes instead of the fractions of a millisecond that are characteristic of chemical reactions in biology.

Advanced Molecular Manufacturing Systems

Ideally, a nanotechnology program with a molecular manufacturing component will be guided not only by what can be worked on in the near future, but by the most profitable and effective destination. Thus, although advanced molecular manufacturing systems cannot be built yet, and can be roadmapped only in outline, it is worth considering some properties and aspects of advanced molecular manufacturing systems, sometimes called nanofactories.

The larger a product the system can build, the more easily its products can be used. No matter how advanced and valuable a cubic-micron device is, it must still be packaged and then integrated into a product. A cubic millimeter device needs far less handling, and may be integrated directly into products if its materials are sufficiently robust not to need packaging. A cubic centimeter device may be a human-usable product without any post-processing.

In order to make grams or kilograms of product, grams or kilograms of molecular manufacturing machines will be necessary. Although large masses of small product might be built with free-floating or bead-adsorbed machines, the ability to build large integrated products implies the ability to build large structures integrating molecular manufacturing machines with larger component-handling systems. Thus, the manufacturing system can be considered as one of the products, and the capability to assemble the products will develop along with the capability to build them. Fastening the machines into a large framework may also make it easier to deliver control and power to the machines. Architectures for feedstock delivery and cooling will be different for rigidly fastened machines than for free-floating machines, but need not be inherently more difficult; passive channels with circulating fluids should work fine, and may be more efficient than mixing and diffusion.

In order to build large products, large numbers of independently controlled operations will be needed. These can be specified with a relatively small blueprint, by designing at multiple levels. For example, a few hundred different molecular components could be combined to make just a few hundred types of sub-micron building blocks, which could be combined to make a few hundred millimeter-scale functional modules, and so on. This approach narrows the design space from completely intractable to comfortably large, and mirrors the approach used to build modern computer hardware and software.

Above 100 nanometers or so, self-assembly will probably be too slow and unreliable, so the parts will have to be handled mechanically. (Binding techniques borrowed from self-assembly may still be used to fasten the parts together after they are placed in mechanical proximity; stronger fastening systems may also be developed.) Thus, even with a relatively constrained design space, building large interesting products probably requires handling diverse components with spatially selected machinery.

Given the vast numbers of molecules required to make even a cubic millimeter product, it is likely that systems of this class will need many onboard computers to unpack an efficiently coded blueprint into locally relevant operation sequences. At the point where massively parallel manufacturing systems can be integrated into a large rigid framework, it may be appropriate to integrate nanocomputers. However, significant attention should be paid to reducing the size and power consumption of the computers.

Control of the system will be much easier if the defect rate of the operations can be reduced in order to reduce feedback and error handling. Ideally, high-level error handling of chemistry mistakes will not be required. There appears to be no theoretical problem with this; mechanically constrained chemistry can use a wider range of conditions than conventional chemistry, including effective concentration, manipulation of energy barriers, and exclusion of undesired competing reactions, to achieve far better yield. Just as enzyme chemistry is more precise, targeted and efficient than conventional chemistry, mechanically constrained chemistry can go quite a bit farther in the same direction. Error rates of 10^{-9} or better seem quite plausible for some operations, and it may well be possible to develop an entire library of fabrication operations, sufficient to do all required chemical manufacturing steps, with extremely low error rates. Although synthesis error rates will never be brought to zero, it may be possible to make them low enough that they

do not have to be addressed in the higher-level planning of nanofactory operation.

When assembling nanoscale components into a product, it is worth noting a useful implication of scaling laws: the linear growth speed of depositing blocks onto a surface via a two-dimensional array of machinery does not change as the scale of the blocks changes, as long as the block handling machinery scales with the block. Thus, this cannot scale all the way to handling individual small molecules, but it can certainly scale to micron-sized blocks, and probably to 100-nm blocks. What this means is that three-dimensional products of any size can (in theory) be assembled from microscopic blocks by a single layer of microscopic machinery. Multiple levels of convergence, as in today's assembly lines, may not be required. Even a 100-nm block has ample room for strong chemical or even mechanical binding or joining apparatus on its surfaces. Thus, even a kilogram-scale nanofactory may require only three top-level architectural units: direct mechanically-guided chemical fabrication of nano- or micro-scale blocks from molecules, transport of those blocks to the product assembly apparatus, and assembly of those blocks directly into the final product. Of course, more complex system architectures may be more efficient.

Biology demonstrates that a mechanosynthesis-based manufacturing system can produce its own weight in product in less than an hour—some bacteria can replicate in as little as fifteen minutes. Thus, for small-product systems, it is not necessary to use dry or highly crosslinked molecules. Architectures might be developed to make large products with wet biopolymer systems, though it is notable that gestation and maturation times of macro-scale organisms tend to be measured in weeks to years, rather than minutes. Be that as it may, the performance of dry crosslinked systems may be substantially higher than wet polymer systems. Dry crosslinked systems may also be more accessible to traditional mechanical engineering, even at the nanoscale—for mechanical systems larger than a few microns, quantum effects are a minor correction, and even the effects of thermal noise (also called Brownian motion) may be largely ignored in some classes of device.

Architecture of a Nanofactory

The architecture of a nanofactory will depend largely on what the nanofactory is capable of building. A molecular manufacturing system that can build only weak molecules such as DNA will need to have most of its structural components supplied by conventional manufacturing. This may be as simple as a conventional reaction vessel with suspended plastic beads to which DNA-based manufacturing aids are bound; these DNA machines would help to construct more DNA-based manufacturing aids (perhaps by guiding and speeding self-assembly of large complexes), as well as other products. Thus, there is a continuum from today's techniques to more advanced nanofactories.

Another approach to molecular manufacturing is to guide the self-assembly of molecular components. This may start by simply increasing the speed of self-assembly by attaching parts to a framework during construction and self-assembly. A framework built of DNA, used to speed the self-assembly of large DNA structures–including improved frameworks–is a natural place to start. Over time, the capabilities of the framework could be improved incrementally, until it uses computer-controlled actuation to control the self-assembly of small molecules, and eventually to control the chemistry to fabricate molecules.

Another approach to a primitive nanofactory is to use a scanning probe array to construct products on a planar surface, via either chemistry or guided self-assembly; the latter may require precision measured only in nanometers, rather than tenths of angstroms; the use of molecular building blocks, substantially larger than atoms, may speed the product construction process. If

those products include arrays of scanning probes, even primitive ones, then the system may be able to produce its own active element. An array of scanning probe microscopes may be scanned by moving the substrate, so the probes only need one degree of freedom apiece (Phoenix C, 2005)

A more advanced nanofactory would use a rigid framework to mount a three-dimensional array of workstations. (A material that is strong, but not very stiff, may be stiffened by holding it in tension; this may be relevant to some biopolymers.) Once it becomes possible to fabricate small fast actuators, control networks, molecule handling machines, and sealed enclosed volumes, a nanofactory that can make three-dimensional products–including more nanofactories–appears to be amenable to today's design capabilities. One architecture would use a branched network of ever-smaller conveyor tubes, terminating in molecular fabrication stations; small components made in those stations would be assembled into larger and larger functional blocks (Phoenix C., 2003). Architecture conveys tiny blocks, made in molecular fabrication stations, directly (or after just one or two stages of component assembly) to a single assembly array that adds the blocks directly to the product (Lizard, 2009).

To a first approximation, the linear growth speed of depositing blocks via a planar array of machines directly onto a product surface does not vary as the size of the blocks and machinery is scaled up or down. This implies that a useful large-product nanofactory could be approached in two mostly-separate projects: one to build precise functional blocks that may be as small as sub-micron; the other to handle those blocks and attach them to their proper place on the product. The block attachment machinery may be built of MEMS, though molecular manufactured machinery has many theoretical advantages over MEMS. Sub-micron blocks are small enough to build molecule-by-molecule in a reasonable time.

DEVELOPING ORGANIZATIONAL COMPETENCE AND EDUCATIONAL OPPORTUNITIES

An organization that intends to contribute to the development of molecular manufacturing may specialize in any of a wide range of fields. Nanoscale actuators, multiscale simulation, self-assembled structures, microscopy and characterization, scanning probe capabilities, lithography, and various types of chemistry are just a few of the fields that may be useful. In fact, almost any nanoscale technology program could be extended to support some aspect of molecular manufacturing development.

At least one area, biopolymer structures, has a smooth continuum of research opportunities from undergraduate teaching to world-class innovation. Any competent undergrad could use Rothemund's DNA staple techniques (Rothemund P.W.K, 2006) to build two-dimensional DNA structures that can easily be observed with a scanning probe microscope. Attaching interesting molecules or nanoparticles to the DNA takes moderate skill in chemistry. Designing new self-assembly techniques or new three-dimensional structures may be an interesting challenge for advanced researchers. Integrating DNA structures into other nanotechnologies spans a range from interesting to world-class. Inventing new families of structure-forming polymers would be a significant contribution to the field.

Thus, molecular manufacturing can be integrated with a nanotechnology program in a wide variety of ways. The educational opportunities provided by molecular manufacturing range from mere tutorial replication of what has already been done, suitable for introductory classes, to world-class innovation. The focus may range from nanoscale technology with incidental relevance to molecular manufacturing, to a program targeted at producing molecular manufacturing systems capable of making commercially relevant products.

CONCLUSION

Molecular manufacturing is a field with much to offer a nanotechnology program. Although its reputation has suffered in the past due to theoretical projections outpacing laboratory capabilities, this gap is rapidly closing, and now is an excellent time to reexamine how to incorporate molecular manufacturing in nanoscale technology programs, or even build a program around it.

Molecular manufacturing can be accessed by many pathways, some with incremental applications, others with immense payoff at the end. The basic concepts–scaling laws, precise mechanically-guided chemistry and/or self-assembly, and nano-building-nano–are easy to understand, and are becoming easier to apply. Any growing nanotechnology program should consider how it might grow in the direction of molecular manufacturing research.

REFERENCES

Abeln, G. C. (1998). Approaches to nanofabrication on Si(100) surfaces: Selective area chemical vapor deposition of metals and selective chemisorption of organic molecules. *Journal of Vacuum Science & Technology B Microelectronics and Nanometer Structures, 16*(6), 3874–3878. doi:10.1116/1.590426

Alivisatos, P. (2007). *Nanotechnology Roadmap for Atomically Precise Nanofabrication and Productive Nanosystems.* Retrieved from http://e-drexler.com/d/07/00/1204TechnologyRoadmap.html

Dienwiebel, M. (2004). Superlubricity of Graphite. *Physical Review Letters, 92*(12). doi:10.1103/PhysRevLett.92.126101

Douglas, S. M. (2009, May 21). Self-assembly of DNA into nanoscale three-dimensional shapes. *Nature, 459*, 414–418. doi:10.1038/nature08016

Drexler, E. (1986). *Engines of Creation: The Coming Era of Nanotechnology*. New York: Anchor Books.

Drexler, E. (1992). *Nanosystems: molecular machinery, manufacturing, and computation.* Hoboken, NJ: Wiley.

Drexler, E. (2009). *Toward Advanced Nanotechnology*. Retrieved from http://e-drexler.com/

Feynman, R. (1959). *There's Plenty of Room at the Bottom.* Retrieved from http://www.zyvex.com/nanotech/feynman.html

Gao, X. (2001). A flexible light-directed DNA chip synthesis gated by deprotection using solution photogenerated acids. *Nucleic Acids Research, 29*(22). doi:10.1093/nar/29.22.4744

Gupta, S., Macala, M., & Schafmeister, C. E. (2006). Synthesis of structurally diverse bis-peptide oligomers. *The Journal of Organic Chemistry,* (71): 8691–8695. doi:10.1021/jo0609125

Jensen, K. (2007). Nanotube Radio. *Nano Letters, 7*(11), 3508–3511. Retrieved from http://dx.doi.org/10.1021/nl0721113. doi:10.1021/nl0721113

Joy, B. (1992). *Why the future doesn't need us.* Retrieved from http://www.wired.com/wired/archive/8.04/joy.html

Kershner, R. J. (2009). Placement and orientation of individual DNA shapes on lithographically patterned surfaces. *Nature Nanotechnology, 4,* 557–561. doi:10.1038/nnano.2009.220

Kuhlman, B. (2003, November 21). Design of a novel globular protein fold with atomic-level accuracy. *Science, 302*(5649), 1364–1368. doi:10.1126/science.1089427

Liao, S., et al. (2004, December 17). Translation of DNA Signals into Polymer Assembly Instructions. *Science, 306*(5704), 2072–2074. Retrieved from DOI: 10.1126/science.1104299

LizardFire Studios. (2009). Retrieved from http://www.lizardfire.com/html_nano/themovies.html

Merkle, R. C. (1996). Design considerations for an assembler. *Nanotechnology, 7,* 210–215. doi:10.1088/0957-4484/7/3/008

Omabegho, T., et al. (2009). A Bipedal DNA Brownian Motor with Coordinated Legs. *Science, 324*(5923), 67. Retrieved from DOI: 10.1126/science.1170336

Phoenix, C. (2003 October). Design of a Primitive Nanofactory. *Journal of Evolution and Technology, 13.*

Phoenix, C. (2005). Large-Product General -Purpose Design and Manufacturing Using Nanoscale Modules. *NASA Institute for Advanced Concepts.* CP-04-01 Phase I Advanced Aeronautical/Space Concept Studies. Retrieved from http://www.niac.usra.edu/files/studies/final_report/1030Phoenix.pdf

Prentiss, M. C. (2006). Protein structure prediction: The next generation. *Journal of Chemical Theory and Computation, 2*(3), 705–716. doi:10.1021/ct0600058

Rothemund, P. W. K. (2006, March 16). Folding DNA to create nanoscale shapes and patterns. *Nature, 440.* Retrieved from doi:10.1038/nature04586

Section 2
Technology Transfer and Innovation

The fields of nanotechnology and microelectronics have evolved out of many years of scientific and technological innovations. In this section, technology innovation, transfer and diffusion are discussed within the contexts of general technology and specificity of nanotechnology. The section also examines the factors that accelerate open innovation, technology creation and why in most cases, some nations exhibit fatigue in technology adoption, lacking resilience in the diffusion process. Kondratieff Cycle, technology trends and knowledge dissemination instruments are discussed.

Chapter 3
Trends in Nanotechnology Knowledge Creation and Dissemination

Nazrul Islam
Cardiff University, UK

ABSTRACT

This chapter explores trends in nanotechnology knowledge creation across scientific disciplines and technology domains, and helps to understand the dissemination of nanotechnology knowledge. In relation to intense global competition in nanotechnology, this study exhibits a forward-looking approach in characterizing nanotechnology research and development trajectories. This research adopts hybrid research methodology, including both quantitative and qualitative methods. The findings imply that nanotechnology knowledge creation and dissemination trends have appeared to bridge divergent disciplines, emphasizing the importance of collaborative research networks among scientists to co-create, share and disseminate nano-knowledge across groups, institutions and borders.

INTRODUCTION

Nanotechnology comprises one of the fastest-growing research and development areas in the world (National Science and Technology Council 2006). Like many areas of scientific and technological exploration, nanotechnology exists on the borders between disciplines including physics, chemistry, materials science, biology, medicine, ICT and engineering. It has been nearly half a century since Nobel Prize winner Richard Feyn-

man advocated widespread nano-scale research by delivering his famous speech "There's plenty of room at the bottom" in 1959, through which the nanotechnology concept first captured the world's attention. Nanotechnology is a field prioritized and promoted by governments worldwide and is a subject of importance in all basic science and engineering fields, for sustainable economic development and has potential to provide comfort and safety for people everywhere. Nanotechnology is regarded as an evolutionary field which emerged from the increased miniaturization of microelectronics and microtechnology in research,

DOI: 10.4018/978-1-61692-006-7.ch003

development and innovation. It can be predicted that in the coming era, nanotechnology will play a central role in a variety of industries and key technologies, in areas of strong societal need, such as energy (e.g. fuel cells, solar, nuclear), biotechnology and medicine (e.g. drug delivery, tissue replacement, directed therapies), information and communications technology (e.g. pervasive computing, data storage, leisure, personal communications) and water and environmental remediation (Roco 2002). In the near future, it is likely that this small scale technology will have the ability to fundamentally change the way almost everything is designed and manufactured, from automobile tires and tennis racquets to air purifiers and life-saving vaccines[1].

As a science based innovation, nanotechnology represents a multi-disciplinary field of research and development since it requires multi-disciplined networked research (Meyer and Persson 1998; Islam and Miyazaki 2009), education and an improvement in the level of human skills performance; it also requires input from, amongst others, chemists, physicists, materials scientists, biologists, engineers and pharmacologists. No clearly defined boundary for nanotechnology has surfaced due to its nano dimension, which randomly affects various technological domains and scientific disciplines. As it concerns a field on the cutting edge knowledge in practice, this study explores trends in the development of knowledge in order to gain a better understanding of the process of knowledge creation and dissemination in the emergent nanosciences and nanotechnologies. In particular, it seeks to ascertain how nano-knowledge has been co-created across disciplines and domains, and seeks to understand the attributes that are likely to enable them to be disseminated.

The rest of the chapter is structured as follows. This introduction is followed by a brief explanation of nanotechnological research approaches and their characteristic differences to other advanced technologies. A hypothesis and a research framework for nanotechnology knowledge creation and

dissemination are proposed and the research has been carried out to prove it by a thorough analysis using both quantitative and qualitative data.

CHARACTERISTIC DIFFERENCES OF NANOTECHNOLOGY

Nanotechnology is a field of scientific constellation and practice in which different scientific disciplines are involved and new knowledge is produced for the sciences involved, as well as for society as a whole. Significant developments in nanotechnology are spreading across the diversified fields of ICT (for faster and smaller processors, higher-density data storage), medicine (for faster drug development, quicker diagnosis, improved drug delivery, better prosthetics), the environment (for pollution control, water purification and clean energy) and materials (for stronger engineering materials, better catalysts, coatings, paints and lubricants, and improved surface properties like scratch-resistance and optical switching) (CRISP/OST Foresight 2001). The decisive factor is that new functionalities and features for the improvement of existing products or the development of new products and application options result from the nanoscalability of the system components alone (Malanowski et al 2006). These new effects and versatile application possibilities are predominantly based on the ratio of surface to volume in atoms and on the quantum-mechanical behavior of the elements of nanomaterials.

It should be noted that all nanotechnological research is geared to two basic strategies, (a) the *'top-down' approach*: miniaturization techniques – decomposition into the smallest manageable entities starting from micro-technology by using techniques (e.g. energy beam and mechanical micro-machining, electro-physical and chem-processes, lithographic methods) that are particularly predominant in physics and physical technology; and (b) the *'bottom-up' approach*: building macrostructures – allowing for the re-engineering of materials at nano-level by using techniques (e.g. assembling, imprinting, electrostatic techniques, laser trapping, sol-gel, colloidal aggregation) that primarily feature in chemistry and biology. Figure 1 illustrates a snapshot of the *top-down* and *bottom-up* processes. The introduction in the market of nanotechnology applications has already occurred and its industry applications are vast and diverse, as materials can be transformed on a very small scale. Virtually all today's commercial nanotechnology is provided by the adoption of the top-down approach, examples of this include manufacturing of computer hard disks, electronic devices and computer chips. The bottom-up approach is still at the research stage and commercial products at least a decade away, for example, making crystal and films for displays, making particles for cosmetics and fuel additives and making atomic or molecular devices. The majority of nanotechnologies, especially those related to electronics and optoelectronics, are seemingly realized by a *top-down* process, where nanostructures are developed through the improvement or advancement of existing technologies. As they are built on cumulative knowledge, *top-down* nanotechnologies are likely to have social and economic impacts in the short and medium terms (Igami and Okazaki 2007). On the other hand, attempts at developing *bottom-up* technologies

have been particularly intense in the past decade and have been fuelled by scientific discoveries, such as carbon nanotubes and fullerenes. At present, *bottom-up* nanotechnology is likely to have a relatively low impact on application fields and it will take time until *bottom-up* nanotechnologies show social and economic impacts.

Nanotechnology provides new possibilities, for manufacturing and control of individual objects on the nano-scale could lead to breakthrough-level innovations. Although biotechnology and nanotechnology have both been regarded as breakthrough technologies with the potential to bring about a wave of radical innovation and industrial restructuring (Peters 2001), nanotechnology also encompasses more distinct areas, such as precision engineering as well as electronics; electrochemical systems (lab-on-a-chip devices) and mainstream biomedical applications in areas as diverse as gene therapy, drug delivery and novel drug discovery techniques. Therefore, in nanotechnology there can be both a *mechanistic version* (more materials science and microelectronics inspired, having new or significantly improved mechanical, electrical and chemical properties or functions) and a *biomimetic version* (more biotechnology inspired, having control of biological systems in order to achieve desired and designed outcomes). In addition, nanotechnology is more dispersed into

Figure 1. Nanotechnology research strategies

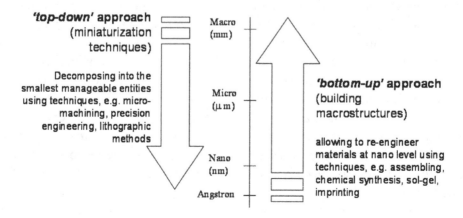

multi-technology domains and multi-disciplines than the initial biotechnology domain. It is evident that, for biotechnology, 90% of the early activity was in the pharmaceutical industry (Peters 2001). In the last few years, applying fundamental discoveries related to nanotechnology has resulted in the development of multibillion-dollar businesses (such as giant magnetoresistance (GMR) for hard disks, nanoparticles for cosmetics and drugs in pharmaceuticals), which has not been the case with biotechnology.

Nanotechnology necessitates thorough spatial control of matter at the level of molecules and atoms, with capabilities to process and rearrange them into customised designs due to the nano dimension. Therefore, nanotechnology differs from traditional chemical manufacturing in that chemical reactions are not left to statistical movements of molecules in solutions, but instead the molecules are brought into appropriate positions with appropriate speeds and orientations to cause desired reactions. Nanotechnology can also differ fundamentally from micro-manufacturing of silicon chips in that the *top-down* approach and repeated refinement of bulk materials into micro

or even nano-scale designs suffers from defects inherent in the original bulk material. In contrast, nanotechnology's *bottom-up* approach can build essentially defect-free structures by manipulating the atoms. Figure 2 presents the relevant characteristics of technical change posed by nanotechnology to make clear the nature of nanotechnological superiority over other advanced technologies.

A FRAMEWORK FOR NANOTECHNOLOGY KNOWLEDGE CREATION AND DISSEMINATION

Due to nanotechnology's multi-disciplinary nature, no single theory or approach can explain nanotechnology's knowledge dynamics, thus the conceptual framework of this study has been drawn by reviewing a number of theories from different angles. Nanotechnology is widely considered as being a general-purpose technology (Bresnahan and Trajtenberg 1995; Helpman 1998) affecting a range of industries in the economy. Technology and knowledge management research related to nanotechnology has been rather limited, since the

Figure 2. Characteristic differences of nanotechnologies

Advanced technologies style	Nanotechnologies style
Control over the microstructure of matter	Control over the nanostructure of matter in the atomic, bit and genomic level
Integrated division between the R&D system	Reform the R&D system so that multi-disciplinary expertise can work together
Relatively smaller markets with faster rate of growth	Wider markets with slower rate of growth
Larger variety of materials with integrated functions	Novel nano materials of specific desired customized properties with new and improved functions
Major consumer: information, telecommunication and pharmaceutical sectors	Major consumer: almost all of the existing sectors
Information intensive (classical)	Quantum and electron wave intensive
Flexible production style and simple miniaturization	Combination with new bottom-up and self-organization manufacturing style
Importance of research for specific market applications	Importance of multi-disciplinary and multi-technological research for all sectors market applications
Academic attitude, firms strategies and behaviour unchanged	Academic attitude, firm strategies and behaviour will change

technology is still in its fluid stage of development. The past few years have seen an explosion of interest in the area of science and technology labelled 'nanotechnology'. There have been few studies on the dynamics of nanotechnology knowledge developments and the emerging convergence of scientific fields in nano-scale research and commercialization. For example, Schummer (2004) compares research collaboration patterns in nanotechnology with those of traditional disciplinary research and concludes that nanotechnology is an aggregation of otherwise disconnected mono-disciplinary fields, rather than multidisciplinary convergence (Roco and Bainbridge, 2003). Meyer (2006) suggests that there are inter-related and overlapping nanotechnologies connected via instrumentation. While Eto (2003) finds evidence of multidisciplinarity in nanotechnology using bibliometric analysis of journals, citations, and authorship patterns. Identifying the trends of how nanotechnology knowledge is co-created and disseminated across scientific disciplines and technology domains may be vital to enrich the literature of nanotechnology knowledge management.

In this study an analytical framework has been chosen to make it easier to identify different instances which may influence or contribute to knowledge creation and dissemination. As economies and society become more knowledge based, the pattern of nano-knowledge generation and nano-information sharing and its exploitation structures become the important determining factor for the success of nanotechnology innovation systems. The author has identified the factors that affect this mechanism. The study has approached the problem of how the technology has been driven by science and technical research using the proposed framework. As nanotechnology is predominantly a science-based and knowledge-intensive sector, it is assumed that nanotechnology evolution is not one directional, rather involves different scientific disciplines and technology domains reversibly interacting to shape the

technology, while government and institutional structures support its successful exploration and exploitation. As a technology, nanotechnology is an emerging field and is characterized by a type of multi-disciplinary approach, which has an impact on various industries. Therefore, *nanotechnology knowledge creation and dissemination can be thought of as the involvement of multiple scientific and technical fields that co-create nano-knowledge and disseminate it across multiple disciplinary boundaries and make a bridge across them fostering a way of working together.* The framework provides a coherent analytical tool for handling the disparate processes of nano-knowledge creation, dissemination and use, including the key obstacles or bottlenecks. Figure 3 is an illustration of this framework.

Generally, nanotechnology research has been developed mainly in universities and public research institutes where scientific disciplines, such as chemistry, physics, materials sciences and biology are extensively involved in the relevant research to explore nano-science as a scientific curiosity and through government initiatives drives. Under the proposed framework, the regulatory infrastructures, public policies and funding allocation system constitute important elements for fostering nanotechnology research across institutions, such as universities, public research institutes and companies. Therefore, institutions are getting involved, initially, for the generation of nano-knowledge to grasp scientific opportunities in several technology domains. The author also argues that nanotechnology evolution is not one directional, rather different technology domains and scientific disciplines reversibly interact, and government and institutional infrastructures are supporting them to evolve. Nanotechnology research transfer can be *direct,* when entities in the public and private sectors form companies with a technological base as spin-offs and direct commercial ventures or *indirect,* when they use a diverse range of agents so that the products and services being innovated are put on the market

Figure 3. A framework of nanotechnology knowledge creation and dissemination

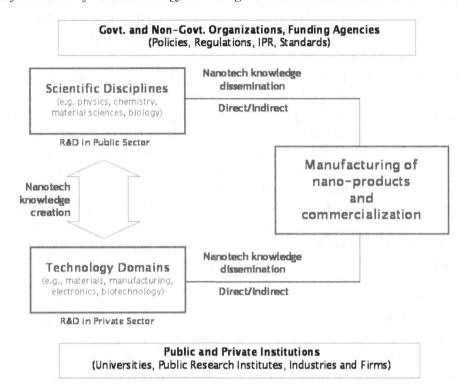

through, for example, the offices of technology transfer labs and science parks. Under the proposed framework, this study focuses on scientific disciplines and technology domains through which nano-knowledge is co-created and disseminated.

RESEARCH METHOD

For nanoscience and technology, increased funding is available worldwide for its rapid development over a wide range of scientific and technological areas. Scientific disciplinary research across institutions (whether it be universities or public research institutions) that were not getting much funding earlier, have recently been taking advantage of increased funding due to the perceived possibilities in nanotechnology rather than for discrete individual disciplines, such as chemistry, physics, material science or biology. Concerning the question of analyzing

the process of nanotechnology knowledge creation and dissemination, this study employs the following approaches: (1) Searching the entire nano-scientific output retrieved through ISI web of science SCI-Expanded database, which covers basic general disciplinary research related to nanotechnology and their reference citations across disciplines over certain periods; (2) Searching the entire nanotechnology applied research output retrieved through Elsevier Engineering Index (EI) Compendex database. The empirical data has been validated by a series of face-to-face interviews carried out in Europe and Japan by which nanotechnology knowledge generation and dissemination trajectories are explored.

The fluid stage of nanotechnology development represents a methodological challenge, conceptually as well as in relation to the data. In this chapter, a hybrid research approach is adopted, analyzing both quantitative and qualitative data. In the quantitative search, the author has used the

Elsevier EI Compendex database which covers a period of over 15 years starting from the 1990s with, first, relevant academic articles and tracking almost the entire lifecycle of the technology's evolution. Altogether 28,559 nanotech-related articles have been retrieved through queries based on 175 specialist keywords[2], derived from Nano Science and Technology Institute (NSTI) publications. Using the ISI web of science SCI-Expanded database over the period of 1995-2005 at 5-year intervals, the study has selected top ranked journals of each discipline, categorized by ESI and JCR in terms of publications. The subsequent analyses were performed by dedicated tech mining software called Vantage Point developed by Alan Porter at the Georgia Institute of Technology (Porter and Cunningham 2005). It carries out automating mining and clustering of terms occurring in article abstracts and article descriptors, such as authors, affiliations or keywords. In addition, the use of specialist computer software helps to statistically and textually analyzes articles, clustering thousands of keywords or specialist terms occurring in abstracts, thus increasing the reliability of the findings and opening up new analytical opportunities for emerging technologies like nanotechnology. The article abstracts from the database were imported to Vantage Point, which has removed duplicates or empty records, typographical errors (typos) in affiliation name and related geo-graphical information, name variations, inconsistency in each field of the entire dataset. The author has double checked the raw records, cleaned the dataset and facilitated the subsequent analyses. In his approach to qualitative methods, the author has used primary data: a series of face-to-face interviews with academic scientists and researchers from universities and public research institutes and with personnel from government bodies and funding agencies in Europe and Japan. Based on both the quantitative and qualitative data analyses, the study has established the trends and patterns of nanotechnology knowledge creation and dissemination trajectories.

TRENDS IN NANOTECHNOLOGY KNOWLEDGE CREATION AND DISSEMINATION

Nanotechnology is affecting various technological domains including advanced materials, biotechnology and pharmacy, electronics, scientific tools and industrial manufacturing processes. It has been of prime importance to first classify nanotechnology domains using Elsevier EI Compendex applied research database to broadly characterize the areas of technology. The quantitative data analysis involves the use of Vantage Point – computer-supported research facilitates processing large amounts of data, including full-text searches, keyword-based profiling and other categorizations. Keywords may not always be reliable indicators for emerging technologies due to the frequent changes of categorization schemes. Therefore, in order to divide whole nanotechnology applied research output into distinctive domains, the author focuses on both article keywords and EI codes to assign specific codes to every article in the database. EI is a standardized and nested structure used in the same way for every Compendex article, moreover, assignment of a specific code to an article is a well-considered decision – therefore this categorization scheme is more reliable than a traditional paper-based bibliometric research. Every Compendex article may be assigned one or more EI code – interdisciplinary research falls into several categories and consequently has several EI codes. Every keyword has also been assigned to a specified domain.

Following the interpretations, the author has divided the entire nanotechnology-related applied research output into four broad categories; nanomaterials, nanoelectronics, nanomanufacturing & tools and bionanotechnology. Figure 4 shows the characteristics of each domain: domain name, short description and examples of EI codes including the relevant keywords. For every domain, detailed lists of relevant EI codes were identified – specific domains corresponded to distinctive EI classes (for example, nanomaterials: EI codes 5.x and 8.x; nano-electronics: EI codes 6.x and 7.x; nanomanufacturing & tools: EI code 9.x; and bionanotechnology: EI code 4.x). There were many individual variances and many of the several hundred sub-classes were excluded from or assigned to other domains. In this way, an exclusive list of EI codes for every nanotechnology domain was generated and used to classify the domains.

NANOTECHNOLOGY KNOWLEDGE CREATION AND DISSEMINATION ACROSS DOMAINS

The variation of scientific output of specific domains related to nanotechnology over time

is illustrated in Figure 5. The results show that materials and electronics domains appear to have achieved superiority in terms of the volume of applied research. The overall output increased slowly from the mid 1990s to 2000 and then increased sharply in the early 2000s, probably due to the announcement of nanotechnology initiatives such as the US government's National Nanotechnology Initiatives, which the rest of world followed. In the case of the other two technology domains, biotechnology and manufacturing & tools, research activities are slowly picking up, which is instructive in that they are emerging fields for nanotechnology (see Figure 5).

Figure 6 indicates that nanotechnology applied knowledge creation is centered on the materials and manufacturing domains. However, there is

Figure 4. Nanotechnologies domains classification

Domain name	Short description	Examples of codes	Examples of keywords
Bionanotechnology	Bionanotechnology concerns with molecular scale properties and applications of biological nanostructures. It can be used in medicine to provide a systematic, as well as a screening, approach to drug discovery, to enhance both diagnostic and therapeutic techniques and to image at the cellular and sub-cellular levels	Compendex EI code 4.x	Biological nanosensors, nanobiomagnetics, nanocantelevers, targeted nano-therapeutics, nanoreplication, nanoencapsulation
Nanoelectronics	Nanoelectronics focuses nanoscale properties and applications of semiconductor structures and devices, and process technology to explore economic and performance benefits in computing, information and communication system	Compendex EI code 6.x and 7.x	Nanodevices, quantum dot lasers, nanosensors, nanocrystal memory, molecular electronics, nanorobotics
Nanomaterials	Nanomaterials concerns with control the structure of materials at nanoscale with great potential to create a range of advanced materials with novel characteristics, functions and applications	Compendex EI code 5.x and 8.x	Nanomaterial, fullerenes, nanocomposite, nanofilms, carbon nanotubes, nanoparticles
Nanomanufacturing /tools	Nanomanufacturing attempts at building more intricate (information-rich, self-assembly) nanostructures. In addition, tools concern its ability to measure and characterize materials at the nanoscale.	Compendex EI code 9.x	Nanoprototyping, nanofabrication, nanolithography, scanning tunneling microscopy, atomic force microscopy

Figure 5. Nanotechnology applied research output by domains

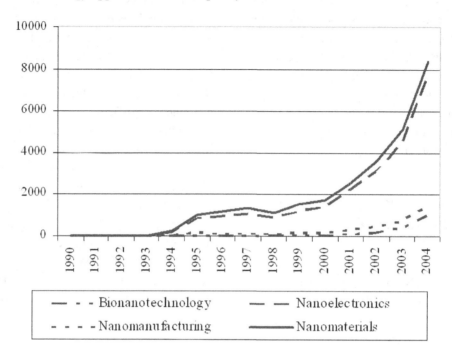

significant and sustained nanotechnology applied research activity in other specific areas of the electronics and biotechnology domains. These data suggest nanotechnology knowledge is co-created through technology domains as the percentage volumes expose a distribution of knowledge through the domains instead of showing a majority in one discrete area. This finding is instructive; confirming that nanotechnology applied research has mastered a diversity of areas that originated in different technology sectors, such as materials science, biotechnology, electronics, engineering and manufacturing. The author has also used the investigation to look into whether any overlapping of nanotechnology knowledge across domains exists, and how much is any overlapping of interest. Figure 7 presents a correlation chart between domains, calculated by software called Vantage Point for the entire dataset. The findings show very low correlations across the domains, suggesting that there are no direct overlaps between the singled out domains, even though many articles are classified as belonging to two domains. If the

analysis were to show a high level of correlation, there might exist the possibility of overlapping nano-knowledge creation and research interests. However, very small values of the finding indicate that nanotechnology knowledge has been created with the divergence of technology domains.

NANOTECHNOLOGY KNOWLEDGE CREATION AND DISSEMINATION ACROSS DISCIPLINES

This section makes an empirical construction of nanotechnology knowledge creation and dissemination across macro disciplines and shows the extent to which diversity of disciplines position each of them across nanotechnology research. Examining snapshots of the prevalent disciplines offers a perspective on the diversity of nano-knowledge flow patterns. Next, the author analyzed the references to nanotechnology research papers extracted from each discipline and investigated the trends of nano-knowledge creation and dissemination

Figure 6. Nanotechnology knowledge creation through divergent technology domains

Nanotechnology Domains	Materials	Electronics	Manufacturing	Biotechnology
Materials	44.42	13.65	32.99	8.91
Electronics	8.33	48.65	34.61	8.40
Manufacturing	10.08	17.34	62.48	10.08
Biotechnology	8.87	13.69	32.78	44.64

Figure 7. Nanotechnology knowledge creation without overlapping domains

Nanotechnology Domains	Materials	Electronics	Manufacturing	Biotechnology
Materials	1.00	0.01	-0.02	0.04
Electronics	0.01	1.00	-0.07	0.01
Manufacturing	-0.02	-0.07	1.00	-0.03
Biotechnology	0.04	0.01	-0.03	1.00

across disciplines over time. The analyses were performed for two domains, materials and electronics. Figure 8 shows that nano-knowledge is almost always created separately through each discipline, early on, rather than through sharing knowledge with other disciplines. Note that the relative shares and the trends in each discipline in nano-knowledge creation have shifted over time, and in 2005, nanotechnology research was disseminating across disciplines. As illustrated in Figure 8, in the case of chemistry, the share of cited references from chemistry-related fields was 64.47% in 1995 and had dropped to 40.65% by 2005. In the case of physics, materials science and biology disciplines, the shares dropped from 74.39% to 53.71%, 39.02% to 29.15% and 54.12% to 36.02% respectively. On the other hand, the shares of cited references across disciplines grew over last ten-year period and made up the gap. For example, in the case of chemistry, the nano-knowledge sharing trend increases from 10.42% to 13.26% in physics; from 7.33% to 19.93% in materials sciences; from 4.63% to 8.86% in biology; and from 1.15% to 6.23% in nanosciences areas. The results indicate that nanotechnology research in chemistry has been disseminated substantially within the materials science, biology and nanosciences areas. This means, for chemists who publish in the area of materials, dissemination seems to have occurred extensively in materials science, biology and nanosciences, and to a lesser degree in physics. Similarly, nanotechnology research dissemination in physics seems to have occurred substantially in the materials science and biological disciplines including nanosciences, and to a lesser extent in chemistry. On the other hand, nanotechnology research dissemination in the materials sciences seems to have taken place in chemistry and biology disciplines including nanosciences, and to a lesser extent in physics. Nanotechnology research dissemination in biology seems to have occurred substantially in chemistry, and to a lesser degree in other disciplines. The analyses could be interpreted in such a way that there was less intention in every discipline to co-create and disseminate nanotechnology knowledge across disciplinary boundaries than in the earlier period. With the evolution of nanotechnology-labelled research pushed by increased R&D funding and the prioritized national initiatives, the basic nanotechnology research trend in natural science disciplines has shifted over time, i.e., from being based on more separate disciplines to being more multidisciplinary. In other words, the trend has moved from a system or culture of specific fields or topics into something that is mixing fields. This may be the cause for the rapid development of nano-instruments (e.g. STM, AFM) that enable control, assembly and manipulation of materials and structures at the nano scale.

For the electronics domain, there is a relatively more noticeable trend towards nano-knowledge creation and dissemination than in the materials domain, as illustrated in Figure 9. The results indicate that each discipline showed less inclination to create and share nano-knowledge across disciplines initially, but this situation changed over time and by 2005 there was a definite trend towards nanotechnology research disseminating across disciplines. The findings show that the share of cited references from chemistry-related areas was 67.85% in 1995 and had dropped to 37.26% in 2005. In the case of physics, materials science and biology disciplines, the shares dropped from 77.69% to 43.16%; 53.26% to 33.69%; and 61.65% to 34.11% respectively. On the other hand, the shares of cited references across disciplines had grown over that ten-year period and filled the gap. For example, in the case of chemistry, the nano-knowledge sharing trend increased from 11.11% to 14.47% in physics; from 7.53% to 20.37% in materials sciences; from 3.96% to 10.18% in biology; from 8.33%

Figure 8. Nanotechnology knowledge creation and dissemination across disciplines in the materials domain

Search period	Macro disciplines	Share of cited references appeared in scientific disciplines					
		Chemistry	Physics	Material Sciences	Biology	Multidisciplinary Sciences	Nanosciences
1995	Chemistry	**64.47**	10.42	7.33	4.63	11.96	1.15
	Physics	11.11	**74.39**	4.34	0	8.21	1.93
	Material Science	26.08	19.92	**39.02**	0.4	12.65	1.89
	Biology	23.19	7.21	0	**54.12**	12.88	2.57
2000	Chemistry	**56.76**	12.05	15.05	5.04	10.25	0.81
	Physics	9.06	**51.1**	5.13	0	33.36	1.32
	Material Science	24.99	13.07	**35.76**	0	25.77	0.4
	Biology	26.38	4.18	7.34	**50.79**	9.62	1.67
2005	Chemistry	**40.65**	13.26	19.93	8.86	11.04	6.23
	Physics	16.05	**53.71**	8.07	1.13	12.9	8.11
	Material Science	22.29	18.87	**29.15**	1.15	16.95	11.47
	Biology	35.57	10.03	5.21	**36.02**	8.6	4.54

to 10.72% in multidisciplinary sciences (general science journals including *Science* and *Nature*); and from 1.19% to 6.97% in nanosciences areas. The results confirm that nanotech research in the chemistry discipline has been disseminated within physics, biology and nanoscience disciplines, i.e., for chemists who publish in the area of nanoelectronics, there seems to have been substantial dissemination in the physics and biology disciplines, and to a lesser extent in materials science. Similarly, nanotechnology research dissemination in physics seems to have involved the chemistry and biology disciplines, and to a lesser extent the materials sciences. Correspondingly, nanotechnology research dissemination in the materials science seems to have involved the chemistry and biology disciplines. Nanotechnology research dissemination in biology seems to have been involved in chemistry, physics and in the materials sciences. This multidisciplinary citation pattern supports the hypothesis that nanotechnology knowledge draws upon widely distributed research knowledge, rather than just individual research fields. One can conclude from this analysis that, for example, nano publications in the chemistry discipline do not just draw narrowly upon chemistry nano-knowledge, but draws from the knowledge in other disciplines as well. The results could be interpreted to signify that nanotechnology knowledge creation and dissemination derive from taking advantage of scientific opportunities that allow researchers to undertake or share a multidisciplinary research. At present, experts in the scientific disciplines need to interact with other researchers across the traditional disciplines to co-create and disseminate their nano-knowledge to explore the opportunities for more efficient outcomes through nanotechnologies. In this way, nanotechnology appears to be leading to a breakdown of the boundaries between scientific disciplines and certainly does not encourage the cultivation of any specific disciplinary knowledge.

Figure 9. Nanotechnology knowledge creation and dissemination across disciplines in the electronics domain

Search period	Macro disciplines	Share of cited references appeared in scientific disciplines					
		Chemistry	Physics	Material Sciences	Biology	Multidisciplinary Sciences	Nanosciences
1995	Chemistry	**67.85**	11.11	7.53	3.96	8.33	1.19
	Physics	3.84	**77.69**	8.46	0.38	7.3	2.3
	Material Sciences	12.74	19.28	**53.26**	0.65	10.78	3.26
	Biology	13.9	6.39	5.26	**61.65**	11.65	1.12
2000	Chemistry	**45.9**	17.27	18.29	2.95	7.3	4.37
	Physics	10.99	**46.23**	13.94	3.12	12.86	6.87
	Material Sciences	20.58	20.96	**39.89**	0.62	10.55	5.07
	Biology	21.12	6.69	11.97	**41.9**	12.32	5.98
2005	Chemistry	**37.26**	14.47	20.37	10.18	10.72	6.97
	Physics	14.48	**43.16**	19.39	7.1	9.01	6.8
	Material Sciences	21.91	19.72	**33.69**	5.47	11.23	7.94
	Biology	21.86	10.49	18.95	**34.11**	7.58	6.99

NANOTECHNOLOGY KNOWLEDGE CREATION AND DISSEMINATION PRACTICES

Advancing an understanding of the trends in nano-technology knowledge dissemination requires an investigation in depth, using both quantitative and qualitative data. In this section, the author uses primary data (a series of face-to-face-interviews with nanotechnology academic scientists and researchers from universities and public research institutes, personnel from government bodies and funding agencies in Europe and Japan) to gather evidence relating to nano-knowledge dissemina-tion across disciplines.

The case of European scientists shows that nanotechnology draws people of different disci-plines towards a hybrid platform that is a mix of multiple fields, primarily because of the merits of utilizing nano-tools. In this sense, an expert in the field needs knowledge of another discipline to work at the nano scale and make cross-links with other researchers, which helps them to move into what is, in effect, a multi-disciplinary system. Nanotechnology knowledge developments and dissemination trends significantly change the traditional culture of knowledge creation both inside and outside the departments. For example, at Oxford University in the department of Phys-ics, which had no Biology department for many years, they now have a program called Biologi-cal Physics, which is one of the largest programs within the department of Physics. An increasing number of undergraduate and graduate students and post-doctoral researchers have been enroll-ing on courses in Biological Physics (a conduit into bionanotechnology). As a result, physicists at all educational stages have the opportunity to learn Biology, biological systems, biological structures, and biological functions as a part of this innovative curriculum. It should be noted that a cultural change in knowledge creation has oc-curred with these developments which has created major changes within the department. Similarly,

outside of the department, it was very unusual to find strong collaboration between different depart-ments in the years before nanotechnology appeared on the scene. Recently, the situation has changed dramatically in this respect, there are major in-terdisciplinary programs, e.g. interdisciplinary research centres and interdisciplinary graduate schools that have been set up and funded by the UK research council. Every year, 75-80 graduate students, who have strong expertise in physical sciences and with no real knowledge and expertise in biological or life sciences, have been admit-ted into these interdisciplinary programmes. The programmes intend to increase knowledge over one year in the fundamental biological sciences and then students begin PhD projects within the IRC, focusing on interdisciplinary research under the umbrella of nanotechnology. These examples have created changes in the educational culture even within the university and have enhanced collaboration across departments and borders.

Some experts from Europe believe that it is now impossible to do research in the nanotechnol-ogy area without there being a strong connection between science and technical disciplines[3]. For example, scientists who were conducting research on simply chemistry or photochemistry, tend to move now into photonics or physics for semi-conductors, as well as into optics and electrical engineering. This technology has appeared purely because of the scientific opportunities presented by working with existing expertise from different fields. Scientists and researchers have argued that nanotechnology holds great promise of providing a new platform where both scientific collabora-tion and opportunities to interact with colleagues across disciplines can take place. Nanotechnol-ogy is simply 'jellying' among different research fields, as Professor Kostas Kostarelos from the University of London puts it. Due to having new material such as CNTs, a physicist (who has no knowledge of its biological, physical or chemical properties) has to interact with biologists, mate-rial scientists, and chemists in order to develop an understanding of its versatile properties and find its applications in different sectors[4]. In this sense, it is necessary to stimulate interaction between multi-disciplinary researchers who need to develop capabilities and knowledge from different disciplines. In other words, in order to make a contribution in nanotechnology, material scientists need knowledge of chemistry, physics and biology. Even within a discipline, several branches are fusing in nanotechnology research. Several examples of evidence of this exist in Europe, for example, in searching for answers to biological questions: Biophysics and Cell Biology are disseminated towards nanotechnology research.

In other cases, people from different research fields have a tendency to work for inter-disciplinary research centers (IRC) that appear to often become excellent nanotechnology research clusters, whether real or virtual. Real inter-disciplinary work is, in a sense, when people from backgrounds specialising in different areas of expertise contribute in part by their physical presence (e.g. centers have chemists to synthesize materials; physicists to carry out instrumental experiments; biologists to do the biological assays; computer scientists to calculate or design structures and explain their function; and engineers to maintain the instruments). Within an IRC, every researcher is assigned projects, discusses or interprets, at a scientific level, the systems, shares their expertise with others, and thus contributes to nanotechnology knowledge creation and accelerates its dissemination. Therefore, the key to achieving this is to make a bridge between different branches of scientific knowledge at the nano-scale and to develop interaction among all practitioners. Virtual inter-discipline is when people with different expertise from different scientific & engineering departments and institutions contribute in part by their virtual presence. For example, IRC Bionanotechnology at Oxford: where researchers from several disciplines are working and collaborating with each other within the institution, as well as virtually with several working groups from dif-

ferent locations through IRCs. Similarly, the IRC Nanoscience Center at Cambridge operates as a real and a virtual centre, sharing knowledge and facilities with a number of university departments and national laboratories across Europe and Asian countries. People can access common equipment, and run facilities where they have specially planned multi-disciplinary projects. Representatives of the departments are, for example, engineers, physicists, chemists and material scientists who are creating an interdisciplinary environment to do nanotechnology research. This trend in the working environment develops gradually. Therefore, to survive the race for nanotechnology, people must change their academic and business attitudes, culture and behavior in this respect of working together within a technology platform of nano-scale[5]. If this is not done, it will decrease the speed of nanotechnology development in general.

Further evidence can be drawn from Japan. Japanese scientists have argued that nanotechnology constitutes a combinatorial aspect of disciplines which enables nano-knowledge creation and dissemination across scientific fields in parallel. A general discipline's research group consists of a variety of experts or researchers – it would include chemists, physicists, engineers, and biologists who are working in parallel. For example, within a scientific discipline, a group of electronic physicist researchers would consist of five sub-groups: one focusing on devices; a second one concentrating on quantum dots and their formation, integration and assembly; a third working on photonics for photonic crystals and LEDs; a fourth sub-group would concentrate on nano-scale thin film oxides; and the fifth sub-group would work on simulation. It can be said that through nanotechnology, scientific research culture has been changing, moving towards making a bridge between multi-disciplinary expertise and experiences. Exchange of nano-information between scientific disciplines seems a crucial phenomenon for nanotechnological systems to evolve.

NANOTECHNOLOGY KNOWLEDGE CREATION AND DISSEMINATION: DEVELOPING MODELS

It appears from the analysis done that the research interest in nanotechnology has been driven by the researchers' own experience in their previous fields of practice. Researchers tend neither to select the nanotechnology field nor diversify into it; rather nanotechnology research has undergone a natural sort of development in an evolutionary way from micro technology, which has been particularly noticeable in the semiconductor and electronics sectors. However, nanotechnology research and development is dispersed through all technology sectors as an evolutionary possibility, except in the biotechnology sector where nanotechnology could really revolutionize the whole spectrum of pharmaceutical and drug discovery processes by changing the whole paradigm of high throughput screening[6].

Both European and Japanese scientists, prior to the emergence of nanotechnology, always believed that research projects in every discipline were very separate and commercialized separately through their distinctive disciplinary features or domain base. Whereas now, nanotechnology knowledge creation and dissemination trends draw researchers from different disciplines towards a hybrid platform that is a mix of multi-fields at the nano-scale, enabling such researchers to work together in parallel. This fostering of its commercialization into the current market drives sustainable innovation as well as creating new markets, which further drives disruptive innovation. In this sense, one expert needs nano-information from another discipline and thus makes cross-linkages among researchers, which helps all involved to innovate in a different way. The essential characteristic of a nanotechnology R&D system is that it involves divergent disciplines to understand the nano-scale systems, sub-systems, and component technologies. The impact between disciplines is also quite

pronounced in that it can initiate the beginning of addressing a new system for practitioners (e.g. a biological system for physicists and engineers; chemistry for biologists, material scientists and engineers). Therefore, it would be a very real and useful reflection of the existing reality to begin purposely bridging disciplines by coming together and taking elements from each of them. From this study it can be deduced that trends in nanotechnology knowledge development and dissemination offer a new way to approach science, which focuses on establishing a new route for practicing science in this decade. The nano-knowledge development and dissemination trend is shown using Chesbrough's (2003) funnel model illustrated in Figure 10.

Nanotechnology is not a new research area; nano-scale research existed previously, but under a different name. Around 50 years ago it was called 'colloid processing' and 20-25 years ago it was called 'sol-gel processing'. In the same way, 'thin films' has been categorized as a nanotechnology area although it has existed for 20-30 years. The most significant change in this field has been brought about by the development of nano-instruments in the early 1980s. Since then, developments in nanotechnology have continued with significant discoveries of new nanomaterials, for example, fullerenes and carbon nanotubes. A revolution in analytical instruments and the introduction of new nanomaterials provided the real breakthrough for the developments of nano-scale technologies. As already identified, people from different disciplines intend to work in parallel at the nano-scale and to help explore this field in such a way that one group of physicists will work on atomic physics while another group works on quantum computing, quantum wires and quantum dots. This knowledge development and dissemination trend will help to evolve a new trajectory by building a bridge between fields. It seems scientists and researchers from various disciplines and technology domains have found new ground to

Figure 10. A model for nanotechnology research and development

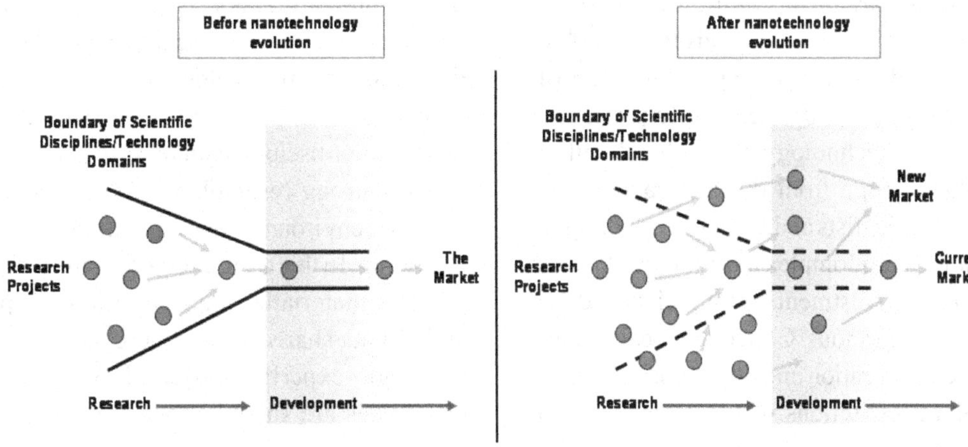

meet at the bottom and synthesize their specializations in new ways, driven by nano-instruments. This further proposes a model of nanotechnology knowledge trajectory as illustrated in Figure 11.

CONCLUSION

This study weighs in with evidence regarding the extent and trend of nanotechnology knowledge creation and dissemination using quantitative and qualitative methods. First, the author used a quantitative approach to visualize nano-knowledge generation and dissemination. These visualizations suggest that nano-knowledge exhibits a high degree of disciplinary diversity, which involves, to a significant degree, many other scientific fields and technology domains. The chapter suggests that every traditional discipline needs to share nanotechnology information with other areas of disciplinary knowledge; this means that nanotechnology researchers do not operate within narrow silos. In this chapter, it has been seen that nanotechnology appears to be not a single field of enquiry, but rather a constellation of several distinct trajectories of scientific advance – a mul-

Figure 11. A model for nanotechnology knowledge trajectory development

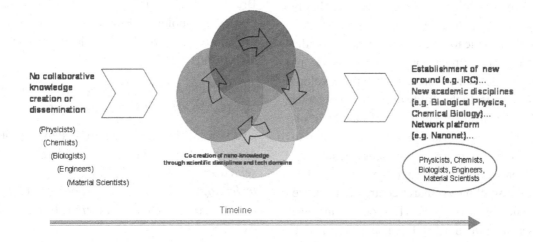

tidisciplinary collection of fields. The findings suggest that as part of the future development of nanotechnology, attention needs to be paid to facilitating the dissemination and absorption of nanotechnology knowledge across disciplines and to assisting nanotechnology researchers' ability to source knowledge from disparate areas. The challenges for scientists and engineers working in nanotechnology are complex and the expectations for the return on investments from society and the government is enormous. Concerted local and international collaboration and co-operation within and outside of institutions would be required to realize the full potential of the technology.

This chapter suggests that every traditional discipline needs to share nano-information with the others. With the evolution of nanotechnology, nanotechnology knowledge creation and dissemination trends have appeared to bridge divergent disciplines. In this respect, it would not be possible to continue nanotechnology R&D, if those concerned lacked relevant background knowledge, lacked links with other experts and possessed a lack of multi-disciplinary knowledge. The author provides examples of how nanotechnology plays a role in bringing scientists from different disciplines together, and of how their impact contributes to collaborative knowledge creation and dissemination. The quantitative studies presented in this chapter analyse the outcomes of the knowledge creation and dissemination process, for example, in the form of publications. Such studies are likely to miss the dynamics of on-going practice and would not be able to assess the role played by the technology in this process or to see how institutions would change their culture in light of the developing knowledge. In this study of on-going practice and the historical development of nanotechnology, using interviews and data analysis helped to reveal the broad trends in knowledge generation and dissemination. The existence of nanotechnology interdisciplinary research centres, both inside and outside of universities in European countries, represent a promising development in the practice

of acquiring multi-disciplinary knowledge through nanotechnology. This study emphasizes the importance of collaborative research networks between divergent scientific fields and institutions in work on the nano-scale. A significant aspect of this is that the interdisciplinary nature of nanotechnology bears a strong resemblance to situations where tools and environments are developed for learning and doing. In this case, everyday nano-knowledge practice materializes in a hybrid, both physical and virtual, shared space where nano-knowledge and nano-expertise acquired from traditional disciplines are shared between participants and locations. The emerging knowledge and experience accumulate in the sphere of nanoscience and technology, enabling scientists and researchers to co-create, share and disseminate nano-knowledge across groups, institutions and borders.

ACKNOWLEDGMENT

The authors would like to show his grateful acknowledgement by giving thanks to European and Japanese nanotech scientists and researchers from universities and public research institutes for sharing their valuable time.

REFERENCES

Bresnahan, T. F., & Tajtenberg, M. (1995). General purpose technologies: Engines of growth. *Journal of Econometrics, 65*, 83–108. doi:10.1016/0304-4076(94)01598-T

Chesbrough, H. (2003). *Open Innovation*. Cambridge, MA: Harvard Business Press.

CRISP/OST Foresight Briefing Paper. (2001). *Nanotechnology: engineering with atoms. Presented in Foresight/CRISP workshop on nanotechnology: What is nanotechnology? What are its implications for construction*. London: The Royal Society of Arts.

Eto, H. (2003). Interdisciplinary information input and output of nano-technology project. *Scientometrics, 58*(1), 5–33. doi:10.1023/A:1025423406643

Helpman, E. (1998). *General Purpose Technologies and Economic Growth.* Cambridge, MA: MIT Press.

Igami, M., & Okazaki, T. (2007). *Capturing Nanotechnology's Current State of Development via Analysis of Patents.* STI Working Paper 2007/4, OECD Directorate for Science, Technology and Industry.

Islam, N., & Miyazaki, K. (2009). Nanotechnology innovation system: Understanding hidden dynamics of nanoscience fusion trajectories. *Technological Forecasting and Social Change, 76*(1), 128–140. doi:10.1016/j.techfore.2008.03.021

Malanowski, N., Heimer, T., Luther, W., & Werner, M. (Eds.). (2006). *Growth Market Nanotechnology – An Analysis of Technology and Innovation.* Weinheim, Germany: Wiley-VCH Verlag.

Meyer, M. (2006). *What do we know about innovation in nanotechnology? Some propositions about an emerging field between hype and path-dependency.* Paper presented at the 2006 technology transfer society conference, Atlanta, Georgia, 27–29 September.

Meyer, M., & Persson, O. (1998). Nanotechnology: Interdisciplinarity, patterns of collaboration and differences in application. *Scientometrics, 42*(2), 195–205. doi:10.1007/BF02458355

National Science and Technology Council. (2006). *The National Nanotechnology Initiative: Research and Development Leading to a Revolution in Technology and Industry, Report-July.* USA: National Science and Technology Council.

Peters, L. S. (2001). Radical innovation and global patterns of breakthrough technology development: an analysis of biotechnology and nanotechnology. In *Proceedings of IEEE Engineering Management Society Proceedings* (pp. 206-212).

Porter, A. L., & Cunningham, S. W. (2005). *Tech Mining. Exploiting New Technologies for Competitive Advantage.* Hoboken, NJ: Wiley-Interscience.

Rocco, M. C. (2002). *International Strategy for Nanotechnology Research and Development Report.* Washington, DC: National Science Foundation.

Roco, M. C., & Bainbridge, W. S. (2003). *Converging technologies for improving human performance: nanotechnology, biotechnology information technology and cognitive science.* Dordrecht, The Netherlands: Kluwer Academic Publishers.

Schummer, J. (2004). Multidisciplinarity, interdisciplinarity, and patterns of research collaboration in nanoscience and nanotechnology. *Scientometrics, 59*, 425–465. doi:10.1023/B:SCIE.0000018542.71314.38

ENDNOTES

[1] For example, nanoparticles are used in automobiles for filler in car tyres, nanoporous filters to minimize the emission of particles on the nanometer scale, catalytic nanoparticles as a fuel additive; carbon nanotubes are used in tennis racquets, field emission displays (FED), transistors, fuel cells and high-performance batteries

[2] The keywords included among others: nanomaterial, nanoparticle, nanocrystal, nanocomposite, carbon nanotubes, fullerenes, nanotubes, nanostructures, nanofiber, plastic nanocomposites, strain-resistant fabrics, nanocoating, nanofilms, thin films, nanorobotics, nanosensor, biological nanosensor

[3] interview with Professor Anne Ulrich, Institut for Organische Chemie and CFN, University of Karlsruhe; interview with

Professor Marcello Baricco, Dipartimento de Chimica IFM and NIS-Centre of Excellence, Universita degli Studi di Torino

[4] interview with Professor Kostas Kostarelos, Center for Drug Delivery Research, University of London

[5] interview with Dr. Karl-Heinz Haas, Spokesman of the Fraunhofer Nanotechnology Alliance and Deputy Director for Fraunhofer ISC, Fraunhofer-Gesellschaft

[6] interview with Professor John Ryan, Director of IRC Bionanotechnology, University of Oxford

Chapter 4
Collaborations in the Open Innovation Era

Annamária Inzelt
IKU Innovation Research Centre, Financial Research Ltd, Hungary

ABSTRACT

Although the impact of open innovation on a global scale on the collaboration between universities and foreign industry is clearly important, empirical evidence from the field is lacking. This chapter investigates the collaboration between Hungarian universities and foreign companies in research and development. The chapter attempts to provide a relevant picture of the research-related linkages of Hungarian universities and foreign companies by employing secondary data processed from various data-banks. The analysis suggests that foreign direct investment and foreign companies play major roles in the internationalisation of research during this second decade of the transition process. Assessing the research and technology products which have originated in university-industry collaboration is no easy task. According to experimental measurements and pilot data-bank, there were more joint publications involving foreign than domestic companies, and the citation value per publication was significantly higher with the former. Data-bank also show that developments in new technology in terms of patent figures rarely involved university-owned or co-owned inventions, although there is some evidence there are more patents which are university-related than owned. Domestic invention and the foreign ownership of patents represent one more sign of Hungarian involvement in global innovation in the development of new technologies.

INTRODUCTION

The new wave of internationalisation is a product of corporate research and development (R&D)

activity. The circulation of international knowledge is critical for the development of innovation performance and for the improvement of national competitiveness in the sense that internationalisation widens the access of companies to academic knowledge and research capabilities. In this pro-

DOI: 10.4018/978-1-61692-006-7.ch004

cess, entities in the international business world influence connections between universities and industry on a cross-border basis. An important question facing policy-makers in Science and Technology asks how this kind of internationalisation affects the universities in the academic host country. The contribution - in terms of inventions of universities to innovation and to economic growth may well become outwardly directed. Conversely, however, without such contracts with foreign companies, universities which are not located in an innovative environment have fewer chances to participate in cutting-edge research activities, and any spillover effects may come much later. The ideal balance between inflow (foreign corporate R&D investment) and outflow (the commercial sales of intellectual property or know-how) is a delicate issue for university administrators, for the corporate sphere and for national policy-makers.

In terms of the internationalisation of university-industry linkages, three fields are currently showing ongoing transformation. The first of these is the changing pattern of innovation which affects the ways in which companies outsource R&D and collaborate commercially; the second relates to the enhancement and globalisation of the Third Mission of Higher Education; the third involves the new wave of internationalisation in which companies' related R&D and innovation activities are globalised. At the same time the policies which stimulate FDI are changing, and the new generations of FDI and other policies focus on FDI-led R&D and innovation. (UNCTAD 2001, Kalotay and Filippov 2009, Guimón 2009)

Although the impact of open innovation on collaboration between universities and foreign industry is clearly important, there is a lack of empirical evidence from this field, and this paper attempts to use of the various data sources available and to develop new indicators to analyse Hungarian involvement in the process

The extent of internationalisation, as reflected by foreign ownership, has increased significantly in Hungary over the last decade, and one of

the consequences of internationalisation is the changing pattern of university-industry relations. The context of the internationalisation of the relationship is distinctive, given that the proportion of domestic invention registered by foreign companies amounts to some 60%.

Hungarian universities do collaborate with foreign-owned companies located in the country - as with companies based elsewhere - and international partners play an important role in linking universities and industry, quite apart from the national environment, which we can describe as moderately innovative.

Following an overview of transformation (on the basis of the literature) the paper offers a number of facts about general foreign involvement in Hungarian business R&D activities and outsourcing. For this part of the analysis we use official statistics on business R&D expenditure. The third and fourth parts of the chapter briefly describe the relationships of Hungarian universities and foreign companies. These sections attempt to illustrate the internationalisation of university-industry relationships using secondary processing from various data sources.

Two different types of foreign business are examined. The first of these relates to entities which are partly or totally foreign-owned but registered in Hungary and the second to foreign entities which are not registered in Hungary and which are only involved in investing in and purchasing R&D.

To describe and analyse relationships, the paper employs certain input and output indicators. The third part is devoted to those inputs where innovation input is represented by R&D expenditure and the fourth investigates the output side of the process by means of publications and patent data.

The available data allow some debate on the specifics of the internationalisation of university-industry partnerships in transition economies, and the paper provides a better understanding of how the open innovation model works and how this affects the triple helix model.

ONGOING TRANSFORMATION

A global transformation is evidenced by the changing structure of innovation, and we can also see the parallel (and closely related) phenomenon of Humboldtian universities being in a similar state of transformation. The new wave of internationalisation touches both. There is a large quantity of published material covering these related issues, and the next sections highlight the new challenges and some new findings in the literature.

The Changing Innovation Model

In the late 19th century a crucial innovation occurred in the generation of useful knowledge for industry: in-house research and development (R&D) laboratories. At that time, when the first company laboratories appeared, this started the trend towards in-house research and to a closed system of innovation. The companies spent majority of their R&D budget in-house in their own laboratories,–until, in fact, the 1960s. Over this period the innovation process was characterised by the internal generation and use of knowledge within a company and by little or nothing which was purchased or could be termed external knowledge. Practice at the dawn of the 20th century leant heavily towards closed or semi-closed innovation.

Starting in the 1960s, however, the (now traditional) closed innovation paradigm was turned around completely and was replaced by the open innovation paradigm. (Chesbrough 2003, Gassmann and Enkel 2004) These linkages lead to a 'network model of innovation' (Callon 1992), 'distributed innovation processes' (Coombs et al., 2003) or 'open innovation' (Chesbrough, 2003). The common, central idea behind these various terms is that, in a world of widely distributed knowledge, companies cannot afford to rely entirely on their own research, but should, instead, buy or license processes or inventions from others.

Companies are again employing extramural activities. Both needs (faster and more efficient innovation) and opportunities (drastic reductions in transport, communication and co-ordination costs, rapid developments in ICT and for greater codification and standardisation of R&D processes) for open innovation have increased the possibilities for segmenting and dispersing R&D over a number of locations and types of organisation (such as other companies and public research organisations).[1] Mainstream companies are increasingly opening their innovation processes and collaborating on innovation with external partners (suppliers, customers, universities etc.) and there is clear movement towards the greater outsourcing of business R&D - either to other companies or to public research organisations.[2] Companies can no longer survive through their own R&D efforts but look for new, more open, methods of innovation.

The external partners chosen by companies differ according to whether the companies wish to collaborate on research or on development. (OECD, 2008a p. 114.) Even if the majority of outsourced business R&D goes to other companies (OECD 2008b p. 19) the demand for public research organisations such as universities has increased significantly.

In the open innovation system, working within R&D and innovation networks is crucial since these networks can contribute to a rational balance between 'intramural' and 'extramural' R&D capacities and activities. In-house capacities are very important for the selection of external partners, for reinforcing decisions on purchasing new knowledge and technology, and for supporting the application of new technology.

The concept of open innovation is closely linked to national and regional innovation systems which emphasise the inter-organisational linkages for knowledge creation and diffusion.[3] (Lundvall, 1992; Nelson 1993) From this complex system the 'triple helix' model describes the linkages between public and private sectors when university, industry and government work together. (Etzkowitz and Leydesdorff (1997), Etzkowitz (2008).

Changes in Research in Higher Education

The university (and academic research organisations) are crucial actors in the knowledge-driven economy as knowledge-generating agents. In the wider innovation process, they are both sources of basic knowledge and potential partners in the open innovation strategy of companies.

The literature is very rich on the changing model of the university in the late 20th century. (Bonacorsi and Dario 2007, Inzelt 2004, Laredo 2007, Martin 2003, Nedeva 2008, Sanchez and Elena 2006, Varga 2000,). Whilst the term "open innovation" explains why it is important for companies to collaborate with universities and public research organisations, the triple helix model focuses on university-industry-government interaction and partnership and studies their changes. (Etzkowitz and Leydesdorff 2000, Etzkowitz 2008) Those researchers who are studying the Third Mission of universities focus on the questions of why it is important and useful for universities to co-operate.

Changes in the Missions of universities and the increasing importance of the Third Mission clearly interrelate with the open innovation system. Sharing university knowledge with the economy and society has many potential benefits for the actors involved. The Third Mission of universities (following Education and Research) is growing in importance (Gulbrandsen and Slipersaeter 2007, Inzelt et al. 2006, Mollas-Gallart et al. 2002). The results of empirical studies and innovation surveys showed that only a small fraction of companies actively demand universities as a partner in innovation. The importance of universities as knowledge sources for companies depends on the characteristics of the sectors (high-technology, advanced, dynamic), the potential to innovate of companies (radical vs. incremental) and the development level of regions.

The available pool of skilled scientists (a critical mass) also influences the capacity of universities to attract lucrative business to the region. (Dőry 2005, Iammarino and McCann 2006, Varga 2005) Last but not least, the local and national environment of universities (e.g., the level of advancement of surrounding industries, the size of a company, the legal framework and intermediaries) also has an impact on collaboration. (A good collection of case studies was published on the role of intermediaries in the Special Section of Research Policy vol. 37, issue 8, 2008)

The strategy of universities toward collaboration has changed due to changes in the size and structure of public funding - parallel to the increased autonomy of universities. (Laudel 2006), and the importance of valorisation in respect of research results, R&D services and collaboration with companies has consequently increased.

The Third Mission depends upon the configuration of activity at a given university activity, how embedded it is in its particular geographical territory and the national institutional framework.[4] The relations of universities with industry have become a major focus for policy-makers, and among these feature contracts with industry, patents taken out directly by universities and the creation of new spin-off and spin-out firms around a university. Similarly, more and more importance has been given to PhD graduates going into industry. (Laredo 2007)

The role of the university in innovation also has a strong impact upon the nature of the knowledge produced by universities.[5] The increased autonomy of universities, the shift towards competitive funding and the changing demand of industry has had an impact not only on the Third Mission of universities but also on education and research. The Bologna process (three degrees in Higher Education, changes to teaching curricula and the introduction of an internationally accepted credits system) is an important adjustment to the needs of the economy and society by institutions of HE within the context of globalised knowledge-driven economies.

As industry-science relationships become global, universities must compete internationally to attract R&D-related FDI, since partners from industry will not finance research into areas in which they are not interested. The ability of universities to compete at world level as well as to join various international networks is important. In addition to an attractive, FDI-related economic policy, the reform of university systems in general, the previous research performances of universities and their accumulated knowledge and capabilities, together with fostering a critical mass in research fields are basic conditions for upgrading international collaboration.

The New Wave of Internationalisation

International co-operation has become more important for companies over the last few decades (Dunning, 2005) characterised by increased interaction between knowledge and globalisation. In this period FDI flows have increased dramatically and continue to be a driving factor of economic globalisation. Corporate innovation activities are increasingly international and tend to favour open innovation–collaborating with external partners, whether suppliers, customers or universities, to maintain their position at the forefront of innovation, and to have new products or services to market before their competitors. More supply-driven factors, such as R&D, have become important.

From the globalisation of open innovation, companies do expect to remain in the forefront of innovation and to outmatch their competitors in introducing new products to the market.

Although the internationalisation of R&D through FDI is not a new phenomenon, its rapid growth and scope have changed dramatically. (Raymond and Taggart 1998, Cantwel and Molero 2003, Narula and Zanfei 2005, Foray 2006) The purchase or outsourcing of R&D (whether domestically or internationally) is now a serious complement to in-house R&D as a part of corporate innovation strategy. (OECD 2008b, EU 2005) The observation of Pavitt (1997, 2002) is still valid–i.e., that a location is attractive for foreign R&D investment if it has a good S&T base (an excellent, or good centre of knowledge, a large pool of skilled S&T workers) and if it provides opportunities to acquire R&D conducted by other companies or institutions and offers more rapid commercialisation.

Traditional cross-border R&D sought to *adapt* products and services to the needs of host countries and to support foreign investors' local operations, and so it was mainly demand-driven. Nowadays, multinational companies (MNCs) seek not only to exploit knowledge generated at home, but also to source technology internationally and to generate new knowledge in other countries - which is why MNCs need access to highly skilled scientific personnel and to tap into worldwide centres of knowledge. (Edler et al. 2002, Edler 2008, Inzelt 2008b, Taggart 1998) More and more companies are responding to increasing global competition and raising R&D costs by internationalising R&D along with other knowledge-intensive corporate functions.[6]

FDI plays a major role in the internationalisation of R&D, and MNCs are the main actors. The literature relating to the internationalisation of innovation systems, which is overviewed by Carlsson (2006), regards inward R&D-intensive FDI as a powerful mechanism of international technology transfer. This mechanism can enable host locations to integrate more advantageously in global value chains. Generally, inward FDI acts as a channel for knowledge-flows and provides opportunities for learning in domestic companies, for establishing regional networks and for involving other foreign- controlled companies.

MNCs have increasingly moved R&D activity across borders within their global value chain and rely on outside innovation for new products and processes. Large companies (mostly MNCs) increasingly adopt innovation networks which link networks of people, institutions (universities, government agencies and other companies) in different countries to solve problems and produce ideas. (Cook 2005) These kinds of internationalised networks generate radical innovation.

Internationalisation relates more to larger companies than to smaller entities and it is more prevalent in certain sectors. Open innovation depends on the technological and industrial context. (Chesbrough and Teece 1996, Gassmann 2006) According to UNCTAD (2005), the activities of the MNCs which spend the most on R&D are concentrated in information technology (hardware), the automotive industry, pharmaceuticals and biotechnology, and in the electronics and electrical industries. OECD adds aviation and aerospace to this list (2008a p. 35). These industries account for over two-thirds of R&D by the world's top 700 spenders (OECD 2008b p.20-21). And so companies or universities active in these fields have a better chance to collaborate internationally than others.

The globalisation of industry-science relationships is also reshaping the triple helix model. (Etzkowitz 2008)

SOME FACTS ON FOREIGN INVOLVEMENT IN HUNGARIAN R&D ACTIVITY

Since the beginning of the transition period, Hungary has attracted a considerable inward flow of FDI and is a typical host country for inward FDI. In 2005 its stock was 56% of GDP, rising to nearly 66% in 2007 - one of the highest levels of foreign ownership in OECD countries. (OECD 2008c, p. 64) Between 1989 and 2008 cumulated FDI per capita in Hungary amounted to US$ 5,314 - third among the countries of Central and Eastern Europe (CEECs), following the Czech Republic (US$ 6,954) and Estonia (US$ 6,749) (EBRD 2005)

Hungary, as a small country, is more dependent on international flows of knowledge and capital than are larger countries with large internal markets

for R&D and innovation. Hungary employs the newest generator of FDI policy that focuses on FDI-led R&D and innovation. In the early 1990s market-based privatization was the main incentive of foreign investment.

No special permit is required to establish a business enterprise in Hungary. Foreign nationals either naturals or legal entities can found companies in Hungary. The only requirement is that the headquarters of such companies must be located within Hungary. Any kind of investors such as foreigners can enjoy the development-tax allowance. That may be utilized for ten tax years following the completion of the development project. Since 1995 Hungary has devoted special attention encouraging foreigners to invest in R&D in Hungary. Hungarian FDI policy backed up with R&D and innovation policy for attracting foreign investment in RDI. These policies together are forming third generation of FDI policy (FDI-led R&D). It is not surprising, therefore, that Hungary should encourage not only general foreign investment but also foreign investment in R&D.[7] The rationale behind this is that globalisation and open innovation provide the country with access to research and innovation networks which will accelerate its own development and better exploit its capabilities. A further assumption is that collaboration in the R&D stage of the value chain makes foreign investment more durable. A long-lasting relationship is an important element in improving competitiveness.

The ratio of Gross Domestic Expenditure on R&D (GERD) to GDP is still under 1% in Hungary, although the proportion of business-funded R&D increased between 1995 and 2005 (OECD 2007, p. 69 and p. 168). In 2006 the ratio of business-funded R&D to GDP was 0.48%, far below the OECD average of 1.56% (OECD 2008c. p. 81.) and also well below the EU's Lisbon targets.

In Hungary, foreign-origin funding for R&D as a percentage of business-funded R&D is, at 18%, significantly above the EU-27 average of 10%. This is also true of business-funded R&D

Figure 1. R&D funding from abroad

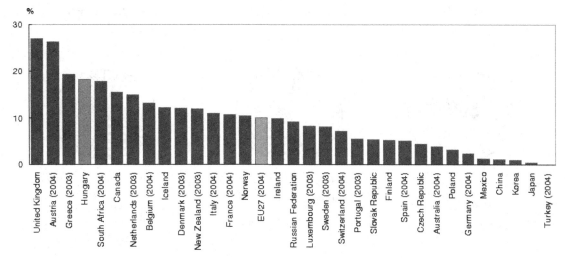

Source: OECD STI Scoreboard, p. 169

in the Higher Education and Government sectors. (Figure 1)

The R&D potential of FDI-funded companies differs from that of domestically owned entities. The indicators give an interesting picture of the innovation potential of domestic and foreign-owned firms, and official statistics provide useful information for such an analysis. The Hungarian Statistical Office groups business founders in 5 categories, which we form into two main groups: (1) the '*domestic business*' group (domestic private business, state-owned and local authority-owned) and (2) the '*domiciled, foreign-owned business*' group (businesses fully- or majority-owned by foreign interests).

As Figure 2 illustrates, of the R&D spending companies only 13% were domiciled foreign-owned. In 2007 these financed 67% of business R&D expenditure and employed 51% of scientists and engineers. R&D expenditure per R&D employee was significantly higher than in domestic companies and the number of R&D personnel per unit was also much higher.

As large companies tend to be foreign-owned, they spend disproportionately more on R&D than do domestic firms, and so, among the R&D-spending companies, MNCs have a particular significance.

The overwhelming role of foreign MNCs in Hungarian R&D has raised the issue of the dependence and vulnerability of the local R&D base. Conversely, the demand of MNCs for R&D plays an important role in preserving and developing R&D capacities and internationalising these activities.

As mentioned earlier, globally open innovation varies by industry. The ratio of outsourced cross-border R&D to total expenditure is more characteristic in certain industries, and the internationalisation of the main R&D-spending MNCs is concentrated on a few sectors. Consequently those companies or universities active in the globally open innovative sectors may be targeted more frequently than others by foreign investors to collaborate internationally in R&D.

Those industries in which the foreign R&D spending of MNCs is concentrated we can refer to as *global R&D spending sectors*. These include the IT, automotive, pharmaceutical, biotechnology, electronic, electrical and aviation and aerospace industries–although the composition of industry varies by region and country.

Figure 2. Key R&D data by company ownership

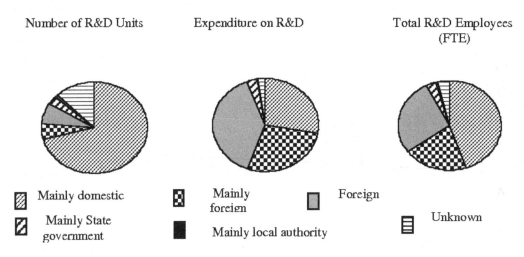

Source: HCSO, R&D Report, 2007

In this context various questions arise, one of which asks how we should characterise the R&D expenditure of the foreign global R&D spending companies if the difference between these and other industries is so clear. A further question asks how purchase and outsourcing of R&D vary according to industrial sector and ownership. In our attempt to examine this issue, we utilise normal R&D survey data, two types of which are significant for us: total R&D expenditure and the proportion outsourced.[8] From the HCSO databank, we selected those industrial sectors which are recognised as global spenders on R&D and segregated the remaining R&D spending sectors into another group. (Table 1) These data give a picture of outsourcing potential which, to some extent, reflects Higher Education, although the data are not sufficiently detailed to assess the real demand of companies for Higher Education.

Table 1. Business R&D expenditure by sector, 2007 (%)

Sectors	Total R&D expenditure			Outsourced R&D expenditure		
	Foreign	Domestic	Total	Foreign	Domestic	Total
Main global R&D spending sectors	75,6	31,2	65,2	96,1	24,3	89,1
Pharmaceuticals	23,9	17,1	22,3	5,4	12,4	6,1
IT	0,5	8,4	2,3	0	10,7	1
Automotive	48,3	0,9	37,2	90,3	0,4	81,5
Electrical/Electronics	2,9	4,8	3,4	0,4	0,8	0,5
Remaining R&D spend-ing sectors	24,4	68,8	34,8	3,9	75,7	10,9
Total	**100**	**100**	**100**	**100**	**100**	**100**

Source: HCSO data-bank, compiled by Zsuzsanna Szunyogh and the author

Note: IT (3002, 7210, 7221, 7222, 7230, 7250, 7260); Automotive industry: (3410, 3420, 3430); Pharmaceuticals and biotechnology (2441, 2442); Electronics and electrical industry (3110, 3120, 3130, 3150, 3161, 3162, 3210)

In Hungary, the main global R&D spending sectors account for two-thirds of business R&D expenditure, and, in the foreign ownership group, no less than three-quarters (Table 1) This sectoral structure is even stronger if we are focus on outsourced R&D, where 96% relates to foreign-owned companies in global R&D spending sectors, the vast majority of which (90%) comes from the automotive sector.[9]

It is evident that the internationalisation of business R&D in Hungary generally follows global trends, and, if we focus on the outsourced R&D proportion of global R&D spending sectors, it is much higher (50%) than for the remaining R&D sectors. In accordance with international trends, partly or totally foreign-owned companies outsource much more in global R&D spending sectors (55%) than do those mainly domestically owned (12%). The opposite trend is visible among the remaining R&D spending sectors. (Table 2) These figures further demonstrate that the internationalisation pattern of business R&D in Hungary is close to the global trends, demonstrating that collaboration serves global open innovation.

The breakdown of global R&D spending sectors shows strong concentrations. The sector shows that the outsourcing activity of foreign-owned companies is more intensive in foreign-owned companies - in the automotive (80% vs. 7%) and in the electrical and electronics (6 vs. 3%) industries. Domestically-owned companies outsource a somewhat higher proportion of their R&D in the pharmaceutical and IT industries - totally against international trends. Foreign companies outsource less than 2%, whilst, for domestically-owned firms the figure approaches 20%. Further investigation is needed to identify the reasons for this deviation and it would also be useful to examine the content of R&D activity.

Official statistics cannot provide any figures relating to the number of universities who are active in performing outsourced R&D. There are, however, a few items of indirect information. One important feature is that the proportion of company- financed R&D to total R&D performed in the HE and government sectors is much higher in Hungary, (11%) than in the EU-27, (6.4%) whilst the ratio of business R&D expenditure to GDP is lower than the EU average. One further fact to be mentioned is that the investment in R&D by foreign-owned companies in Hungary has its own significance. As foreign business is responsible for the lion's share of Hungarian business R&D, we can assume that FDI-related R&D is important for Higher Education Institutions (HEIs)—or at least in global innovation-related fields of science. We

Table 2. Business outsourced R&D in Hungary by international character of sectors 2007

Sectors	Foreign	Domestic	Total	Foreign	Domestic	Total
	Outsourced R&D to total R&D expenditure (%)			Outsourced R&D to in-house R&D expenditure (%)		
Main global R&D spending sectors	55,1	12,0	50,3	122,8	13,6	101,1
Pharmaceuticals	9,8	11,1	10,0	10,8	12,5	11,1
IT	1,8	19,5	16,7	1,8	24,3	20,0
Automotive	81,1	6,8	80,6	427,8	7,3	416,6
Electrical/electronics	6,3	2,6	5,1	6,8	2,7	5,4
Remaining R&D spending sectors	6,9	16,8	11,5	7,4	20,3	13,0
Total	**43,4**	**15,3**	**36,8**	**76,6**	**18,1**	**58,2**

Source and note: see table 3.

attempt to investigate this issue further using new data and indicators.

COLLABORATION BETWEEN FOREIGN COMPANIES AND HUNGARIAN UNIVERSITIES

Companies commonly seek partners for collaboration in development, (industrial partners, clients, suppliers). (OECD 2008a p.91) Universities are natural partners for profit-oriented strategies although less important for asset-exploiting strategies in many countries. Whether we approach industry-university collaboration from the perspective of the company or of the university, both prefer to collaborate with one another in the pursuit of profit

According to the experience of advanced countries or regions, companies search for universities which could be potential partners in research collaboration, whilst in less advanced countries (or regions), where appropriate innovative business partners are rare and there are limited opportunities for foreign investors to find business partners for innovations, this shortage may lead to a division of collaboration among a handful of universities prominent in research and industry. In less innovative circumstances, the second-best solution for innovative business is to contract with universities not only for profit-seeking but also for asset-exploiting R&D tasks. For foreign investors, universities may to some degree be a substitute for non-existent local innovation actors and can also act as magnets, attracting attention to the region. If the local ecology is not rich enough, internationalisation reshapes the university-industry relationship.[10] (Inzelt 2004, Inzelt 2008a, Kállay and Lengyel 2008, Inzelt and Csonka 2008)

Companies build especially close relationships with certain universities, and the academic excellence of universities and public research laboratories attract the R&D departments of large firms. The attractiveness of a university to external actors depends on many factors such as the university's research potential, the way in which the university is equipped to develop knowledge jointly with companies and the capabilities for technology transfer offices. Another crucial factor is "subject-mix" and the universities' existing research fields, since open innovation is more frequently found in those industries which target faculties in their own fields.

Contracts with Industry

As discussed in the literature, there are many dimensions and forms of collaboration. (Gulbrandsen and Slipersaeter 2007, Inzelt et al. 2006, Molas-Gallart et al. 2002) So-called purchasing-based innovative companies interact with Institutions of HE as they purchase inputs for their innovation process. Companies active in creating new knowledge collaborate on research. These companies establish partnerships to innovate jointly with a common goal in view. (OECD 2008a, p. 22) These contracts may cover various forms of relationship. Contracts with industry cover an institution's revenues from private companies for undertaking research, providing research services or for carrying out testing for industrial partners. The contracts may have a 'soft' dimension: large companies may pay (directly or through the university) faculty members' membership fees to professional associations, the cost of travel to participate in conferences, or funds to cover the cost of professional publications. (Inzelt et al. 2006) Contracts may also include several other activities besides research and research services. For example, PhD students might be supported by industry, or a company may contract with an university to provide training courses for their employees.

From these linkages we focus only on those which are R&D-related. To demonstrate international collaboration, we employ one of the input indicators of innovation practice - R&D expenditure funded by the business sphere. Two types of output indicator are employed: scientometrics which are generally used for measuring scientific performance and patent indicators to characterise technology creation. Indicators are designed to measure collaborative performances.

Most relations between Hungarian universities and companies are covered by research contracts and we can examine some official statistics concerning the role of business in funding university research.

In financing Higher Education R&D, the major source is still the government sector. (Table 3) Business represents a smaller (but growing) proportion. In 2007 business financed 14% of total HE expenditure on R&D (HERD). This figure was similar in 2005, although it had been only 6% in 2000. This increase indicates that more and more national (and EU) programmes have initiated public-private partnerships encouraging private business demand. These incentives have affected the 'contracts with industry'. (Table 3)

Companies which are working in Hungary (domestic and domiciled foreign firms) financed 2% of HERD in 1995 when FDI scarcely featured in R&D. Around the turn of the century business became much more important in financing (12% in 2002) as FDI-led R&D activity reached the universities.

Investigating the funding structure of HERD, we can also see another type of foreign corporate player: some foreign companies with no production or commercial investment in Hungary are contracting with universities. Since 2002 these companies have provided 0.2-0.3% of HERD. This tiny percentage is easy to overlook, but the source has its own importance for the university departments involved and for the field of science. The source represents an opportunity to collaborate in the development of advanced technology and to break into the network of pioneering companies since companies target partners from among the cutting-edge departments of universities.

As the official statistics cannot provide information on the owners and origins of those companies funding Higher Education, we must employ another database to learn more about foreign HERD sources. An administrative databank at the Ministry of Education which existed between 1995 and 2005 contained detailed information on various research contracts of universities.[11] (See the description in Inzelt 2004, p. 979) Employing the figures and value of business funding can provide information on the size of the business sources which were attracted. The main advantage of this databank compared to official statistics is that it allows us to break down university-industry research contracts according to the owners and origins of business organisations for so-called large income contracts, meaning that the sum of the contract exceeds HUF 5m (€19-20,000). HE Institutions have many contracts below this

Table 3. Distribution of the sources of HERD (%)

Funding sources	1995	2000	2002	2004	2006	2007
Government	89.8	85.8	83.8	80.9	77.1	76.8
Business (domestic and domiciled firms)	2.1	5.7	11.8	12.9	13.0	13.7
Non-profit	-	1.0	0.8	0.6	1.7	1.8
Foreign sources	3.8	5.4	3.7	5.7	8.2	7.7
Foreign firms working abroad	-	-	0.3	0.2	0.2	0.3
Others	4.3	2.1	-	-	-	-
Total	100.0	100.0	100.0	100.0	100.0	100.0
Source: compiled from HCSO databank, June 2009						

threshold, but information on them is either missing or not detailed.

To investigate the origin of large contracts and of contracts with foreign industry we selected 12 Hungarian universities (6 large and 6 in the medium-to-small bracket). The main selection criterion was good research performance as demonstrated by publications. (The 12 provide 90% of ISI publications.)[12]

The so-called foreign partners of universities belong to two different groups: (1) partly or totally foreign-owned domiciled firms and (2) foreign firms with no local manufacturing or service activity. Categorisation was carried out manually.[13]

Between 2000 and 2005 more than 50% of contracts (by number) and two-thirds of business-derived income came from foreign-owned and foreign-located companies. Table 4 shows contracts with industry according to the origin of the commercial partner.

Similarly to the general picture given in the previous section on domestic and domiciled foreign firms, the number of contracts with domestic firms is high, but University income from contracts is lower from this source. The average income per contract with domestic companies was HUF 23.5m, a sum which was 54% higher when foreign-located firms were involved and 66% higher in the case of contracts with domiciled foreign-owned firms.

Available data do not allow us to separate these contracts by purchasing-based activities or collaborative relationships. However, the importance of the university as a source of knowledge is different for the two types. According to anecdotal evidence, the bulk of contracts are more purchasing-based than collaborative in character.

Further information is needed to assess the impact of contract research on HE research activities and the innovativeness of companies. For example, if we wish to know the aims of university-industry collaboration, the character of the knowledge purchased, the information needed, the number and total value of contracts according to purpose we need much more detailed information. It would be very useful if we could break down the number and the value of contracts by their aim (research, research services, training, consultancy and expert advice to industry, university faculty development, supported research chairs) and by types of partner (MNCs, large companies, SMEs) and investigate these data on the level of faculty, university and scientific field. Further useful information would be the size and level of internationalisation of the collaborating firms.

In addition to information on partner companies, it is also important to learn the capabilities of universities to produce useful knowledge for the outside world and also to understand university departments and which fields of science are best

Table 4. Overview of contracts with industry, 2000-2005 (12 universities)

Owners and Origin	No. of Contracts	Income	Income / contracts	Distribution by	
				No. of Contracts	Income
		M HUF			
Mainly Domestic, Private	143	3367.4	23.5	47	36
Domiciled Foreign (majority shareholding)	66	2578.6	39.1	22	28
Foreign-located	92	3320.6	36.1	31	36
Total	*301*	*9266.6*	*30.8*	*100*	*100*

Source: IKU's compilation from the databank of the Ministry of Education (and Culture)

Note: the table contains only those contracts whose total value exceeded HUF 5m. Each university has had many other minor contracts.

suited to collaboration with business. We need further statistics in all of these areas.

Information on the duration of contracts and renewed partnerships is also important. The regular appearance of a partner as an outsourcer or collaborator in joint research is a sign that the partner is innovative and has a strong motivation to acquire new knowledge regularly. Although longer contracts can provide more stability in research agendas and in financing, the actual duration of contracts may well differ from that originally contracted. In Hungary, as in several other countries, the regulatory framework may make it more advantageous for industry to break comprehensive collaboration down into smaller units and to renew contracts annually. This means that "cleaned data" are important if we are to be able to identify the real duration of a contract - and also that "renewed partnerships" may be less significant.

MEASURABLE OUTPUT FROM COLLABORATION

Collaboration with industry has a variety of output. In addition to publications and patents there are other valuable products of collaboration such as grey literature and confidential expert reports to industry. No systematic information is available on these categories.

The most traditional output indicators are scientific publications and patents. These are discussed in this section and have also been developed further by disaggregating them in order to gain a closer insight into the main products of collaboration. Joint university-industry publications are characterised by scientometric indicators and data extracted from the "Web of Science" on Hungarian universities. For an analysis of co-patenting and collaboration leading to patenting, we employed two data sources: nationally registered patent data and the OECD Triadic Patent Family -and we extracted data for Hungary from both to create indicators.

Co-Publications

Scientific publication has intensified worldwide. The share of co-authored papers by industrial and academic scientists grew rapidly. (Calvert and Patel 2002, Hicks and Hamilton 1999) Indicators of international co-authorship (the number of articles by two or more authors from different countries) point to increasing cross-border collaboration, and international co-authorship has increased in most countries in the past decades. (Glänzel et al. 2006) The vast majority of these publications have originated from academic circles in different countries, although an (as yet, much smaller, but growing) number of internationally co-authored papers come from international collaboration between academia and industry.

In Hungary changes corresponding to these world trends and the number of internationally co-authored papers are increasing significantly. This process was supported by the transition, a process which opened up the country in the '90s. During the second half of this transition period (2001-2005) the proportion of internationally co-authored papers seems to have stabilised.

The total number of Hungarian scientific publications (according to WoS ISI data) has increased more than 30% whilst the parallel figure for internationally co-authored papers was slightly below 30% during the period investigated at the selected universities.[14] Increased domestic and international co-authorship indicates the crucial role of interaction among researchers with different backgrounds for diversifying their sources of knowledge, and internationally co-authored papers accounted for two-thirds of all scientific publications.

As the publication pattern varies greatly according to the fields involved, we divided universities into three subgroups: (1) Universities with Faculties of Medicine, Universities with Faculties of Technology and (3) Universities with Faculties other than Medicine and Technology (= Others) (Table 5)

Table 5. Ratio of co-authored scientific articles to total publications (12 universities, 2001-2005, %)

Universities with Faculties of	Total		Total international		With international business		Growth rate of international business co-authored publications
	co-authored papers						
	2001	2005	2001	2005	2001	2005	2005 to 2001
Medicine	89.6	88.7	66.8	63.7	2.0	2.3	1.48
Technology	99.5	98.6	61.7	58.0	2.2	3.0	1.44
Others	95.9	97.2	61.4	65,1	1.5	0.6	0.67
Total (12)	92.7	92.4	64.7	63.4	1.9	1.9	1.31

Source: extracted from background documents to Inzelt and Schubert. 2009

Papers co-authored with international business (foreign-located companies) represent a minute fraction (2%) of the total of scientific publications. The characteristics of this minor group are investigated here when a foreign-located company is the co-author. The importance of papers co-authored with business is the generation of useful knowledge for practice and for knowledge diffusion.

Of the 12 universities examined, it is mainly those with Faculties of Medicine and Technology who regularly produce joint publications with industry, and, of these, it is those with Faculties of Technology who more frequently (albeit not by a large margin) publish in collaboration with international business than those with Faculties of Medicine. It is, however, remarkable that universities with Faculties of Medicine have been able to increase their number of joint publications with international business faster than those with Faculties of Technology. At the same time the ratio of internationally co-authored papers is not only lower at universities without these specific Faculties but significantly so (in favour of academic partners).

Figure 3 compares university-industry co-authored papers by domestic and foreign business partners.

Figure 3. Co-authorship between faculty members and industrial researchers by number of co-authored publications by origin of firms (2001-2005)

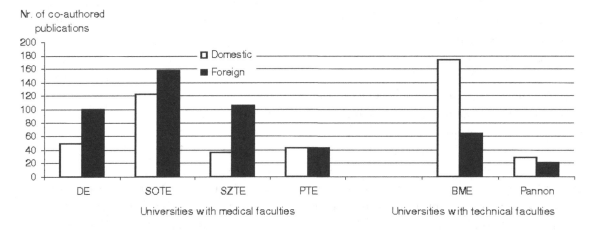

Source: extracted from background documents to Inzelt et. al. 2009

The number of co-publications involving foreign business partners was larger at 3 of the 4 universities with Faculties of Medicine than with domestic partners. In the case of universities with Faculties of Technology, the situation was the opposite: the number of domestic co-authored papers was much larger. We may assume that foreign-owned, domiciled firms (FDI-led) are among the important co-authors with universities with Faculties of Technology, but, in the absence of precise data, we can only guess the FDI-led R&D role in 'domestic' co-authored papers. In any case the output of universities with Faculties of Medicine supports the hypothesis that foreign companies are dynamos of university-industry collaboration in a less innovative environment.

To judge innovation potential embedded in scientific results and the robustness of research findings, it is important to know the value of new knowledge originating from papers co-authored with industry. For this evaluation the most widely used indicator is the citation index, which is a guide to the scientific importance of the paper.[15]

According to our earlier study on Higher Education publications (Inzelt et al., 2009) international co-authorship has positive effects on the citation rate of publications. This general picture is true if we focus on co-authorship with industry. (Figure 4)

Figure 4 illustrates the citation performances by domestic and foreign co-authored publications. Of the 6 universities investigated, 5 have a higher citation rate per publication where co-authors were foreign rather than domestic companies. This picture varies from the norm at Hungary's largest University of Technology. (Figure 4)

As Figure 4 illustrates, except for the largest university in Hungary (comprising several Faculties of Technology), the other 5 universities have papers (co-authored with international business partners) which are heavily cited. This figure suggests that an international business partnership has a positive effect on the citation-level of publications. This general picture may not only be a sign of the novelty value of the papers, but of the greater weight accorded to foreign partners in the

Figure 4. Citations per publication of senior academics' and industrial researchers' co-authored papers by company origin (2001-2005)

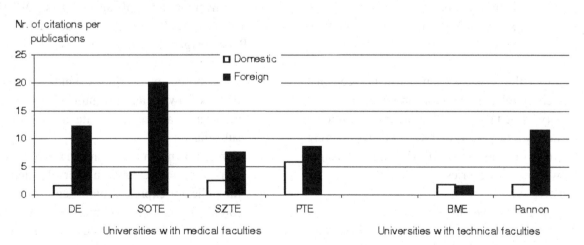

Source: extracted from background documents to Inzelt et. al. 2009

scientific world, or, again, of scientific marketing capabilities which are stronger than the domestic.

Invention and Patent Application

Cross-sectoral (referring to the business and academic sectors) collaborations in inventions are crucial for the capitalisation of knowledge. Universities as a site of invention are subject to their individual national and internal regulations on patent ownership. They differ not only by their patent regulations but in their capabilities to manage intellectual property. The various national innovation systems usually offer 3 main approaches to patent ownership for universities (Inzelt et al. 2006):

- Institutional ownership
  a. University-owned
  b. Jointly-owned by university and other organisation(s)
  c. Owned by research funding company or agency
- 2. Individual ownership
  a. Patent owned by inventors (Faculty members)
  b. Jointly-owned by academics and other individuals or organisations.
- 3. Mixed individual and institutional ownership (see all above)

From this classification it can be seen that university-related inventions are becoming university-owned patents in the case of institutional ownership (1a and 1b), and so, when investigating university patents it is worthwhile distinguishing between two categories:

1. Patents *owned* by the university
2. Patents *invented* (or co-invented) in the university (so-called indirect university patent)

In the first category are those patents for which the patenting university applied alone or with co-applicants and became the owner or co-owner of registered patents. The second category contains those patents which are not owned by the university but where all or some of the inventors are Faculty members.

The difference in size of these two categories is quite significant. If the regulations were to permit ownership of the patent by the university, by the company or by the inventors as individuals, the number of patents produced by the university could be 10 or 20 times higher than the number of university-owned or co-owned patents actually shown.

The first category may be calculated from patent statistics, although it is more difficult to obtain reliable data on the second category. (See the discussion in Inzelt et. al. 2006, pp. 139-147)

According to international experience, the increasing volume of R&D investment abroad is matched by the increasing importance of the home and host country's role in patenting. However, the increasing volume of FDI-led R&D investment has raised important questions for both home and host countries where patents are being applied for.

The OECD distinguishes 3 important categories of patenting international R&D activity; these help to characterise cross-national relationships between inventors and applicants/owners. (OECD STI Scoreboard 2007, p. 162, 164, Guellec, D. and van Pottelsberghe de la Potterie 2001)

1. Cross-border ownership: Country of residence of owner and inventor differs. (e.g. multinational conglomerate and foreign subsidiary);
2. Foreign ownership of domestic invention: Compared to the total number of patents, the indicator expresses the extent to which foreign firms control domestic inventions and reflects the importance of a country's inward R&D investments;
3. Domestic ownership of inventions made abroad refers to the property of a country, but requires that at least one inventor be

located in a foreign country out of the total number of domestic applications. This indicator evaluates the extent to which domestic firms control inventions made by residents of other countries.

We may add another category: (4) Joint ownership of co-invention when domestic and foreign actors together own the patent.

In the context of Hungarian universities and foreign firms the second type is most marked, and the others scarcely exist.

Measuring the technology innovation performance of Hungarian universities is not an easy task even if we do not focus on internationalisation.[16] Hungarian HEIs belong to the third group of patent ownership by universities (i.e., mixed individual and institutional ownership). Hungarian universities rarely own the inventions of their faculty members.[17]

Data is available on university-owned patents but not on indirect university patents. The latter still has to be created.

University-Owned Patents

The share of patents filed under PCT and owned by universities is 1.2% in Hungary while the OECD average was 4.3 and the EU average 3.1% in 2002-2004. (OECD 2007, p. 75.)

Among the main reasons for the low level of Hungarian university-owned patents were the lack of resources for patenting, the poor management of inventions and the confused regulations on university-related intellectual property in effect prior to the enactment of new legislation in 2004-2005.[18]

Taking into account the small number of university-owned patents, it is not surprising that only two applications by joint university-foreign-owned companies (registered in Hungary) can be seen in the Hungarian Patent Office databank for the period between 2000 and 2007. In respect of the patent performance of universities, the rev-

enues from the licensing of patents were almost negligible.

Indirect University Patents

Knowledge of the university-linked inventions owned by others is important for several reasons. The information on indirect patents gives a more reliable picture of the technology creation capabilities of universities than that on owned patents if the ownership is not strictly institutional. In addition, information on the organisations of inventors and of applicant organisations helps us to understand collaboration linkages.

Among the OECD's 36 members, Hungary ranks fourth in terms of the foreign ownership of domestic inventions, a proportion which is relatively high in Hungary compared to other OECD economies. Almost 60% of domestic inventions were under foreign ownership in 2003. This proportion was around 30% in 1993. (OECD 2008b. 33-36)

Universities do not differ from this general picture. Although there are no data on the foreign ownership of domestic university inventions, all indirect information suggests that foreigners own a considerable part of domestic university inventions, even if they covered only the minor part of the research expenditure.

Due to the shortage of data on indirect university patents, we utilise data on the cross-border ownership of inventions. This method is acceptable as a first attempt at measuring, and we can assume that the university picture on inventor-applicant relationships is very close to the general picture.

For our analysis of the inventor-applicant relationship we used the OECD developed *Triadic Patent Family*[19] which has been available since 2008. The advantages of this data-set '... only patents applied for in the same set of countries are included in the family ... patents included in the family are typically of higher value, as patentees only take on the additional costs and delays of extending protection to other countries if they deem it worthwhile.' (OECD 2009, p. 71-72)

This international resource of patent statistics can provide information only on relationships between Hungarian inventors and foreign applicants and not specific information on relationships between universities and international actors.[20]

The Triadic Patent Family shows 58 patent applications listing Hungarian inventors for the period 2000-2004.[21] Table 6 gives an overview of Hungarian related patents in the Triadic Patent Family.

There were no Hungarian universities among the applicants.[22] Neither academic nor corporate collaboration produced a Hungarian university as a co-applicant. Foreign laboratories were characterised by the foreign ownership of domestic inventions (the number of patents relating to domestic inventions and owned by non-residents). Among the applicants involving Hungarian inventors, 2 foreign universities (USA) and 2 foreign laboratories (USA and France) are found. When US universities were the applicants, the inventors were from both countries (Hungary and the US) as the invention was based on collaborative research.[23]

Two-thirds of the inventors were Hungarian when the applicant was a foreigner. The majority of applicants were foreign companies (36), and,

out of this group, the cross-border ownership of the invention is clear in the case of 14 foreign companies (all inventors are Hungarian but the applicants belong to different countries). The majority of applicants are foreign investors in Hungary and, in these cases, some collaboration can be presumed between the foreign-owned company and faculty members. The other 22 applicants list both Hungarian and foreign inventors, signifying cross-border ownership based on cross-border collaboration.

Although we do not have exact figures on domestic university inventions controlled by foreign firms, all indirect information suggests that foreigners control a considerable part of the universities' inventions, even if foreign funding only covered a small fraction of the research costs.

In Hungary, as well as in many European countries, specific measures are important to protect the proprietary knowledge created by domestic universities. In addition to the low level of patenting activities in universities, the weakness of intellectual asset management may encourage companies to become applicants for patents rather than to purchase licences. The effect of new regulations (2004, 2005) on managing intellectual

Table 6. Hungarian related inventions, by applicant, in the Triadic patent family (2000-2004)

Applicants	No. of Applicants by origin of Inventor			Number of Inventors		
	only Hungarian	Hungarians & foreigners	Total	Hungarian	Foreigners	Total
Institutional applications						
Foreign university		2	2	4	12	16
Foreign research laboratory	1	1	2	4	3	7
Foreign companies	14	22	36	112	51	163
- with Hungarian location	11	10	21	70	23	93
- no Hungarian location	3	12	15	42	28	70
Mainly Hungarian-owned	12	1	13	90	2	92
Hungarian laboratory	1	-	1	6	-	6
Individual applications						
Individuals	1	3	4	10	6	16
Total	29	29	58	226	74	300

Source: compiled on the base of OECD Triadic Family Patent Databank property by universities and interactions with firms in this field will be measurable in the years to come, and, one day, universities may become, along with foreign companies, co-applicants in respect of their own inventions.

FINAL REMARKS

Globalisation has reduced the barriers to entry into global networks and has also created opportunities for new players to enter. The penetration of FDI-led R&D in Hungary has involved Hungarian universities in international university-industry collaboration, and, as foreign business holds the lion's share of Hungarian business R&D, it is clear that FDI- related R&D is important for HEIs, at least in those fields of science relevant to global innovation.

In accordance with international trends, partly or totally foreign-owned firms outsource much more in global R&D spending sectors than do companies primarily in domestic ownership. Inward FDI and outsourcing R&D from abroad are now playing a major role in university research and are crucial in several fields. The internationalisation of universities' business research contract portfolio has speeded up the development of university-industry collaboration.

The fact that foreign business accounts for a relatively large proportion of the financial resources of universities may indicate that the S&T capabilities of Hungarian HE are suitable for foreign business. Contracts with industry have their effect on both the input and output sides.

Universities make up a relatively large proportion of partners in FDI-led R&D. That may be either a sign of the attractiveness of universities or of the limited availability of suitable domestic companies. A further alternative is that this symptomises an imbalance between publicly- and privately-funded researches since it is the case that private funds penetrate into sectors earlier occupied by public funds.

In the context of open innovation, business-funded university research is crucial, providing opportunities for universities to diversify their sources of funding and for society as a whole to become more innovative and competitive. However, an appropriate balance between public and private funding must be found, especially since companies are generally reducing their focus on basic and longer-term research in response to competition and shorter product cycles.

The present economic crisis inevitably has an effect on FDI-led R&D and innovation. In an environment where economic crises are so threatening, the strong sectoral concentration of outsourced R&D is dangerous. Although the presence of the automotive industry in the economy and in business-financed R&D was one of the country's strengths in the first part of the 21st century, it leaves Hungary's economy and business-funded university research fragile, as a narrow research portfolio is problematic.

The increasing worldwide competition among HE organisations creates a greater demand for detailed information on university collaboration, and it is vital to know the impact of collaborative (and contract) research on the university research agenda and performance, since a high level of FDI also creates risks that national resources may be diverted from the country's needs to meet the short-term objectives of foreign interests.

The scarcity of data and indicators were obstacles to a thorough analysis, and the indicators which permit a little analysis (at least testing ideas and metrics) were prepared from various databanks. Our statistical analysis showed that, to some extent, the internationalisation of business R&D in Hungary matches those global trends which are serving global open innovation. The so-called global R&D spending sectors have outsourced a much higher proportion of R&D to Hungary than have other sectors.

A very important question for the future of the triple helix model is whether collaboration for asset-exploitation can be transformed into profit-driven partnerships, and a further issue is how local businesses can expand their development partnerships with global players and benefit from spillovers from global innovation and foreign investors.

Our analysis demonstrates the importance of this topic for Hungary and for similarly situated countries. Increasing competition for R&D-related foreign direct investment is making the catching-up process more difficult and more risky for Hungary as a small emerging economy. Yesterday's successes in attracting investors and collaborators give no guarantees for the future.

Our analysis also highlighted the importance for policy-makers to investigate further the linkages between production and the exploitation of new knowledge if these are separated in spatial terms by globally open innovation.

REFERENCES

Bonaccorsi, A., & Daraio, C. (Eds.). (2007). *Universities and Strategic Knowledge Creation: Specialization and Performance in Europe.* PRIME Series on Research and Innovation Policy in Europe.

Callon, M. (1992). The Dynamics of Techno-Economic Networks . In Coombs, R., Saviotti, P., & Walsh, V. (Eds.), *Technical Change and Company Strategies* (pp. 72–102). London: Academic Press.

Calvert, J., & Patel, P. (2002). *University-Industry Research Collaborations in the UK.* SPRU working paper, University of Sussex, Brighton, UK.

Cantwell, J., & Molero, J. (Eds.). (2003). *Multinational Enterprises, Innovative Strategies and Systems of Innovation.* Cheltenham, UK: Edward Elgar Publishing Limited.

Carlsson, B. (2006). Internationalization of innovation systems: A survey of the literature. *Research Policy, 35,* 56–67. doi:10.1016/j.respol.2005.08.003

Chesbrough, H. (2003). *Open Innovation: the new imperative for creating and profiting from technology.* Cambridge, MA: Harvard Business School Press.

Chesbrough, H., & Teece, D. J. (1996). When is virtual virtuous? Organizing for innovation. *Harvard Business Review, 74*(1), 65–73.

Cook, P. (2005). Regionally asymmetric knowledge capabilities and open innovation . *Research Policy, 34,* 1128–1149. doi:10.1016/j.respol.2004.12.005

Coombs, R., Harvey, M., & Tether, B. (2003). Analysing distributed processes of provision and innovation. *Industrial and Corporate Change, 12*(6), 1125–1155. doi:10.1093/icc/12.6.1125

Dőry, T. (2005). *Regionális innováció-politika. Kihívások az Európai Unióban és Magyarországon.* Budapest, Pécs: Dialóg Campus.

Dunning, J. H. (2005). The evolving world scenario . In Passow, S., & Runnbeck, M. (Eds.), *What's Next? Strategic Views on Foreign Direct Investment* (pp. 12–17). Stockholm, Sweden: Invest in Sweden Agency.

EBRD. (2005). Transition Report: Business in Transition. London.

Edler, J. (2008). Creative internationalization: Widening the perspectives on analysis and policy regarding international R&D activities. *The Journal of Technology Transfer, 33*(4). doi:10.1007/s10961-007-9051-1

Edler, J., Mayer-Krahmer, F., & Reger, G. (2002). Changes in the strategic management of technology: Results of a global benchmarking study. *R & D Management, 32*(2), 149–164. doi:10.1111/1467-9310.00247

Etzkowitz, H. (2008). *The Triple Helix: University-Industry-Government Innovation in Action.* London: Routledge. doi:10.4324/9780203929605

Etzkowitz, H., & Leydesdorff, L. (Eds.). (1997). *Universities in the Global Economy: A Triple Helix of University–Industry–Government Relations.* London: Cassell Academic.

Etzkowitz, H., & Leydesdorff, L. (2000). The dynamics of innovation: from national systems and 'mode 2' to a Triple Helix of university-industry-government relations. *Research Policy, 29*(2), 109–123. doi:10.1016/S0048-7333(99)00055-4

EU. (2005). *The Handbook on Responsible Partnering–Joining forces in a world of open innovation. A guide to better practices for collaborative research and knowledge transfer between science and industry.* EUA, ProTon Europe, EARTO and EIRMA. Retrieved from http://www.responsible-partnering.org/library/rp-2005-v1.pdf

Foray, D. (2006). *Globalization of R&D: linking better the European economy to 'foreign' sources of knowledge and making EU a more attractive place for R&D investment.* Technical Report, Expert Group 'Knowledge for Growth.'

Gassmann, O. (2006). Opening up the innovation process: Towards and agenda. *R & D Management, 36*(3), 223–228. doi:10.1111/j.1467-9310.2006.00437.x

Gassmann, O., & Enkel, E. (2004). Towards a Theory of Open Innovation: Three Core Process Archetypes, In *Proceedings of the R&D Management Conference* (RADMA), Sessimbra, Portugal July 8-9, 2004.

Glänzel, W., Debackere, K., & Meyer, M. (2006). *Triad or Tetrad? On Global Changes in a Dynamic World.* Paper presented at the 9th International Conference on S&T Indicators Leuven (Belgium), September.

Guellec, D., & van Pottelsberghe de la Potterie. (2001). The internationalisation of technology analysed with patent data. *Research Policy, 30*(8), 1253–1266. doi:10.1016/S0048-7333(00)00149-9

Guimón, J. (2009). Government strategies to attract R&D-intensive FDI. *Journal of Technology Transfer.*

Gulbrandsen, M., & Slipersaeter, S. (2007). The 3rd Mission and the Entrepreneurial University Model . In Bonaccorsi, A., & Dario, C. (Eds.), *Universities and Strategic Knowledge Creation. Specialization and Performance in Europe* (pp. 112–143). Cheltenham, UK: Edward Elgar.

Hicks, D., & Hamilton, K. (1999). Does University-Industry Collaboration Adversely Affect University Research? *Issues in Science and Technology Online.* Retrieved from http://www.nap.edu/issues/15.4/realnumbers.htm

Iammarino, S., & McCann, P. (2006). The structure and evolution of industrial clusters: Transactions, technology and knowledge spillovers. *Research Policy, 35*(7), 1018–1036. doi:10.1016/j.respol.2006.05.004

Inzelt, A. (2004). The evolution of university-industry-government relationships during transition. *Research Policy, 33*, 975–995. doi:10.1016/j.respol.2004.03.002

Inzelt, A. (2008a). Strengthen and upgrade regional capabilities (Regional University Knowledge Centre Programme in Hungary). *Romanian Journal of Economics, 26*(1), 133–154.

Inzelt, A. (2008b). The inflow of highly skilled workers into Hungary: a by-product of FDI. *The Journal of Technology Transfer, 33*, 422–438. doi:10.1007/s10961-007-9053-z

Inzelt, A., & Csonka, L. (2008). Strengthening and Upgrading Regional Knowledge Capabilities in Hungary . In Filho, W. L., & Weresa, M. (Eds.), *Fostering Innovation and Knowledge Transfer in European Regions* (pp. 109–132). Frankfurt, Germany: Peter Lang.

Inzelt, A., Laredo, P., Sanchez, P., Marian, M., Vigano, F., & Carayol, N. (2006). Third mission in Methodological Guide, Observatory of European University. *PRIME NoE*. Retrieved from http://www.prime-noe.org

Inzelt, A., & Schubert, A. (2009). *Collaboration between Professionals in Academia and in Practice (in the light of scientometric indicators for 12 universities)*. Minerva.

Inzelt, A., Schubert, A., & Schubert, M. (2009). Incremental citation impact due to international co-authorship in Hungarian higher education institutions. *Scientometrics*, *78*(1), 37–43. doi:10.1007/s11192-007-1957-8

Kállay, L., & Lengyel, I. (2008). The Internationalisation of Hungarian SMEs . In Dana, L., Han, M., Ratten, V., & Welpe, I. (Eds.), *A Theory of Internationalisation for European Entrepreneurship* (pp. 22–36). Cheltenham, UK: Edward Elgar.

Kalotay, K., & Filippov, S. (2009). *Foreign Direct Investment in Times of Global Economic Crisis: Spotlight on New Europe*. UNU-MERIT Working Paper, 2009-021.

Laredo, P. (2007). Revisiting the third mission of universities: toward a renewed categorisation of university activities? *Higher Education Policy*, *20*(4), 441–456. doi:10.1057/palgrave.hep.8300169

Laudel, G. (2006). The art of getting funded: how scientists adapt to their funding conditions. *Science & Public Policy*, *33*(7), 489–504. doi:10.3152/147154306781778777

Lengyel, B., & Leydesdorff, L. (2007). *Measuring the knowledge base in Hungary: Triple Helix dynamics in a transition economy*. Paper presented at the 6th Triple Helix Conference, 16-19 May 2007, Singapore.

Lundvall, B.-A. (Ed.). (1992). *National Systems of Innovation–Towards a theory of innovation and interactive learning*. London: Pinter Publishers.

Martin, R. B. (2003). The changing social contract for science and the evolution of the university . In Geuna, A., Salter, A., & Steinmueller, W. E. (Eds.), *Science and innovation: Rethinking the rationales for funding and governance* (pp. 1–29). Cheltenham, UK: Edward Elgar.

Molas-Gallart, J., Salter, A., Patel, P., Scott, A., & Duran, X. (2002). *Final Report to the Russell Group of Universities, SPRU*. Unpublished paper.

Narula, R., & Zanfei, A. (2005). Globalization of Innovation: The Role of Multinational Enterprises . In Fagerberg, J., Mowery, D., & Nelson, R. (Eds.), *The Oxford Handbook of Innovation* (pp. 318–345). New York: Oxford University Press.

Nedeva, M. (2008). New tricks and old dogs? The 'third mission' and the re-production of the university . In Epstein, D., Boden, R., Rizvi, F., Deem, R., & Wright, S. (Eds.), *The World Yearbook of Education 2008: Geographies of Knowledge/ Geometries of Power–Higher Education in the 21st Century*. New York: Routledge.

Nelson, R. (Ed.). (1993). *National Innovation Systems*. New York: Oxford University Press.

OECD. (2006a). *Science, Technology and Innovation Outlook*. Paris: OECD Publishing.

OECD. (2007). *OECD Science, Technology and Industry Scoreboard 2007, Innovation and Performance in the Global Economy*. Paris: OECD Publishing.

OECD. (2008a). *Open Innovation in Global Networks*. Paris: OECD Publishing.

OECD. (2008b). *The Internationalisation of Business R&D. Evidence, impacts and implications* (Guinet, J., & De Backer, K., Eds.). Paris: OECD Publishing.

OECD. (2008c). *Review of Innovation Policy, Hungary*. Paris: OECD Publishing.

Pavitt, K. (1997). National Policies for Technological Change: Where are the Increasing Returns to Economic Research? In *Proceedings of the National Academy of Sciences*, Washington DC.

Pavitt, K. (2002). Public policies to support basic research: What can the rest of the world learn from US theory and practice? (And what they should not learn). *Industrial and Corporate Change, 11*, 117–133. doi:10.1093/icc/11.1.117

Raymond, S., & Taggart, J. H. (1998). Strategy shifts in MNC subsidiaries. *Strategic Management Journal, 19*(7), 663–681. doi:10.1002/(SICI)1097-0266(199807)19:7<663::AID-SMJ964>3.0.CO;2-Y

Sanchez, M. P., & Elena, S. (2006). Intellectual capital in universities: Improving transparency and internal management. *Journal of Intellectual Capital, 7*(4), 529–548. doi:10.1108/14691930610709158

Taggart, J. H. (1998). Determinants of increasing R&D complexity in affiliates of manufacturing multinational corporations in the UK. *R & D Management, 28*(2), 101–110. doi:10.1111/1467-9310.00086

UNCTAD. (2001). *World Investment Report: Promoting Linkages*. New York: United Nations.

UNCTAD. (2005). *World Investment Report, Transnational Corporations and the Internationalisation of R&D*. New York: United Nations.

Varga, A. (2000). Local academic knowledge spillovers and the concentration of economic activity. *Journal of Regional Science, 40*, 289–309. doi:10.1111/0022-4146.00175

Varga, A. (2005). Localized knowledge inputs and innovation: The role of spatially mediated knowledge spillovers in the new EU member countries from Central Europe: The case of Hungary. In *The impact of European integration on the national economy* (pp. 118–133). Cluj-Napoca, Romania: Babes Bolyai University Press.

KEY TERMS AND DEFINITIONS

Business Research and Development: Business R&D comprise creative work undertaken on a systematic basis in order to increase the stock of knowledge, including knowledge of man, culture and society, and the use of this stock of knowledge to devise new applications The financier and/or performer of these activities are business organisations. (Frascati Manual, 2002 p. 30.)

Co-Publications: A co-publication is the result of co-operation between representatives of each entity and each country taking part in a particular joint research programme. Such research forges links between the parties (scientists, laboratories, institutions, countries, etc.) that have worked together to produce a scientific paper. (Okubo, 1997, p. 28.)

Global R&D Spending Sectors: In certain industries outsourced cross-border R&D is more characteristic than in others. The main R&D-spending multinational companies are concentrated on these few sectors. These include the IT, automotive and pharmaceutical industry, biotechnology, electronic, electrical and aerospace industries.

Indirect University Patents: The owner of patent is the inventor (faculty member) not the university as an institution itself or the patent is jointly-owned by faculty members and other individuals or organisations.

Open Innovation: Open innovation is a paradigm that assumes the firms can and should use external ideas as well as internal ideas, and internal and external paths to market, as the firms look to advance their technology (Chesbrough, 2003, p. 24.)

Patent Inventor: Those individuals are inventors who attended in producing novelty that has industrial applicability. Country of residence of the inventor may differ from country of patent applicant/assignee.

Patents Owned: Patent is a legal title protecting an invention. The legal protection gives its owner the right to exclude others from making, using, selling the patented invention for the term of the patent. The applicants (or assignees in the US) will be the owners of the patent if it is granted. The owner may be the same person as inventor or employer company of inventor. But owner can be different from inventors and their respective organizations.

Third Mission of Higher Education: This means the university's relationship with the non-academic outside world: industry, public authorities and society. The 3rd mission includes several different activities such as the commercialization of academic knowledge through collaboration with industry, patenting/licensing, creation of spin-off companies, participation in policy-making, involvement in social and cultural life. (OEU Guide, 2006 p. 127.)

Transition Economies: The term *"transition economy"* is frequently used to refer to the countries of Central and Eastern Europe after the fall of the communist or socialist regimes in the end of the 1980's. Thereby, transition means the status of those countries during the evolution from a command economy to a market-based economy. This movement is usually characterised by the changing and creating of institutions, particularly private enterprises; changes in the role of the state, thereby, the creation of fundamentally different governmental institutions; and the promotion of private-owned enterprises, markets and independent financial institutions. (Falke, 2001 p. 1-2.)

ENDNOTES

1   "Open" does not mean "free" (as with some software); the payments of licence fees as well as other financial arrangements are a feature of open innovation.

2   "Open innovation" is broader than pure outsourcing, but this paper does not deal with other forms.

3   The concept of open innovation relates not only to the importance of knowledge-sourcing, but also to the exploitation of internal innovation together with external partners.

4   Instead of three "Missions", Laredo (2007) suggests that universities carry out three "Functions" which can be categorised as: (1) Mass Tertiary Education (leading to a Bachelor degree), (2) Professional Specialised Higher Education and Research (leading to a "professional" MA or MSc as the core degree, with "problem-solving research" as the core activity) and (3) "Academic Training and Research" (leading to a PhD as the core degree and involving publications as a core output). While the first and the third of these are clearly found already at local and international level, the second is focused on professions and follows their internationalisation.

5   The role of business in financing university R&D is studied in relation to issues such as the danger for the basic ethos of a university of how the choice of topic and the input of the university can contribute to the advancement of science. These important issues are beyond the scope of this chapter.

6   According to anecdotal evidence, the world economic crisis has affected corporate R&D less than production. The withdrawal of FDI, and closed factories have resulted in a reduction in FDI-led R&D, but, at the same time, we can observe strong innovation activity applied to try to break out of the economic

crisis.. Naturally, withdrawal and increasing FDI-led R&D investment occur in different companies and sectors.

[7] Both innovation and FDI policies focus on this issue. The emerging vision of the modern, innovative Hungarian economy, able to compete successfully in the global arena, produced policy which encourages companies to be innovation-oriented and universities to develop, beyond their traditional teaching mission, both their research performance and their capacity to transfer research results and new knowledge in order to convert these into commercially relevant innovations.

Despite many efforts to launch relevant programmes, the competence and attractiveness of universities for strategic research partnerships with the private sector has remained heterogeneous and somewhat unsatisfactory due to shortcomings in their knowledge base and their capability to act as high-performing research partners in collaborative projects.

[8] Outsourced R&D expenditure cannot be broken down further, and no information is available on the proportion of R&D outsourced to Higher Education.

[9] The automotive industry is a typical example of where the borderline between experimental and simple development is not very strong. If companies wish to avoid an innovation tax or benefit from an R&D tax credit system, they 'extend' experimental development, and so these figures must be interpreted with caution. Due to the R&D tax credit system there are always problems with figures relating to the automotive industry. Even if R&D expenditure and contracting-out figures were lower after revision, this sector actually invests significant amounts in R&D and outsourcing to universities.

[10] Lengyel and Leydesdorff (2007) observed that 'foreign-owned firms may have had determining roles on triple helix mechanisms in Hungary.' During the first and early second phase of the transition process 'the internal linkages were weakened and external linkages asynchronously reinforced. ... Universities could further develop international relations ... and FDI became a major factor in the transformation process.' (pp. 22-23)

[11] During the course of this exercise, the report form was modified several times and the availability of this source for research purpose changed. The new law on HE which guaranteed the autonomy of HE Institutions abolished this data collection on the R&D finance sources of HE Institutions. For our analysis, the disadvantage of this is that there are no figures after 2005 following Hungary's accession to the EU (in May 2004).

[12] The databank was prepared for the Verinekt project. Here we use only one part of it to discuss the relationship between foreign businesses and universities.

[13] Three graduate students, Gábor Csizmazia, Vilmos Klein and Szabolcs Szőke collected the names of the contractors from the report forms and classified them by sector (public and private) and by origin (domestic-owned, domiciled foreign-owned, foreign-located, EU and other international sources). Web searches were an important tool for classification.

[14] See the description of the sample in Inzelt et. al. 2009.

[15] In addition to the values of this indicator, its shortcomings are also well known.

[16] The patent documents list both inventors and applicants. The inventor is the person who invents something which did not exist before, but this novelty has to have industrial applicability to obtain a patent. The applicant is the holder of the legal rights and obligations of a patent application. Applicant may be the same individual as inventor or may be an organisation (company or university). If

the applicant is an organisation it can be the employer of an inventor or another organisation in the same or in different countries. Patent documents allow inventors and applicants to be identified by their country of origin, by the organisations of applicants and, in many cases, by the organisations of inventors. 'Through the applicant's and inventor's addresses it is possible to track the patterns and the intensity of international co-invention ... foreign ownership of domestic inventions and vice versa.' (OECD 2009, p. 32.)

[17] If the research funding company is the sole owner of the patented university-related invention, the contracts between university/faculty members and company may regulate the role of the inventors in two different ways: include or exclude the name of the faculty members from the list of inventors in the patent application. The compensation of inventors is usually generous if they are excluded from the list of inventors. However, the university itself is usually not compensated. The patent statistics are not able to follow these 'indirect' university patents. (See the discussion on measurement in Inzelt et. al. 2006 pp. 139-147)

[18] Data are not yet available on how the Law on Higher Education (2005) affects the processes.

[19] According to the OECD definition (2009), the Triadic Patent Family is a set of patent applications filed at the European Patent Office (EPO), the Japan Patent Office (JPO), and granted by the US Patent and Trademark Office (USPTO), sharing one or more priority applications.

[20] In the present stage of the databank, inventors are not identified as employees of universities. Presumably the general picture on the relationship between inventors and applicants does not differ from the relationship between universities and foreign applicants. The relevant sources and resources are missing pair-match the list of inventors with faculty members.

[21] The locations of inventors and applicants were identified by the addresses, and so those with Hungarian addresses are listed as Hungarians. The Hungarian sample was selected by Zoltán Benke, graduate student, from the "Triadic Patent Families".

[22] Following the period investigated, legislation changed. Under the new regulation (Law on Innovation 2004 and Law on Higher Education 2005) HE Institutions have to set up or strengthen their technology transfer organisations, re-regulate ownership and share licence income from patents between university and inventor. More transparent regulation and better management of intellectual property may make HE Institutions more attractive for business organisations and also may result in the better treatment of inventions.

[23] In given cases the information was enough to link Hungarian inventors to universities as the inventors were well-known Hungarian scientists.

Chapter 5
Towards the Sixth Kondratieff Cycle of Nano Revolution

Jarunee Wonglimpiyarat
Thammasat University, Thailand

ABSTRACT

Nanotechnology is currently seen as a paradigm shift towards scientific revolution or 'nano revolution. This chapter discusses the nano revolution within the global context. It is interesting to see that the governments around the world have formulated policies to manage the research and development (R&D) efforts and exploit the potential of nanotechnology to increase industry's ability in the global economy. The chapter analyses the successive waves of technological change based on Kuhn's model of scientific change and Schumpeter's model of Kondratieff cycles. As nanotechnology would have significant impacts on virtually every commercial sector, many countries commit to foster nanotechnology developments. This chapter will focus on nanotechnology framework policy recommendations. The policies and research activities of the most preeminent nations discussed in this chapter represent global research trend towards nano revolution in the next decades.

INTRODUCTION

Nanotechnology is widely perceived as one of the key technologies of the 21st century that would transform the world's economy (Roco and Bainbridge 2002). The supramolecular architectures represent a new revolutionary approach in research and production. The nature of interdisciplinary technology research makes it useful in many applications. Nanotechnology has been recognized as a promising new growth technology, opening up a floodgate of opportunities for developing viable applications (Roco 2001, Luther 2004). In other words, this field of technology offers the possibility of transforming the international science and technology policy landscape and making significant impacts on the direction of research and development for a wide range of nations and companies (Michelson 2008). Given that nanotechnology is one of the fastest-growing

DOI: 10.4018/978-1-61692-006-7.ch005

research areas in scientific and technical fields in the world, it is expected that nano revolution would create a wealth of new materials and manufacturing possibilities (Ikezawa 2001, Wilson 2002, National Science and Technology Council 2006). It is now a science and technology priority area for many countries with the governments' efforts to put the results of nanotechnology development to commercialization. The national policy for nanotechnology is to change the existing technology system and bring about an industrial revolution (the nano revolution). Under the pressure of competition, the key to a success would lie in how each country could find the right application to focus on in order to survive through international competitions.

At present, the limit of mono-disciplinary science to reach a solution to a particular problem sets the stage for the possibility of scientific revolution - the progress towards broad research areas such as physics, biology, materials and engineering sciences. Within the global economy, there is a large potential given by the opportunities of nanostructures for the commercialization. The scientific and technical challenges of working at the nano scale are huge as nanotechnology is expected to cause discontinuous progress and provide massive industrial applications. Many industrialists see that the commercial potential of nanotechnology will have at least the same magnitude as biotechnology. However, while previous research studies have focused on improvements of advanced materials and manufacturing techniques, the policy perspective of nanotechnology has received little attention. The study attempts to fill this gap by looking into a global perspective of nano revolution with an aim to understand developments in nanotechnology innovations and the extent to which the nanotechnology would affect the whole economy. The focus of this chapter is on policy recommendation to assist the science and technology-based economic development in the global economy.

The structure of this chapter is as follows. Section 2 presents the literature review on the models of technological change. Section 3 discusses nanotechnology as a revolutionising technology bringing about a paradigm shift in industrial research. Section 4 discusses the structural crisis and technological forecasting of nanotechnology. It also reviews nanotechnology policies and research activities of some of the most preeminent nations in nanotechnology initiatives - USA, China, Germany, South Korea, France, Taiwan. The policy recommendations to encourage the undertakings of nanotechnology research and development towards the revitalisation of the global economy as well as conclusions are drawn in Section 5.

MODELS OF TECHNOLOGICAL CHANGE

There is a clear implication that technological change is the phenomenon of structural crises of adjustment (Freeman and Perez 1986, 1988; Dosi 1982, 1988). From reviewing the common ground in relation to the progress of business cycles, the technological change is argued to be a fundamental driving force bringing about economic growth. In respect of evolutionary economics, technological change reflects the innovative efforts with varying degrees of appropriability and uncertainty about the technological and commercial outcomes. According to Dosi's models of technological paradigms and trajectories, the term 'technological paradigm' is defined as a pattern for solution of selected techno-economic problems based on highly selected principles (Dosi 1982, 1988). Similarly, Freeman and Perez (1986, 1988) used the term 'techno-economic paradigm' to refer to an innovation that affects the whole economy e.g. steam power, electric power, electronic computer. Tushman and Anderson (1987) further argued that there are 2 processes of technological change: competence-enhancing and competence-destroying. The latter is regarded as a technological shift, representing discontinuity.

Kuhn (1970) used the term 'paradigm shift' to refer to the mark of maturity of a science. Kuhn's model of scientific change explains the progress of scientific development as coming from the paradigm shift. Kuhn argues that the successive transition from one paradigm to another via revolution is the usual developmental pattern of mature science. In other words, there are periods of stability (normal science)[1] punctuated by periods of crisis[2], leading to a revolution and a new normal science. As the acquisition of a paradigm provides model problems and solutions to a community of practitioners, the essence of Kuhn's process of scientific development (the development of paradigms) can be shown in the figure below.

Figure 1 shows the development of paradigm (Glass and Johnson 1989). The normal science represents a period of puzzle-solving activities in response to a mismatch between the paradigm and reality. The puzzles that resist solution are seen as anomalies of a paradigm. When there is a failure to solve the puzzles within the current paradigm, these anomalies would further produce disorder or crisis in the process of scientific development. According to Kuhn (1970), the symptoms of the crisis encourage the willingness to try anything new which would bring revolution (Brown 1977, Daneke 1998, McCraw 2007).

Schumpeter's (1939, 1967) work on business cycles shows the phenomena of Kondratieff waves of technological change. The long-wave theory explains the technological revolutions underlying the Kondratieff cycles. The Kondratieff cycle is the long wave of technology which help advance an economy. Schumpeter's long cycle reflects the changes in technology systems which have effects on the behavior of the entire economy. Each business cycle, according to Schumpeter, is unique because of the variety of technical innovations. The long waves in economic development or the so-called Kondratieff waves are the structural changes in techno-economic paradigm (technological revolutions) (Freeman and Perez 1986, 1988). Such changes (the process of structural changes) can be seen as the main engine of capitalist growth. In other words, the long waves of economic development bring about the technological revolutions leading to an advancement of the economy.

The successive waves of technological change bringing about economic growth are shown in Table 1. Regarding Schumpeterian conception of long waves on technological transformations, cotton is a key factor of textile innovations in the first Kondratieff cycle; coal and iron for the industries

Figure 1. The process of scientific development

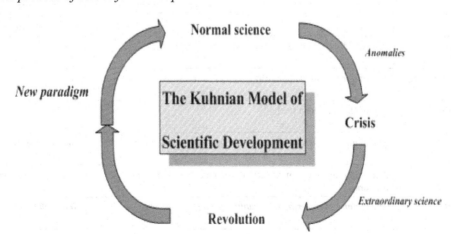

associated with steam power and railways in the second Kondratieff; steel for the industries based on electric power, chemicals manufacture in the third Kondratieff; energy (oil) for industries such as consumer electronics, synthetic materials and pharmaceuticals in the fourth Kondratieff; and chips (integrated circuits) for innovations based on information and communication technology (ICTs) in the fifth Kondratieff (Freeman 1982, Freeman and Soete 1997).

It is argued that nanotechnology is entering the sixth Kondratieff to change the industrial system (molecular manufacturing) of the world. Investment in the technology of nanometer scale has suggested the progress of theoretical concepts from macro- and micro- to nano-system. Nanotechnology offers unprecedented challenges and opportunities to generate new applications fostering the development of whole new sectors of industrial activities. Walker (2000) stresses the importance of embedded institutional, political and economic commitments to building a particular technological paradigm. From the previous 5 Kondratieff cycles based on mechanisations, it seems that nanotechnology is now the starting point of the scientific breakthroughs which would drive the next industrial revolution called 'nano revolution'.

NANO REVOLUTION: A PARADIGM SHIFT IN INDUSTRIAL RESEARCH

Nanotechnology deals with the nanometer scale- a scale at one-thousandth of a micrometer or one-billionth of a meter (10^{-9} or 0.000000001 m). The scale of molecules enables the development of many new materials and devices with wide-ranging applications such as medicine, electronics, energy production and any other field[3]. Currently, the scientific development is moving towards the convergence of technologies: biotechnology, information technology and nanotechnology (represented in Figure 2 by each circle). The overlapping area of these three circles represents the interactions of sciences bringing about wide-ranging applications. Based on Kuhn's theory of scientific revolutions, the limit of mono-disciplinary science to reach a solution to a particular problem sets the stage for the possibility of scientific revolution.

There are many subfields of nanotechnology: electrical, structural, biomedical, energy, environment where the potential of nanotechnology lies in its molecular components. Seen from Schumpeter's perspective of technological change, the nanoscale has the potential to bring

Table 1. Kondratieff cycles in industrial revolution

Period	Description	Key factor of economic development
First Kondratieff (1780s - 1840s)	Industrial Revolution: factory production for textiles	Cotton
Second Kondratieff (1840s – 1890s)	Age of steam power and railways	Coal
Third Kondratieff (1890s – 1940s)	Age of electricity and steel	Steel
Fourth Kondratieff (1940s – 1990s)	Age of mass production of automobiles and synthetic materials	Energy (especially oil)
Fifth Kondratieff (late 1990s)	Age of information, communication, communication and computer networks	Chips (micro-electronics)
Sixth Kondratieff (2000s onwards)	*Age of nano engineering and manufacturing*	*Nanotechnology*

Source: The author's design (adapted from Schumpeter 1939, 1967)

Figure 2. Convergence of technologies (Source: Catching the next Wave in the Corridor, IC2 UT Austin)

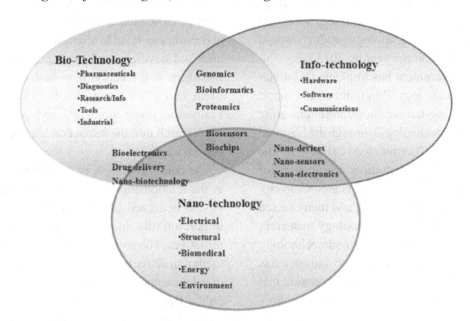

about scientific and technological revolutions. In other words, nanotechnology promises to bring about new products that would have been impossible with macro-sized materials. It can be argued that the world industry is now entering the sixth Kondratieff of nano revolution. Nanotechnology represents a new revolutionary approach in fundamental research moving from a macrocentric to nanocentric system.

Figure 2 also shows the coming era of nanotechnology and nanosystems where technology has emerged as a discipline of nanoscale science. Nanotechnology constitutes an interdisciplinary field that integrates principles and theories inherent to computer science, biology, physics, chemistry and mathematics. It has transformed scientific research in such a way that separate fields have converged under a single engineering paradigm umbrella (Nordmann 2004). The interdisciplinary nature of nanotechnology can be seen from the European Union's NaPolyNet project involving 15 partners from 10 European countries. The NaPoly-Net project team is set up to map the competences in the different fields of characterization of polymer nanostructures. The aim of the NaPolyNet is to

network at regional, national and international level with experts on the characterization of polymer nanostructured materials in the field of packaging, textiles and membranes for improving the structure-performance of polymer devices[4].

Recent years have seen an explosion of nanotechnology research. More than 40 countries have developed programmes in nanotechnology since 2005, illustrating the importance of this field of research. Research and Development (R&D) activities in nanotechnology have been strengthened worldwide to provide a foundation for technological advancement in the future. Figure 3 compares the amount of nanotechnology investment in various countries. Japan is the most preeminent nation in supporting nanotechnology investment. Nanotechnology is one of the priority areas in the Science and Technology Basic Plan that Japan aims to improve the national competitiveness. The Japan Nanotechnology Business Creation Initiative drafted the roadmap (projection up to Year 2020) to focus on strategic applications of nanotechnology in electronics, materials, biomedical, fuel cells and environment. Currently, the Japanese government attempts to enhance the efficiency of the technology transfer process (technology transfer to innovation). The Japanese government expects that the nanotechnology products would reach USD 1 trillion per year by 2015[5].

The US government has implemented a National Nanotechnology Programme to support investment in long-term scientific and engineering research in nanotechnology (through the National Nanotechnology Coordination Office). In the President Bush administration, the Houses of Congress passed the 21st Century Nanotechnology Research and Development Act and then enacted the law to support nanotechnology research. The policy of cost-sharing for nanotechnology research between government and industry has been initiated by President Obama to ensure that US is a competitive leader in the nanotechnology revolution[6]. In the European Union countries, the European Commission announced a five year plan (Years 2005-2009) to keep Europe at the forefront of the fast-moving field of nanotechnol-

ogy in a safe and responsible way. This Action Plan proposes measures to be taken at national and European level to strengthen research in this area and develop useful products and services. Measures in the action plan include a doubling of the budget for nanotechnology in the Seventh Framework Programme (FP7) and specific support for research into the impact on human health and the environment[7].

As nano revolution is a fundamental change in the relationship between society and technology, it is argued that the governments play an important role in organizing nanotechnology initiatives. The government is in the best position to play an active role in speeding nanotechnology developments and applications by coordinating public R&D activities with those of the private sector, increasing investment and creating necessary infrastructures to foster nanotechnology innovations. Importantly, the government would act as a convener of different sectors in build-

Figure 3. Comparison of nanotechnology investment in various countries (amount in million euros, Year 2007) (Source: 'Current status and direction of nanotechnology in Europe', Professor Geoffrey Mitchell, Director of the Centre for Advanced Microscopy, University of Reading, UK.)

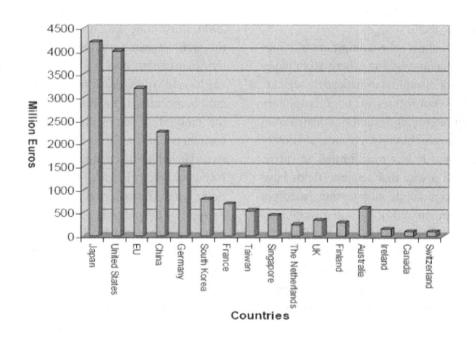

ing bridges across disciplines and setting policy mission-oriented R&D to turn R&D for practical and industrial use. Further, the government plays the key role to enforce effective coordination of R&D programmes at the national level for minimizing duplication and achieving greater efficiency. For example, the US government establishes nanotechnology as a central investment and policy priority. The government support for nanotechnology investment has increased from USD 450 million 2001 to USD 1.5 billion in 2008 as the US aims to be the world leader in the discovery, development, and commercialization of nanotechnology. The government agencies participate in the National Nanotechnology Initiative to fund nanotechnology research in order to create a physical and intellectual infrastructure for sustaining America's economic viability in the high-tech and manufacturing sectors. China has emerged as a fast follower in nanotechnology. The rapid development of China's nanotechnology industries is due in large part to the intervention of the central government. The nanotechnology industries have enjoyed state funding since the 1990s through National 863 Hi-Tech R&D Plan.

Today, the major research centers in Beijing, Shenyang, Shanghai, Hangzhou and Hong Kong account for 90 percent of all nanotech research and development. The Chinese government provides investment supports for nanotech projects with an aim to transform the nanotech industry by 2010 (developing the nanotechnology industry to achieve high performance like the microelectronics, telecom, and other high-tech industries in China).

It is interesting to see that nanotechnology will have an impact on countries both in the developed and developing world by transforming the commodities market, global production, value chains, and the nature of scientific collaboration (Lux Research 2006, Michelson 2008). Figure 4 shows the potential impacts of nanotechnology on the industries. It can be seen that nanotechnology would have high commercial impacts on the industrial sectors of healthcare (genomics and biotechnologies for health), information society technologies (telecom, computers and electronics).

Figure 4. The impacts of nanotechnology on different industries (Source: Rohit et al. (2003))

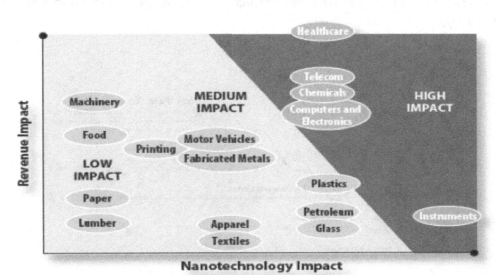

STRUCTURAL CRISIS AND TECHNOLOGICAL FORECASTING OF NANOTECHNOLOGY

This study uses Schumpeter's (1939, 1967) model of technological change to explain the structural developments and impacts of nanotechnology. Figure 5 illustrates the dominant infrastructure and applications for successive waves of technological change. Whereas cotton, coal, steel, energy and chips led the progress and development of preceding Kondratieff waves, scientists, industrialists and researchers currently see nanotechnology as the source of new business cycle and a main engine towards capitalist growth. Nanotechnology research began slowly in the 1960s; however, the world increasingly knows the capacities and potential of this technology since Richard Smalley, Professor of chemistry, physics, and astronomy at Rice University, Houston, USA., received the Nobel Prize for chemistry and nanotechnology research. The potential of nanotechnology is pervasive in many applications, for example, nanomedicine, nanoelectronics, nanotechnology in batteries, solar cells and fuel cell.

Currently, there is a mismatch between the normal science and reality. Nanotechnology is a scientific movement that can solve many of the world's current problems. Taking into account the problem of energy crisis (scarce fossil fuels like oil, coal, and gas) and high oil prices at present (oil prices have surpassed USD 130 per barrel in 2008), nanotechnology opens up the possibility of moving towards new technologies that are efficient, inexpensive, and environmentally friendly like solar power and electric cars. The technology provides improved energy storage materials for hydrogen and fuel cells, better building insulations and combustion engines. Nanostructured membranes, nanohorn electrodes, and nanocatalysts have made fuel cells smaller, lighter, and more affordable. The nano renewable energies help improve energy efficiency and reduce the adverse effects of energy production for the natural environment. Nanotechnology has been applied widely to molecular manufacturing bringing about improvements in computing, medical diagnosis and treatment and many other technological applications such as healthcare and life sciences, electronics, communications and computing.

Figure 5. The successive waves of technological change (Source: The authors' design, based on Schumpeter (1939, 1967))

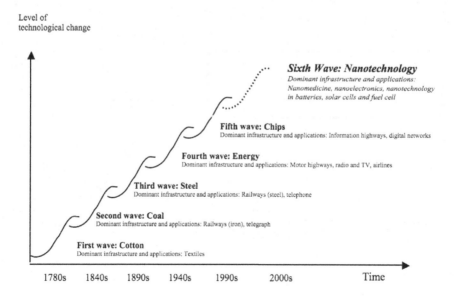

Nanotechnology is regarded as new approaches to manipulation of materials and structures. Many countries regard nanotechnology as an important area of future exploitation and have set up national initiatives to prepare for this technological challenge. As a result, the world is struggling to fulfill the conditions of a new techno-economic paradigm (Freeman and Perez 1988) leading to the new wave of nanotechnology or nano revolution:

(i) Falling Cost

Nanotechnology can be seen as a cost effective manufacturing technology as its benefits include the ability to manufacture at practically no materials' cost. For example, advanced materials can help reduce costs of clean energy whereas nanoelectronics components can help accelerate the development of nanoelectromechanical/ nanodevice innovations. Nanotechnology can be used as a substitute for gold and silver in printable electronics and Radio Frequency Identification (RFID) tags which require unit costs below one cent and many other low-cost applications such as blood-testing kits, solar cells, electronic memory.

(ii) Unlimited Availability of Supply

At present, it is not evident whether nanotechnology has reached a stage of unlimited supply of resources. However, this new technology clearly has helped create additional sources of supply to the industry. For example, nanotechnology provides alternate sources of energy, from solar cells to hydrogen fuel cells and nano-efficient fuel batteries. Nanotechnology contributes to resource saving in industrial production since less materials are required to generate the same amount of production outputs. Many companies are now using nanotechnology to reduce manufacturing costs and compete to get their share of the marketplace.

(iii) Pervasive Use in Many Products and Processes

Nanotechnology has opened up new avenues for producing molecular and nanoscale innovations. It is an enabling technology that improves many types of manufactured products and electronic applications products. For example, nanotechnology coating/ nanocrystalline coating helps improve the efficiency of automobile engines, nano-enabled finishes help protect electronic devices, and nanoparticulate reformulations help make cholesterol-reducing drugs more effective. Nanotechnology can be used in many other applications such as new material, medical, pharmaceutical and agricultural products.

Many industrialists see that nanotechnology would bring discontinuous progress and create the sixth Kondratieff cycle of industrial revolution (European Commission 2004, Macnaghten et al. 2005, Miller 2008). In other words, nanotechnology is seen as a promise of revolution over industry and society (technology that would revolutionize and transform every sector of industry). Currently, nanotechnology has been high on the global scientific agenda. Governments of more than 80 countries plan major investments in nanotechnology and have formulated policies to manage the R&D efforts and exploit the potential of this technology. Researchers in global communities are embarking on new nanotechnology path. As can be seen from Table 2, many countries incorporate nanotechnology as national plans and/or establish national nanotechnology centres within the context of the country's strategy. The national laboratories have been set up and technology areas have been prioritised to prepare for this technological challenge.

POLICY RECOMMENDATIONS AND CONCLUSION

This chapter discusses the nano revolution based on Kuhn (1970)'s model of scientific change and Schumpeter's model of Kondratieff cycles (1939, 1967). The sixth Kondratieff wave of technological change can be seen from its interdisciplinary nature. A budget set to invest in nanotechnology research in many countries illustrates an increased level of commitment by local governments towards this paradigm shift (the nano revolution). The governments of many countries see nanotechnology as key to rebuild their economies and set policy agendas to encourage national competitiveness. This chapter offers key policy recommendations to encourage the undertakings of nanotechnology research and development towards the revitalization of the global economy as follows:

Table 2. Nanotechnology research and policies of some of the most preeminent countries

Country	Research policies and activities
USA	The National Nanotechnology Initiative (NNI), advocated by Former President Bill Clinton, provides a multi-agency framework to ensure US leadership in nanotechnology. In 2007, the National Nanotechnology Initiative released a Strategic Plan outlining updated goals and programme component areas to strengthen fundamental nanoscience and engineering research. The NNI sets aside the budget of USD 1.5 billion for Fiscal Year 2009 to support basic research, infrastructure development and technology transfer.
China	The Chinese Academy of Sciences, jointly with the National Natural Science Foundation of China and the State Science and Technology Commission actively support nanoscale research since 1980s. The Chinese Academy of Science's Knowledge Innovation Programme gives priority to nanotechnology so as to bridge the gap between the research lab and the marketplace. The government policy includes the National High Technology and Development Programme which sets goals for the short-term (nanomaterials development), medium term (bionanotechnology and nanomedical technology development), and long term (nanoelectronics and nano chips development).
Germany	The Ministry for Education and Research plays an important role to support pre-commercial nanotechnology R&D investment. In 2007, the German government spent USD 515 million (representing 40% of public EU nanotechnology funding) in nanotechnology research, making Germany a leader in European Union in nanotechnology research and investment. The Ministry of Education and Research has launched the Nano-Initiative Action Plan 2010 to provide a single framework for the nanotechnology goals and activities of Germany's government departments. The federal government's high-tech strategy, with the help of eight ministries, puts together the package of actions focusing on turning nanotechnology research into commercial products.
South Korea	The Korean government launched the 10 Year Plan (Years 2001-2010) for the Promotion of Nanotechnology with the budget amounting USD 1.485 billion aiming to become the top 5 Nanotechnology countries in the world. The Korean nanotechnology programmes focus on funding the nanotech research areas in nanoelectronics (IT) and nanomaterials. The Nanotechnology Investment Plan supports the creation of new R&D programmes in nanotechnology through nano research networks of Korea Institute of Science and Technology Evaluation and Planning (KISTEP) and Korea Science and Engineering Foundation (KOSEF).
France	France's priority research areas in nanotechnology are nanoelectronics and position system. The National Programme Nanosciences was launched by the Ministère de la Recherche with the CNRS (Centre National de la Recherche Scientifique), the CEA (Commissariat à l'Energie Atomique) and the DGA (Délégation Générale à l'Armement) to coordinate and develop fundamental research in nanosciences. The Nano 2012 nanotechnology programme was launched as a strategic investment programme to foster the development of the electronics industry. The programme is defined by STMicroelectronics, IBM, CEA and local authorities with the budget of USD5.65 billion that would create 650 new jobs.
Taiwan	The Industrial Technology Research Institute (ITRI) serves as Taiwan's research and development hub undertaking nanotechnology research in science foundation, platform technologies and application technologies. Taiwan launched the industrialisation programme with the budget of USD 412.8 million to support research in nanomaterials, nanoelectronics, nanodevices, nanopackaging technology, nano optical communication, nano biotechnology and nanotechnology for energy applications. Currently, the Taiwan government has approved the second phase of the Taiwan National Nanotechnology Programme (Years 2009-2014) to strengthen the industrialisation of nanotechnology.

Cooperative Learning and Integrative Research Networks

Nanotechnology is seen as the 21st century technology where its commercial potential is enormous. Nanotechnology presents a new paradigm. In other words, nanotechnology can be seen as a rational process to solve problems and pave the way for a revolution in materials, information and communication technology and industrial processes (improving products and production processes with better characteristics or new functionalities). At present, many countries have set strategic priorities in nanotechnology aimed at smoothing the commercialization of outputs of nanotechnology R&Ds. In developing nanotechnology economy, the technological innovation could be done by cooperative learning and integrative R&D networks. The research consortia should link up with the incubators and deployment organizations in order to accelerate the speed of R&D towards commercialization and improve innovative nanotechnology applications.

Setting Nanotechnology Priorities for Competition

Seeing nanotechnology as a major factor in the social-economic development, many governments will face international competition. In such a situation, the governments should set nanotechnology priorities in terms of finding the right areas of nanoscience with specific expertise to promote industrialisation of R&D outcomes. A strong national policy agenda is needed for this thematic prioritisation. To keep up competitiveness, the governments should take efforts in allocating funding to activities that would lead to knowledge advancement and/or commercialization, depending on the national conditions. In linking nanotechnology capabilities to nanotechnology opportunities, this requires a formulation of the national programmes that can facilitate the development of nanotechnology. The governments

should therefore consider formulating the detailed programmes and building upon their core strategy to strengthen the nanotechnology economy.

Developing the Financial Infrastructure to Support Nanotechnology Research

An initial seed stage of funding nanotechnology research is difficult because the timeframe to market is large (moving from laboratory discovery to a commercial product) and the research projects require much funding. To improve nanotechnology performance, the financing infrastructure to support innovation is needed, particularly for small- and medium-sized enterprises (SMEs). For any country to develop and improve nanotechnology economy, the governments should support venture capital funding to accelerate nanotechnology becoming a commercial opportunity. The governmental support in terms of funding should be on focused research areas with R&D priority consideration to ensure that basic nanotechnology research can be converted to new commercial products, new companies and jobs. Furthermore, institutional investors may take over deals from start to finish which would enhance the effectiveness in turning research into useful and commercial innovations.

The above points present the nanotechnology framework policy recommendations for improving the global competitiveness on nanotechnology research. For developing countries to catch up with developed countries, they should consider using a combination of cooperative learning and research networks among universities and companies to share research excellence and strengthen nanotechnology innovations between all actors. In the course of their networks, developing countries would be able to increase the ability to innovate and generate new nanotechnology applications. Developing countries might consider investing in the nanotechnology research through academic funds and subsidies in order to explore nanotechnology opportunities. China and India are examples of developing countries focusing their efforts on high technology growth as a means of fueling economic development. China and India have initiated their science, technology, and innovation policies around funding for nanotechnology R&D amongst other high-tech approaches with an aim to leapfrog development in the key field of nanotechnology. This chapter has shown that nanotechnology is seen as the next technological revolution by many scientists, industrialists and researchers. The technological advancement of nanotechnology has attracted developed and developing country governments to make ambitious investments. However, there remains an important question of whether all nations would be ready to meet the readiness challenges that most certainly will lie ahead.

REFERENCES

Brown, H. I. (1977). *Perception, Theory and Commitment: The New Philosophy of Science.* Chicago: Precedent Publishing.

Daneke, G. A. (1998). Beyond schumpeter: Nonlinear economics and the evolution of the U.S. innovation system. *Journal of Socio-Economics, 27*(1), 97–115. doi:10.1016/S1053-5357(99)80079-1

Dosi, G. (1982). Technological paradigms and technological trajectories. *Research Policy, 11,* 147–162. doi:10.1016/0048-7333(82)90016-6

Dosi, G. (1988). The Nature of the innovation process . In Dosi, G., Freeman, C., Nelson, R., Silverberg, G., & Soete, L. (Eds.), *Technical Change and Economic Theory.* London: Pinter.

European Commission. (2004). *Nanotechnology - Innovation for tomorrow's world.* Luxembourg: European Commission Community Research.

Freeman, C. (1982). Schumpeter or Schmookler? In Freeman, C., Clark, J., & Soete, L. (Eds.), *Unemployment and Technical Innovation.* London: Pinter.

Freeman, C., & Perez, C. (1986). *The Diffusion of Technical Innovations and Changes of Techno-economic Paradigm.* Paper prepared for the Venice Conference, March 1986. Science Policy Research Unit, University of Sussex.

Freeman, C., & Perez, C. (1988). Structural crises of adjustment, business cycles and investment behaviour . In Dosi, G., & Freeman, C. (Eds.), *Technical Change and Economic Theory.* London: Pinter.

Freeman, C., & Soete, L. (1997). *The Economics of Industrial Innovation* (3rd ed.). London: Pinter.

Glass, J. C., & Johnson, W. (1989). *Economics: Progression, Stagnation or Degeneration?* Hemel Hempstead, UK: Harvester Wheatsheaf.

Ikezawa, N. (2001). Nanotechnology: Encounters of Atoms, Bits and Genomes. *NRI Papers 37.*

Kuhn, T. S. (1970). *The Structure of Scientific Revolutions.* Chicago: University of Chicago Press.

Luther, W. (2004). *International Strategy and Foresight Report on Nanoscience and Nanotechnology.* VDI Technologiezentrum.

Lux Research. (2006). *The Nanotech Report* (4th ed.). New York: Lux Research.

Macnaghten, P., Kearnes, M., & Wynne, B. (2005). Nanotechnology, governance, and public deliberation: what role for the social sciences? *Science Communication, 27*(2), 1–25. doi:10.1177/1075547005281531

McCraw, T. K. (2007). *Prophet of Innovation: Joseph Schumpeter and Creative Destruction.* Cambridge, MA: Harvard University Press.

Michelson, E. S. (2008). Globalization at the nano frontier: The future of nanotechnology policy in the United States, China, and India. *Technology in Society*, *30*, 405–410. doi:10.1016/j.techsoc.2008.04.018

Miller, G. (2008). Contemplating the social implications of a nanotechnology revolution. In Fisher, E., Selin, C., & Wetmore, J. (Eds.), *Yearbook of Nanotechnology in Society* (pp. 215–225). Berlin: Springer. doi:10.1007/978-1-4020-8416-4_19

National Science and Technology Council (NSTC). (2006). *The National Nanotechnology Initiative: Research and Development Leading to a Revolution in Technology and Industry*. NSTC Report, July 2006.

Nordmann, A. (2004). *Converging Technologies: Shaping the Future of European Societies. Report of the High Level Expert Group: Foresighting the New Technology Wave*. Brussels, Belgium: European Commission Research.

Roco, M. C. (2001). International strategy for nanotechnology research and development. *Journal of Nanoparticle Research*, *3*, 353–360. doi:10.1023/A:1013248621015

Roco, M. C., & Bainbridge, W. S. (2002). Converging technologies for improving human performance: Integrating from the nanoscale. *Journal of Nanoparticle Research*, *4*(4), 281–295. doi:10.1023/A:1021152023349

Rohit, S., Hwang, V., Sood, K., Klein, J., & Cohn, K. (2003). *Nanotechnology What to Expect*. A Larta White Paper. Los Angeles, CA: Larta

Schumpeter, J. A. (1939). *Business cycles: A Theoretical, Historical and Statistical Analysis of the Capitalist Process (2 Vols.)*. New York: McGraw-Hill.

Schumpeter, J. A. (1967). *The Theory of Economic Development* (5th ed.). New York: Oxford University Press.

Tushman, M., & Anderson, P. (1987). Technological Discontinuities and Organization Environments. In Pettigrew, A. (Ed.), *The Management of Strategic Change*. Oxford, UK: Blackwell.

Walker, W. (2000). Entrapment in large technology systems: institutional commitments and power relations. *Research Policy*, *29*(7-8), 833–846. doi:10.1016/S0048-7333(00)00108-6

Wilson, M. (2002). *Nanotechnology: Basic Science and Emerging Technologies*. Boca Raton, FL: Chapman & Hall/CRC. doi:10.1201/9781420035230

ENDNOTES

[1]   In Kuhn's terms, 'normal science' is a strenuous and devoted attempt to force nature into the conceptual boxes supplied by professional education that often suppresses fundamental novelties because they are necessarily subversive of its basic commitments … the very nature of normal research ensures that novelty shall not be suppressed for very long (Kuhn 1970).

[2]   Kuhn uses the term 'crisis' to refer to the scientific revolution. According to the Kuhnian paradigm concept, a paradigm shift occurs when the existing paradigm comes into a state of crisis as an explanatory model, that is too many phenomena have accumulated in the normal science phase that the existing paradigm cannot explain. This crisis then leads to the emergence of the new paradigm (Kuhn 1970).

[3]   Sustainable Development: A Challenge for European Research, Janez Potočnik, European Commissioner for Science & Research.

[4]   NaPolyNet Project (www.napolynet.eu)

[5]   US National Science and Technology Council

6   State of the Union, President Barack Obama, the White House.

7   The current status and direction of Nanotechnology Development in the European Union (EU), Professor Geoffrey Mitchell, Director of the Centre for Advanced Microscopy, University of Reading, UK - Workshop on Nanotechnology and the Industrial Development organised by the National Innovation Agency, Ministry of Science and Technology, 20 January 2009, Bangkok, Thailand.

Chapter 6
Technology Resilience and Diffusion:
Capability Formation Dilemma in Non-Agile Economies

Yves Ekoué Amaïzo
Austria Afrology Think Tank, Austria and International Business and Project Management, Austria

ABSTRACT

Economic zones are facing difficulties to control their endogenous process of technology creation. Technology poverty does exist. How to ensure that fragile economic zones become more resilient in capturing technology? How to achieve a competitive advantage in non-agile economies and get a larger category of their populations benefit from it? These challenges have taken all serious developmentalists by storm. Unfortunately, the so-called 'technology transfer' does not yield to the outcomes it promotes resulting to mitigated results in non-agile countries. This chapter argues that in an unpredictable and changing environment, technology diffusion is part of a complex change process which needs to be dealt with in a comprehensive manner. It cannot take place without revisiting the capability formation dilemma in non-agile economies. The technology-driven capability should therefore be supported by strategic policies and public private institutions. Technology diffusion may appear as the systemic link between shared wealth creation and collective efficiency.

INTRODUCTION: TECHNOLOGY CONTENT AS THE MISSING LINK

The paradigm of catching-up or falling behind industrialized countries is an endless issue in the development paradigm. The global trend towards a sustainable wealth creation is a technology-innovation driven process. With reference to this

debate, the world is basically split into three main groups of countries: industrialized countries (IC), emerging or newly industrialized countries (NIC) and least industrialized countries (LIC). Selected criteria such as the gross domestic product (GDP) per capita or the share of manufacturing or industry in the GDP appear as partial indicators while trying to restore the crucial role of science and technology in determining the wealth's creation of countries or economies. The correlation be-

DOI: 10.4018/978-1-61692-006-7.ch006

tween the lack of technology, the level of poverty and the pace of development is more and more obvious. Some countries have organized and structured their technology capability to respond positively and efficiently to change. Based on a clear commitment of governments, it is possible to support an endogenous process towards industrialization. As agility becomes an issue in the wealth creation process, States could be split into two broad categories: agile and non-agile economies. Economies whose selected economic drivers are below the world average will be considered as non-agile economies. Priority will be given to gross domestic product (GDP) (or gross national income (GNI)) per capita and the share of manufactured value-added in GDP.

Highlighted by many development institutions in the 1970s and 1980s, the naive approach of transferring technology to people without taking into consideration the existence of a comprehensive technology-driven policy embedded in local culture appears today as a strategic mistake. Any effective policy scheme should include the mastering of a local capacity of absorption, ensure the presence of support institutions, offer smart partnership approaches favoring investors and innovators, both local and foreign, and contribute to capture and diffuse knowledge. It is not surprising to register mitigated failures of past radical top-down precepts, wrongly considered as solutions which were supposed to make the difference for people in regions which are lacking technology infrastructure and leaders' effective commitments. Billions of US dollars (US$) of development aid[1] transferred to non-agile economies hardly improve the lives of those who are at the bottom of the wealth's ladder although there is no bottom to poverty. With the structural decline and mitigated efficiency of development aid, it is not surprising that a growing number of concerned people are clearly advocating against the effective benefit of aid (Moyo, 2009). Aid does not contribute to an effective transfer of technology. It often contributes to increase the number of people falling into

dependency and poverty, an impressive number of African people which cannot be limited to the 'bottom million[2]' (Collier, 2007). In fact, with improved relations between effective democratically elected leaders and the population in fragile, failing and failed States (Reinert, Amaïzo & Kattel, 2010 forthcoming) and the renewed interest to foster entrepreneurship at all levels, the road to economic prosperity is possible. As recalled by the President of Rwanda, Paul Kagame, *'We know the road to prosperity is a long one. We will travel it with the help of a new school of development thinkers and entrepreneurs, with those who demonstrate they have not just a heart, but also a mind for the poor'* (Kagame, 2009). Alternative economists and new developmentalists' views may become more prominent in the future (Chang & al., 2010, forthcoming).

With reference to the mainstream neo-liberal economics schools of thoughts, the principle of effective technology transfer is simply restricted to those who can afford the high cost of transferring technology. For many governments with budgets constraints, technology transfer is simply an undeliverable option in a competitive globalized unpredictable environment. The importance of the State intervention was often underestimated in modern liberal economics. The recent financial crisis and its negative collateral effects have reversed that perception. In order to ensure technology diffusion in countries whose immediate objectives are to sustain or support an industrialization process, a smart and cross-cutting policy to improve the level of technology content has more chances to succeed if promoted by a government. Industrialization without clear linkages with the entrepreneurial community and a re-orientation of skills in line with global market evolution may not lead to the creation of decent jobs. In a competitive environment, it could even become a trap for countries whose policy combines low-level skills and high transaction costs while pursuing trade without value-added or trade limited to no- or low-technological content goods.

There is no such thing as a forced consensus on trade at the expense of industry (Soludo, Ogbu and Chang, 2004). This situation is worsened when those countries face additional difficulties in mastering the outsourcing of economy of scale's solutions. For these reasons, the diffusion of technology content process cannot exclusively be left to the market forces. It does not mean that smart public interventionism (UNCTAD, 2009) should not be combined with private entrepreneurships in a competitive environment. Thus, alternative approaches in wealth creation may appear as incomplete if they are not part of a structuring of the local/regional economy's strive to develop and sustain productive capacities with the involvement of the concerned population (Reinert, Amaïzo and Kattel, 2010, forthcoming). Accordingly, public private partnership solutions and technology diffusion cannot take place without:

- An endogenous reform of the infrastructure of innovation;
- A conducive business environment favoring technology absorption, creation, diffusion, promotion and resilience;
- A watch system structured around regular benchmarking of key technology drivers in line with the existing skills infrastructure and the overall policy objectives;
- An agglomeration process linking skills, knowledge and Diaspora to inject technology content at all level of the production system.

Technology is a fundamental instrument in the generation and the sustaining of both wealth and shared economic growth for all economies, non-agile economies in particular. However, technological development could generate externalities, such as costs, in terms of air and water pollution and other undesirable environmental effects.

The overall challenge of the introduction of technology is to improve individual and collective economic performances using productivity as a benchmark instrument. Otherwise differences in economic performance may hamper weak industrialized countries' progress (Ocampo, Jomo and Vos, 2007). Without focusing explicitly on nano-technologies and micro-electronics, various interdependent issues are preconditions for the emergence of a rapid diffusion of modern and advanced technologies. Nano- and micro-technologies may require more skilled resources as well as additional capital to acquire appropriate support equipments.

As technology transfer has never taken place for free, nor has been sustained without competition, technology transfer is becoming an obsolete concept and should be replaced with *technology diffusion*. This concept is closer to the effective practice of actors in a competitive and global market. Shared wealth creation is directly correlated with the effective commitment of decision makers, both in public and private sector, to diffuse technological content broadly. Technology diffusion should contribute to the generation of wealth and returns for the State through taxes. Thus, four interdependent issues will be developed with the objective to contribute to open new avenues for partial benchmarking of economies:

- Performance and progress based on the positive interdependency of people, processes and techniques are better achieved through the introduction of technological content in all forms and at all levels of economic production. Thus, nano-technologies appear as one form among others of the diffusion of technology content.
- Based on selected drivers, technology disparity between countries is clearly obvious (see Appendices). The pace of technology diffusion partly explains the economic growth divide. How to achieve technology diffusion in a global and unpredictable competitive environment is a challenge for both agile and non-agile economies.

- Without a clear and flexible strategic policy favoring the development of technology content, it would be difficult to create and enforce public-private institutions that should promote technology diffusion. The problem of transfer of technology cannot be considered as a simple exchange of goods and services but as part of a process of knowledge capturing and diffusion. With the rise of the environmental concern, selected domains of technology will gradually become de facto public goods and should be treated with preferences in non-agile economies.

- The building of an agile, diversified and specialized technology-driven capability formation should be part of a renewed vision of new developmentalists. Building and keeping skilled human resources is a real development challenge for non-agile economies. It could become a vicious circles' trap as most of the skilled human capital in non-agile economies are attracted by better prevalent conditions in agile economies. Dynamic and constructive linkages between R&D, business community and States need to be explored further. Weak and newly industrialized countries are particularly concerned as late comers in the process of injecting technology content in their economies. They should be prepared to compete and take advantage of new opportunities in a global production networks system while ensuring an endogenous process of technological change.

Non-agile economies are missing technology content at several levels of their productive structures. The catch-up process syndrome should gradually be replaced by a resilience culture. Leveraging performance in such circumstances is a real challenge.

TECHNOLOGY DIFFUSION: A PERFORMANCE'S LEVERAGE

Technology diffusion is a complex issue which needs interdisciplinary collaboration and agglomeration. How to contribute efficiently to the improvement of the predictability of the resilience and the sustainability of performance of an economic area (region, state, town, economic zone, etc.)? How to promote technology diffusion with a comprehensive approach based on the upgrading of technology content at all levels (policy, programs, institutions, projects, process, skills, and collaborative actors)? The answers must be found in a cultural agile mindset. Achieving technological change while ensuring or improving technology diffusion, becomes de facto the emerging challenge for all countries.

Technology diffusion is a matter of urgency for newly industrialized countries (NIC) and even a matter of emergency for least industrialized countries (LIC). Countries with a share of manufactured value-added in gross domestic product (MVA in GDP) below the world average should be considered as weak or least industrialized countries (LIC). It is therefore the building of a system of fragile and uneven diffusion of technology which usually leads to a weak-performing dynamism of the industrialization process.

Even while trying through several iterations to offer a *"predictive model of technology transfer"* and highlighting difficulties in the acceptance of new and innovative technology (Jolly & Creighton, 1977), it is still unclear today whether the technology transfer process could be reduced to a concept, a framework or a methodology. The correlation between technology transfer and the specific method used in transferring technology as part of an international partnership is questionable because of the imbalance of power in global relationships. The *transfer of technology* concept has a special cultural bias. The dominant culturally led approach of technology transfer is often conceived as the transfer from *those who have* to *those who do not have*. It is a one-way relationship and should be considered as a serious preconceived notion structured as a static approach in a technology diffusion paradigm. It should be replaced by a more dynamic, complex and systemic understanding of technology diffusion based on the interdependency of skills, productivity, performance, shared wealth creation, and development progress.

As stated by R. Wright, *'Society becomes increasingly non-zero-sum as it becomes more complex, specialized, and interdependent'* (Wright, 2001). Non-agile economies cannot escape this development path. Selected economic zones are facing difficulties to master their endogenous process of creating and mastering technology. As part of cascading fragilities, a static approach generates weaknesses in terms of wealth creation and productivity enhancement. How to ensure that weak economic zones become more resilient in capturing technology and transform it into a competitive advantage for a larger category of their populations is the challenge at stake? This change process is a complex issue which needs to be dealt with in a comprehensive manner especially at the policy level. While the root problems are well identified by serious analysts, the policy precepts often diverge between decision-makers and appear irreconcilable on the ground. Strategies and policies adopted converge on the fact that the improvement of technology content does support technology diffusion. Failing to promote technology content is de facto a clear policy against the interest of non-agile economies. It should open rooms for investigations as this leads to a deindustrialization process (UNCTAD, 2009).

In the past, mercantilism, slavery, colonialism and cultural neutralization did impede the innate path of technology development in numerous non-industrial countries. The issue of technology transfer cannot anymore and exclusively be explained using a cultural bias introduced by countries which first master technology, productivity and wealth creation. Improvement in local endogenous technology could be more relevant for technology diffusion than the transfer of advanced technology itself. One of the common mistakes lays in the difficulty for some past and present decision-makers to differentiate, and sometimes to consider as equal, the concept of *'transfer'* and *'resilience'* in technology. Resilience appears as a better descriptive approach of what an economy needs. Technology transfer appears therefore as a simple sub-group of resilience.

In some cases, technology transfer could impeach technology diffusion. The generation process, the selection of technologies and the ability to absorb and innovate at local level for local needs are crucial. It should not be split when defining priority challenges. Improving the well-being for all implies that comprehensive approaches and exogenous priorities are not bypassed otherwise it could simply lead to a deviation from the original goal of introducing technologies. As a result of difficulties faced in transferring technology, palliative transfer of technology has gradually become the common law in countries with weak economic and political influence. For non-agile economies such as many African countries or countries which were victims of a series of internal or external neutralization of technological processes, resilience means the ability of an economic zone under constraints or influences to recover its capability formation in technology (Amaïzo, 2004), especially after several years of technology deformation or eradication. Resilience is therefore the capability of a strained economic zone to recover its original proactive generation and to master a regular and smooth introduction of technology content as a means to improve, in all sectors, its shared wealth creation process. It is presumed that the overall objective is to build and sustain economic returns for social development and progress.

This agility process is called technology diffusion and could support technology resilience. The power of leapfrogging using the original endogenous technological capabilities while taking advantages of the opportunities offered by the globalization process should be considered as part of the agility process. It is a mix of local knowledge diffusion, flexibility, resistance, adaptability and proactive innovation. How to recover and reinstate this self-generating process in a complex, competitive and unpredictable global environment appears as a real challenge for lately industrialized countries. The commitment of a State to improve productive structures is partially measured by the share of an economy's manufactured value-added in gross domestic product. Both the self-technology generation process and the progress in creating value-added as an outcome of technology upgrading are opening new avenues for resilience consciousness and ethical behavior while diffusing technology.

Technology diffusion could finally be defined as the propension of an economic zone to develop its capability absorption possibilities with the immediate objectives to enable, in a glocal (global and local) networked production system environment, a rapid resurgence of productivity and wealth diffusion. Consequently, technology diffusion and resilience turn out to be the foundation of agility, a dynamic process towards future competitive advantages' capabilities.

The methodological approach adopted is part of an attempt to shed a new light on the interdependency of various concepts and features. The usual North-South straightjacket approach of *transfer of technology*, often supported by dominant cultural conceptual bias, has not resulted in the effective transfer of technology in the interest of the large majority of the population in non-agile economies. It is often restricted to the interest of major transnational companies or local powerful decision-makers. These two groups of actors often even refrain from diffusing fully both knowledge and technology as a means to keep power and preserve their technology advance, limit wages distribution and therefore purchasing power improvements. As a result, these groups of actors master dependencies over the rest of the population. Competition is either ruled out or simply restricted under window-dressing sweet statements. Effectiveness of technology diffusion is a sign of the existence of a competitive environment. It is an acknowledgement of an endogenous political maturity and appears as better suited to modern management behavior striving to achieve collective efficiency. As management was qualified by S. Berkun (2008) as an art of *'how to make things happen'*, technology diffusion is in fact a comprehensive process of managing technology content in developing economic zones with a voluntary attempt to progress evenly and proactively towards collective efficiency, while limiting social and environmental externalities. The management of technology supposes a complete re-organization of knowledge management in non-agile economies (Hosni and Khalil, 2004)

Technology diffusion is an issue of technology change and resilience. It takes place with the identification and the ownership of the problem to be solved with clear inputs of technological contents. The contribution and support of appropriate skills, institutions and partners and the integration of the local culture constitute agile preconditions which may delay the dynamic process if missing. The immediate objective of technology diffusion must clearly be focused on how productivity and performance improvement could be achieved in the interest of a large number of people. In order to secure resilience on a sustainable basis, the introduction of ethic, social and environment considerations cannot anymore be separated from a comprehensive process of technology diffusion.

It would be unwise to list the large number of existing definitions of technology transfer because of various cultural bias related to the level of the on-going industrialization process. As technology diffusion has been preferred as a more appropriate terminology to describe the effective technological shift phenomena leading to a change process, it is therefore redefined as follows:

Technology diffusion is happening in a globalized and unpredictable context and cannot be delinked from the process of sharing, the imperative of solving a problem, the crucial need to ensure a wider diffusion as well as a voluntary acceptance to let the technology be further develop or exploit in all forms without any restrictive conditions. Technology diffusion has more chance to spill-over in a scientific and technological knowledge environment and institutions. Leapfrogging through technology content improvement in a competitive and conducive environment is facilitated through the exchange and the adaptability of strategies, policies, processes, projects, products, applications, materials and services. Failures should be considered as part of the learning process and are usually reduced in a competitive environment. In light of the above, technology diffusion is part of the process of converting knowledge advance into marketable or non-marketable features which are supposed to improve the way things are done.

As technology diffusion is crucial to leverage performance, means to measure technology diffusion are difficult to aggregate.

MEASURING TECHNOLOGY DIFFUSION'S DISPARITIES

The technology diffusion disparity is defined as the combination of:

- the lack of proactive and voluntary commitments to take advantage of technology in all human activities gearing towards improvement and productivity, and

- the constraints faced in attracting, developing, diffusing broadly and democratically technology content in each segmented part of policies, processes, instruments and equipments, which are contributing to the improvement of human well-being.

- It is therefore different from the common definition of technology gap limited to *"those who have access to technology and are using it effectively, and those who*

do not" or *"the difference in technologies developed and used in countries and firms where one is more advanced than the other"* (UNCTAD, 2005: 5). As a consequence, narrowing the technological gap without entering a technology catch-up race is possible at the entrepreneurial level (firms) and at the governance level (governments). Two main groups of activities should be promoted:

- Generation of technology: the ability to *'acquire, master, adapt and improve upon scientific and technical knowledge'*; Based on existing data which is considered as credible, a list of drivers will be suggested in the first category called: *Technological capability formation*.

- Diffusion of technology: the ability to *'keep up with technological advances'* (UNCTAD, 2005: 5) through innovation, spreading at large knowledge as a public good and opening avenues for collective improved capabilities. A second list of drivers will be suggested and grouped in a second category called: *Technological absorption and diffusion capacity*.

Because of the negative cultural bias present in the definition of fragile States, it was decided to differentiate between two categories of economies: agile and non-agile in achieving technology diffusion and resilience. An agile economy in this context is an economy with leaders committed to generate and sustain a systemic approach to technology content diffusion as part of public, private or public-private initiatives. It is one way to improve wealth creation and sharing with the objective to achieve democratically collective efficiency in a complex, unpredictable globalized world environment. The 'there is no alternative' (TINA) syndrome of technology catch-up could become a misleading concept. It should be revisited in the light of local culture, concerned stakeholders' interests with special reference to their effective absorption capabilities and the level of external unstated barriers. Non-agile economies, which are not exclusively fragile economies, are defined here as economies which are facing major constraints to generate and diffuse technology with negative consequences on their ability to support shared economic growth for all.

The measurement of technology generation is usually related to the benchmarking of a series of specific drivers. The drivers mentioned below are relevant for nano-technology and micro-electronics but also for any advanced technologies which cannot be promoted in a vacuum or could be sustained with identified constraints.

The measurement of technology diffusion is more difficult to highlight. It often results in the observation of a lack of technology content or knowledge and its collateral consequences on the improvement of the human well-being.

The following selected and non-exhaustive drivers might be of interest while benchmarking agile and non-agile economies as to the effective commitment of governments to promote technology content. Two major groups of drivers were identified as a crucial leverage for competitive advantages, collective efficiency and *in fine* the improvement of populations' well-being:

- Technological capability formation (see Appendices 1 and 2)
    - Mean year schooling (Average years of schooling of adults by country, for most recent years)
    - Primary school pupil-teacher ratio which is an indication of education efficiency especially in comparison with the level of the public expenditure on education (as % of total government expenditure);
    - Public expenditure per student in the tertiary (as % of Gross Domestic Product (GDP) per capita);
    - Expenditures for Research & development (as % of GDP);

    - Gross enrollment ratio in the Tertiary (as % of the relevant age group);
    - Researchers in R&D, (in millions of people);
    - Qualitative innovation in Business signaled by the International Organization for Standardization (ISO) certification ownership (as % of quality certified firms);
- Technological absorption and diffusion capacity (see Appendices 3 and 4)
    - High-technology exports (in $US million);
    - Share of manufactured exports in High-technology exports (in %);
    - Royalty and license fees (receipts and payments, in $US million);
    - Scientific and technical journal articles (in numbers);
    - Patent applications filed (residents versus non-residents, in numbers);
    - Trademark registered with national or regional trademark offices (residents versus non-residents, in numbers);
    - Electric power (consumption per capita in KWh);
    - Affordability and efficiency of fixed line and mobile telephones (in $US per month);
    - Internet users (number per 100 people);
    - Affordability of Internet (average price in $US per month);
    - Information and communication technology expenditures (as % of GDP);
    - Secure Internet servers (per million people);
    - Agricultural productivity (agriculture value-added per worker), 2003-2005 (in 2000 $US).

Most of the statistical figures should be considered with great caution as the accuracy, the lack of data, the variation in years of data collection are some of the problems encountered. It is also difficult to combine those data with the objective to offer one reliable aggregated index. Based on those constraints, technology diffusion can be analyzed with:

1. Data trends over the years;
2. Benchmarking national drivers with regional or global ones;
3. Using sectoral average as a best practice yardstick.

The mean year schooling[3] is a powerful driver in measuring chances of a country to promote technology-driven economies. The knowledge gap between USA with 12 years schooling as compared to Guinea-Bissau with 0.8 mean year of schooling is obvious (see Appendix E). Countries below the world average of 6.2 years of formal schooling have less opportunity to achieve quickly technology diffusion and resilience. Those countries are considered as non-agile economies. It is not surprising that the number of qualified human resource is low, and the level of technology-knowledge weak. There are hardly any available institutions or a proactive technology-driven policy. The generation of technology from an endogenous viewpoint becomes a serious challenge. Negative collateral effects on performance and productivity improvements are numerous. It is also advisable to review this driver with the share of tertiary enrolment in Science and technology[4]. From that perspective, the domestic capability formation and the ability to create, absorb, diffuse, and innovate are simply hampered. The various Information, Communication and Technology (ICT) drivers help to highlight the digital divide between non-agile and agile technology/innovation-driven economies.

The problem of technology diffusion and resilience cannot find a sustainable solution if the concept underlying any technology development process in a country is based on an aid syndrome. Those who have a technology or knowledge should transfer *'what they want'*, *'when they want'* and *'to whom they want'* without discussing the effective impact or effective improvement on beneficiaries. Technology should be delinked from aid. For example, food aid should be replaced by technology content support to farmers in non-agile economies. G8 members[5] came to a consensus on this conceptual shift. Food security without charity could improve performance in a given cultural and economical context if implemented at the development pace of the economy and the population's absorption capacity. Nano-technology may open new opportunities in the food production value chains.

If the globalization of technology has open new opportunities areas, it is unclear whether these opportunities are not in fact favoring agile technology-driven economies. The advantages provided to foreign productive capacity facilities in non-agile technology-driven economies (usually through foreign direct investments), and the low level of local absorption capacities limit the potential benefits which could result from intra-firm collaboration in the growing global production and knowledge sharing networks system. Besides, foreign best practices may not necessary favor the generation of technologies at the local level.

To enable the emergence of modern technology, a combination of human expertise, supportive institutions, appropriate equipments, conducive policies and a system of diffusion structured as a sustainable self-financed process must be in place. Effective commitment of the Government and a proactive role of private sector representatives supporting technology capability formation are required. Nano-technology is a typical example of high-technology content where the lack of any of those conditions may result in complex obstacles and sometimes failures.

It is therefore quite essential to diverge from the TINA syndrome that one can find 'the best transfer of technology paradigms' across the globe. Such common statements have led to misleading blueprint technology policies or copy-paste approaches which ended in dramatic, time consuming and non-cost effective achievements in technology diffusion and resilience. The TATA syndrome meaning *there are thousands of alternatives* should be preferred as they are many recipes in economics (Rodrik, 2007). The issue at stake is the management of a self-conducive environment which should favor technology content diffusion partly as public goods, partly as a result of the outcomes of a competitive market environment. The management of technology in non-agile economies is often simply forgotten (Hosni and Khalil, 2004).

How to achieve technology diffusion without revisiting the role of technology content development in countries' strategic policies is becoming a new area of interest. It has a direct implication on the quality of support institutions and the importance of the local leadership in driving the technological ladder on the right path despite existing constraints and unpredictable changes. In fact, the lack of technology content promotion in non-agile economies leads to wealth divide.

TECHNOLOGY CONTENT DISPARITY: ECONOMIC GROWTH DIVERGENCES

The measurement of technology content disparity is a difficult task. The problem of data availability, the selection of appropriate drivers, and the high-level of secrecy in information circulation on technology are some of the constraints faced while comparing economic growth among countries. Although the correlation between industrialized economies and technology-driven policies is not under question anymore, it is unclear why developing countries with large financial possibilities are still not on the path of upgrading the technology ladder. One of the most obvious explanations seems to be a clear neutralization of the protection of infant industries in non-agile

economies (Chang, 2002). Another explanation might be that technology development is not embedded in a systemic comprehensive approach. Two interdependent groups of requirements need to be proactively established to ensure a sustainable process towards technology content diffusion with direct consequences on shared economic growth. While highlighting some of the available drivers, it is suggested to focus on: technological capability formation and technological absorption and diffusion capacity.

Technological Capability Formation

Education especially in the tertiary group appears as a pertinent driver to access the technological capability formation of a region or a country. In 2006, the gross enrollment ratio in the tertiary level of education in High-income countries (HIC) was 13.4 times higher than in Sub-Saharan Africa (see Appendix A). When comparing countries, the same ratio for the United States of America (USA) was 16.4 times higher than Ghana and 13.6 times higher than Bangladesh. It is therefore not surprising to take note of major disparities in education efficiency. As an example in 2006, the primary school pupil-teacher ratio which is the number of pupils per teacher in High-income countries (HIC) was 2.93 times higher than in Sub-Saharan. Between USA and Ghana, the disparity is 2.26 times higher as compared to 3.65, the disparity level between USA and Bangladesh. Between USA and China, the disparity was lowered to 1.29. Between USA and Cuba, it is the USA which lags behind as the disparity of 1.4 clearly indicated that they are more teachers per pupils in Cuba. Amazingly, the gross enrollment ratio in the tertiary level of education in Cuba is also higher with 88% of the relevant age group than in the USA with 82% (see Appendix B). With 14.2% of the total Government expenditure on education in 2006, Cuba seems to do more than the USA with 11.7% in terms of Government commitment to support the technological capability formation[6].

This is confirmed with the higher level of public expenditure per student in the tertiary in 2006 in Cuba (34.5% of GDP per capita) than in the USA (27.6% of GDP per capita).

As it is the case for many African countries, Ghana did invest heavily in the public expenditure per student in the tertiary (209.4% of GDP per capita) but has a poor gross enrollment ratio in the Tertiary, 5% of the relevant age group in 2006. There is clearly a lack of consistency and coherence in the planning of technology diffusion and resilience in this country. Too much investment in physical infrastructure (school building without regular maintenance of appropriate equipments) and the brain drain pandemics could also be advanced as acceptable explanations of this contradictory situation. Transfer of technology does not take place smoothly in such an environment. It is of upmost importance to revisit the process of technological capability formation in Ghana. For Cuba and USA, it could be the appropriate entry point for renewed win-win cooperation between an agile and a non-agile economy.

Technological Absorption and Diffusion Capacity

Most of the poor people in non-agile economies are living in rural areas and depend heavily on agriculture. It is crucial to focus on the role of technological absorption and diffusion in that particular sector. The agricultural productivity driver enables to benchmark economies using the agriculture value-added per worker during 2003 and 2005 and measured in 2000 $US to the number of workers in agriculture. It is obvious that any inputs of technology contents or technology processes in the agricultural sector would drastically boost the level of productivity. With reference to the available data compiled by the World Bank (2008:138-141), the share of agricultural value-added (Appendix D) takes into consideration agriculture production and outcomes of forestry and fishing but does not properly reflect the land

productivity. The interpretation of this driver should be limited to comparison purposes.

The technology content disparity between regions and countries is huge and could partly explain the economic growth differences between non-agile and agile economies. In high income countries (HIC), agricultural productivity is 35.3 times higher than in MIC, 71.6 times higher than in LIC and 80.4 times higher than in Sub-Saharan Africa (Appendix D). There is a clear lack of technology content and processes in the agricultural sector in non-agile economies (low- and middle-income countries). The transfer of technology did not obviously take place in non-agile economies during the last five decades, even with the development aid. At the local level, the constraints of technological absorption and technological diffusion capacities could also explain the disparity between world regions. At the country level, Singapore, with a ratio of 40,323 in agricultural productivity per worker, is about to 'catch-up' with the USA, 41,797 (World Bank, 2008b:140). France with 44,017 has taken the lead among HIC. It must be put into perspective with the lowest level of two of the poorest countries in agricultural productivity: Burundi with 70, Eritrea with 61 (World Bank, 2008b:138). As part of humanitarian aid, transfer of technology is often not mentioned at all. It could be explained by many objective reasons such as war, civil internal unrest, lack of democracy and legitimacy, unclear land reforms, etc. In fact, technology diffusion was obviously not on the development aid agenda either. Sometimes, technology transfer was often mentioned but was hardly implemented on a sustainable basis. There is a donor's bias in defining technology diffusion, often mixed with palliative technology transfer. The donor's definition is radically different from the one of the people on the ground. It is too often limited to the transfer of donors' money to international consultants, preferably from the donors' country.

Nevertheless, it is the lack of technological absorption and diffusion capacities in those countries which should be primarily addressed in explaining these technological disparities. Among African countries, there are also important disparities. The relative positioning of Togo as compared to Ghana, Tunisia, South Africa or Nigeria (see Appendix D) are outcomes of strategies and policies. Countries which focus their development on exports of raw materials or minerals without structuring a productive capacity infrastructure to move the ladder up in the value chain production are also those where the technology absorption and the diffusion capacities are fragile.

Commitment and pro-active policies and strategies towards upgrading competencies and capability at local level were often simply bypassed. The main reason for that originates from the *copy-paste* syndrome of the obsolete myth of import-substitution and the transfer of technology bias considering technology as a gift. The today's new myth in non-agile economies focuses on 'trade creating sustainable wealth' without a proactive development of productive capacity and technological capability formation. The diffusion of technology appears also as another major difficulty. Based on the agricultural productivity ratio, China with 401, India with 392 did attract an impressive number of transnational companies in the agricultural sector. Nevertheless, the diffusion of technology seems to be limited to the boundaries of those firms or their supportive and productive networks. As an example, technology diffusion in India has been based on a comprehensive approach including education, agglomeration of productive structures (industrial clusters) and competencies, smart state intervention preceded by a watch system policy and improvement in wages distribution (Lall, 2000).

Technology Diffusion: From Public Domain to Public Goods

To enable a large impact, the diffusion of technology already in the public domain should gradually be considered as a public good. Both public and private decision-makers should integrate it culturally as a collective efficiency issue. A proactive role of the Government and a clear policy of interventionism (UNCTAD, 2009) should not be considered anymore from a dogmatic approach especially after the unsolicited role devoted to Government as *last resort interventionist* in limiting the negative collateral consequences of the 2008 financial crisis. Countries such as Tunisia (with 2,719 in terms of agricultural productivity ratio in 2006) or Venezuela (6,292 in agricultural productivity, in 2006) should be scrutinized in order to extract some best practices on agricultural productivity (see Appendix D) and strengthen cooperation on technology diffusion starting with those already in the public domain.

There is an indirect correlation between the consumption of electrical power and the rapid diffusion of technology in regions or countries. Consumption per capita in KWh in HIC 2005 was 5.06 times higher than in MIC, 24.96 times higher than in LIC and 18 times higher in Sub-Saharan Africa. The low level of Nigeria (127 KWh per capita) is a clear indication of mismanagement and a lack of effective commitment towards energy production and its use for technology production, absorption and diffusion. It was acknowledged that the transmission and distribution losses could be up to 24% as compared to Sub-Saharan average (542 KWh per capita) with only 9% of transmission and distribution losses. Nigeria's level is even lower than Bangladesh with 136 KWh per capita in 2005 (see Appendix D). It is interesting to see how Costa Rica with 1,719 KWh per capita is almost at the same level as China with 1,781 KWh per capita (World Bank, 2008b:304).

With those constraints in mind, it is important to mention that statistics focusing on High-technology exports expressed in million of US $ do not favor countries with a large internal market such as the USA, India or Nigeria. Nevertheless, it appears that regions and countries which are considered as the main producers and exporters in the global production networks system are also those who have or are in the process of mastering their technological absorption, production and diffusion process. From that angle, unit cost of labor and labor productivity could favor emergent countries. It is against those background that China had achieved $US 271,170 million in 2006 as compared to USA with $US 219,179 million in High-technology exports. It is not surprising that the share of manufactured exports in high-technology exports of both countries in the same year reaches 30%.

So, technology absorption, production, transfer, and diffusion seem to spread on a more rapid pace in China than in the USA. It is obvious that it cannot take place without an agile manufacturing sector. Thus, promoting trade without production as it has been advised to Low income countries especially African countries appears in the light of these examples as a bad advice in terms of technology upgrading. In fact, focusing on exporting goods and services without technological content or with low technology content appears as being a major mistake in the long run. It could become an unworkable barrier to overcome, at least within a five year policy orientation program. Ghana has achieved only $US 1 million of High-technology exports in 2006 as compared to South Africa with $US 1,799 million. These performances should directly be correlated to the lack of manufactured exports in High-technology exports (0% for Ghana in 2006) as compared to South Africa with 6% (see Appendix D).

The direct correlation between technology content and industrialization cannot be ignored in non-agile economies and should be addressed to secure sustainability in socio-economic progress.

Technology Content and Industrialization

One could assume that future emerging economies are those who clearly have a technology content-driven policy with the objective to achieve 20% of their share of manufactured exports in High-technology exports. Emerging economics are those whose share of MVA in GDP exceeds the average of their respective economic regions. Manufacturing value-added is the total of gross output less the value of intermediated inputs used in production for productive capacities. To achieve the suggested 7% economic growth to reach the United Nations Millennium Development goals targets (reducing extreme poverty by half in 2015), an average of 17% of MVA in GDP is required (Amaïzo, Atieno, McCormick & Onjala, 2004). For non-agile economies, those goals cannot be achieved without a proactive and technology-driven policy and technology diffusion implementation scheme. Manufacturing should be redefined in a broader sense especially in non-agile economies as the process of building productive structure for long-term resilience. It cannot happen without a cumulative technology content process.

Manufacturing is the proactive process of injecting technology content at all levels of the value chain production while transforming raw materials, components and processes, full or in parts, into finished products that is of value to the market. The introduction of technological contents is part of the division of labor process and cannot take place without a clear and measurable productivity effect. The manufacturing process cannot be limited to the objective of taking advantages of economy of scale and low wages in non-agile economies. Otherwise, it will favor the obsolete and wrong transfer-substitution approach of technology which is partly responsible of the 1980's debt trap in Africa. Basically, turn-key-factories or technologies were bought and implemented with no clear terms of reference on technology absorption, diffusion and capability formation. This danger does exist for the transfer of technology in the nano- and micro-technologies sectors. Export and trade in non-agile economies without a local operational technology content program are clearly not advisable as a long term policy. Wealth creation on a sustainable basis is a systemic and dynamic process which cannot bypass technology content spill-over and resilient productive structure. Wealth creation without an industrialization process based on technology content can only lead to palliative measures (Reinert, 2007). Additional drivers which are taking into consideration the diffusion of technology within an existing market (national large market, regional trade free areas, etc.) should complement the two drivers suggested in this paper, namely high-technology exports and share of manufactured exports in High-technology exports.

Selected drivers such as the number of scientific and technical journals available in a country could provide an idea of the free circulation of knowledge and favor knowledge/technology capturing and diffusion. With the internet, nobody is stopped to access all 'free to air' knowledge or technology available in the public domain. So, it is the infrastructure of capturing technology in the public domains which needs to be created or strengthened. In addition, a system of alert becomes unavoidable in the promotion technology absorption and diffusion. Using a shortcut, the level of royalty and license fees received and paid, the number of patent applications filed by residents and non-residents actors, the number of trademarks registered with residents and non-residents actors are some of the drivers which illustrate the agility of an economy in acquiring and diffusing technological knowledge and applications. Royalties and license fees received in LIC are 392.8 times less important than in HIC (see Appendices 3 and 4). This is clearly an acknowledgement that transfer of technology does not take place as a donation, nor as a top-down aid-driven phenomenon. It must be embedded in a real proactive and aggressive government interventionism in promoting technology content at all level of the productive system. It requires the full participation of vibrant active private sector actors. Fiscal policy could help ensuring that the diffusion of technology of non-resident actors benefits on a sustainable basis the country or region.

Nevertheless, fiscal policy should not become, as it is the case in several imperfect democratic countries, an additional burden on those economic actors active in transforming patent applications into technological applications or trademarks into productive businesses.

When the production of knowledge is limited or the knowledge is already available (with or without costs), all forms of incentives geared to attract technologies should be promoted. A real conducive business environment is required in terms of communication infrastructure. Selected drivers focusing on basic information technology requirements are also mentioned as indicators of an economy's ability to attract freely and diffuse rapidly technology and related knowledge. The affordability and efficiency of the energy and the communication networks system (telephone, Internet) is another important criteria of technology diffusion and diffusion. The related cost and the security of the system are some of the specific criteria which could by nature discourage any collective development based on technology diffusion in an economy. Although there is no clear convergence in the price of mobile telephone, cost to access the Internet between HIC and LIC, the issue of securing internet servers and protecting data content are real challenges. Only two in LIC versus 445 Internet servers per million people in 2007 in HIC are considered as secured and registered as such (Appendix C). Any progress in this area is directly linked to the improvement of freedom and democracy in the countries. Technology divide does exist and technological poverty should be considered as a new area for consideration by development stakeholders.

One needs to ascertain that technology diffusion and resilience is then considered as an area of intervention for international public donors. The disparities between Ghana or Nigeria and USA (Appendix D) should be benchmarked with Ghana's improvement over the years in the West African regions. For many non-agile economies, the lack of availability and accuracy of data is acknowledged. Absorption and diffusion capabilities become reality when policies, institutions and learning process generate a unique capability formation environment.

TECHNOLOGY-DRIVEN POLICIES, INSTITUTIONS AND CAPABILITY FORMATION

In order to promote all forms of support activities and institutions favoring technology-driven capabilities, a new approach in technology diffusion is required. Generation and diffusion of technologies have more chances to be implemented in a Public-Private Partnership environment. Policy instruments supported by clear and concerted immediate objectives should gradually become the basis for

the building of agglomerated knowledge. Those policies should be supported by the introduction of technological processes and contents at all level of an innovative development approach focusing on sustainable wealth creation.

Four main layers of policy responses could be offered to promote technology diffusion and resilience:

It is important not to rely exclusively on transnational corporations and foreign direct investment as the main source of technology diffusion in non-agile economies. This was yet the main channel of innovation in high-tech industries such as nano-technologies, electronic and ICT in general. The direct correlation between the improvement of High-technology exports and especially the share of manufactured exports in High-technology exports for several newly-

Table 1. Technology diffusion

Technology Diffusion Policy guiding principles for the diffusion of technology content		
Layers of Public-private partnerships in technology	**Immediate objectives**	**Policy catalyst instruments**
1. Technology generation and improvement	Exploitation and incentives for endogenous technologies, innovations and improvement processes	Selective incentives: Favoring the learning process, the identification of free available technologies and knowledge
2. Attracting and diffusing Foreign technology capabilities	Identification and organization of local capabilities and competencies as a source for external productive capacities to ensure resilience and diffusion at local both local and global levels	Tax facilities and negotiation for technologies that are upgrading local technology infrastructure and innovative institutions. Introduction of medium and long term concessional arrangements with special incentives for fair contractual proposals
3. Building of Techno-scientific Public-private productive communities focusing on technology content development	Upgrading technology content in productive capacities and productivity enhancement processes based on clear linkages between knowledge centers (including universities), productive ventures, support institutions and Government	Contractual tax forgiveness arrangement versus commitment to the creation of decent jobs and entrepreneurship entities
4. Institutionalization of technological and scientific linkages among ethics actors	Enabling economic stakeholders to participate in the agglomeration and the clustering of productive networks in support to technology enhancement and resilience Specializing local competences and promote aggressive marketing policy towards the identification of an Anchor region, country, sector, firms and knowledge centers	Ensuring an effective competitive conducive environment for regular cost reduction and improved services for users Reorientation of the training and education programs towards technology acquisition, entrepreneurship and innovation

industrialized and emerging countries may explain the overall consensus on the diffusion of this approach as a policy best practice.

The development path of weak-technology driven countries cannot be a duplication of the path of newly industrialized countries. The comparative advantages and the excessive specialization in production and exports might lead to severe difficulties during economic crisis with serious consequences on job destruction and decent job creation. The overall principle of adapting to change in an unpredictable environment should be supported by an institutionalized system of capability formation. The immediate objective is to enable systemic adjustment of the technology infostructure and infrastructure and its innovation process regularly with the market requirements. It is argued that this is possible through the adjustment of the technology content defined here as the smallest building object in technological enhancement.

The challenge is to avoid a window-dressing approach usually published in the media and move more deeply in the absorption and integration process at the local level. It is also important not to forget about the primarily objective of increasing productivity at and for the structuring and resilience of the local or regional technological capabilities. It is therefore crucial to limit the brain-drain of the Diaspora while keeping technology knowledge at local level. Special incentives to attract the Diaspora back home should become an integrated part of comprehensive negotiated circular migration agreements between agile and non-agile economies. Specific legal agreements must be in place, if necessary as part of global networks with global international lawyers, to avoid that the exploitation of local generated technologies and innovations by international transnational companies is not simply bypassing the local economies in terms of knowledge sharing. From that perspective, a clear legal framework should be prepared and anticipated, if necessary as part of international global agreements,

when innovation and technology diffusion takes place at global level. It is a matter of reducing or limiting the inequality divide while pursuing scientific and technological collaboration between agile technology-driven economies and non-agile technology-driven economies. It is crucial to focus the diffusion of technology through small and medium entrepreneurs and enterprises (SMEE). Often, the technological linkages between SMEE in technology advanced and less technology advanced countries are poor and must be institutionalized. The role of high profile global buyers is becoming contraproductive for poor economies whose infant industries' development is simply frozen. Because of competitive pressures, many fragile technology-driven economies can only compete at the margin (UNCTAD, 2005:12) and sometimes keep exporting no- or low technological content goods.

The overall export oriented approaches need to be re-discussed on a case- by-case basis in non-agile States, most of them no-technology-driven economies. The building of national technology capabilities as well as national innovation capability systems should become an integrated part of a compulsory development package. Acquiring, anticipating and mastering technology as part of the productivity and performance improvement at national and regional levels need a real commitment from decision-makers. It needs to be approved by democratically elected policy makers to ensure legitimacy. The Government cannot be neutral as claimed in the 'invisible hand economic theory[7]' (Aydinonat, 2008). A proactive role of States is required especially in providing incentives for the creation of sustaining technology institutions. Those technology intermediaries do not need to be created by the Government (UNCTAD, 2005:13). It needs regular Public-Private negotiation arrangements between decision makers (Speser, 2006). As part of a system of protection of local people's interests, those arrangements may be developed as a national or regional innovation observatory or watch system. It cannot bypass the promotion, the attracting of R&D activities, the building of appropriate infrastructure including equipment, people, information and freedom. It is obvious that State interventions must be smart interventions. Policy instruments are numerous but are not all are required. At last the capabilities of decision makers to select wisely the appropriate approach, technology and the prospect to generate resilience and improve performance locally in the interest of the population need to be strengthened.

Succeeding in technology diffusion is not a matter of a few top-down political decisions or the goodwill of selected transnational companies. It needs the interdependency of the proactive and agile dimensions of innovation while using wisely all natural and human resources to achieve productivity and performance with the objective to benefit a large majority of concerned populations.

Any alternative development road map should consider productivity and performance improvement as part of the technology diffusion and resilience system. A regular benchmarking system should be established to ensure convergence with the regional average situation and progress while benchmarking against neighbors' economies and selected anchor economies. These are practical ways to support progress in diffusing technology content at local level.

Paradoxally, technology diffusion may not necessary have a compulsory development path into high-technology. Sustainability could simply arise with the move into innovative, dynamic and permanent technology content upgrading process with qualified competencies including those of the Diaspora. Furthermore, it is not obvious how a country with a poor budget and a lack of human capital in appropriate magnitude could access appropriate technology symbols (royalties, licenses, patents, trademarks, etc.). The dream of leapfrogging into dynamic high-tech activities such as nano-technologies and electronics must be thoroughly prepared and sustained with regular policy incentives and fiscal advantages. Without this overall conducive environment, weak tech-

nology-driven economies could take advantages from existing technologies as they may not even have the appropriate infrastructure to capture and prospect, adapt and diffuse technology available for free in the public domain. Using emerging advanced technology might be easier, but costly. It may not necessary favor the overall process of capability formation supported by an adequate technology-driven policy on education and continuous learning.

Social considerations are necessary to ensure that outcomes from technology diffusion are widely, wisely and equally shared among all citizens.

CONCLUSION: TECHNOLOGY DIFFUSION: COLLECTIVE EFFICIENCY AND SHARED WEALTH CREATION

Technology diffusion cannot be reduced to the subcontracting or outsourcing of technology acquisition even if means of payment are available. It is a mix of policy commitments in support to the promotion of technology content at all levels of development activities. It is important to have an innovation strategy based on home-led interdependency while competing in a global environment. An institutionalization of public-private technology negotiation processes and a voluntary peer-review benchmarking mechanism could foster collective efficiency. Technology diffusion is a multi-stakeholder partnership coupled with international and endogenous processes of capturing and implementing best practices. It favors an effective and smart transfer of brain gains between knowledge centers and productive structures. As a result, interdependency between on one side, private and public actors and on the other side, institutions and enterprises which are promoting low, middle and high-tech content technologies implementation processes and productivity could result in collective efficiency.

It is therefore recommended to move away from the catch-up syndrome or the closing-gap approach which appears more as a symptom of the problem than a solution to the problem. The alternative approach is to look for convergence in a given sector or in a region as a means to strengthen regional integration first. The concept of transfer of technology should be replaced by technology diffusion and resilience. The latter concept may have more chances to promote interdependency and favor a democratic process of technology diffusion. The capability formation environment needed should be centered around a technology-driven capability and embedded in strategic policy and institutionalized institutions' support.

Technology diffusion cannot really take off without clear technology absorption capacities and capabilities. Even, if the diffusion process is in place, it does not necessary guarantee a successful result (Criscuolo & Narula, 2001) because of risks of bias and misuses towards non-resilience goals. The role of information, communication, training and learning is crucial and should be embedded in local culture. The tolerance and the time made available for the learning by mistakes and experimentation as well as the commitments of local authorities to establish and strengthen dedicated institutions should also be an integrated part of the technology diffusion package. A conducive policy of incentives on technology diffusion including rewards and recognition for the human capital including the Diaspora should be clearly advocated. It is in fact this knowledge agglomeration process which could unlock to the brain-drain drama (Brinkerhoff, 2006b). Technology diffusion could become a smart solution if included in rotation migration policies and contribute to attract both remittances and returning migrant people (Brinkerhoff, 2006a).

It was acknowledged that for non-agile economies, '*weak absorptive capacity may constrain further technological progress*' (World Bank, 2008a). The integration process is also an agglomeration process. Pro-active measures favor-ing technology diffusion should strive to create and strengthen linkages between the knowledge institutions (R&D, universities, innovation centers, etc.) and productive structures namely firms or entities generating productive employments. For these reasons, the learning and knowledge upgrading institutions should be considered as focal points for knowledge sharing and spreading.

The challenge for non-agile economies in succeeding in their technology diffusion strategy supposes a cultural change in mindsets on performance and collective efficiency. Whenever major strategic decisions are at stake, leaders should be requested to focus on introducing technology content inputs as a mean to ensure positive results out of interactions between development processes, institutions and responsible individuals. Diffusion technologies at national level should be part of a new strategic management in non-agile economies (Cohen, 2004). This supposes the development of local capabilities, able to respond on time to change through an improved and transparent management process, especially for the negotiation of long-term concessional arrangements contracts. In order to ensure resilience, a workable cumulative approach based on the diffusion of knowledge and technology must be gradually established and developed as a means to transfer to the next generation effective development building blogs. Innovation is part of an economic prosperity strategy (Nastas, 2007) and is determinant in structuring the economy's competitive advantage (Porter, 1990).

Technology diffusion and resilience is in principle eased when the GDP growth rate as well as the share of manufactured value-added in GDP are both high, usually above the regional average (Amaïzo, Atieno, McCormick & Onjala, 2004). If the role of foreign direct investment is fundamental in achieving high productivity gains, it is the technology diffusion and resilience which made it sustainable. It cannot take place without a comprehensive support to both entrepreneurship (Libecap, 2005) and intrapreneurship and an endogenous-driven technology 'transfer' process (Jolly & Creighton, 1977). With those structures in place, technology diffusion in return improves the level of the investment and the economy's diversification with a positive impact on the country's performance (Lall, 2000). It could be substantiated with a rapid credit expansion, a better involvement of foreign and local financial institutions in risk taking in the productive sector and a larger remittance inflow from workers overseas. Technology diffusion made countries less vulnerable to exogenous shocks or crisis (World Bank, 2008a) when strategic priorities are focusing on productive structures (Reinert, Amaïzo & Kattel, 2010, forthcoming) rather than on speculative investments. From that perspective, the improving level of South-South cooperation may slowly move from the usual first generation of trade relations with little value-added and low level of technology content to a second generation of South-South win-win cooperation based on a clear commitment to industrialization and development of productive structures. Technology diffusion among emerging agile economies and non-agile economies will depend heavily on government effective commitment to technology content driven investment, low level of corruption and effective participation of local people in the process. The agglomeration process will then become borderless. A systemic approach on collective efficiency may gradually replace the existing industry-centered-analysis founded on direct and indirect effects (Schiff & Yanling, 2006).

Technological progress and economic prosperity are directly correlated with an increase in total factor productivity and gross national income per capita over the years (at least 10 years). The catch-up process between non-agile economies and agile economies could be compared in terms of convergence or divergence in per capita income gains over the years. Should the technology diffusion infrastructure takes roots, the per capita income of non-agile economies should rapidly grow faster than in agile economies. It is nevertheless not sure

that this good performance is well shared among people in the same country or between countries (World Bank, 2008).

The slowness of technology diffusion from cities to rural areas, as well as from the rich to the poor (UNCTAD, 2007b), is the challenge which needs to be considered seriously as available indicators and drivers only provide a partial view of the technology progress and its impact on the population. Available indicators offer a quantitative approach on technology whereas the qualitative approach and the effective impact and acceptance by the population are often what is critical. Over the longer term, the economic resilience of non-agile economies, most of them being weak industrialized countries and late industrial economies (Amsden, 2001), will be tested through their ability and commitment to promote a comprehensive technology diffusion and resilience policy. Based on a clear vision and strategy and under such circumstances, poverty reduction has some chances to be replaced by wealth creation as technology diffusion and resilience spread and spill-over. Using new and modern technology to accelerate growth in non-agile economies should not be discredited because of past numerous failures such as *'bad luck, poor governance, flawed policy design and poor implementation'* (Chandra and Kolavalli, 2006).

In fact, *'catching up, forging ahead and falling behind'* (Abramovitz, 1986) is not a zero-sum development competition. Governments may have to review the way they treated the private sector as it contributed to limit pro-active initiatives dedicated to productivity. Transnational companies may face difficulties in sustaining their profit if they do not proactively revise their *'business as usual'* model along more social and ecological considerations. They should be prepared to focus more on ethics and international best practices and avoid underestimating the value of local people. For many multinational firms, deviations from ethical behaviors are simply considered as *'normal'* in non-agile economies as possible sanctions could easily be neutralized with corruption. Technology diffusion approaches could be of help in reducing corruption. With the rapid technological changes, more possibilities to access new advanced technologies exist. The absorption capability should not exclusively be the responsibility of the public sector. Capability formation could also be proactively generated by private actors, foreign as well as local (Amaïzo, 2004). The leapfrogging across several stages of technology development is achievable and requires an effective structuring of the technology diffusion process. As already stated in 1962, '*differences in nations' ability to innovate technology and to adapt it to their particular circumstances were the primary causes of differences in per capita income between countries, and that the ability to appropriate what other had innovated was the essence of the latecomers' advantages*'Gerschenkron (1962). Non-agile economies have the opportunities to capture these advantages if they collectively opt for smart technology content driven policies.

How to increase resilience in non-agile economies could become a new area of research which cannot be limited to the traditional and deregulated '*opening of market*' dogma sometimes known as '*market fundamentalism*' (Kozul-Wright and Rayment, 2007). Technology diffusion could prevent non-agile economies from becoming addicted to the market as a slave (Chang et Al., 2010 forthcoming). Regulatory barriers could impede technology diffusion but it could also improve technology diffusion. Restrictions on labor mobility and reallocation could slowddown the process of acquiring and accessing new technologies. But, they could also take away all available skilled labor from a non-agile economy to an agile economy. This contribute to structure '*cascading fragilities*' (Reinert, 2007), and limiting room for endogenous innovation and progress.

Last but not least, existing indicators and drivers do not take into consideration techno-

logical content which was part of the production process. Benchmarking maize production and export between non-agile and agile economies could be misleading as the former still produce with almost no- or low-content technology while the latter produces with advanced sophisticated nano-technologies which enable increasing returns on productivity[8]. From that perception, resiliency could be redefined as how a non-agile economy plans and remains in control of productivity and shares under a competitive process, increasing returns among its population while achieving technology progress, well-being enhancement and a convergence in shared economic growth per capita.

Resilience should contribute to end the vicious circles of cascading fragilities in non-agile economies, enable a transformation change from divergence to convergence with agile economies. This should not anymore be considered as a catch-up process. The dynamic of failures might then be neutralized. The strategic commitment of leaders to promote shared wealth creation and technology diffusion may then produce collective efficiency. The best transfer paradigm is unique for each form of agglomeration, but it does not prevent from collecting best practices across the globe while adapting, innovating and spreading knowledge with reference to each local or particular circumstance. Speed, efficiency and capacity building have no future in cultural environment where the ownership process of technology diffusion by concerned people has no endogenous roots. For non-agile economies, the use of alliances with appropriate anchor economies and smart strategic partnerships in technology diffusion are critical to leapfrogging with a minimum of risks. The beyond-technology-transfer era is emerging and appears as a change transformation process which needs alternative economic considerations. Non-agile economies should not miss it.

REFERENCES

Abramovitz, M. (1986). Catching up, forging ahead, and falling behind. *The Journal of Economic History, 46*(1), 385–406. doi:10.1017/S0022050700046209

Amaïzo, Y. E. (1998). From Dependency to Interdependency. Globalization and marginalization: any chance for Africa. Paris, France: L'Harmattan Publisher.

Amaïzo, Y. E. (2004). *Global Value Chains and Production Networks: Promoting Capability formation in South Africa. Discussion Paper.* Dept of Trade and Industry, Ministry of Industry, South Africa: Competitiveness Conference in cooperation with UNIDO. Pretoria, South Africa, 7-11 June. Retrieved July 2, 2009, from http://www.thedti.gov.za/invitations/unidoinvitation.htm

Amaïzo, Y. E., Atieno, R., McCormick, D., & Onjala, J. (2004). *African Productive capacity Initiative: From vision to action.* Vienna, Austria: UNIDO Policy Papers.

Amsden, A. (2001). *The Rise of the Rest: Challenges to the West from Late-Industrializing Economies.* New York: Oxford University Press.

Aydinonat, N. E. (2008). The Invisible Hand in Economics: How Economists Explain Unintended Social Consequences. New-York: Routledge.

Barro, R. J., & Lee, J.-W. (2000). International Data on Education Attainment: Updates and Implications. *NBER Working Paper 7911.* Cambridge, MA: National Bureau of Economic Research.

Berkun, S. (2005). Making Things Happen. Mastering Project Management: Mastering Project Management (Theory in Practice) (Rev. Ed.). Sebastopol, CA: O'Reilly Media, Inc.

Brinkerhoff, J. M. (2006a). Diasporas, Mobilization Factors, and Policy Options . In Wescott, C., & Brinkerhoff, J. (Eds.), *Converting Migration Drains into Gains: Harnessing the Resources of Overseas Professionals* (pp. 127–153). Manila, Philippines: Asian Development Bank.

Brinkerhoff, J. M. (2006b). Diasporas, Skills Transfer, and Remittances: Evolving Perceptions and Potential . In Wescott, C., & Brinkerhoff, J. (Eds.), *Converting Migration Drains into Gains: Harnessing the Resources of Overseas Professionals* (pp. 1–32). Manila, Philippines: Asian Development Bank.

Chandra, V., & Kolavalli, S. (2006). Technology, Adaptation, and Exports: How Some Developing Countries Got It Right: How Some Countries Got It Right . In Chandra, V. (Ed.), *Technology, Adaptation, and Exports* (pp. 1–47). Washington, DC: World Bank Publications. doi:10.1596/978-0-8213-6507-6

Chang, H.-J. (2002). *Kicking away the ladder. Development Strategy in Historical Perspective.* London: Anthem Press.

Chang, H.-J. (2010). *Market as means rather than master: Towards new developmentalism.* London: Routledge Ed.

Cohen, G. (2004). *Technology Transfer. Strategic management in Developing countries* (1st ed.). New Delhi, India: SAGE Publications Pvt. Ltd.

Collier, P. (2007). The Bottom Billion: Why the Poorest Countries are Failing and What Can Be Done About It (1st Ed.). New-York: Oxford University Press.

Criscuolo, P., & Narula, R. (2001). *A novel approach to national technological accumulation and absorptive capacity: Aggregating Cohen and Levinthal.* Oslo, Norway: University of Oslo.

Gerschenkron, A. (1962). *Economic Backwardness in Historical Perspectives*. Cambridge, MA: Belknap Press.

Hosni, Y. A., & Khalil, T. M. (Eds.). (2004). *Management of technology. Internet Economy: Opportunities and Challenges for Developed and Developing Regions of the World.* Selected Papers from the 11th International Conference on Management of Technology, Oxford, UK: Elsevier Ltd.

Jolly, J. A., & Creighton, J. W. (1977). The technology transfer process: Concepts, framework and methodology. *The Journal of Technology Transfer, 1*(2), 77–91. doi:10.1007/BF02622191

Kagame, P. (2009, May 7). Africa has to find its own road to prosperity. *Financial Times*.

Kozul-Wright, R., & Rayment, P. (2007). *The Resistible Rise of Market Fundamentalism: The Struggle for Economic Development in a Global Economy*. London: Zed Books.

Lall, S. (2000). The technological structure and performance of developing country manufacturing exports, 1985–98. *Oxford Development Studies, 28*(3), 337–369. doi:10.1080/713688318

Libecap, G. (Ed.). (2005). *University Entrepreneurship and Technology Transfer: Process, Design, and Intellectual Property. Advances in the Study of Entrepreneurship, Innovation. Innovation and Economic Growth*. London: Jay Press.

Moyo, D. (2009). *Dead Aid: Why Aid Is Not Working and How There Is a Better Way for Africa*. London: Penguin Group.

Nastas, T. (2007). *Scaling up Innovation: The Go-Forward Plan to Prosperity*. Washington, DC: World Bank Institute.

Ocampo, J. A., Jomo, K. S., & Vos, R. (2007). *Growth Divergences: Explaining Differences in Economic Performance*. London: Zed Books.

Porter, M. (1990). *The Competitive Advantage of Nations*. London: Macmillan Publications.

Reinert, E., Amaïzo, Y. E., & Kattel, R. (2010). The Economics of Failed, Failing and Fragile States: Productive Structure as the Missing Link. In Chang, H.-J. (Eds.), *Market as means rather than master: Towards new developmentalism*. London: Routledge Ed.

Reinert, E. S. (2007). *How Rich Countries Got Rich... and Why Poor Countries Stay Poor*. London: Constable & Robinson Ltd.

Rodrik, D. (2007). *One Economics, Many Recipes*. Princeton, NJ: Princeton University Press.

Schiff, M., & Yanling, W. (2006). North–South and South–South trade-related technology diffusion: An industry-level analysis of direct and indirect effects. *The Canadian Journal of Economics. Revue Canadienne d'Economique, 39*(3), 831–844. doi:10.1111/j.1540-5982.2006.00372.x

Soludo, C. C., Ogbu, O., & Chang, H. J. (2004). The Politics of Trade and Industrial Policy in Africa: Forced Consensus? Ottawa, Canada: Africa World Press and the International Development Research Centre, Speser, P. L. (2006). The Art and Science of Technology Transfer (1st Ed.). Hoboken, NJ: John Wiley and Sons.

UNCTAD. (1995). *Bridging the technology gap*. CSTD Issues Paper, Inter-sessional panel, UNCTAD Commission on Science and Technology for Development, 10-12 November. Retrieved April 16, 2009, from http://www.unctad.org/en/docs/ecn162006crp1_en.pdf

UNCTAD. (2007a). International Investment Rule-setting: Trends, Emerging Issues and Implications. Note by the UNCTAD Secretariat, Commission on Investment, Technology and Related Financial Issues, Geneva, 8-14 March 2007, TD/B/COM.2/73. New York: United Nations.

UNCTAD. (2007b). *The Least Developed Countries 2007: Knowledge, Technological Learning and Innovation for Development*. Geneva, Switzerland: UNCTAD.

UNCTAD. (2009). *The Least Developed Countries 2009: The State and Development Governance*. Geneva, Switzerland: UNCTAD.

World Bank. (2008a). *Global Economic Prospects: Technology diffusion in the Developing World*. Washington, DC: World Bank.

World Bank. (2008b). *World Development Indicators 2008*. Washington, DC: The World Bank.

Wright, R. (2001). *NonZero: The Logic of Human Destiny*. New-York: Vintage Books.

ENDNOTES

[1]   Moyo D. (2009). *Dead Aid: Why Aid Is Not Working and How There Is a Better Way for Africa*. London. UK: Penguin Group.

[2]   Collier, P. (2007). *The Bottom Billion: Why the Poorest Countries are Failing and What Can Be Done About It*. 1st edition. New-York, USA: Oxford University Press.

[3]   It is the years of formal schooling received, on average, by adults over age 15 (Barro & Lee, 2000).

[4]   World Bank, *World Development Indicators 2005*. Washington. D.C.

[5]   The G8 meeting in L'Aquila, Italy ended in July 2009, just like the summit at Gleneagles in 2005, with promises for Africa: A pledge of US$ 20 billions from the world's industrialized economies to support food production for farmers through 'investment aid' in non-agile economies over three years.

[6]   The financing of education in USA is largely taken over by private entities. Thus, one should be cautious while comparing USA and non-agile economies.

[7]   Also referred to as 'neoclassical economics', the invisible hand economic theory was first proposed by Adam Smith in the 18th century.

[8]   For example, maize is produced in agile economies with *'sophisticated crop rotation methods, enhanced irrigation and fertilization strategies based on satellite imaging, and bioengineered seeds'* (World Bank, 2008a:60).

APPENDIX A: TECHNOLOGICAL CAPABILITY FORMATION

Table 2. Selected world regions

Selected drivers	Sub-Saharan Africa	Low-income countries (LIC)	Middle-income countries (MIC)	High-income countries (MIC)
Education efficiency: Primary school pupil-teacher ratio (pupils per teacher), 2006	47	41	20	16
Education efficiency: Public expenditure on education as % of GDP), 2006	4.2	..	4.3	5.4
Education efficiency: Public expenditure on education (as % of Total Government expenditure), 2006	..	..	..	12.5
Public expenditure per student in the tertiary (as % of GDP per capita), 2006	..	..	25.9	29.0
Expenditures for R&D (as % of GDP), 2000-05*	..	0.57	0.85	2.38
Gross enrollment ratio in the Tertiary (as % of the relevant age group)**	**5**	**9**	**27**	**67**
Researchers in R&D (per million people), 2000-05*	..	..	803	3,731
Qualitative innovation in Business signaled by the ISO certification ownership (as % of quality certified firms)	..	..	..	..
.. Not available; *Most recent years available between 2000 and 2005; ** Provisional Data Source: World Bank, *World Development Indicators 2008*, pp. 76-78, 312-314; 80-82; 272-274.				

APPENDIX B: TECHNOLOGICAL CAPABILITY FORMATION

Table 3. Selected countries

Selected drivers	Africa					Latin America and Caribbean			Asia			North America
	Ghana	Nigeria	South Africa	Togo	Tunisia	Brazil	Cuba	Vene-zuela	Bangla-desh	China	India	USA
Education efficiency: Primary school pupil-teacher ratio (pupils per teacher), 2006	32	37	36	34	20	21	10	17	51	18	40	14
Education efficiency: Public expenditure on education (as % of GDP), 2006	5.4	..	5.4	..	7.3	8.7	9.1	3.7	2.5	..	3.8	5.4
Education efficiency: Public expenditure on education (as % of Total Government expenditure), 2006	..	..	17.6	..	20.8	..	14.2	..	14.2	..	..	11.7
Public expenditure per student in the tertiary (as % of GDP per capita); 2006**	209.4	..	50.1	..	56.4	32.6	34.5	34.3	49.4	..	61.0	27.6
Expenditures for R&D (as % of GDP), 2000-05*	..	..	0.87	..	1.03	0.91	0.56	0.25	..	1.34	0.61	2.68
Gross enrollment ratio in the Tertiary (as % of the relevant age group), 2006**	5	10	15	..	30	24	88	52	6	22	11	82
Researchers in R&D (per million people), 2000-05*	..	..	379	..	1,450	462	..	..	..	708	..	4,605
Qualitative innovation in Business signaled by the ISO certification ownership (as % of quality certified firms)***	6.0 in 2007	..	42.4 in 2003	..	..	19.1 in 2001	..	12.5 in 2006	7.8 in 2007	35.9 in 2003	22.5 in 2006	..

..: Not available;
*Most recent years available between 2000 and 2005;
** Provisional Data.
*** Representative sample of the non-agricultural economy, excluding financial and public services;
Source: World Bank, *World Development Indicators 2008*, pp. 76-78, 312-314; 80-82; 272-274.

APPENDIX C: TECHNOLOGICAL ABSORPTION AND DIFFUSION CAPACITY

Table 4. Selected world regions

Selected drivers	Sub-Saharan Africa	Low-income countries (LIC)	Middle-income countries (MIC)	High-income countries (MIC)
High-technology exports (in $US million), 2006	..	..	478,215	1,322,714
Share of Manufactured exports in High-technology exports (%), 2006	..	6	20	21
Royalty and license fees (receipts (R) and payments (P) in $US million), 2006	R=1,417 P=1,471	R=334 P=1,163	R=3,743 P=22,719	R=131,201 P=124,636
Scientific and technical journal articles (number of), 2005	3,563	16,711	112,719	578,656
Patent applications filed (Residents (R) versus Non-Residents (NR)), 2005	R=16 NR=5,554	R=364 NR=267	R=132,662 NR=137,246	R=782,572 NR=415,654
Trademark registered with national or regional trademark office (Residents (R) versus Non-Residents (NR)), 2005	R=439 NR=28,750	R=1,157 NR=2,884	R=898,687 NR=200,348	R=684,902 NR=217,497
Electric power (consumption per capita in KWh), 2005	542	391	1,928	9,760
Fixed line (F) and mobile telephones (M): Affordability and efficiency (Price basket in $US per month), 2006	F=11.6 M=12.3	F=6.1 M=10.0	F=9.2 M=10.2	F=26.6 M=17.0
Internet users (number per 100 people), 2006	3.8	4.2	14.1	59.3
Affordability of Internet (Price basket in $US per month), 2006	15.9	12.0	11.2	13.7
Information and communication technology expenditures (as % of GDP), 2006	..	6.1	5.1	7.2
Secure Internet servers (per million people), Dec. 2007	2	0	5	445
Agricultural productivity (agriculture value-added per worker), 2003-2005 in 2000 $US	335	376	763	26,940

.. Not available;
Source: World Bank, *World Development Indicators 2008*, pp. 138-140, 304-306, 308-310, 312-314.

APPENDIX D: TECHNOLOGICAL ABSORPTION AND DIFFUSION CAPACITY

Table 5. Selected countries

Selected drivers	Africa					Latin America & Caribbean			Asia			North America
	Ghana	Nigeria	South Africa	Togo	Tunisia	Brazil	Cuba	Venezuela	Bangladesh	China	India	USA
High-technology exports (in $US million), 2006	1	..	1,799	0	344	8,426	59	80	21	271,170	3,511	219,179
Share of manufactured exports in High-technology exports (%), 2006	0	..	6	0	4	12	12	2	0	30	5	30
Royalty and License fees (receipts (R) and payments (P) in $US million), 2006	R=0 P=..	R=.. P=45	R=46 P=1,282	R=0 P=3	R=14 P=11	R=150 P=1,664	R=.. P=..	R=0 P=..	R=0 P=5	R=205 P=6,634	R=112 P=949	R=62,378 P=26,433
Scientific and technical journal articles (number of), 2005	81	362	2,392	..	571	9,889	261	534	193	41,596	14,608	P=205,320
Patent applications filed (Residents (R) versus Non-Residents (NR)), 2005	R=.. NR=..	R=.. NR=..	R=.. NR=5,554	R=.. NR=..	R=56 NR=282	R=3,821 NR=2,560	R=94 NR=191	R=.. NR=..	R=.. NR=..	R=93,172 NR=80,155	R=6,795 NR=10,671	R=202,776 NR=187,957
Trademark registered with national or regional trademark office (residents versus non-residents), 2005	R=.. NR=..	R=.. NR=..	R=.. NR=28,331	R=.. NR=..	R=.. NR=..	R=83,117 NR=15,981	R=301 NR=482	R=.. NR=..	R=.. NR=..	R=593,382 NR=63,902	R=.. NR=..	R=224,269 NR=28,359
Electric power (consumption per capita in KWh), 2005	266	127	4,847	94	1,194	2,008	1,152	2,848	136	1,781	480	13,648
Fixed line (F) and mobile telephones (M): Affordability and efficiency (Price basket in $US per month), 2006	F=9.8 M=7.0	F=.. M=10.7	F=22.7 M=13.8	F=15.4 M=12.1	F=2.9 M=5.3	F=15.6 M=26.2	F=13.1 M=22.6	F=.. M=1.2	F=4.0 M=2.6	F=.. M=2.9	F=3.3 M=2.5	F=25 M=5.2
Internet users (number per 100 people), 2006	2.7	5.5	10.9	5.0	12.8	22.5	2.1	15.3	0.3	10.4	5.5	69.5
Affordability of Internet (Price basket in $US per month), 2006	11.9	11.3	11.6	0	3.1	10.1	30	12.5	24.0	10.0	6.6	15.0
Information and communication technology expenditures (as % of GDP), 2006	..	3.4	10	..	6.0	6.4	..	3.7	2.7	5.4	6.1	8.7
Secure Internet servers (per million people), Dec. 2007	0	0	23	0	2	16	0	5	0	0	1	868
Agricultural productivity (agriculture value-added per worker), 2003-2005 in 2000 $US	320	950	2,484	347	2,719	3,126	..	6,292	338	401	392	41,797

.. : Not available; Source: World Bank (2008). *World Development Indicators 2008*, Washington D.C., pp. 312-314, 304-306, 308-310, 138-140.

APPENDIX E: MEAN YEAR OF SCHOOLING

Table 6. Mean year of schooling

Selected driver	Africa					Latin America and Caribbean			Asia			North America
	Ghana	Nigeria	South Africa	Togo	Tunisia	Brazil	Cuba	Venezuela	Bangladesh	China	India	USA
Mean Year Schooling	3.9	..	6.1	3.3	5	4.9	..	6.6	2.6	6.4	5.1	12

Source: UNESCO. (2001) *Statistical Yearbook 2001*. Paris: Available from http://www.nationmaster.com/red/graph/edu_ave_yea_of_sch_of_adu-education-average-years-schooling-adults&b_map=1.[Accessed 27 April 2009]

Chapter 7
Adopter Fatigue Phenomenon in Diffusion of Innovations

Augustine O. Ejiogu
Imo State University, Nigeria

ABSTRACT

This chapter examines the adopter fatigue phenomenon in the diffusion of nanotechnology and micro-electronics innovations. It is hypothesized that innovations spread through a social system in an s-curve and that the speed of technology adoption is determined by two variables p and q where p represents the speed at which adoption takes off and q the speed at which later growth occurs. However, this two-variable model has been criticized as an over-simplification of a complex reality hence the need to examine adopter fatigue phenomenon defined as the hesitation, delay, or refusal by an individual to adopt an innovation on account of prevailing circumstances including the rapid evolution of new tech-nologies. This phenomenon is particularly relevant to nanotechnology and microelectronics products and processes which are characteristically continuously refined and upgraded. Because the phenom-enon is a symptom of poverty, it is recommended that overcoming adopter fatigue be achieved through multidisciplinary approach including empowering individuals by subsidizing the cost of adoption, and developing ancillary infrastructure such as electricity. Both governments and non-governmental orga-nizations, in the spirit of public-private-partnership, should act in synergy in solving the poverty-linked problem of adopter fatigue.

INTRODUCTION

Any society can be construed as consisting of organized system in which individuals and all artifacts are complementary parts connected in a myriad of ways (Hagerstrand, 1972). If in a sub-region of the system, a hitherto unknown concept is brought in, for example, a new technology, this forms a disturbance. Under certain conditions, this disturbance can be spread to the conterminous regions and be propagated. In the course of time, the entire social system will become permeated and to some extent be transformed.

DOI: 10.4018/978-1-61692-006-7.ch007

A permeation of this nature, whether total or partial, is central to the socio-economic development of nations. The attendant new technologies play catalytic role in the development process. Nanotechnology and microelectronics technologies which engender new technologies are also a player in the development process. Nanotechnology generally involves developing materials or devices or structures of sizes equal to or less than the size 100 nanometers. Nanotechnology can create many new materials and devices in electronics and energy production. Some of the products of nanotechnology include, but not limited to electronics and computers, health and fitness items, food and beverage goods for children and home and gardening products. A subfield of electronics, microelectronics is the study and manufacture of electronic components which are very small (Economic Expert, 2009). Nanotechnology and microelectronics processes and products have wide-ranging applications in information and communication technologies. Hence they aid in economic growth and development.

In spite of the pervasive impact of these innovations on daily lives and businesses, the technologies have not yet diffused globally (Ekekwe, 2009).

In most instances, innovations spread through a social system in an S-curve. This is brought about by early adopters selecting the technology first followed by the majority, until the technology or innovation is well diffused in the society or system (see Figure 1). It is further theorized that the speed of technology adoption is determined by two variables p, which is the speed at which adoption takes off, and q, the speed at which later growth occurs. A cheaper technology might have a higher p, for example, taking off more quickly, while a technology that has network effects (like a fax machine, where the value of the item increases as others get it) may have a higher q (MSU, 2009). The S-curve (see Table 1) is a robust yet flexible framework to analyze the introduction, growth and maturation of innovations and to understand the technological cycles (Scocco, 2006).

However, this model has been criticized as being an over-simplification of a complex reality. Hence, there is the need to critically examine the adopter fatigue phenomenon in the adoption process. Many developing nations, being adopters, are in most cases associated with this fatigue. Adopter fatigue may be defined as the lack of motivation on the part of an adopter to acquire

Figure 1. S-Curve (Source: BCT Partners, 2009)

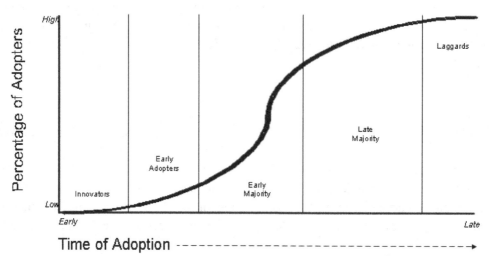

new technologies on account of prevailing circumstances including the rapidity of the evolution of new technologies. This paper argues that adopter fatigue engendered by weak to near non-existent infrastructure and high cost of maintenance and up-grading on account of the rapid evolution of new technologies should be of major concern to policymakers.

Table 1. The S-curve framework

The S-Curve Framework
The S-Curve emerged as a mathematical model and was afterwards applied to a variety of fields including physics, biology and economics. It describes for example the development of the embryo, the diffusion of viruses, the utility gained by people as the number of consumption choices increases, and so on
In the innovation management field, the S-Curve illustrates the introduction, growth and maturation of innovations as well as the technological cycles that most industries experience. In the early stages, large amounts of money, effort and other resources are expended on the new technology but small performance improvements are observed. Then, as the knowledge about the technology accumulates, progress becomes more rapid. As soon as major technical obstacles are overcome and the innovation reaches a certain adoption level, an exponential growth will take place. During this phase, relatively small increments of effort and resources will result in large performance gains. Finally, as the technology starts to approach its physical limit, further pushing the performance becomes increasingly difficult, as Figure 2 shows.
Consider the supercomputer industry, where the traditional architecture involved single microprocessors. In the early stages of this technology, a huge amount of money was spent in research and development, and it required several years to produce the first commercial prototype. Once the technology reached a certain level of development the know-how and expertise behind supercomputers started to spread, boosting dramatically the speed at which those systems evolved. After some time, however, microprocessors started to yield lower and lower performance gains for a given time/effort span, suggesting that the technology was close to its physical limit (based on the ability to squeeze transistors in the silicon wafer). In order to solve the problem supercomputer producers adopted a new architecture composed of many microprocessors working in parallel. This innovation created a new S-curve (Figure 3), shifted to the right of the original one, with a higher performance limit (based instead on the capacity to co-ordinate the work of the single processors).
Usually the S-curve is represented as the variation of performance in function of the time/effort. Probably that is the most used metric because it is also the easiest to collect data for. This fact does not imply, however, that performance is more accurate than the other possible metrics, for instance the number of inventions, the level of the overall research, or the profitability associated with the innovation.
One must be careful with the fact that different performance parameters tend to be used over different phases of the innovation, as a result the outcomes may get mixed together, or one parameter will end up influencing the outcome of another. Civil aircraft provides a good example; on early stages of the industry fuel burn was a negligible parameter, and all the emphasis was on the speed aircrafts could achieve and if they would thus be able to get off the ground safely. Over the time, with the improvement of the aircrafts almost everyone was able to reach the minimum speed and to take off, which made fuel burn the main parameter for assessing performance of civil aircrafts. The fuel burn metric was also compounded by cost and climate protection factors.
Overall we can say that the S-Curve is a robust yet flexible framework to analyze the introduction, growth and maturation of innovations and to understand the technological cycles. The model also has plenty of empirical evidence; it was exhaustively studied within many industries including semiconductors, telecommunications, hard drives, photocopiers, jet engines and so on.

Source: adapted from Scocco (2006)

Figure 2. Performance index vs. Time

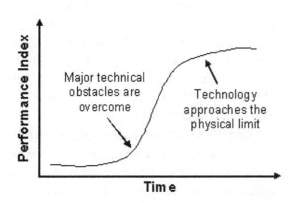

Figure 3. Performance vs. Time

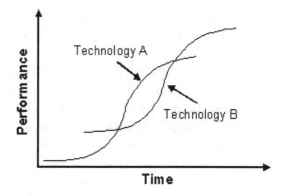

THE DIFFUSION PROCESS

In its common usage, diffusion refers to the spreading out or scattering. In the study of social systems, it refers to the transmission of elements or features of one culture to another. In all, the diffusion process is the spread of new technical device or innovation from its source of invention or manufacture to end users or adopters (Ekong, 2003). It should be pointed out that this definition is bereft of such critical elements of diffusion process as time and space and channels of communication. A more comprehensive definition is given by Rogers (1964) who stated that diffusion is a process in which an innovation is passed through certain channels over time among members of a social system. As a concept, diffusion is a flow and not a stock. It involves a series of events that occur over time through certain channels finding expression in natural changes in a society.

Four important elements in the diffusion process have been identified (Pemberton, 1936; Ekong, 2003). They are innovation, communication channels, social system and time or rate of adoption. These elements typically characterize a diffusion process. A closer look at these characteristic parts of diffusion process is very important to this discussion. Innovation is an idea, method or object perceived as new by an individual or group (van den Ban and Hawkins, 1996), in any given social system. Nanotechnology and microelectronics have enabled the realization and development of innovative products and services in the global economy. The communication channels refer to the media through which the innovation is extended to individuals or groups in their own social environment. Examples of these channels include personal contacts, the print and electronic media.

According to Ekong (2003), a social system can be defined as a group of individuals who interact and influence the behaviour of one another almost on a permanent basis. As an element in the diffusion process, it is a social environment within which innovation spreads. This element goes a long way in emphasizing the human dimensions

of diffusion of innovations. This is accentuated by the realization that the social system is the matrix in which innovation decision process is moulded. But then the individuals or groups in a social system and their entire inanimate cultural milieu are not mutually exclusive. Hence social system defined as the scene within which innovation spreads (Eze, 2005) enables us to interpret the social system as essentially encompassing the individuals and groups interacting among themselves as dictated by the existing cultural milieu within a given time period (Katz, 1955). In the diffusion process not all individuals exert equal amount of influence over all others. Hence there are leaders also called opinion leaders who have a lot of influence in spreading news about a new technology.

Finally, the time or rate of adoption refers to the speed with which individuals in a social system adopt or reject an innovation. Also known as adoption period (Ekong, 2003), this varies from one innovation to another and from culture to culture.

ADOPTION PROCESS

Diffusion of innovation takes place through a five-stage process. It is referred to as resembling a normative decision-making model. Referring to the five-stage process as a type of decision-making tends to be justified based on the fact that it enables the individual or group to accept or reject an innovation (Rogers, 1964). Within diffusion of innovation, there are three types of innovation-decisions: optional innovation decisions, collective innovation decisions and authority innovation decision. Innovation-decisions itself may be construed as the decision to adopt/reject an innovation arrived at by an individual or an organization (Veneris, 1990).

Optional innovation decision is reached by an individual distinguished in some ways from others in a social system. Collective innovation decision is collectively arrived at by sundry individuals in a social system. And authority innovation decision is made by a few influential or powerful individuals for an entire social system. An individual or a group of individuals bases the type of decision on whether an innovation is adopted/rejected. In all, diffusion of an innovation occurs via a five-stage-process which includes awareness, interest, evaluation, trial and adoption. Explaining the different stages, van den Ban & Hawkins (1996) stated that the potential adopter first hears about the innovation (awareness), makes relevant enquiry about it (interest), weighs up the costs and benefits of using the innovation (evaluation), applies the innovation on a small scale (trial) and large scale and subsequent application of the innovation in preference to other methods (adoption). Hence, the changes that take place within individuals from when they become aware of an innovation to the decision to use it or not are the adoption process. The time it takes for such changes to occur is also relevant in the adoption process.

It should be noted that although both diffusion and adoption are a process they are not synonymous. As a social phenomenon, more is known of the results of diffusion process than of the process in action. Based on valid and reliable quantitative data, measurement of the results of the process gives information on the decision to apply an innovation and to continue to use it (adoption). Thus in a given sub-region, diffusion process tells us how innovations spread among individuals overtime. Adoption process tells us about the changes that take place within individuals as regards an innovation from when it was first heard to the trial decision to accept it or not. Simply put, measurement of a diffusion process is captured by the individuals' decision to accept or reject a given innovation in a social system. The adoption process does not rigidly follow this sequence in practice. Furthermore, the use of such products depends on the availability of such supporting infrastructure as skills and power supply which the adopter must possess. Also interest may precede awareness such as when individuals are looking for a method of communicating from the comfort of the rural areas with friends and family in faraway places.

CHARACTERISTICS OF INNOVATIONS

Some intrinsic characteristics of innovations influence an individual decision to adopt or reject them. Rogers (1964), van den Ban and Hawkins, (1996) and Ekong (2003) agree on the following characteristics: relative advantage, compatibility, complexity, trialability and observability. Relative advantage refers to how improved an innovation is over the previous one. Ekong (2003) points out that the relative advantage may be expressed either in economic or social terms. Ayichi (1995) argues in terms of the financial and economic viability of acquiring the technology. Economic viability measures adoption of the technology in terms of social costs and benefit. Financial viability on the other hand measures the profitability of the innovation to the individual adopter. It must be emphasized that cost should be evaluated in all its ramification including what has to be given up (opportunity cost) along with the cost of other innovations on which utilization of that under consideration depends. For instance, some ten years ago, information and communication technologies are thin on the ground in rural areas in Nigeria. The few that had cellphones, also acquired individual masts to the bargain. So purchase of a cellphone had a derived demand for mast, pole and wires with all the cost implications. Nowadays, individual masts have virtually come down and replaced by giant masts provided by the service providers, yet network availability fluctuates in the rural areas.

Other characteristics include the compatibility of the innovations, socio-cultural values and beliefs, and complexity. Complexity refers to the degree to which an innovation is relatively difficult to understand or use. Compatibility and complexity of innovations are particularly critical to adoption of nanotechnology and microelectronics technology. This is so because the scientists involved in their development are driven more by market forces and rightly quest for innovation. The issues of compatibility and complexity are in most cases not critically examined. Following compatibility and complexity are triability and observability. Triability as a characteristic refers to how easily an innovation may be experimented with in the process of its being adopted while observability refers to the extent to which an innovation is visible to others.

ADOPTER CATEGORIES

Not all individuals adopt innovations at the same rate. Some people accept new technology years before others. According to Rogers (1964), adopter category is a classification of individuals within a social system on the basis of innovativeness. Ekong (2003) defines innovativeness as the extent to which an individual is relatively earlier in adopting a new technology than others in a social system. The five adopter categories are the innovators, the early adopters, the early majority, the later majority and the laggards. Youngest in age, willing to take risks and great in financial lucidity, innovators are the first to adopt new technology or an innovation. The early adopters are the second fastest category of adopters. The early adopters are usually regarded as opinion leaders. Although they show some degree of opinion leadership, early majority adopter category adopts an innovation after a significantly longer time than the innovators and the early adopters. Late majority adopter category is made up of individuals who approach an innovation with skepticism even after the majority of the society has adopted it. Individuals in this group have very little financial resources and very little opinion leadership. The laggards are the last to adopt an innovation. Laggards are typically the oldest of all adopters, lowest in financial resources and of little or no opinion leadership

Division of individuals into these categories depends not only on the extent to which individuals have adopted an innovation but also on the assumption that distribution of adoption over time is normal. However, technologies including nano-technology and microelectronics are not static. There is continuous innovation, which engenders new products and processes together with new signals in the diffusion process that tend to take the steam off the previous innovation. Furthermore, continuous innovation tends to lend itself to continuous adoption. This tendency is capable of creating different versions of a technology being available for adoption with equally different sub categories of adopters within a given category. As stated earlier, adoption of innovation follows an S-curve when plotted over a time period. However, the S-curve does not just come about. The S-curve can rather be seen as composed of a series of bell-shaped curves. These curves depict different segment of a population adopting different versions of a generic innovation. This is particularly relevant to nanotechnology and microelectronics whose products and processes are inherently dynamic.

If we have gone to a great extent in focusing on the diffusion and the adoption processes, it is to bring out the idea that the spread of dynamic and evolutionary innovation as is the case with nanotechnology is capable of overwhelming an average adopter and hence affect the receptivity of the new technology. This brings us to the phenomenon of adopter fatigue.

ADOPTER FATIGUE PHENOMENON

It is hypothesized that the speed of technology adoption is determined by two characteristics: p and q. p denotes the speed at which the adoption takes off. q denotes the speed at which subsequent growth occurs. However, there are other factors that affect the rate of adoption of innovation. The following phenomena can influence the rate of adoption of innovation. These include the change in the nature of a given innovation from the early adopters to the majority of users on account of adapting the innovation to the needs of the adopters. With respect to nanotechnology and microelectronics, changes in a given innovation arising from modification to local needs are in turn dependent on the technological development of the social system. For less developed countries, these changes tend to be non-existent. Other factors that can influence adoption include disruptive technology and path dependence.

Disruptive technology refers to an innovation that upgrades a product or service in ways that the market does not envisage. Typically, this comes about by being low priced or produced for a different community of consumers. Disruptive technologies are a factor in the rate of diffusion and adoption in nanotechnology and microelectronics. Different sets of electrical and electronic products are known to be manufactured for different segments of the market. Such products are sometimes tagged "for export", "adapted to tropical conditions", and the like. Some are actually without obvious labels but are intrinsically characterized by being cheaper, and of lower capacity. A disruptive technology can dominate an existing market by serving a purpose in the new market which the older technology did not serve. Alternatively, disruptive technology can through performance improvement gradually move up market and eventually displace the incumbents. In this way, disruptive technology can fundamentally change the diffusion patterns for any existing technology by starting a different and competing S-curve.

Path dependence explains the way a set of decisions one faces for any given circumstance is restricted by earlier decisions even where past circumstances are no more relevant. For example, several alternatives to QWERTY, the most widely used modern-day keyboard layout on English language computer, have been developed over the years. These alternatives, the designers claim, are more efficient. However none of the alternatives has witnessed widespread adoption due partly to the sheer dominance of available keyboard and training. The foregoing factors that can affect the rate of adoption are not in the least exhaustive. In fact it cannot be overemphasized that an aggregate of convoluted and complicated circumstances causes eagerness, hesitation, delay or refusal to adopt innovations. Part of the picture, to all intents and purposes, is the personal characteristics of the individual made manifest in the adopter fatigue phenomenon.

As stated before, adopter fatigue is the hesitation, delay or refusal by an individual to adopt innovations suggested by signals through the communication network on account of prevailing circumstances including the alarmingly rapid evolution of new technologies. This definition gives at least three variables that give expression to the phenomenon. These are: individuals' personal characteristics, rapidity of evolution of the technology and interaction of the two variables. First let us look at the rapidity of evolution of new technology.

Nanotechnology products require upgrading on account of rapid evolution of the products. For nanotechnology to be adopted especially in developing countries the products have to be affordable and operable. While affordability has to do with economic and financial realities of the individual, operability pertains to technology issues. Affordability and operability concerns both come into play in adoption of nanotechnology products. The plot thickens for the low income technology consuming - as opposed to technology producing-individual- who is confronted with rapidly evolving software and hardware products. For instance, in June, 2009, Samsung produced 16GB NAND flash memory described as the highest capacity memory now available. Samsung said it will manufacture the devices in 51 nanometers (nm). The 51nm NAND flash chips can be manufactured 60 percent more efficiently than those manufactured with 60nm process technology. This new migration milestone was achieved by Samsung just eight months after the manufacture of its 60nm 8GD NAND flash last August. Some product and processes even enter the market at faster rates.

As of August 21, 2008, the project on Emerging Nanotechnologies estimated that more than 800 manufacturer–identified nanotech products are available to the public and new ones were entering the market at a rate of 3-4 per week. The individuals' personal characteristics (e.g. socio-economic status and educational attainment) are intricately linked to the overall social system in which one ordinarily resides. The social system is in turn defined and delineated by the cultural milieu including the level of technological development. So in the diffusion process, the individuals' personal characteristics interact with the fleeting realities of a rapidly evolving technology. The rapid rate at which nanotech products and processes evolve is capable of making an innovation somewhat obsolete before an average adopter can get a handle on the innovation. Thus, an adopter being overwhelmed by the cost of adoption, challenged by the incompatibility of the existing systems with the ever changing products and processes, hamstrung by weak, inefficient and sometimes non-existent supporting infrastructure, and confronted by unique obstacles with different cultural contexts, fatigue results. This phenomenon may create a shadow effect over a wider region and to that extent constrict the adoption process. It should be appreciated that the phenomenon is a symptom of a larger and more complex problem – that of debilitating deprivation and inequality. Recognition of the adopter fatigue phenomenon as having a killing effect in the diffusion process should lead scholars, policy-makers and the public in devising ways round the problem. While scientists are genuinely encouraged to continuously create values in processes and products, overcoming the adopter fatigue in the diffusion of such innovations should inform and influence policies especially in developing countries.

OVERCOMING ADOPTER FATIGUE IN THE DIFFUSION PROCESS

The best point of departure in overcoming adopter fatigue phenomenon in the diffusion process is to examine the conditions which give rise to it. Because poverty is implicated as precipitating the phenomenon, and also poverty is essentially of a multi-dimensional nature, overcoming adopter fatigue must then be achieved through multi-dimensional fronts. Some solutions are as follows:

- The role of government in overcoming the adopter fatigue phenomenon is critical. Once a technology is certified economically viable, the government exercising its authority innovation-decision should accept the innovation and additionally empower the citizens for adoption. Empowerment can come in various forms including subsidizing the cost of adoption of such technology, development of supporting infrastructure such as electricity, especially as it concerns information and communication technologies. With particular reference to African Union member countries measures agreed for eradicating poverty under the institution of NEPAD (New Partnership for Africa Development) should be adhered to.
- In this era of public-private-partnerships, non-governmental organizations should also play a role in solving the poverty-linked problems of adopter fatigue. This the NGOs can do in partnership with the government in effort aimed at overcoming the fatigue.
- Furthermore, developing countries should map out workable strategies to be a part of the global player in the manufacture, distribution and use of nanotech products. This is expected to reduce or even eliminate any inertia in diffusion process as the average adopter is likely to be a part of the manufacturing process.

CONCLUSION

This paper has identified adopter fatigue as a factor in the diffusion of innovations. Because the diffusion of innovation is a complex phenomenon, it is recommended that other interactive factors especially in relation to nanotechnology and microeconomics be empirically studied. Also, the notion that the speed of technology adoption is determined by two variables has been criticized as being an unacceptable over-simplification of a complex reality. The need therefore arises to explore other factors that also determine the adoption of technology. Furthermore, in exploring other factors, this paper focuses on the adopter fatigue phenomenon in the diffusion of innovations. This phenomenon has been discussed in some details. Ways for overcoming it in the adoption process have also been discussed. This contribution to knowledge is expected to inform and influence relevant policies. With respect to ways of overcoming the adopter fatigue phenomenon, no single approach is sufficient; a multifaceted methodology is vital especially in emerging and developing economies.

REFERENCES

Ayichi, D. (1995). Agricultural Technology Transfer for Sustainable Rural Development in Nigeria . In Eboh, E. C., Okoye, C., & Ayichi, D. (Eds.), *Rural Development in Nigeria: Concepts, Process and Prospects* (pp. 126–134). Enugu, Nigeria: Auto Century Publishing Company.

Economic Expert. (2009). *Microelectronics*. Retrieved from http://www.economicexpert.com/a/Microelectronics.htm

Ekekwe, N. (2009), Towards competitiveness and global outsourcing: practical model for microelectronics diffusion in Africa, Int'l Conference on Industry Growth, Investment and Competitiveness in Africa (IGICA), 8-10 June 2009, Abuja, Nigeria.

Ekong, E. E. (2003). *Rural Sociology an Introduction and Analysis of Rural Nigeria*. Uyo, Nigeria: Dove Educational Publishers.

Eze, S. O. (2005). Diffusion and Adoption of Innovation. In Nwachukwu, I., & Onuekwusi, G. C. (Eds.), *Agricultural Extension and Rural Sociology* (p. 249). Enugu, Nigeria: Snaap Press Ltd.

Hagerstrand, T. (1972). Diffusion of Innovations. In D. C. Sills (Ed.), International Encyclopedia of Social Sciences (Vol. 3 & 4). New York: The Free Press.

Katz, E., & Lazarsfeld, P. (1955). *Personal influence: The part played by people in the flow of mass communications*. Glencoe, IL: Free Press.

MSU. (2009). *Diffusion of innovations*. Retrieved from http://www.educ.msu.edu/epfp/meet/02-06-06files/Diffusion_Innovations.pdf

Partners, B. C. T. (2009). *S-curve*. Retrieved from http://www.bctpartners.com/

Pemberton, H. E. (1936). The curve of culture diffusion rate. *American Sociological Review*, *1*(4), 547–556. doi:10.2307/2084831

Rogers, E. M. (1964). *Diffusion of Innovations*. New York: Free Press.

Scocco, D. (2006). *Innovation Zen*. Retrieved from http://innovationzen.com/

van den Ban, A. W., & Hawkins, H. S. (1996). Agricultural Extension. Cambridge, MA: Blackwell Sciences Inc.

Veneris, Y. (1990). Modeling the transition from the industrial to the informational revolution. *Environment & Planning A*, *22*(3), 399–416. doi:10.1068/a220399

Section 3
Industry, Policy and Experiences

In this section, industry, policy and case studies for both nanotechnology and microelectronics, and technology in general are discussed. It looks at how nanotechnology and microelectronics are shaping and transforming industries and how policies continue to influence their penetrations across the globe. The issues surrounding technology licensing and R&D, barriers stalling the global diffusion of emerging technology to developing nations along with the mechanism for assessing maturity and performance of both nanotechnology and microelectronics are examined. Case studies on national policies as they pertain to developing knowledge economies and a study evaluating university-industry network on nanotechnology are also presented.

Chapter 8

Nanotechnology, Firm Innovation and University-Industry Networks:
The Case of the UWS[1] Nanotechnology Network in Sydney

Cristina Martinez-Fernandez
University of Western Sydney, Australia

ABSTRACT

Nanotechnology is becoming a transformative element for the manufacturing sector into the knowledge economy. Access to relevant knowledge is a critical factor in this transformation as manufacturing firms cluster in peripheral suburbs away from the knowledge intensive ring of central business districts. Results from a project conducted in South-West Sydney shows that informal university-industry networks raise the awareness of firms to the potential of nanotechnology applications, their willingness to invest in nanotechnology R&D and the number of university-industry cooperation initiatives and business-to-business partnerships. Results from the project also suggest that, despite the importance for firms of being involved in global networks, access to local knowledge and local networks is significant for the innovation process of small and medium enterprises (SMEs).

INTRODUCTION

Technological development is one of the critical factors on innovation and firm competitiveness (Maskell 2001; Smith 2000; OECD 1999) and the absorption of General Purpose Technologies (GPT) is a significant factor in the continued competitiveness of the firm (Shea, 2005). The impact of nanotechnology in innovation as a frontier technology and a general-purpose technology is only starting to be analysed in detail. Many firms, from a broad variety of sectors, are unaware of the transformation effect nanotechnology might have in the competitiveness of their business. In this context of early path development of nanotechnology, access to specialised knowledge and to knowledge infrastructure is critical for firms to evaluate nanotechnology investments.

The significance of knowledge infrastructure

DOI: 10.4018/978-1-61692-006-7.ch008

for innovation is becoming increasingly acknowledged in the literature. Today, more research than ever is exploring the relationships of creation and diffusion of knowledge, the impact of knowledge and skills, the commercialisation of knowledge, knowledge flows and firm competitiveness, and new enablers of the knowledge society (OECD, 2005; OECD 1999; Maskell 2001; Smith 2000; AEGIS 2005; Audrestsch, 1995; TIAC, 2002; Turpin and Martinez, 2003; Acs, 2002; Martinez-Fernandez et al 2005a). It is now widely recognised that global competitiveness is dependent on the capacity of economies to acquire knowledge capital and to apply new knowledge through a highly trained and specialised workforce.

The role that universities and other knowledge institutions play in the generation, sharing and transferring of knowledge is being scrutinized by governing bodies of universities and other research organizations (AVCC, 2005; OECD 2005). In addition to training new talent and providing research outputs, 'Third Stream Activities' such as strategic engagement with industry and the community (Molas-Gallart, 2002) are becoming an important part of universities role in society. Universities are increasingly active in influencing regional competition and recent literature points to a conscious, robust strategy by some universities of driving regional knowledge development (Garlic, 2000; Faulkner & Senker, 1995; Sproats, 2003). While firms are increasingly embedded in global networks, access to local knowledge is still a determinant factor for firm innovation (Martinez-Fernandez & Potts, 2007).

It is generally agreed today that participation in networks, clusters and alliances proves a powerful learning mechanism for firms, especially if the relationships are in geographical proximity to enable extensive informal knowledge sharing (Martinez-Fernandez, 2004). It has also been discussed elsewhere that regions should maximise the value of their knowledge-generation institutions through linkages with the different actors in their innovation system (Maskell, 2001) to fa-

cilitate transfer of knowledge between knowledge providers and specialised industry users (Teece, 1987). However, there is still a lack of industry-university cooperation in many fields while at the same time the benefits of universities to their regions' knowledge intensity is firmly advocated (Acs 2002, Martinez-Fernandez & Leevers 2004, Martinez-Fernandez 2004).

This paper discusses the effects of a university-industry network on the absorption of nanotechnology applications by SMEs in the region of South-West Sydney. A survey conducted among South Western Sydney firms shows that the role of universities in the dissemination and transfer of knowledge of frontier technologies such as nanotechnology is more significant than in established technologies such as ICT.

NANOTECHNOLOGY: DEFINITION, GENERAL POLICY AND INDUSTRY CONTEXT

There is not just one definition of 'nanotechnology' although it is agreed that it has to do with the science of the very small. The OECD defines nanotechnology as the "range of new technologies that aim to manipulate individual atoms and molecules in order to create new products and processes: computers that fit on the head of a pin or structures that are built from the bottom up, atom by atom" (OECD 2003). Basically, nanotechnology is the design, characterization, production and application of structures, devices and systems that entails controlling the shape and size at the nanometer scale (Royal Society 2004). The size range of nanotechnology is often delimited to 100 nm down to the molecular level (approximately 0.2nm) because this is where materials have significantly different properties. Nanoscience refers to the scientific analysis of materials at atomic, molecular and macromolecular scales while nanotechnology applies scientific developments to commercialization (Shea, 2005). The field of nanotechnology is incredibly diverse, and many of the applications represent radical innovations in the market, such as an invisible zinc cream, self-cleaning materials, unbreakable fabrics or the development of cheap, portable, personal solar cells for recharging laptops and mobile phones.

The interest on 'nano' is not only a firm issue it is also a national priority in a worldwide race. The increase of governments' awareness in the promising economic potential of nanotechnology can be observed in the increase R&D expenditure between 1997 and 2000. R&D funding for nanotechnology grew from USD 114.4 million to more than USD 210.5 million in the European Union and from USD 102.4 million to USD 293 million in the US during this period (OECD 2003). The US government invested nearly USD 1 billion into nanotechnology research during 2004, and will add USD 3.7 billion more between 2005 and 2008. Clayton Teague, Director of the US National Nanotechnology Coordinator Office, a government department that facilitates cooperation between academic researchers, corporations and other government offices said at the 2004 Boston conference 'Nanotech', "the US federal government is committed to the promise of nanotechnology. With all that support, the government and lawmakers are really looking to this field to be a major contributor to our economy over the coming years." The trend is not unique to the US;

countries around the world are increasing their R&D investment in nanotechnology. Figure 1 shows public investment R&D in nanotechnology since 1997.

The US, the EU and Japan are far ahead in R&D investment in nanotechnology since 1997, although other countries such as Italy, Sweden, Finland and France have a high average annual growth rate. According to the US National Nanotechnology Initiative (NNI) worldwide government funding has increased to about five times what it was in 1997, exceeding 2 billion dollar in 2002 (Knop, 2005). The US National Science Foundation has predicted that the market for nanotechnology will reach 1 trillion dollars in 10-15 years (OECD 2004) with an estimated 7 million jobs needed to support the industry worldwide (Knop 2005:8). Venture capital is still small; less than 2 percent of the investments made by venture capital firms in the US in 2003 where in nanotechnology (Lux Research 2004). Analysts consider investments in nanotechnology difficult because the investment landscape is not yet established, there is not much scope for risk diversification and the interdisciplinary, multi-industry nature of nanotechnology means that the investment is not a single, well-defined part of the stock market (Custer, 2005). Large companies with exposure to nanotechnology, companies investing in nanotechnology research and smaller companies with

Figure 1. Public investment in nanotechnology 1997-2003 (Source: National Science Foundation, 2003)

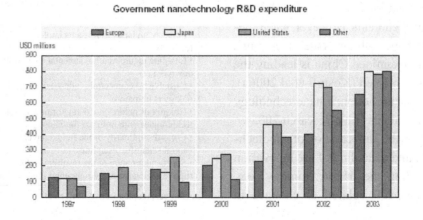

high exposure to nanotechnology seem to be the best options for high-risk investors at the moment (Custer, 2005). The Australian venture-capital sector has shown little interest in nanotechnology investments, in part for the reasons outlined above but also because the sector doesn't seem to have the depth of knowledge and confidence to move forward in nanotechnology (BRW 2005:24).

Patents are led by the private sector; the three top patents assignees by 2004 were L'Oreal (266), IBM (125) and Regents of the University of California (107) (Lux Research 2004). Most of the patents are assigned to large multinational corporations[2] with a large share of investment in nanotechnology (Shea, 2005). Ten countries (US, Japan, France, UK, Switzerland, Taiwan, Italy, South Korea, Netherlands, Australia) account for almost 90 percent of worldwide nanotechnology patents, the US is leading with over 80 percent and as much a 57 000 patents (Shea, 2005). At the same time there has been an increase in scientific output as measured by the number of scientific publications, which increased from 10 575 in 1997 to 15 667 in 2000. The publication of papers on nanotechnology was largely dominated by the US, Japan and Germany (OECD 2003). A recent bibliographic analysis of nanotechnology citations (Kostoff et al 2006a,b) has found that three Western countries (USA, Germany, France) have about eight percent more nanotechnology publications (for 2004) than the three from the Far Eastern group (China, Japan, South Korea). China, despite the high level of government investment, has minor representation in the most highly cited nanotechnology documents. However, when specific nanotechnologies sub-areas such as coatings and powders are examined China is leading the USA in articles published (Kostoff et al 2006a). Australia ranks near the bottom in nanotechnology publications and patents (Lux Research 2005).

Understanding the path creation of nanotechnology requires understanding of the innovation system surrounding the nanotechnology; e.g. organizational aspects related to changes in the organisation of nano-knowledge generation and diffusion, identifying search trends of the nanotechnology community and their emergent strategic research and technology agendas (Andersen, 2005). The application of nanotechnology at the firm level is still in very early stages which is understandable as per the complexity of nanotechnology dynamics (see Table 1).

These dynamics govern emergent nanotechnology activities. Firms are not always aware of the possible radical or even disruptive effects of nanotechnology and of the implications of its possible transformative effects in society (Shea 2005). As a General Purpose Technology the application to industry sectors is exceptionally varied and requires a different approach to industry analysis as the sectors are blurred and the boundaries are unclear. The use of 'nanotechnology ecosystems' might constitute a better term and analytical framework for the study of industry applications of nanotechnology. Within each ecosystem more than one industry coexists, as well as research labs and other organisations. Figure 2 shows seven nanotechnology ecosystems that can be differentiated from the applications so far (Shea 2005, The Royal Society 2004, Wood et al 2003, Luther 2004, Andersen 2005, Lux Research 2004).

The industry sectors within each ecosystem are varied, in many cases inter-related and changing at a rapid pace as more discoveries and applications are released in the market. Nano applications also

Table 1. Nanotechnology dynamics

• Fuzzy boundaries as it is a technological field not a single technology
• Enabling nature as a fundamental General Purpose Technology (GPT)
• Unstable base, uncertain industrial applications as the technology is immature
• Ubiquitous nature on existing or new materials or converging technologies with multiple functions
• Cross disciplinarity of the field, convergence of several natural scientific disciplines
• Spectacular nature of the resulting technology. Possible disruptive societal and environmental effects of the technology

Source: Adapted from Andersen 2005, Shea 2005

Figure 2. Nanotechnology ecosystems

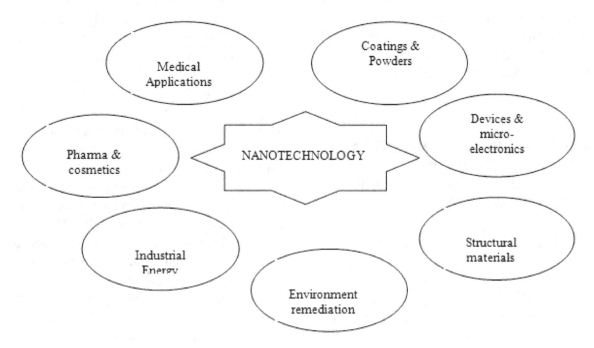

reflect the impact of other technological advances from complementary technologies such as ICT and biotechnology. This strong, changing nature of innovative activity in nanotechnology is both an advantage and a barrier for the firm. For example, disruptive nanotechnology based innovations are more likely to result in decline in economic performance of the firm than non-disruptive innovations (Shea 2005: 196). However, even with the risk involved in disruptive innovation this will eventually improve and become the mainstream technology and those companies ahead of the game will have better bases for sustained competitiveness in the market place.

It is unclear from the literature which mechanisms can be put in place from a policy perspective to assist the transition from emerging technologies to new industries. Firms entering the market either grow and survive or exit the market and at the centre of this process is the complexity of the commercialisation process. The speeding up of the transition from emerging technologies to new industries is central to successful economic growth, employment, competition and sustainability (Hung & Chu 2006). So understandably this constitutes a policy priority for modern economies. Three policy mechanisms are pointed out by Hung & Chu as means for speeding up this process.

The first is *encouraging partnerships in the commercialisation process* to link discoveries to market opportunities. The second is *fostering entrepreneurship and venture initiatives* in the innovation system to encourage the expansion of entrepreneurial behaviours in the community. The third is *sustaining commercialisation and the creation of new firms*, promoting the survival and growth of those firms that are able to adjust in the market. In the case of Taiwan, these functions are developed as a whole by the Industrial Technology Research Institute (ITRI) - a non-profit, national research institute that undertakes applied research, produces patents, and transfers its nanotechnology to local industry (Hung & Chu, 2006). One of the activities developed by the ITRI was the 'knowledge-based NanoNetwork', which has a positive effect in improving the domestic nanoresearch environment. A similar initiative, though focused at the local level, was undertaken by the University of Western Sydney with the formation of the 'UWS Nanotechnology Network' in South-West Sydney. An analysis of this case is presented next.

NANOTECHNOLOGY IN AUSTRALIA: THE CASE OF THE UWS NANOTECHNOLOGY NETWORK IN SOUTH-WEST SYDNEY

Australia is not a leading country in the nanotechnology race. An analysis of 14 countries by Lux Research (2005) shows four different rankings. *Dominant* countries scoring high on both nanotechnology activity and technology development strength are the US, Japan, South Korea and Germany. *Niche players* that score low in nanotechnology activity but high on the technology development strength needed to convert that activity into jobs and products. In this group are Taiwan, Israel and Singapore. *Ivory Towers* nations, high on nanotechnology activity but low on technology development are the UK and France. And finally the *Minor League*, scoring low on both axes are China, Canada, Australia, Russia and India. Australia has high-profile life science nanotechnology successes but the level of government expenditure in nanotechnology is at the bottom of the rank as well as the number of patents and publications (Lux research 2005).

The Australian nanotechnology ecosystem is very small, characterised by emerging businesses, most of them undercapitalized, and a growing academic research capability. A National Nanotechnology Strategy has been advocated to address specific needs of coordination across all levels of government and to provide a regulatory framework for the sector (PMSEIC, 2005). Recommendations from this Taskforce address the need to overcome fragmentation in the sector, linking business and researchers and enhancing industry application

of nanotechnology (PMSEIC, 2005:5).

Australian businesses show limited interest on nanotechnology if compared with US firms. A government commissioned report investigating adoption of nanotechnology by Australian businesses found that companies are interested in commercial solutions to problems, not the technology itself. Local companies, in particular, see themselves as users of the technology, rather than developers (Dandolopartners, 2005a).

A survey of 105 businesses investigating investments and attitudes towards nanotechnology (Dandolopartners 2005b) found that 92 percent of respondents were aware of nanotechnology with 40 percent indicating they have detailed knowledge in the subject. Companies appear to be investing in nanotechnology or expect to do so in a significant way in the next five years. 21 percent of companies were already investing in nanotechnology. Companies in the manufacturing sector were the heaviest investors (45%). The size of the firm seems to be important when investing in nanotechnology being the very large (32%) and the small (26%) the bigger investing groups. The majority of businesses do not perceive strong barriers for investing in nanotechnology apart from acquiring the right skills in-house and finding information about products and latest research. Businesses see the main roles of government as providing funding to public institutions for research (66%), encouraging private sector investment in nanotechnology (56%) and providing information to the public about nanotechnology (53%). This report is much more detailed than the previous one by the same company so some of the differences are due to the different sample size.

The University of Western Sydney Nanotechnology Network Project

Business attitudes towards nanotechnology as outlined in the Dandolopartners' reports contrast with findings of the University of Western Sydney Nanotechnology Network Project that reports the majority of firms were unaware of the implications of nanotechnology for manufacturing SMEs. The project started in 2003, funded by the Department of Transport and Regional Services (DoTaRS) Sustainable Regions Programme with AUD 255,000 specifically for the Campbelltown/Camden Local Government Areas in South-West Sydney; a region with a significant concentration of manufacturing firms from the metals, glass and building and construction industries.

The objective of the project was to identify and build nanotechnology business potential in the Campbelltown/Camden Region, specifically in the area of nano-materials. The University aim was to work with existing organisations including local and regional peak industry associations to:

- identify existing enterprises with potential for application of nano-materials technology;
- facilitate and support development of networks and new enterprises applying nano-materials technology; and to
- enable access to funding opportunities for industry and product development with concomitant research and training, including research and development granting schemes.

The project was designed as a 'Knowledge Intensive Service Activity' (KISA)[3] where the UWS Office of Regional Development would organise activities providing specific knowledge, specialised information and opportunities to discuss nanotechnology and possible applications to manufacturing processes. The UWS Nanotechnology Network is formed by academics, research students, industry, business people, government representatives and community groups that meet quarterly to discuss, attend lectures or seminars and showcase products. Attendance is around 50 people at each meeting. The UWS Nanotechnology Network has a part-time facilitator that compiles a newsletter, maintains a mailing list with over 300 members and a dedicated website.[4] Networking at the quarterly meetings is carefully planned with quality catering and preparation to provide opportunities for people to meet and discuss in an informal setting. The mix of private, public and community sector is also strategically planned so as to facilitate emergence of partnerships. To ensure that the project positively contributed to the region there was a strong emphasis placed upon identifying and evaluating the opportunities that arise from the network activities and the soft infrastructure developed.

The industry responses to the UWS Nanotechnology Network and its effects were measured via two small surveys in 2003 and 2005 and three in-depth case studies in 2005. The industry break up in the 2003 survey of the network (279 members) shows that the majority of business are in manufacturing (32%), followed by business services (19%), government (17%), university and research and education organisations (29%) and other businesses in the area of biotechnology, ICT, health or packaging (3%). Only 27 percent of the firms new about nanotechnology and only 6 percent were using nanotechnology. Six percent had plans to introduce nanotechnology, 24 percent were in partnership with a university conducting a particular project and 42 percent of the firms wanted to network with other members of the network.

In the 2005 survey, 46 percent has participated regularly in the network activities and 51 percent had gained significant new knowledge as a result of their participation. Of the respondents, 26 percent had plans to introduce nanotechnology into their companies and 14 percent had already invested funds on nanotechnology R&D. Up to 26 percent of the companies have initiated new partnerships during their participation in the network. These partnerships were significantly more frequent with other companies rather than with universities, regional organisations or industry associations (see Figure 3).

Although partnerships appear to be more fre-

Figure 3. Partnerships for nanotechnology development/applications (Source: UWS ORD Nanotechnology Survey (2005))

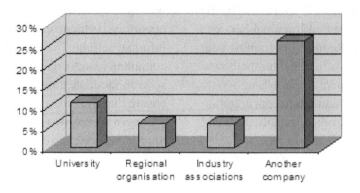

quent among companies than with universities or other public or non-for-profit organisations, universities and research and technology organisations (RTOs) are top providers of knowledge and information for nanotechnology (see Table 2) indicating the role of RTOs in diffusion of nano knowledge to industry.

These findings contrast with previous research on established technologies such as ICT and software that rely more on knowledge produced by their network of contacts: customers, competitors, and conferences/professional meetings. The role of universities as providers of knowledge is very small in this sector (see Table 3).

Companies applying frontier technologies seem to rely more in scientific knowledge pro-

Table 2. Sources of nanotechnology knowledge (No firms; n=38)

Sources of Knowledge	Not relevant/ small importance	Medium/Great Importance
UWS	8	**20**
Other Universities	10	**14**
RTOs	10	**13**
Internet, databases	8	**17**
Business networks	9	**15**
Suppliers	12	13
DSRD	11	12
Customers	13	11
Industry associations	14	10
Within the firm	16	8
Other firms within same industry group	15	8
Competitors	16	7
KIBS	16	8
Conferences, journals	16	8
Fairs & exhibitions	16	7

Source: UWS ORD, Nanotechnology survey 2005

Table 3. Important sources of knowledge for software firms

Sources	No of firms (n=54)
Within the firm	47
Customers	47
Databased information networks (Internet)	32
Competitors	22
Conferences, meetings or periodicals	20
Other firms within the same industrial group	17
Suppliers of equipment, material or components	17
Fairs and exhibitions	15
Industry associations	9
Universities and colleges	6
Public/private non-profit research centers	6
Consultancy firms	5
Public patent documents	2

Source: Martinez-Fernandez et al 2005

duced by universities and public/private research institutions (RTOs). In-house knowledge has small relevance for frontier technologies but is the greatest source of knowledge for established technologies (Martinez-Fernandez et al, 2005b). Consultancy firms (characterized as knowledge intensive business services), industry associations and government departments have only a small role in the co-production of knowledge in both cases. The survey of South-West Sydney firms found that firms applying nanotechnology were more reluctant to trust in-house sources of knowledge and information while in the case of software, tourism or mining technology services internal sources of knowledge were more important for the firm (Martinez-Fernandez et al, 2005b,c,d; Martinez-Fernandez & Leevers, 2004). This difference might relate to the early path creation of nanotechnology if compared with more established technologies. Thus, firms have limited experience with the technology and they turn to RTOs to contrast and complement information.

An interesting result of the survey is that the perceived barriers for the introduction of nanotechnology in the firm's product development have more to do with issues of 'relevance and information' than with the market or the expertise (see Figure 4).

While funding is seen as a barrier to the introduction of nanotechnology by 29 percent of the respondents, issues of relevance for the firm and of information seen to be far more important

(72% of respondents). Network approaches such as the UWS dissemination of nanotechnology information are a good vehicle to address this point because companies are able to have a quality well planned informal context where they are confident to discuss nanotechnology with other interested parties and they are also able to see ready applications by other companies, new prospects, and are exposed to discoveries by university researchers. The case studies further confirm this argument.

Case Studies: Three Manufacturing Approaches to Nanotechnology

Three of the UWS Nanotechnology Network companies from South-West Sydney were selected for in-depth case studies, based on their involvement with nanotechnology applications. These three companies were considered best practices of rapidly adopting or developing nano-applications in the western Sydney region. The analysis focused on the firm's innovative activity, the type of nanotechnology knowledge intensive service activities (KISA) undertaken, the transforming internal processes used by the firm for the adoption of nanotechnology and their involvement on the UWS nano network. The three firms were new to the nanotechnology world so they cannot be considered yet under a specific 'nanotechnology ecosystem' but as companies operating in the manufacturing or consulting sector and with some interest in developing nano applications.

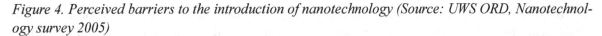

Figure 4. Perceived barriers to the introduction of nanotechnology (Source: UWS ORD, Nanotechnology survey 2005)

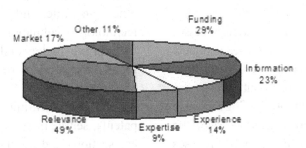

A) R.J. Walsh & Son Ltd

R.J.Walsh & Son Ltd is a 45 year old company, which mainly produces are animal powered vehicles. The company has 8 staff, approximately AU$ 1 million annual revenue and is currently in expansion. The company is characterised by incremental innovations (products/services and organisational). Recently the company has introduced 'Nanoflex', a new steel material that is stronger but lighter which constitute an incremental change to the materials used before in their sulkies. Nanoflex enables lighter products of equivalent strength therefore influencing harness racing and dog carting buggy development because they are easier to lift and move. A lighter dog carting buggy is especially important because 80 percent of this market is composed of women. So far the barriers the company has encountered when innovating are the legislative burden in Australia for introduction of the modified buggy and the lack of information or collaboration related for the application of Nanoflex to its buggies. The most significant barrier has been access to the material itself and the patent which has delayed production for about 18 months.

Knowledge intensive service activities associated with nanotechnology are not very frequent; most of them are undertaken in-house, with other consulting businesses or with UWS, industry associations and government departments. Financing the technology has been an issue, with the owner investing 10 percent of his time on related nanotechnology R&D. The participation in the UWS Nanotechnology Network has provided a reference point for assistance and also has facilitated the introduction to other companies and their products. The company considered it important to link to UWS for their innovation processes because of UWS's diffusion of knowledge of new technologies and facilitation of meetings with other companies looking for similar solutions.

The application of nanotechnology is also transforming internal operational processes of the firm, in small, opportunistic and informal bases and mostly led by the owner of the business. One constrain is lack of expertise so at least one engineer with nanotechnology knowledge will be needed for further developments. Collaboration partners in the innovation process are found locally, in a radius of 20 kilometers, especially with UWS, suppliers and clients.

B) Broens Industries

Broens Industries is a 25-year-old company specialising in technology development and special purpose machinery to be used in the automotive, aerospace and biotechnology industries. The company has 132 fulltime employees, is currently in expansion and has a total annual revenue of more than AU$ 25 million. The innovative activity of the firm is regarded as incremental at the product/service and organisational level. Latest innovations for the Australian market are new machines for the aerospace and automotive industries. Barriers to innovation were mainly the cost or availability of finance and the changing currency conditions. The management of cash flow is difficult when venture capital is scarce and there is not much government assistance leaving limited margins for investment in technology developments.

The use of nanotechnology KISA is limited to research & product development with other businesses and UWS. The application of nanotechnology is also limited and will be guided by client demand. The investment in nanotechnology is small at the moment, mostly in-kind time by the owner, as it is hard to see the short-term benefits. The company is participating with 3 other companies and UWS on a proposal for a technology roadmapping project on toolmaking to improve understanding of the market and focus areas for a more comprehensive applied research project.

The transforming internal processes are patents, seminars and meetings between different sections and a continuous training, mostly customer driven. There is also a mix of people working together; for example a group of designers or engineers work together with a group of tradespeople to work out ideas. The firm believes innovation is necessary to survive, and it must be commercialised, so teams are very focused on product development and manufacturing. The firm will need at least one person (part-time) with nanotechnology specific knowledge for the future. Collaboration partners are all from the local area, within 20 kilometres and still relating to the more traditional activities of the firm.

C) KIRK Group

KIRK Group is a 32-year-old firm specialising in high resolution printing for corporate plans. The firm has 90 staff members, the business structure is in expansion and the annual revenue is more than AU\$2 million. The types of innovations are mostly incremental (product/service and organisational) but they also have a radical process innovation when they standardised international brand colours circular plates for their printing cylinders.

The use of nanotechnology KISA is small and limited to industry development, business planning advice and information sharing and new knowledge. These activities are mainly performed in-house, with other businesses and UWS. They have also gained access to government funding for nanotechnology R&D through an Australian Research Council Linkage grant with UWS. This project provides for the development of specific printing technology needed by the firm and it will constitute a radical innovation. Transforming internal processes are through project folders that can be consulted by all employees, feedback from employees on particular aspects of the project and from allocating current staff to new projects. In the case of the nanotechnology being developed with UWS, current staff will move into the new technology instead of hiring new specialised people. The critical aspect is that technology moves too fast and the time in getting the tech-

nology (sooner better than later) is an important competitive advantage. Collaboration partners are found in the local area within 20 kilometres and, as in the previous case, still around traditional activities of the firm.

The surveys conducted on the UWS Nanotechnology Network over a 2 year period suggest that knowledge intensive service activities facilitated by local universities have a high impact on firms' absorption of nanotechnology influencing partnerships in co-production of innovation. It also suggests that suburbs/regions with manufacturing business have the potential to apply nanotechnology in the medium term as nanotechnology greatly applies to many manufacturing processes related to fabrics, cosmetics, coatings, metals etc.

The case studies show that *partners in the innovation process of nanotechnology are from the local area,* therefore innovation policies and programs need to also have a focus at the regional and local level. Local networking strategies and especially those linking industry and university research departments are needed to advance the field in the medium to long term in a way that is relevant for industry.

Is Local Knowledge Infrastructure a Factor Impacting University-Industry Networks?

A further question relates to industry access to knowledge infrastructure in the region covered by the UWS Nanotechnology Network and whether the geography of knowledge matters so that the firms involved in the project had limited access to other sources of knowledge. An analysis of the state of New South Wales shows two bands of knowledge institutions. The first along the coast and the second in the regional areas to the west of the first band. No institutions are located in the far west of the state. Within the coastal band, there is an obvious concentration bordered by the cities of Newcastle, Wollongong, Penrith and the coastline. This can be called the 'Sydney Knowledge Corridor' stretching from Newcastle region to Wollongong with multi-centres around Newcastle, Sydney and Wollongong (AEGIS, 2005). There are clear concentrations of knowledge producing institutions within this knowledge corridor and the highest concentrations are in the eastern and central suburbs of Sydney. These are also areas of high concentration in terms of the number of ICT and KIBS companies (Martinez-Fernandez et al, 2005e).

In contrast, few knowledge institutions are located in the western section of the corridor at any of the city centres. The University of Western Sydney (UWS) has 6 central campuses in the west, the University of Sydney has a campus in the south and there is a concentration of research and teaching at and around Westmead Children's Hospital but no CSIRO[5] units in the west. UWS is the only large tertiary level teaching university operating in the western part of the knowledge corridor, serving a very large urban population as well as the rural fringes of the city such as the Blue Mountains, Wollondilly Shire and the Southern Highlands. While this area also supplies other Sydney universities with many students, UWS remains the closest for many and the tertiary institution of their choice. In most cases, therefore, businesses both in the west of the State and in the west of Sydney do not have the same access to knowledge intensive institutions as those located further east, closer to the business and financial centre of Sydney. The challenges of accessing knowledge were actually one of the determinants for UWS formation of the nanotechnology network and it would be difficult to create and maintain this network outside the University.

University expertise (skilled personnel and high level equipment) available in a region is important in two main ways. Firstly, it serves to 'capture' knowledge generated elsewhere. Secondly, it enables the region to participate in creating specialised expertise and developing it further to meet local needs. Thus, local, national and international knowledge is translated by university players and *transferred* into locally useful knowledge for supporting existing industries, generating new industries, informing public policies and meeting other community needs such as health, urban planning, environmental control, education, and aged care.

Universities have an important role to play in the generation, transmission and transfer of knowledge but they are not alone and a strong interdependent relationship between universities and the other knowledge producing and using sectors that contribute to knowledge development is also vital (Tornatzky et al. 2001:3). The presence of universities or evidence of the generation and transmission of knowledge does not necessarily mean that a *transfer* into industry, business and commercial activity is taking place. Well-functioning university-industry transfer requires input from both producers and users of knowledge. These interactions between 'users' and producers of knowledge have two functions. They create new knowledge, through implementation of technologies, prototyping, pilot plant trials and other forms of testing, and they generate the critical mass needed for sustained innovation. The simple linear model of knowledge flows as between RTOs and companies has long ago been replaced by an interactive model (Kline and Rosenberg 1986).

Therefore, university-industry networks have the most to offer to the transfer of knowledge but this needs to be field or sector specific within a region as 'knowledge needs' and firm behaviour vary depending on industry specialisation and the level of university research strengths. Our study suggests that *access to knowledge infrastructure and linkages to universities facilitates radical innovation of local firms*. The case of the UWS Nanotechnology Network indicates[6] that universities impact is greater in the case of frontier technologies such as nanotechnology than in the case of established technologies where firms have developed expertise over an extended period of time. It can also be argued that competitiveness of innovative firms need to be seen within the network of relationship with other parts of the regional innovation system in order to understand firm's access and use of cutting edge technology. Informal networks seem to be a powerful instrument for transfer of new knowledge and its impact seems to be more powerful when the main players are found in close geographical proximity and the network involves players from different parts of the value-chain.

CONCLUSION

Nanotechnology constitutes a breakthrough in science with the capacity to control matter at the atomic level. Industry applications are vast and diverse as materials can be transformed at a very small scale. The introduction in the market of nanotechnology applications has occurred at a very high speed in many OECD countries and the scientific community is growing in strength and industry relevance. However, nanotechnology dynamics related to fuzzy industry boundaries, immaturity of the technology, cross-disciplinarity of the field and the possible radical or even disruptive effects of nanotechnology complicates the absorption of nanotechnology by the firm.

As a General Purpose Technology, the application to industry sectors is exceptionally varied and requires a different approach to industry analysis as the sectors are blurred and the boundaries are unclear. The use of 'nanotechnology ecosystems' constitute a better analytical framework for analysis as within each ecosystem more than one industry coexists, as well as research labs and other organizations. Seven nanotechnology ecosystems can be identified: medical applications, coating and powders, devices & micro-electronics, pharmaceuticals & cosmetics, industrial energy, environment remediation and structural materials.

International competitiveness of manufacturing industries is impacted by implementation of nanotechnology in their innovation processes.

However, the awareness and participation of manufacturing companies in the nanotechnology race still limited. In particular, Australia has weak R&D investments, patents and citation measures. This paper discussed the case of a university-industry network, which has a positive impact on manufacturing firms in a peripheral area of Sydney Metropolitan Region. The following conclusions can be extracted:

First, *companies applying Frontier Technologies seems to rely more on scientific knowledge produced by Universities and public/private research and technology institutions (RTOs) than on knowledge generated by the private sector.* The survey of South-West Sydney firms also found that firms applying nanotechnology were more reluctant to trust in-house sources of knowledge which contrast with other established sectors. This difference might relate to the early path creation of nanotechnology if compared with more established technologies such as ICT. Consultancy firms (KIBS), industry associations and government departments have only a small role in the co-production of nanotechnology knowledge.

Second, results of the survey indicate that the *perceived barriers for the introduction of nanotechnology in the firm's product development have more to do with issues of 'relevance and information' than with market demand or lack of skills in the firm.* Network approaches are a good instrument to provide relevant information in a well planned context where to discuss nanotechnology with other interesting parties, showcase current commercializations and discuss new business prospects and discoveries by university researchers.

Third, the surveys conducted on the UWS Nanotechnology Network over the 2 year timeframe suggest that suburbs/regions with manufacturing business can develop smarter and more sustainable industry as nanotechnology greatly applies to many manufacturing processes involving materials, cosmetics, coatings, metals etc.

Four, the case studies show that *partners in the innovation process of nanotechnology are from the local area* and therefore policies and programs to develop networked nanotechnology innovation systems need to also have a focus at the regional and local level. Local networking strategies and especially those linking industry and university research departments are needed to advance the field in the medium to long term in a way that is relevant for these firms.

Five, access to knowledge infrastructure and linkages to universities facilitates innovation of local firms. Universities impact on firm innovation is even greater in the case of frontier technologies such as nanotechnology. *Competitiveness of innovative firms need also to be seen in the context of their access to relevant knowledge and network of relationships with other parts of the regional innovation system; not just as a set of productive resources.*

Six, the *location of knowledge institutions does matter for promoting interactions with industry.* Even within Sydney itself, where there is an overall concentration of knowledge institutions, there are many areas with relatively little access, notably the greater west, in spite of the growing population in Western Sydney. Firms in these areas are therefore faced with barriers to accessing knowledge and might be in more need to participate in activities oriented to sharing knowledge.

Finally, university strategies for knowledge generation, diffusion and transfer matter for the development of our cities and regions. Networking projects that facilitate well planned knowledge intensive service activities for those areas of expertise that are cutting edge, such as nanotechnology, are fundamental for maintaining industry competitiveness in areas of commercial uncertainty. These networks of innovation require public support to be sustainable in the short term and experiences such as the UWS nano-network suggest their effectiveness as innovation policy instruments. The results of the network are also bidirectional; at the same time that industry benefits from research discoveries and product showcase, research departments benefit from open dialogue with industry in relation to opportunities for commercialization of new technological developments. The strength of the UWS nano-network also have to do with the lack of knowledge institutions in the region, leaving this network as the only relevant meeting point for SMEs willing to embrace radical innovation. The different roles of universities in the co-production of knowledge for industry nanotechnology applications and the analysis of its effects in society need further exploration.

REFERENCES

Acs, Z. (2002). *Innovation and the Growth of Cities*. Cheltenham, UK: Edward Elgar Publishing Ltd.

AEGIS. (2005). *Stocktake of NSW as a Knowledge Hub*. Sydney, Australia: UWS.

Andersen, M. M. (2005). Path Creation in the Making – the Case of Nanotechnology. In *DRUID 2005 Conference*.

Audretsch, D. (1995). *Innovation and industry evolution*. Cambridge, MA: MIT Press.

Australian Vice-Chancellors Committee-AVCC. (2005 December). *Engagement with Business and Community: Enhancing Universities' Interaction.*

Business Review Weekly (BRW). (2005, August 4). Big Hopes for Tiny Floats. *BRW*, 24.

Custer, M. (2005 June). Investing in Nanotechnology is its Own Science. *Bulletin Special, Credit Suisse*, 32-33.

Dandolopartners, 2005a. *Nanotechnology and the Business Community: A study of business' understanding of and attitudes towards nanotechnology.* Report for Nanotechnology Victoria. 11 July 2005.

Dandolopartners. (2005b). *Nanotechnology: National Business Interviews*. Detailed Report. Report commissioned by Nanovic and DITR.

Dandolopartners. (2005c). *Nanotechnology: A National Survey of Consumers*. Detailed Report. 11 July 2005.

Faulkner, W., & Senker, J. (1995). *Knowledge Frontiers*. Oxford, UK: Oxford University Press.

Garlic, S. (2000). *Engaging Universities and Regions: Knowledge contribution to regional economic development in Australia*. Occasional Paper Series 00/15, Higher Education Division, Department of Education, Training and Youth Affairs, Canberra. Retrieved from http://www.detya.gov.au/highered/eippubs/eip00_15/00_15.pdf

Hung, S.-Ch., & Chu, Y.-Y. (2006). Stimulating new industries from emerging technologies: challenges for the public sector. *Technovation, 26*, 104–110. doi:10.1016/j.technovation.2004.07.018

ISR. (2001). *Nanotechnology in Australian Industry. Proceedings and outcomes report*. Canberra, Australia: ISR.

Kline, S. J., & Rosenberg, N. (1986). *The Positive Sum Strategy, Harnessing Technology for Economic Growth An Overview of Innovation*. Washington, DC: National Academy Press.

Knop, K. (2005 June). Nanotechnology: A Big Future for Small Things. *Bulletin Special, Credit Suisse*, 9-11.

Kostoff, R. N., Murday, J. S., Lau, C. G. Y., & Tolles, W. M. (2006b). The seminal literature of nanotechnology research. *Journal of Nanoparticle Research, 8*(1).

Kostoff, R. N., Murday, J. S., Stump, J. A., Johnson, D., Lau, C. G. Y., & Tolles, W. M. (2006a). The structure and infrastructure of the global nanotechnology literature. *Journal of Nanoparticle Research, 8*(1).

Luther, W. (2004). *Industrial Applications of Nanomaterials – chances and risks. Technology Analysis*. VDI Technologiezentrum.

Lux Research. (2004). *The Nanotechnology Report*.

Lux Research. (2005). *Ranking the nations: Nanotech's Shifting Global Leaders*.

Martinez-Fernandez, M. C. (2004). Regional Collaboration Infrastructure: Effects in the Hunter Valley of NSW. *Australian Planner, 41*(4).

Martinez-Fernandez, M. C. (2005b). *Knowledge Intensive Service Activities (KISA) in Innovation of Mining Technology Services in Australia*. Sydney, Australia: University of Western Sydney.

Martinez-Fernandez, M. C., & Leevers, K. (2004). Knowledge creation, sharing and transfer as an innovation strategy: The discovery of nano-technology by South-West Sydney. [IJTM]. *International Journal of Technology Management, 28*(3-6), 560–581.

Martinez-Fernandez, M. C., & Potts, T. (2006). *Innovation at the Edges of the Metropolis: An Analysis of Innovation Drivers in Peripheral Suburbs of Sydney*. Opolis.

Martinez-Fernandez, M. C., & Potts, T. (2007 March). Innovation at the Edges of the Metropolis: An Analysis of Innovation Drivers in Peripheral Suburbs of Sydney. *Opolis*.

Martinez-Fernandez, M. C., Potts, T., Receretnam, M., & Bjorkli, M. (2005a). *Innovation at the Edges: An analysis of Innovation Drivers in South West Sydney*. Sydney, Australia: University of Western Sydney. Retrieved April 28, 2005, from http://aegis.uws.edu.au/innovationedges/main.html

Martinez-Fernandez, M. C., & Rerceretnam, M. (2006). *The Role of UWS as a Knowledge Hub*. Sydney, Australia: UWS.

Martinez-Fernandez, M. C., Soosay, C., Krishna, V. V., Turpin, T., & Bjorkli, M. (2005d). *Knowledge Intensive Service Activities (KISA) in Innovation of the Software Industry in Australia*. Sydney, Australia: University of Western Sydney.

Martinez-Fernandez, M. C., Soosay, C., Krishna, V. V., Turpin, T., Bjorkli, M., & Doloswala, K. (2005c). *Knowledge Intensive Service Activities (KISA) in Innovation of the Tourism Industry in Australia*. Sydney, Australia: University of Western Sydney.

Martinez-Fernandez, M. C., Soosay, C., & Tremayne, K. (2005e). *Learning Spaces, Co-production of Knowledge and Capacity Building in the Service Firm*. AEGIS Working Paper 2005-06. Sydney, Australia: AEGIS, UWS.

Maskell, P. (2001). Towards a knowledge-based theory of the geographical cluster. *Industrial and Corporate Change, 10*(4), 921–943. doi:10.1093/icc/10.4.921

Mnyusiwalla, A., Daar, A. S., & Singer, P. A. (2003). Mind the gap: science and ethics in nanotechnology. *Nanotechnology, 14*, R9–R13. doi:10.1088/0957-4484/14/3/201

Molas-Gallart, J., Salter, A., Patel, P., & Scott, A. (2002). *Measuring Third Stream Activities – Final Report to the Russell Group of Universities*. University of Sussex, Science & Technology Policy Research.

New South Wales Department of State and Regional Development. (2002 December). Retrieved from http://ats.business.gov.au

OECD. (1999). *Benchmarking Knowledge-based Economies*. Paris: OECD.

OECD. (2001). *Cities and Regions in the New Learning Economy*. Paris: OECD.

OECD. (2003). *OECD Science, Technology and Industry Scoreboard*. Paris: OECD.

OECD. (2004, May). Nanotech is not small. *The OECD Observer. Organisation for Economic Co-Operation and Development*, 243.

OECD. (2005). *University Research Management. Developing Research in New Institutions*. Paris: OECD.

Prime Minister's Science, Engineering and Innovation Council (PMSEIC). (2005). *Nanotechnology: Enabling Technologies for Australian Innovative Industries*. Canberra, Australia: DEST.

Senker, J., Faulkner, W., & Velho, L. (1998). Science and technology knowledge flows between industrial and academic research: a comparative study . In Etzkowitz, H., Webster, A., & Healey, P. (Eds.), *Capitalizing Knowledge: New Intersections of Industry and Academia* (pp. 111–132). New York: State University of New York Press.

Shea, C. M. (2005). Future management research direction in nanotechnology: A case study. *Journal of Engineering and Technology Management, 22*, 185–200. doi:10.1016/j.jengtecman.2005.06.002

Smith, K. (2000). Innovation as a systemic phenomenon: Rethinking the role of policy. *Enterprise and Innovation Management Studies, 1*(1), 73–102. doi:10.1080/146324400363536

Sproats, K. (2003). *The role of universities as economic drivers in developing their local environment*. Presented at ACU General Conference, Belfast.

Teece, D. (1987). Profiting from technological innovation: Implications for integration, collaboration, licensing and public policy . In Teece, D. (Ed.), *The competitive challenge: Strategies for industrial innovation and renewal* (pp. 185–219). Cambridge, MA: Ballinger.

The Royal Society and the Royal Academy of Engineering. (2004). *Nanoscience and Nanotechnologies: Opportunities and Uncertainties*. Retrieved from http://www.nanotech.org/uk

TIAC. (2002). *The Organisation of Knowledge: Optimising the Role of Universities in a Western Australia 'Knowledge Hub*. Australia: TIAC.

Tornatzky, L., Gray, O., & Waugaman, P. (2001). *Making the Future: Universities, Their States and the Knowledge Economy*. Raleigh, NC: Southern Growth Policies Board.

Turpin, T., & Martinez-Fernandez, C. (2003). *Riding the Waves of Policy*. AEGIS Working Paper Series 2003-02. Sydney, Australia: AEGIS.

Wood, S., Jones, R., & Geldart, A. (2003). *The Social and Economic Challenges of Nanotechnology*. Swindon, UK: Economic & Social Research Council.

ENDNOTES

[1]   University of Western Sydney

[2]   Top ten firms are IBM, Xerox, Micron, Eastman Kodak, Motorola, Texas Instruments, NEC, Canon, Advanced Microdevices, GE.

[3]   Knowledge-Intensive Service Activities (KISA) are defined as production and integration of service activities undertaken by firms in manufacturing or service sectors, in combination with manufactured outputs or as stand-alone services. KISA can be provided by private enterprises or public sector organisations. Typical examples include: R&D services, management consulting, IT services, human resource management services, legal services such as IP-related issues, accounting and financing services, and marketing services. (Martinez-Fernandez et al, 2005b,c,d).

[4]   http://www.uws.edu.au/nano

[5]   Australia's public research institution

[6]   Based on interview data

Chapter 9
Licensing in the Theory of Cooperative R&D

Arijit Mukherjee
University of Nottingham, UK & The Leverhulme Centre for Research in Globalisation and Economic Policy, UK

ABSTRACT

The literature on cooperative R&D did not pay much attention to knowledge sharing ex-post innovation through technology licensing, which is a common phenomenon in many industries. The author shows how licensing ex-post R&D affects the incentive for cooperative R&D and social welfare by affecting R&D investment and the probability of success in R&D. Licensing increases both the possibility of non-cooperative R&D and social welfare.

INTRODUCTION

Cooperation in R&D is a common phenomenon in this contemporary world.[1] A large body of literature has emerged to explain the rationale for cooperative research and development (R&D). Current literature has identified knowledge spillover, uncertainty in the R&D process, cost sharing in R&D and unsuccessful patent application as the main motives for cooperative R&D (see, e.g., d'Aspremont, and Jacquemin, 1988, Kamien et al. 1992, and Suzumura, K., 1992, Marjit, 1991, Combs, 1992, Mukherjee and Marjit, 2004, Mukherjee, 2005, Kabiraj, 2006 and 2007 and

DOI: 10.4018/978-1-61692-006-7.ch009

Mukherjee and Ray, 2009). However, the previous works have generally ignored knowledge sharing ex-post innovation through technology licensing, which is an important element of conduct in many industries.[2] In this chapter, we show how licensing ex-post R&D affects the incentive for cooperative R&D and social welfare by affecting R&D investment and the probability of success in R&D.[3]

Considering knowledge sharing as the benefit from cooperative R&D, we show that licensing ex-post R&D does not encourage the firms to do cooperative R&D, while the firms do cooperative R&D in the absence of licensing provided the cost reduction through R&D is small. It must be clear from our analysis that if cooperative R&D provides other benefits such as cost sharing and synergies, along with knowledge sharing, cooperative R&D may occur even in the presence of licensing, yet licensing reduces the incentive for cooperative R&D. Hence, one would expect less cooperative R&D in industries where knowledge sharing through licensing is easy.

If there is no licensing, we show that there can be a situation where the firms prefer cooperative R&D but welfare is higher under non-cooperative R&D, thus creating a conflict of interest between the firms and the society. However, even if licensing reduces the incentive for cooperative R&D, it increases welfare irrespective of its effect on R&D organization.

The remainder of the paper is organized as follows. Section 2 provides the model with non-cooperative and cooperative R&D without licensing. We extend this model in section 3 by incorporating licensing ex-post R&D. Section 4 shows welfare implications. Section 5 concludes.

THE CASE OF NO LICENSING

Consider an economy with two firms, called firm 1 and firm 2. Assume that these firms can produce a homogeneous product with a technology corresponding to the constant marginal cost of production $\bar{c}$. Both firms do R&D and each of them tries to reduce the cost of production to c, which is assumed to be zero for simplicity. However, success in R&D is uncertain. Assume that p and $(1-p)$ show respectively unconditional probability of success and failure in R&D. We assume that both firms face the same probability of success in R&D. Assume that each firm affects the probability of success in R&D through its own R&D investment, i.e., $p(x_i)$, $i = 1,2$, where x_i is firm i's R&D investment. We consider that $p'(x_i) > 0$, $p''(x_i) < 0$, $p'(0) = \infty$ and $p'(\infty) = 0$ for $i = 1,2$. We further assume that there are no fixed costs related to production and R&D.

The assumption of no fixed cost of R&D implies that both firms always find it profitable to do R&D compared to no R&D. If there are fixed costs of R&D, firms might not find it profitable to do R&D if the probabilities of success in R&D were sufficiently low. Since our purpose is to examine the impact of licensing on R&D organization, i.e., non-cooperative and cooperative R&D, and social welfare, we abstract the possibility of no R&D by assuming no fixed cost of R&D.

Assume that the inverse market demand function is

$$P = 1 - q, \tag{1}$$

where, P is price of the product, q is the industry output and $1 > \bar{c}$. We further assume that $\bar{c} < .5$ which implies that if one firm has a technology corresponding to the marginal cost of production 0 and the other firm has a technology corresponding to the marginal cost of production $\bar{c}$, the outputs of both firms are positive.

We consider the following game in this section. At stage 1, both firms decide whether to do non-cooperative R&D or cooperative R&D. At stage 2, they compete like Cournot duopolists in the product market. We solve the game through backward induction.

The expected profit of the ith firm under non-cooperative R&D and under cooperative R&D are respectively

$$\begin{aligned}
V_i(NC) &= p(x_i)p(x_j)\pi_i(0,0) \\
&+ p(x_i)(1 - p(x_j))\pi_i(0,\bar{c}) \\
&+ (1 - p(x_i))p(x_j)\pi_i(\bar{c},0) \\
&+ (1 - p(x_i))(1 - p(x_j))\pi_i(\bar{c},\bar{c}) - x_i
\end{aligned} \tag{2}$$

and

$$V_i(C) = p(x_i)p(x_j)\pi_i(0,0)$$
$$+p(x_i)(1 - p(x_j))\pi_i(0,0)$$
$$+(1 - p(x_i))p(x_j)\pi_i(0,0) \tag{3}$$
$$+(1 - p(x_i))(1 - p(x_j))\pi_i(\bar{c},\bar{c}) - x_i$$

where $i,j = 1,2$ and $j \neq i$.

The respective first order conditions for profit maximization with respect to R&D invests are

$$p'(x_i)p(x_j)\pi_i(0,0)$$
$$+p'(x_i)(1 - p(x_j))\pi_i(0,\bar{c})$$
$$-p'(x_i)p(x_j)\pi_i(\bar{c},0) \tag{4}$$
$$-p'(x_i)(1 - p(x_j))\pi_i(\bar{c},\bar{c}) = 1$$

and

$$p'(x_i)p(x_j)\pi_i(0,0)$$
$$+p'(x_i)(1 - p(x_j))\pi_i(0,0)$$
$$-p'(x_i)p(x_j)\pi_i(0,0) \tag{5}$$
$$-p'(x_i)(1 - p(x_j))\pi_i(\bar{c},\bar{c}) = 1$$

The optimal R&D investments of the firms for the particular regime can be found by solving the respective first order conditions. Symmetry of the firms implies that the equilibrium R&D investments of the firms are the same for a particular situation, though these equilibrium values may be different under non-cooperative R&D and under cooperative R&D. We assume that the probability function is such that it generates unique equilibrium R&D investments.

To find out closed form solutions for our analysis, let us assume that the probability function of the ith firm, $i = 1,2$ is $p(x_i) = x_i^{\frac{1}{2}}$. Given this probability function and due to symmetry, we find from (4) and (5) that the equilibrium R&D

investments of the ith firm, $i = 1,2$ are respectively $x_{NC} = \left(\dfrac{4\bar{c}}{18 + 4\bar{c}^{-2}}\right)^2$ and

$x_C = \left(\dfrac{\bar{c}(2 - \bar{c})}{18 + \bar{c}(2 - \bar{c})}\right)^2$ under non-cooperative

R&D and under cooperative R&D. The corresponding equilibrium probabilities of success are

$$p_{NC} = \frac{4\bar{c}}{18 + 4\bar{c}^{-2}} \quad \text{and} \quad p_C = \frac{\bar{c}(2 - \bar{c})}{18 + \bar{c}(2 - \bar{c})}.$$

Hence, the equilibrium expected payoffs of the ith firm, $i = 1,2$ under non-cooperative R&D and under cooperative R&D are respectively

$$V_i(NC) = \frac{[16\bar{c}^{-2} + 4\bar{c}(18 - 4\bar{c}(1 - \bar{c}))((1 + \bar{c})^2 + (1 - 2\bar{c})^2) + (18 - 4\bar{c}(1 - \bar{c}))^2(1 - \bar{c})^2 - 144\bar{c}^{-2}]}{9(18 + 4\bar{c}^{-2})^2} \tag{6}$$

and

$$V_i(C) = \frac{[\bar{c}(2 - \bar{c})(36 + \bar{c}(2 - \bar{c})) + 324(1 - \bar{c})^2 - 9\bar{c}^{-2}(2 - \bar{c})^2]}{9(18 + \bar{c}(2 - \bar{c}))^2} \tag{7}$$

The firms do cooperative R&D compared to non-cooperative R&D if neither of them is worse-off under cooperative R&D compared to non-cooperative R&D. So, the firms do cooperative R&D instead of non-cooperative R&D if

$$V_i(C) > V_i(NC). \tag{8}$$

Figure 1[4] plots the difference $V_i(C) - V_i(NC)$

Figure 1 shows that the firms prefer cooperative R&D compared to non-cooperative R&D if $\bar{c} < .35$ (approx.).

Figure 1. The relationship between $\bar{c}$ and $(V_i(C) - V_i(NC))$ for $\bar{c} \in (0,.5)$

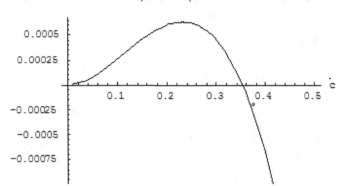

THE CASE OF LICENSING

Now we extend the model of the previous section by incorporating the possibility of licensing ex-post R&D. Consider the following game. At stage 1, the firms decide whether to do non-cooperative R&D or cooperative R&D. At stage 2, they decide on technology licensing. At stage 3, they compete like Cournot duopolists. We solve the game through backward induction.

It is clear from our framework that licensing is not an option if the firms do cooperative R&D since, if successful, they share information about new innovation through cooperative R&D. However, under non-cooperative R&D, firms may have the incentive for licensing ex-post R&D if there is unilateral success in R&D. We assume that, if the firms do non-cooperative R&D, under unilateral success in R&D, the successful firm can give a take-it-or-leave-it licensing offer to the unsuccessful firm, which accepts the offer if the offer does not make it worse off compared to no licensing. Under licensing, licenser can charge a non-negative up-front fixed-fee, F and a per-unit output royalty, r.[5]

Before examining the firms' preference over cooperative R&D and non-cooperative R&D, we first determine the optimal licensing contract. Since, the firms are symmetric, without loss of generality, we assume that, under unilateral suc-

cess in R&D, firm 1 succeeds in R&D and decides whether to license the technology to firm 2 or not. Therefore, in this situation, firm 1 maximizes the following expression:

$$\underset{F,r}{Max} \; F + rq_2 + \pi_1(0, r) \tag{9}$$

s.t., $\quad F + rq_2 + \pi_1(0, r) \geq \pi_1(0, \bar{c}) \tag{10}$

$$\pi_2(0, r) - F \geq \pi_2(0, \bar{c}) \tag{11}$$

$$F \geq 0, \; q_i \geq 0, \; i = 1, 2.[6] \tag{12}$$

Solution of the problem given by (9)-(12) will provide the following result.

Proposition 1:*If the firms do non-cooperative R&D, under unilateral success in R&D, the successful firm will always license its technology to the unsuccessful firm. The licensing contract will involve only output royalty equal to* $\bar{c}$.

Since the proof of the above proposition is similar to Mukherjee and Balasubramanian (2001), we are leaving the details here. Ignoring the constraint $F \geq 0$ and constraint (11), the optimal output royalty is $r^* = .5$ and this amount of royalty implies zero production by the licensee. In this situation, the up-front fixed-fee, i.e., F, will be negative to satisfy constraint (11). The non-negativity constraint on F implies that the optimal output royalty cannot exceed $\bar{c}$. Therefore, the optimal royalty is $\bar{c}$, since $r^* = .5 > \bar{c}$, and the optimal up-front fixed-fee is 0.

Licensing has two opposing effects on the profit of the licenser. On one hand, licensing makes the licensee more cost efficient and reduces the profit of the licenser. On the other hand, licensing helps the licenser to raise its profit by charging a price for its technology. The optimal royalty $\bar{c}$ makes the effective marginal cost of the licensee the same under licensing and under no licensing. Hence, this amount of royalty eliminates the negative impact of licensing on the profit of the licenser but increases its profit through royalty income.

Though we have considered linear demand function, it is easy to understand that even for non-linear demand function, licenser can charge the output royalty in a way that eliminates the negative impact of licensing on the profit of the licenser and hence, the optimal output royalty will be equal to the difference between the marginal cost of the licensee before licensing and the marginal cost of the licenser. Rockett (1990) shows this as the optimal output royalty for non-linear demand functions.

Under licensing, the expected payoff of the ith firm is

$$V_i(NCL) = p(x_i)p(x_j)\pi_i(0,0)$$
$$+p(x_i)(1 - p(x_j))(\pi_i(0,\bar{c}) + rq_j^*)$$

$$+(1 - p(x_i))p(x_j)\pi_i(\bar{c},0)$$
$$+(1 - p(x_i))(1 - p(x_j))\pi_i(\bar{c},\bar{c}) - x_i, \quad (13)$$

where $i, j = 1, 2$ and $j \neq i$.

The respective first order conditions for profit maximization with respect to R&D invest is

$$p'(x_i)p(x_j)\pi_i(0,0)$$
$$+p'(x_i)(1 - p(x_j))(\pi_i(0,\bar{c}) + rq_j^*)$$
$$-p'(x_i)p(x_j)\pi_i(\bar{c},0)$$
$$-p'(x_i)(1 - p(x_j))\pi_i(\bar{c},\bar{c}) = 1$$
$$(14)$$

Proposition 2: *The equilibrium R&D investment of the ith firm, i =1,2, under non-cooperative R&D with licensing is higher compared to R&D investment under non-cooperative R&D without licensing (except for the situation where the probability of success becomes 1), which is greater than the equilibrium R&D investment under cooperative R&D.*

Proof: Suppose, x_{iNC}^* and x_{jNC}^* are the equilibrium R&D investments of the ith and jth firms under non-cooperative R&D without licensing, where $i, j = 1,2$ and $i \neq j$. So, x_{iNC}^* and x_{jNC}^* satisfy condition (4). Straightforward calculation shows that, given the R&D investments x_{iNC}^* and x_{jNC}^*, left hand side (LHS) of (5) is less than 1. Therefore, given the R&D investment x_{jNC}^*, the optimal R&D investment of the ith firm under cooperative R&D is lower than x_{iNC}^*. Since the firms are symmetric, similar logic implies that, given x_{iNC}^*, the optimal R&D investment of the jth firm under cooperative R&D is lower than x_{jNC}^*. Given that we have unique equilibrium in R&D investments, this implies that the optimal R&D investments are lower under cooperative R&D compared to non-cooperative R&D without licensing.

Similar argument proves that the optimal R&D investments of the firms under non-cooperative R&D with licensing are greater than the optimal R&D investments under non-cooperative R&D without licensing, except for the situation where the probability of success in R&D reaches 1.[7] Q.E.D.

If the probability of success in R&D takes the form $p(x_i) = x_i^{\frac{1}{2}}$, the equilibrium R&D investment of the ith firm, $i = 1,2$ is

$$x_{NCL} = \left(\frac{\bar{c}(7 - 6\bar{c})}{18 + \bar{c}(3 - 2\bar{c})} \right)^2 \text{ under non-cooperative}$$

R&D with licensing. The corresponding equilibrium probability of success is

$$p_{NCL} = \frac{\bar{c}(7 - 6\bar{c})}{18 + \bar{c}(3 - 2\bar{c})}.$$ Hence, the equilibrium

expected payoff under non-cooperative R&D with licensing is

$$V_i(NCL) = \frac{\begin{array}{c}[\bar{c}^2(7 - 6\bar{c})^2 + \bar{c}(7 - 6\bar{c})(18 - 4\bar{c}(1 - \bar{c}))((1 + \bar{c})^2 \\ + 3\bar{c}(1 - 2\bar{c}) + (1 - 2\bar{c})^2) + (18 - 4\bar{c}(1 - \bar{c}))^2(1 - \bar{c})^2 - 9\bar{c}^2(7 - 6\bar{c})^2]\end{array}}{9(18 + \bar{c}(3 - 2\bar{c}))^2}$$

$$(15)$$

Figure 2 plots the difference $V_i(C) - V_i(NCL)$

Figure 2 shows that $V_i(NCL)$ dominates $V_i(C)$ for all $\bar{c}$ between 0 and .5. This result can be explained in the following way. If the firms do cooperative R&D, under unilateral success in R&D, the successful firm needs to share knowledge with the unsuccessful firm. Hence, the successful firm faces higher competition from the unsuccessful firm compared to the situation where

the firms do non-cooperative R&D and do not share knowledge with the competitor. If the firms do non-cooperative R&D and can license afterwards then, under unilateral success in R&D, the successful firm always benefits by licensing its technology to the unsuccessful firm. In this situation, the successful firm can eliminate higher competition from the unsuccessful firm by designing a suitable licensing contract. Since the firms are symmetric and have the similar probability of being the licenser or the licensee, both firms always prefer non-cooperative R&D compared to cooperative R&D under licensing.

The comparison of Figures 1 and 2 shows that if there is no licensing, cooperative R&D occurs for $\bar{c}$ between 0 and .35 (approx.), while cooperative R&D does not occur if there is licensing. Hence, the following result immediate.

Proposition 3:*The possibility of licensing ex-post R&D increases the incentive for non-cooperative R&D compared to cooperative R&D.*

The above analysis shows that the firms always do non-cooperative R&D in the presence of licensing. However, one should take this result with a caution. In this paper, cooperative R&D only helps the firms to benefit from knowledge sharing. We have examined relative profitability of knowledge sharing at the time of R&D compared to knowledge sharing after innovation. However, in our framework, there is no benefit from cost

Figure 2. The relationship between $\bar{c}$ and $(V_i(C) - V_i(NCL))$ for $\bar{c} \in (0,.5)$

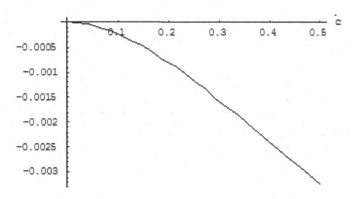

saving or synergies under cooperative R&D. If R&D involves fixed costs as in Marjit (1991), the cost saving provides another incentive for doing cooperative R&D. The possibility of getting benefit from synergies as in Kabiraj and Mukherjee (2000) can also provide an incentive for doing cooperative R&D. With sufficiently higher benefits from cost saving or synergies under cooperative R&D, firms might do cooperative R&D for some parametric configurations even if they have the option for licensing ex-post R&D. However, even in this situation, our basic argument holds, i.e., the possibility of licensing ex-post R&D reduces the incentive for cooperative R&D. The possibility of knowledge sharing through licensing increases the expected profits under non-cooperative R&D and reduces the firms' incentive for cooperative R&D compared to the situation with no licensing.

Welfare Implications

Now we examine how licensing ex-post innovation affects social welfare, which is the sum of consumer surplus and the net industry profit. Like the above analysis, we assume that the probability of success in R&D takes the form $p(x_i) = x_i^{\frac{1}{2}}$.

If the firms do non-cooperative R&D without licensing, cooperative R&D and non-cooperative R&D with licensing, welfare are respectively

$$W(NC) = \frac{\begin{array}{c}[64\bar{c}^{-2} + 4\bar{c}(18 - 4\bar{c}(1 - \bar{c}))(8 - 8\bar{c} + 11\bar{c}^{-2}) \\ +4(18 - 4\bar{c}(1 - \bar{c}))^2(1 - \bar{c})^2 - 288\bar{c}^{-2}]\end{array}}{9(18 + 4\bar{c}^{-2})^2}$$

(16)

$$W(C) = \frac{[4\bar{c}(2 - \bar{c})(36 + \bar{c}(2 - \bar{c})) + 1296(1 - \bar{c})^2 - 18\bar{c}^{-2}(2 - \bar{c})^2]}{9(18 + \bar{c}(2 - \bar{c}))^2}$$

(17)

and

$$W(NCL) = \frac{\begin{array}{c}[4\bar{c}^{-2}(7 - 6\bar{c})^2 + \bar{c}(7 - 6\bar{c})(18 - 4\bar{c}(1 - \bar{c}))(8 - 2\bar{c} - \bar{c}^{-2}) \\ +4(18 - 4\bar{c}(1 - \bar{c}))^2(1 - \bar{c})^2 - 18\bar{c}^{-2}(7 - 6\bar{c})^2]\end{array}}{9(18 + \bar{c}(3 - 2\bar{c}))^2}$$

(18)

We have shown that if there is no licensing, the firms do cooperative R&D in the absence of licensing if $\bar{c} \in (0, .35)$. We plot the difference $W(C) - W(NC)$ in Figure 3 for $\bar{c} \in (0, .35)$.

It is immediate from Figure 3 that, though the firms prefer cooperative R&D for $\bar{c} \in (0, .35)$, welfare can be higher under non-cooperative R&D. Hence, in the absence of licensing, there

Figure 3. The relationship between $\bar{c}$ and W(C)–W(NC) for $\bar{c} \in (0,.35)$

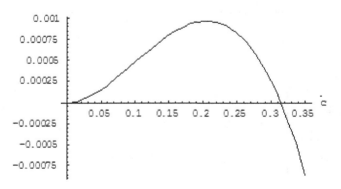

Figure 4. The relationship between $\bar{c}$ and W(C)–W(NCL) for $\bar{c} \in (0,.35)$

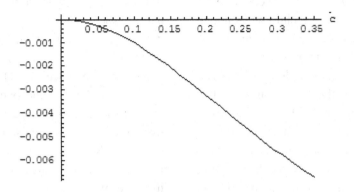

may be a conflict of interest between the firms and the society and cooperative R&D may reduce welfare compared to non-cooperative R&D. Even if cooperative R&D helps the firms to share the new technology and to increase the probabilities of success in R&D by increasing R&D investments, the higher R&D investments under cooperative R&D compared to non-cooperative R&D create higher costs to the society under the former compared to the latter. The negative effect of higher R&D costs may reduce welfare under cooperative R&D compared to non-cooperative R&D.

However, if there is licensing, the firms do non-cooperative R&D even for $\bar{c} \in (0,.35)$. Hence, if $\bar{c} \in (0,.35)$, licensing changes R&D organization from cooperative R&D to non-cooperative R&D. We plot *W(C)–W(NCL)* in Figure 4 for $\bar{c} \in (0,.35)$.

Figure 4 shows that welfare is higher under non-cooperative R&D with licensing than under cooperative R&D. Hence, even if licensing reduces the incentive for cooperative R&D, it helps to increase welfare. Non-cooperative R&D with

Figure 5. The relationship between $\bar{c}$ and W(NC)–W(NCL) for $\bar{c} \in (.35,.5)$

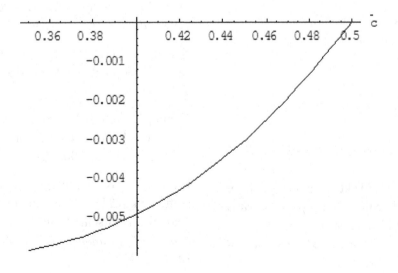

licensing increases the industry profit under unilateral success in R&D compared to cooperative R&D where the firms share technologies under unilateral success in R&D.[8] Further, the R&D investments are higher under non-cooperative R&D with licensing than under cooperative R&D. These effects together help to outweigh the negative effects of higher R&D costs and lower consumer surplus in the case of a unilateral success in R&D under non-cooperative R&D with licensing compared to cooperative R&D, thus increasing welfare under the former than the latter.

The following figure shows that even if licensing does not affect R&D organization (i.e., the firms do non-cooperative R&D irrespective of licensing), which happens for $\bar{c} \in (.35, .5)$, welfare increases with licensing.

Figure 5 plots $W(NC) - W(NCL)$ for $\bar{c} \in (.35, .5)$ and shows that welfare is higher under $W(NCL)$ than under $W(NC)$. Licensing increases the probabilities of success in R&D and the industry profits under unilateral success in R&D. These benefits together outweigh the negative effect of higher R&D investments under non-cooperative R&D with licensing compared to non-cooperative R&D without licensing, thus creating higher welfare under the former than the latter.

We summarize the above discussion in the following result.

Proposition 4:*The possibility of licensing increases welfare, whether or not it affects R&D organization.*

CONCLUSION

We show the effect of licensing on R&D organization and social welfare. If there is licensing, firms do non-cooperative R&D keeping the option for licensing open, while they do cooperative R&D without licensing if the cost reduction through R&D is small. Hence, licensing reduces the incentive for cooperative R&D.

If there is no licensing, we show that there may be a conflict of interest between the firms and the society with respect to the R&D organization. There can be a situation where the firms prefer cooperative R&D but welfare is higher under non-cooperative R&D. However, licensing increases welfare whether or not it affects the R&D organization. Though licensing increases R&D investments, thus increasing the probabilities of success in R&D and the costs of doing R&D, the former effect dominates the latter and licensing increases welfare.

REFERENCES

Combs, K. L. (1992). Cost sharing vs. multiple research projects in cooperative R&D. *Economics Letters, 39*, 353–371. doi:10.1016/0165-1765(92)90273-2

d'Aspremont, C., & Jacquemin, A. (1988). Cooperative and non-cooperative R&D in duopoly with spillovers. *The American Economic Review, 78*, 1133–1137.

DeCourcy, J. (2001). *International trade policy and the National Cooperative Research Act.* PhD dissertation, Michigan State University.

Gallini, N. T., & Winter, R. A. (1985). Licensing in the theory of innovation. *The Rand Journal of Economics, 16*, 237–252. doi:10.2307/2555412

Kabiraj, T. (2006). On the incentive for cooperative and non-cooperative R&D in duopoly. *Arthaniti, 5*, 24–33.

Kabiraj, T. (2007). On the incentives for cooperative research. *Research in Economics, 61*, 17–23. doi:10.1016/j.rie.2006.12.003

Kabiraj, T., & Mukherjee, A. (2000). Cooperation in R&D and production: a three firm analysis. [Zeitschrift fur Nationalokonomie]. *Journal of Economics, 71*, 281–304. doi:10.1007/BF01228744

Kamien, M. (1992). Patent licensing. In Aumann, R. J., & Hart, S. (Eds.), *Handbook of game theory*. Amsterdam: Elsevier.

Kamien, M. I., Muller, E., & Zang, I. (1992). Research joint ventures and R&D cartels. *The American Economic Review, 82*, 1293–1306.

Katz, M. L., & Shapiro, C. (1985). On the licensing of innovations. *The Rand Journal of Economics, 16*, 504–520. doi:10.2307/2555509

Marjit, S. (1991). Incentives for cooperative and non-cooperative R&D in duopoly. *Economics Letters, 37*, 187–191. doi:10.1016/0165-1765(91)90129-9

Mukherjee, A. (2005). Innovation, licensing and welfare. *The Manchester School, 73*, 29–39. doi:10.1111/j.1467-9957.2005.00422.x

Mukherjee, A. (2008). Technology licensing. In Rajan, R. S., & Reinert, K. A. (Eds.), *Princeton Encyclopaedia of the World Economy*. Princeton, NJ: Princeton University Press.

Mukherjee, A., & Balasubramanian, N. (2001). Technology transfer in horizontally differentiated product-market. [Richerche Economiche]. *Research in Economics, 55*, 257–274. doi:10.1006/reec.2001.0254

Mukherjee, A., & Marjit, S. (2004). R&D organization and technology transfer. *Group Decision and Negotiation, 13*, 243–258. doi:10.1023/B:GRUP.0000031079.32373.a4

Mukherjee, A., & Ray, A. (2009). Unsuccessful patent application and cooperative R&D. *Journal of Economics, 97*, 251–263. doi:10.1007/s00712-009-0071-1

Rockett, K. (1990). The quality of licensed technology. *International Journal of Industrial Organization, 8*, 559–574. doi:10.1016/0167-7187(90)90030-5

Rostoker, M. (1984). A survey of corporate licensing. *IDEA, 24*, 59–92.

Suzumura, K. (1992). Cooperative and non-cooperative R&D in an oligopoly with spillovers. *The American Economic Review, 82*, 1307–1320.

Vonortas, N. (1997). Research joint ventures in the U.S. *Research Policy, 26*, 577–595. doi:10.1016/S0048-7333(97)00032-2

ENDNOTES

[1] DeCourcy (2005) documents that there have been over 800 cooperative R&D ventures that have registered in the U.S. since the passage of the National Cooperative Research Act (NCRA) in 1984. See Vonortas (1997) and DeCourcy (2001) for detailed discussions of the cooperative R&D ventures registered in the NCRA.

[2] Given the vast literature on technology licensing, we do not try to review that literature here. Instead, we refer to Rostoker (1984), Kamien (1992) and Mukherjee (2008) for surveys on technology licensing.

[3] Gallini and Winter (1985) and Katz and Shapiro (1985) show how technology licensing affects the incentive for innovation. However, neither of these papers considers the effect on cooperative R&D. Mukherjee (2005) considers the effects of licensing on R&D organization and welfare. However, unlike this paper, Mukherjee (2005) considers that R&D investment and the probability of success in R&D are exogenously given, thus ignoring the effects of licensing on R&D investments, which, in turn, affect the incentive for cooperative R&D and welfare.

[4] We use 'The Mathematica 4.2' for the figures of this paper.

[5] One may think that firms can make similar contract under cooperative R&D where the

contract is contingent upon R&D outcome, saying that, in case of unilateral success in R&D, successful firm will give the technology to the unsuccessful firm but against a payment consisting of fixed-fee and output royalty. However, if possibility of coordination in the R&D stage under cooperative R&D helps the unsuccessful firm to acquire knowledge about new innovation then this contingent contract under cooperative R&D becomes ineffective.

[6] Antitrust law may be responsible for non-negative constraint on licensing contract.

[7] Given x^*_{iNC} and x^*_{jNC}, LHS of (13) is greater than LHS of (4) if $p'(x^*_{iNC})(1 - p(x^*_{jNC}))rq^*_j > 0$.

[8] Under unilateral success in R&D, the industry profit under non-cooperative R&D with licensing is $\dfrac{2 + \bar{c}(1 - \bar{c})}{9}$, while it is under cooperative R&D is $\dfrac{2}{9}$.

Chapter 10
Entry Barriers to the Nanotechnology Industry in Turkey

Neslihan Aydoğan-Duda
İzmir University of Economics, Turkey

İrge Şener
Çankaya University, Turkey

ABSTRACT

Nanotechnology is the science that focuses on the control of matter at the atomic scale. It has the potential to create many new materials and devices with wide-ranging applications, such as in medicine, electronics and energy production. There are many entry barriers which can affect nanotechnology penetration in developing and emerging nations. This chapter discusses such barriers for Turkey. Despite about 10 universities having nanotechnology programs, the number of nanotechnology firms in the country is still low. Using combinations of interviews, surveys and literature, these issues that continue to stall the commercialization of discoveries in Turkey are examined.

INTRODUCTION

Nanotechnology is the study of the control of matter on an atomic scale. Generally nanotechnology deals with structures of the size 100 nanometers or smaller, and involves developing materials or devices within that size. Nanotechnology is very diverse, ranging from extensions of conventional device physics, to completely new approaches based upon molecular self-assembly, to developing new materials with dimensions on the nanoscale, even to speculation on whether we can directly control matter on the atomic scale. Nanotechnology has the potential to create many new materials and devices with wide-ranging applications, such as in medicine, electronics and energy production.

It is obvious that entry barriers to the nanotechnology industry should be discussed in detail if it is the case that such entry is barred for a variety of reasons and that the public can not benefit from the nanotechnology products. Turkey is a significant example to such occurence. In particular it appears that there are ten nanotechnology research centers in ten different universities in Turkey. These centers are furnished with strong infrastructure both physically and research wise. Most of the

DOI: 10.4018/978-1-61692-006-7.ch010

researchers have their doctorate degrees from the prominent universities in the U.S. and they are working on cutting edge issues on nanotechnology. However, it appears that there are only thirteen companies in the entire country which can be classified as operating in the nanotechnology sector. Such observation is as alarming as it might be perceived as natural. For example, one could easily claim that in a developing country such as Turkey an industry which requires a long-term R&D investment might not develop as the fruits of such investment would take years to materialize. However such argument can easily be refuted as there are 41 nanotechnology firms in China and 17 in İndia and 195 for example in Germany. The country size and development levels do not seem to be the main determinants here. So what drives this lack of nanotechnology commercialization?

In Aydoğan, Chen(2008) and Aydoğan (2009), some insights were provided on the above question. Extending those insights, and in this piece we survey ten academicians from four different nanotechnology Centers in Turkey (interviews were conducted at Marmara University, İstanbul Technical University, Koç University and Sabancı University) to gather their thoughts on this pressing issue of the lack of commercialization in the nanotechnology industry in Turkey. In the below section we conduct the literature review, we summarize the ideas of the interviewees and we conclude by some policy suggestions along with some directional throughts on research.

COMMERCIALIZATION OF NANOTECHNOLOGY: LITERATURE REVIEW

Although many countries have been developing strategies for the development of the nanotechnology industry since the beginning of the 1990's, Turkey has been lagging behind such efforts. Not until year 2000 that some steps were taken in this direction. Particularly nanotechnology has been identified as one of the critical sectors for the development of the Turkish economy. In particular a very well-equipped nanotechnology center at Bilkent, one of the most prominent universities in Turkey has been funded by the Prime Ministry State Planning Organization (Devlet Planlama Teşkilatı). Following this a number of centers among the ones we have interviewed in this study have been established. However, the commercialization process of the nanotechnology industry in Turkey has been stalled majorly. As mentioned earlier currently there are only thirteen nanotechnology companies in Turkey.

Although nanotechnology is at its nascent stages, it has become apparent that it will cause in stark changes in every area of our lives. Nanotechnology has its basis in many different sciences, and this makes the basic difference when compared with the impact of other sciences. According to Niosi and Reid (2007), the many different underlying pro-genitor technologies, of which have bases in molecular biology, electronics, materials science, physics (optics and quantum) and others, contribute to the composition of nanotechnology and hence this makes nanotechnology as inherently complex and diverse with diverse applications.

The broad spectrum of nanotechnology has lead to the development of various materials. For example, widespread commercial adoption of nanotechnology is growing rapidly; where early commercial applications are focused on the improvement of cosmetics, coatings, textiles and displays *(Bozeman, Hardin & Link, 2008)*.

Since nanotechnology is related to many different fields, it has received considerable attention among researchers from all over the world. Many countries have mobilized their universities and laboratories to conduct research and development activities in nanotechnology. However, as it is in the case of other technological developments, nations differ in their standing related to nanotechnology. Niosi and Reid (2007) suggest that due to the complexity of technologies involved in the development of nanotechnologies, training and support for researchers are necessary, and for these high levels of training, public funding and support for infrastructure are required. Among the developing countries, the authors indicate China, India and Brazil are the three main contenders for catching up in the nano-space. For other developing countries, like Turkey, the authors propose two strategies that can be used to promote the development of emerging technologies. These strategies are early patenting in areas with the potential to attract foreign venture capital and formation of clusters and alliances.

Cluster formation is especially important for the commercialization of nanotechnology products. Uranga, Kerexeta and Campas-Velasco (2007) state that one part of the dynamics prior to the commercialization of a science takes place at the local geographical level such as local clusters, mega-centers and regions. The importance of regions that give rise to the formation of clusters in the nanotechnology field is also addressed by other authors; Zucker, Darby, Furner, Liu, and Ma (2007), for example, conduct an empirical analysis for which authors measure the regional growth of new knowledge in nanotechnology. They do this by article and patent counts where they find that geographical regions gain cumulative advantage in the production of nanotechnology research. The authors argue that the production of nano-technological knowledge is embedded in the wider social context of institutional organization, cross-institutional collaboration, and national structures of incentives and rewards. According to the authors, this embeddedness is both constraining and enabling. It is constraining since the range of possible actions is narrowed, and on the other hand since the flow of tacit knowledge deepens and the flow between organizations becomes more differentiated, it becomes enabling.

Until now, there had been a disconnection between the nano-technological advancement undertaken in the research centers or laboratories, and the commercialization of these products. Successful commercialization of developed products is still

problematic in many countries. Since nanotechnology is still at its infancy, there exist few research on the commercialization problems. According to Bozeman, Hardin & Link *(2008)*, given the recent origins of nanotechnology research, there is heretofore no systematic research on barriers inhibiting the diffusion of nanotechnology from the laboratory to commercial application. Bozeman, Hardin & Link (2008) undertook a research project on nanotechnology firms in North Carolina region. The results of this research indicate a market failure in the capital markets because of asymmetric information on the risk and return associated with the adoption of nanotechnology and its impact on commercialization. In this piece we offer a thorough account of this disconnection from an academic angle by speaking with several university researchers.

In the commercialization of nanotechnology, knowledge plays an important role. Rothaermel and Thursby (2007) argue that an incumbent firm's ability to exploit new methods of invention in biotechnology and nanotechnology fields initially depends on its access to tacit knowledge with regard to the employment of new methods. However, over time, as firms learn and/or the knowledge becomes codified in routine procedures or commercially available equipment, inventive output, becomes more dependent on traditional R&D investments.

Shapira and Youtie (2008) indicate that path-dependent stocks of knowledge, capabilities, finance and other resources, business or policy induce capital investments in facilities, institutional strategies and linkages, and the availability of talent and human capital may influence the places where the nanotechnology research and commercialization gravitate towards; and for development in the nanotechnology sector although all the factors are relevant, a greater impact of one of them over the others will have regional implications. According to the results of their analysis, due to the concentration of research either in universities or government laboratories, nanotechnology emerges in nontraditional places of technological development. This finding indicates the prominent role of universities in nanotechnology research.

University spin-offs and new start-ups also contribute to the diffusion of nanotechnology. Uranga, Kerexeta and Campas-Velasco (2007) believe that the origin of start-ups in the fields of biosciences and nanotechnologies has to be linked to research in laboratories and universities. These companies' main function is to commercially exploit research results. In other words, these firms work in the field of exploiting intellectual property rights.

Libaers, Meyer, and Geuna (2006) find that although university spin-off companies in the field of nanotechnology in UK have an important role, they do not necessarily play a dominant role in the development of nanotechnology. According to the results of the research, while university spin-off companies are important drivers of technological change in the subfields of nanomaterials, nanodevices, nanobiodevices and nanoinstruments; they do not play a part in the formal commercialization of nanoservices or nanofabrication facilities. The overwhelming majority of nanopatents are owned by multinational corporations (MNC) and indirectly through academic technology transfer offices by university spin-off companies. Therefore, the increase in the commercialization of nanotechnology is not just related to university spin-offs but also MNCs.

Turkish Context: Nanotechnology Sector and Problems with its Commercialization

There appears to be several issues that are related to the inability of the nanotechology sector to blossom in Turkey. For example the key researchers at Marmara University in Istanbul although strongly think that the Nanotechnology Center at Marmara has a practical focus that targets commercialization they also add the following issues for the lack of the number companies in this area.

First of all these scholars think that the potential number of researchers who would engage in such ventures is limited. This is because there is not enough demand for scientists with Ph.D. degrees to work in such companies. Most R&D start-ups companies appear to be owned by individuals that are not as highly educated and hence lack vision. On the other hand, the researchers claim that financing a nano-tech venture is basicly formidable in Turkey as there is no developed venture capital market and the banks would not invest in technology related ventures that require such large funds. Also TÜBİTAK (The Scientific and Technological Research Council of Turkey) although is able to fund selected research projects the amount of the extended funds would certainly be not enough to develop products in the nanotechnology sector as the budget of this agency is way too narrow to this end. Researchers suggest an inventive solution to the lack of interest in such funding and the underdeveloped capital markets. They claim that a ministry of technology would be able to set the vision, handle the groundwork and hence solve the funding problem.

Apart from the financing problem researchers also mention that there are no incubation centers of nanotechnology in universities partly because nanotechnology products require much longer periods of research time. Another issue related to the commercialization of nanotechnology products is a possible lack of demand to Turkish produced nanotechnology products due to the lack of trust in their reliability. A related issue which the researchers raise is the weak patent and copyright laws in Turkey which curb the researchers' incentives to engage in researching and developing nanotechnology products. Researchers also mention that there is a lack of trust between the industry and the university as private sector professionals in Turkey do not have enough confidence in the ability of the university professors in handling the practial matters of the market. Finally, researchers at Marmara argue that the whole country is not as developed as the main cities like Ankara,

İzmir and İstanbul, particularly the east side of the country is underdeveloped and somewhat disconnected from the west. Hence the scholars stress this creates a difficulty for research results to spread across the entire country.

Following this interview we spoke with the researchers from İstanbul Technical University who propose an inception of a National Nanotechnology Platform to connect industry with the university researchers as in Turkey they claim there is a lack of knowledge transfer between the industry and members of universities. Such platform they suggest would include TÜBİTAK (The Scientific and Technological Research Council of Turkey), Ministries, public institutions, universities and industry members. Researchers also mention that the technopark at Istanbul Technical University is built to bridge universities with the industry members. Researchers also suggest the inception of technology cities to be build around universities.

Another interview with a researcher from Koç University reveals that their research is mostly theoretical but sets a pre-stage for quantum computers and some of their research involve surfaces that do not hold water and hence have some practical applications. Researchers emphasize the lack of venture capital for the development of nanotechnology products. In fact in the context of nanotechnology the narrow possibility of research success reveal the difficulty of obtaining funds in an environment where there is no risk capital. Researchers also emphasize the lack of skilled labor for the development of the nanotechnology sector. Researchers claim that products that appeal to the Turkish economy, such as nano-farming, would possibly make it to the market. Researchers propose a nanotechnology center which is subsidized by companies who receive consulting services from the center researchers and they eventually buy the research output to turn it into a product. These companies might also generate new ideas which they could request feasibility reports from the university center and hence both sides could benefit from such cooperation. Researchers

emhpasize that if a cluster project would be put into action they would have to be built around major cities with good universites. Researchers also emphasize Turkey's candidacy to Europian Union which channels alot of R&D funds into universities and the industry and possibly this would include the nanotechnology sector as well.

Another interview with the researchers at Koç University reveal that most of the nanotechnology research centers in Turkey are involved in theoretical research rather than research that targets product development. In fact the researchers add even though there might be some studies on nanotechnology which are practially oriented, commercialization of such work is stil a problem due to the lack of venture funds and the fact that risk of failure is too high. Hence the authors suggest partnering with a foreign firm with deep pockets could be the way to go. Researchers also suggest that the government can offer step-wise grants that are extended based on the success of a project. Researchers also argue that nanotechnology products that are produced in Turkey might be difficlt to export as quality standards might not match to for example U.S. standards. Researchers suggest that there could be a funding consortium where firms give seminars on their future products and companies jointly extend funds. Another obstacle for the Turkish firms is that most firms are not institutionalized which makes it difficult for firms to take risks.

Researchers from Sabancı University suggest the inception of application oriented centers inside Universities which would help the commercialization of research ideas. They also suggest that poor quality imports should be prevented to help the domestic nanotechnology sector to evolve. Researchers also suggest that in universities, particularly in nanotechnology research, research is driven by academic success and hence this also could curb practical focus. They implicate hence that research side is done with success but researchers need an interface to commercialize what they have been working on. In addition researchers claim academicians in Turkey have not been supported to start-up companies and hence it is not in the culture. Therefore researchers suggest that there could be a sabbatical system where academicians can work in industry and members of industry in turn could work in universities so that two sides could understand each other.

Also all scholars indicate the importance of clusters or to begin with technoparks but they also mention that although there are some formations as such, a full-fledged nanotechnology cluster or even a technopark does not exist in Turkey yet.

The major issues behind the lack of commercialization and diffusion of the nanotechnology sector in Turkey are listed below:

- Strong theoretical focus in university research
- The significance of theoretical research for the promotion of university professors
- Lack of interface between the university and industry
- Lack of a supportive culture for academician-entrepreneurs
- Lack of skilled labor
- Prejudice on university researchers from the practitioners
- Lack of venture capital
- Lack of funding agencies for projects that require large funds
- Lack of incubation centers
- Weak patent and copyright laws
- Lack of trust in Turkish produced goods
- Competition from cheap imports
- Lack of fully-operational industry clusters
- Disconnection between the industrial and the underdeveloped part of the country

CONCLUSION AND SUGGESTIONS

It appears that in reality Turkey needs to cover some mileage before research shows itself in the marketplace. The issues range from cultural as-

pects of the matter to the lack of venture funds for such occurrence. Scholars have some important ideas to build an interface between universities and the industry. For example, a ministry of technology, a national nanotechnology platform, a funding consortium, a nanotechnology center which all would solve the severe funding problems in Turkey and bridge the gap between the university and the industry. Turkish researchers could also solve a possible lack of demand to their products by producing nanotechnology goods in the agricultural sector which the country definitely is in the severe need of. In addition the funding problem can be somewhat solved by channelling the increased R&D funds to the nanotechnology industry. This can be done by informed lobbying as if funds are not provided for the development of the nanotechnology products particularly in a strategic industry such as the defense sector, the security of the country could be compromised. European Union membership is likely to help with such lobbying as the funds for R&D are increased substantially. In addition, a nanotechnology cluster formation would substantially solve the time to market, financing and risk issues.

The research on the commercialization of the nanotechnology industry appears to be rather scant. However, given its significance a thorough theoretical and empirical work is urgently needed. In particular, issues that are laid out in this piece need to be systematically analyzed within an empirical framework.

REFERENCES

Aydoğan, N. (2009). *Innovation Policies, Business Creation and Economic Development: A Comparative Approach*. New York: Springer. doi:10.1007/978-0-387-79976-6

Aydoğan, N., & Chen, Y. P. (2008). *Social Capital and Business Development in High-technology Clusters: An Analysis of Contemporary U.S. Agglomerations*. New York: Springer. doi:10.1007/978-0-387-71911-5

Bozeman, B., Hardin, J., & Link, A. N. (2008). Barriers to the Diffusion of Nanotechnology. *Economics of Innovation and New Technology*, *17*(7&8), 749–761. doi:10.1080/10438590701785819

Istanbul Technical University. (n.d.). Retrieved from http://www.itu.edu.tr/

Koç University. (n.d.). Retrieved from http://www.ku.edu.tr/

Libaers, D., Meyer, M., & Geuna, A. (2006). The Role of University Spinout Companies in an Emerging Technology: The Case of Nanotechnology. *The Journal of Technology Transfer*, *31*, 443–450. doi:10.1007/s10961-006-0005-9

Marmara University. (n.d.). Retrieved from http://www.marmara.edu.tr/

Niosi, J., & Reid, S. E. (2007). Biotechnology and Nanotechnology: Science-based Enabling Technologies as Windows of Opportunity for LDCs? *World Development*, *35*(3), 426–438. doi:10.1016/j.worlddev.2006.11.004

Rothaermel, F. T., & Thursby, M. (2007). The Nanotech versus The Biotech Revolution: Sources of Productivity in Incumbent Firm Research. *Research Policy*, *36*, 832–849. doi:10.1016/j.respol.2007.02.008

Sabancı University. (n.d.). Retrieved from http://www.sabanciuniv.edu.tr/

Shapira, P., & Youtie, J. (2008). Emergence of Nanodistricts in the United States: Path Dependency or New Opportunities? *Economic Development Quarterly*, *22*, 187–199. doi:10.1177/0891242408320968

Uranga, M. G., Kerexeta, G. E., & Campas-Velasco, J. (2007). The Dynamics of Commercialization of Scientific Knowledge in Biotechnology and Nanotechnology. *European Planning Studies*, *15*(9), 1199–1214. doi:10.1080/09654310701529136

Zucker, L. G., Darby, M. R., Furner, J., Liu, R. C., & Ma, H. (2007). Minerva Unbound: Knowledge Stocks, Knowledge Flows and New Knowledge Production. *Research Policy*, *36*, 850–863. doi:10.1016/j.respol.2007.02.007

Chapter 11
Micro and Nanotechnology Maturity and Performance Assessment

Nazrul Islam
Cardiff University, UK

ABSTRACT

This chapter aims to provide a new readiness matrix called 'innovative manufacturing readiness levels (IMRLs)' to evaluate and assess the areas of micro and nanotechnology maturity including their performance. The study employs a case study approach through which the practicability and applicability of the IMRLs conceptual matrix were verified and confirmed. A case study with laser-based manufacturing technologies explores the stages of micro and nano technologies (MNTs)' maturity, including the key issues and performances that contributed to the development of a new assessment tool. Concerning intense global R&D competition in MNTs, this study exhibits a forward-looking approach in assessing MNTs maturity and performance. A generic conclusion is reached by which product designers and technology managers position themselves and take into account risk reduction exercises related to MNTs. The novelty of the research could be that organizations, which develop and use MNTs, have an opportunity in applying such a specific assessment matrix to quantify the technology readiness of unreleased MNTs.

INTRODUCTION

Nanotechnologies have been attaining wide recognition in recent years and are expected to bring significant changes to a wide range of technologies due to having novel and improved properties in their materials in relation to the nanometre scale. Because of their potential applications, nanotechnologies are considered as one of the emergent fields having the potential for important social and economic impacts in the future. A turning point in science and technology policy in relation to nanotechnology has been the launch of the National nanotechnology Initiatives (NNI) in the United States in 2000. Since then, developed countries and many developing countries have launched initiatives or have prioritised research in nanotechnology. Nanotechnology as it appears today is not

DOI: 10.4018/978-1-61692-006-7.ch011

a single technology but a constellation of diverse and unfocused technological fields. Research into technologies based on nano scale structures is nothing new. What has led to a nanotechnological breakthrough is the development and application of new sophisticated instruments that have been applied to observe and manipulate processes at the nano scale level. The so-called breakthroughs were mainly made with the development of the STM (scanning tunneling microscope) and the AFM (atomic force microscope) in the mid 1980s by IBM for the semiconductor industry. Since then a range of other tools and advancements have appeared in how to measure and interpret at the nano scale (Royal Society 2004).

At the present time, manufacturing industries see the benefits of micro and nano technologies (MNTs) for industrial production, since R&D activities in the emergent technologies have been strengthened worldwide, and it is thought their potential is likely to be pervasive in wider applications. Micro and nanomanufacturing technologies are increasingly employed in a huge variety of consumer products and industrial components (Rooks, 2004) incorporating machine tools and processes. In addition, the demand for nano-products and components has been increasing rapidly with versatile applications in electronics, optics, medicine, biotechnology, automotive and communications industries (Alting et al, 2003). Researchers in academia and industry worldwide are striving to get involved in the development of innovative (micro and nano) manufacturing technologies to meet the demand of miniaturized products (Islam, 2009). The mission for micro and nanotechnology-based companies, both small and large, is to adopt or implement the technology successfully. However, to achieve this, companies face challenges rooted in technological and process uncertainties and risks which demand new managerial approaches. With the fast moving pace of emerging MNTs and their disruptive nature, coping with their maturity and performance measures has become the foremost

issue for manufacturing design engineers as well as technology managers. Therefore, it is crucial for successful tracking and controlling of an innovative manufacturing technologies life cycle.

The importance of technology readiness and capability has been recognized as an important driver for the successful deployment of technological as well as manufacturing systems. To implement the technologies successfully, companies face challenges rooted in technological and process uncertainties which demand new and improved managerial approaches. In addition, in the management of micro and nano technologies' life cycles, challenges still remain in its practice and implementation which demand novel or improved approaches. The main objective of this chapter is to build and use a new readiness matrix called 'innovative manufacturing readiness levels (IMRLs)' to evaluate and assess technology maturity, including the performances of the emergent MNTs. It also has the practical purpose of supporting the companies related to micro and nanotechnologies to develop their capabilities in technology management. For assessing the maturity of generic technology and manufacturing systems several concepts and scales currently exist. For example, technology readiness levels originally developed in the 1980s by the National Aeronautics and Space Administration and further adopted in the 1990s by the United States Air Force, and manufacturing readiness levels as defined by the Department of Defense in 2005. These existing concepts have limitations, specifically in embracing MNTs maturity. Therefore, a comprehensive understanding in the implementation of MNTs has provided the basis on which an improved matrix can be developed.

Since micro and nanotechnologies have been regarded as a major driver of future global economic development, small and medium-sized companies and industry increasingly seek more effective ways to manage MNTs. This chapter explores how MNTs are managed throughout their lifecycle in a more explicit way than has so far been accomplished by this research. The research presented here sets out to examine how scientific and technological readiness, product and process readiness, market and business readiness, and other associated aspects of MNTs readiness can be depicted over the life cycle. Also, it seeks to provide the IMRLs readiness framework as a tool for product designers and technology managers to position themselves and to enable them to take into account the key elements relating to MNTs. This can be addressed by the following two specific objectives:

1. To develop a generic readiness framework that can be abstracted and applied to managing the whole process of micro and nanotechnology innovation.
2. To establish generic issues and criteria for each phase of the micro and nanotechnologies life cycle

As this research is exploratory, strategic and theory building in nature, it embraces an interpretative position and employs a case study approach through which the practicability and applicability of the conceptual framework of the IMRLs can be confirmed. The study uses the IMRLs matrix and applies this method to the case of micro and nano scale laser processing technologies to build the IMRLs readiness framework. This framework can be used in the capture and assessment of performance measures along the MNTs maturity phases, ranging from concept development to production environment.

In the past decade, the quality of conventional laser manufacturing has increased dramatically and extended its reach even further into versatile application sectors. The need for extremely precise ultra-small micro and nano scale components and devices with high quality finishes and tolerances is continuing to promote growth in the area of precision laser machining. Laser micro and nano manufacturing technologies have thus been developed into the most flexible and powerful technique for rapid prototyping, which is allow-

ing micro engineers and new product designers to take advantage of an immense range of new manufacturing routes for developing micro and nano products. Hence, companies are benefitting from the significantly reduced time-to-market offered by micro and nano scale laser processing, improving their flexibility and versatility. For easy understanding of a wide range of technologies related to MNTs, a brief overview and a refined classification of micro and nano manufacturing technologies, specifically the laser processing technologies, are presented here.

AN OVERVIEW OF MICRO AND NANO MANUFACTURING TECHNOLOGIES

Innovative manufacturing process technologies offer better and sufficiently improved quality not only for conventional products but also for making possible entirely new products, while also emphasizing the issues of high product and process performance, miniaturization and conflation of technologies. Therefore, development of process technologies is closely linked to advances in micro and nano manufacturing technologies and their management. The micro and nano manufacturing industry faces substantial challenges, for example, producing smart products with more functions containing fewer component parts, which may use new or improved materials and processes with high precision. Since micro and nano manufacturing is a scale-integrated production technology based on high accuracy and quality, the whole micro and nano scale machining system is required in the manufacture and assembly of component technologies in ultra-small regimes. At present, many advanced technology products necessitate manufacturing processes and machines operating in the micro and nano regimes for successful competitive development. On the other hand, innovative manufacturing processes need to be evaluated on their quality and capability, including in the areas of accuracy and reproducibility, as well as the achieved precision of the related equipment.

Micro and nano manufacturing technologies cover a broad range of technologies. Several MNTs researchers have proposed classification schemes to categorize these technologies, focusing on micro machining processes (Masuzawa, 2000), micro fabrication techniques (Madou, 2002), micro and nano engineering technologies (Brinksmeier, 2001) and materials and the micro process dimension (Dimov et al, 2006). In recent years, manufacturing industries have witnessed a rapid increase in demand for micro and nano components and products in a wide range of industrial sectors, including the electronics, optics, medical, biotechnology and automotive sectors.

As a result of the current trend towards product miniaturization, product designers and technology managers require a comprehensive classification of micro and nanomanufacturing technologies. A refined and comprehensive classification scheme for micro and nanomanufacturing technologies and processes has been developed based on relevant literature reviews and the author's accumulating experience in researching the area of nanotechnologies. As demonstrated in Table 1, micro manufacturing technologies are classified into six main categories, (1) Energy-assisted micro machining, (2) Mechanical micro machining, (3) Electro-physical and chemical processes, (4) Lithographic methods, (5) Shaping and/or replication processes, (6) Additive processes. The current developments in nanotechnologies can be divided

Table 1. Classification of micro and nanomanufacturing technologies

Classification of micro manufacturing technologies	
Categories	**Key technologies and/or processes**
1) Energy beam micro-machining	Laser processing, Focused ion beam, Electron beam, X-ray
2) Mechanical micro-machining	Micro-milling, Micro-drilling, Micro-turning, Micro-grinding, Chemical mechanical polishing, Lapping, Micro-abrasive jet machining, Micro-abrasive blasting
3) Electro-physical and chemical processes	Micro-EDM, ECM, Ultrasonic micro-machining, Hybrid micro-machining, Etching, Electro-chemical polishing, Electroforming
4) Lithographic methods	Photo/UV lithography, Electron beam and Ion beam lithography, X-ray lithography, Laser lithography
5) Shaping processes (Replication)	Micro-molding, Hot/UV embossing, Casting, Electroplating, Spin coating, Extrusion, Punching, Bending, Blanking, Injection molding, Chemical vapor deposition, Physical vapor deposition
6) Additive processes	Micro-sintering, Electroplating, Welding, Gluing, Rapid prototyping, CVD, PVD, Micro-spark coating
Classification of nanomanufacturing technologies	
Categories	**Key technologies and/or processes**
1) Top-down technologies and/or processes	Energy beam machining, e.g., laser, electron beam, ion beam, x-ray
	Lithography, e.g., scanning probe, e-beam, ion beam,
	Erosive processes, e.g., chemical, mechanical, electrical, ultrasonic
2) Bottom-up technologies and/or processes	Nano imprinting, Contact printing, Replica molding
	Assembling, e.g., self-and directed assembly, LISA
	Electrostatic techniques, e.g., xerography of crystals, coating
	Laser trapping or tweezers
	Sol-gel, Colloidal aggregation
	Spinoidal wetting or dewetting

into two different manufacturing approaches, (1) The top-down approach (more incremental), and (2) The bottom-up approach (more radical). The top-down approach aims at the miniaturisation of current technological solutions, involving decomposition of materials into the smallest manageable entities. Thus, the majority of contemporary applied R&D in top-down nanotechnology builds on prior knowledge and is thus more incremental by nature (Igami and Okazaki, 2007). The radical nature of nanotechnology is most likely linked to bottom-up approaches, where the innovations rely on assembling or molecular level manipulation rather than the current advances in miniaturisation. Bottom-up processes aim at building macrostructures from nanostructures by allowing the re-engineering of materials. The R&D efforts related to bottom-up approaches are still mostly research orientated, while more applied research is predicted to emerge in the mid-to long-term (Islam and Miyazaki, 2009).

Classification of Laser Manufacturing Technologies and Their Application Domains

The intense energy that a laser may produce on materials enables several types of ultrafast, novel and economical processing. These types of processing are advantageous from the point of view of quality, productivity and efficiency, and superior to those possible with their conventional counterparts. In this section, we will present a brief classification of laser processing. The classification of laser materials processing can be grouped into two broad categories based on laser power density and materials phase transition, e.g. high power density lasers (involving change of phase/state, for example, vaporization and melting) and low power density lasers (involving no change of phase/state, for example, heating). It is evident that materials transformation without phase change (for example, hardening, bending, and lithography) requires a low power density

laser which relies on heating. Materials transformation with phase change (e.g. welding, cutting, drilling, sintering, coating, rapid prototyping), on the other hand, requires a high power density laser which relies on melting and vaporization. Table 2 represents this classification, including key processes and examples of application domains involved with each category. Demonstrating the relevance of laser micro and nano manufacturing, application examples where laser processing is in use for production are also presented. The key micro and nano applications include micro and nano sensors and actuators, micro and nano fluidics, lab-on-a-chip, micro and nano optics and micro-nano tooling (Table 2).

DEVELOPMENT OF A NEW READINESS MATRIX

Businesses in micro and nanotechnologies are constantly seeking tools to analyze their efforts, since they may lose substantial competitive advantage in bringing emerging and disruptive technologies to market. Despite the recognized benefits that result from MNTs, whether directed at products or production processes, very little attention has been paid to proposing methods for evaluating and assessing technology maturity consistently between organizations. Although many different scales are common in industry, each criterion has a specialized nomenclature that varies based on the specific industry sector or organizational cluster. Most organizations apply specific assessment scales to quantify the technology readiness of an unreleased technology. Thus, the objective of this research is to present an alternative methodology that would simplify the evaluation and assessment of MNTs maturity by different organizations. The term 'maturity' is employed throughout the chapter as 'readiness' tends to refer to the suitability of using a given technology in a particular context.

For a decade, technology readiness levels (TRLs) (Mankins, 1995) have become a tool that

Table 2. Classification of laser processing technologies

Classification of laser materials processing				
Categories	**Sub-categories**	**Key processes**	**Key application domains**	**Key micro and nano applications**
High power density lasers (involving change of phase/state)	Phase changes from solid to vapor (vaporization)	-Laser machining (e.g. cutting, drilling, scribing) -Deposition/coating (e.g. marking) -Laser-assisted purification (e.g. cleaning) -Laser spectroscopy (e.g. isotope separation)	-Medical (e.g. angioplasty, tumor therapy, skin/dental/eye surgery); -Chemical (e.g. photochemical deposition, pollution control, spectroscopy); -Military (e.g. atomic fusion, weapon guide); Surface engineering/treatment; -Manufacturing	-Micro and nano sensors and actuators (e.g. low noise amplifiers, memory module, RF switches); -Micro and nano fluidics (e.g. flow pumps, valves); -Lab-on-a-chip; -Micro and nano optics (e.g. photonic crystal devices); -Micro and nano tooling (e.g. stripping, cleaving, lensing)
	Phase changes from solid to liquid (melting)	-Joining (e.g. welding, brazing) -Surface alloying (e.g. cladding, glazing) -Rapid prototyping -Reclamation/repair -Laser sintering/soldering		
Low power density lasers (involving no change of phase/state)	No phase changes (heating, i.e. solid to solid)	-Bending or forming -Transformation (e.g. surface hardening) -Laser lithography Semiconductor annealing	-Communication (e.g. optical fibers, telecommunication, optical data storage and computation); -Reprography (e.g. scanning, printing) -Entertainment (e.g. audio-video recording, laser light beam show, pointer) -Metrology (e.g. inspection)	

has added clarity to technical assessments and the discussion of technology maturity and risk. TRLs are a measure used by some US government agencies (e.g., NASA, DOD) and many of the world's major companies to assess the maturity of evolving technologies (for example, materials, components, devices) prior to incorporating that technology into a system. Since new technologies are not suitable for immediate application when first conceptualized, they are usually subjected to experimentation, refinement, and increasingly realistic testing. Once the technology is sufficiently proven, it can be incorporated into a system or subsystem. These types of assessment (for example, TRL scales) take into account a variety of factors that can describe the current state of the technology in relation to, readiness and risk assessment.

The manufacturing readiness levels (MRLs) were developed by the Department of Defense to provide a similar understanding of generic manufacturing risk and maturity, aiming to find out how assessments fit into the defense acquisition process (Technology Readiness Assessment Desk book, 2005). The MRL definitions and descriptions are based on the integration of existing industry, government, and technical coalition standards and recommendations.

The innovation readiness levels (IRLs) give an overall idea of how companies and regions can sustain their ability to innovate and how communication can contribute to this (Zerfass, 2005). The IRLs take into account the relevance of internal as well as external stakeholders within the innovation process and considers the relevance of regional cluster development, as well as highlighting the importance of communication for the implementation of new ideas, products, and services for managing the process of incremental innovation over its life cycle (Tao et al, 2008).

Figure 1 illustrates the definitions of existing readiness scales.

Several theories and concepts exist regarding technology and business development as well as their maturity and life cycles. For example, Stage-gate (Cooper, 2001), S-curve (Foster, 1986), TRL (Mankins, 1995) and MRL (DOD report, 2005) provide guidance of maturity and/or performances for generic technology and manufacturing. While the technology diffusion theory (Rogers, 1995), market adoption model (Moore, 1999) and product/service life cycle (Beacham, 2006; Genaidy and Waldemar, 2008; Niemeyer and Whitney, 2002) help understand market evolution trajectories. Recently, innovation readiness levels

(Tao et al, 2008) have depicted the development of an innovation over its life cycle. Measurement of the technology's maturity and performance is an indicator of technology management capacity and assessing this indicator is of significance for emergent MNTs companies. Assessment of this indicator can help organizations confirm the level their capacity of management of nanotechnology has reached, help managers identify the problems and shortcomings, and could offer a guiding framework for the improvement of technology management practices.

The existing theories or concepts possess some limitations, specifically in embracing the micro and nanomanufacturing technologies life

Figure 1. Definitions of existing readiness levels

Technology Readiness Levels	Definitions
TRL 1	Basic principles observed and reported
TRL 2	Technology concept and application formulated
TRL 3	Analytical and experimental proof-of-concept
TRL 4	PoC validation in laboratory environment
TRL 5	PoC validation in relevant environment
TRL 6	System model or prototype demonstration in a relevant environment (ground or space)
TRL 7	System prototype demonstration in a space (operational environment)
TRL 8	System completion, test and demonstration (ground or space)
TRL 9	Actual system prove through successful mission operations

Source: Mankins, 1995

Innovation Readiness Levels	Definitions
IRL 1 (*TRL 1-3*)	Basic scientific principles of the innovation observed and reported
IRL 2 (*TRL 4-6*)	Components developed and validated, and a prototype is developed
IRL 3 (*TRL 7-9*)	Technological development completed and the complete system is proven
IRL 4	Challenges/difficulties that innovation may encounter when first introduced in the market
IRL 5	Maintain and enhance the position of innovation and to cope with competition
IRL 6	Re-innovation of technology, and obsolescence and exits of innovation

Source: DOD report, 2005

Manufacturing Readiness Levels	Definitions
MRL 1	Manufacturing Feasibility Assessed
MRL 2	Manufacturing Concepts Defined
MRL 3	Manufacturing Concepts Developed
MRL 4	Capability to produce the technology in a lab environment
MRL 5	Capability to produce prototype components in a production relevant environment
MRL 6	Capability to produce a prototype system in a production relevant environment
MRL 7	Capability to produce systems, subsystems in a production representative environment
MRL 8	Pilot line capability demonstration. Ready to begin low rate production
MRL 9	Low rate production demonstration. Capability in place to begin full rate production
MRL 10	Full rate production demonstration and lean production practices in place

Source: Tao et al, 2008

cycle. The TRLs and MRLs concepts have some limitations which prevent their straight-forward application to a specific domain where a wide range of technologies are being developed concurrently. For example, the terminology and assessment method is specific and is difficult to apply to different technological domains, is strictly subjective as it relies on questionnaires completed by experts, which are time consuming to fill out. Micro and nanomanufacturing technologies are interdisciplinary in nature, where multiple component technologies can be developed in parallel. Therefore, a comprehensive understanding and assessment of the maturity of micro and nanomanufacturing technologies requires a new readiness matrix and a new classification tool. In this section, the author constructs a matrix and selects a laser-based micro and nanomanufacturing processing industry to participate in a technology maturity and performance assessment case study. These are the research steps undertaken in this study:

1. Construction of an assessment matrix,
2. Design of the questionnaire and carrying out of the case study,
3. Data capture and analysis, as, well as providing recommendations for continuous management of nanotechnology practice improvement.

Micro and Nanotechnology Maturity and Performance Matrix

The proposed IMRLs readiness matrix is a four part 'performance' model (e.g. science and technology, product and process, market and business, firm and industry performance) which separates the life cycle of micro and nanomanufacturing technologies into five phases or levels for capturing the main attributes in implementing the technologies. The IMRLs is a descriptive assessment matrix used for guiding the development of the capacity for management of nanotechnology practices and for providing methods to facilitate

self-improvement in organizations. The questionnaire has been worked out based on the proposed IMRLs readiness matrix and has been carried out at a laser-based company in the UK in order to confirm the validity of the matrix. Figure 2 illustrates the IMRLs matrix. A basic comparison between readiness scales is illustrated in Figure 3.

Defining the Key Levels of the IMRLs Matrix

1. **Micro and nanotechnology concept development and refinement (IMRL 1)** Innovative (micro and nano) manufacturing concepts and ideas are generated and refined at this stage. This level involves the characterization of micro and nanostructures and understanding of the properties of materials. Technical and manufacturing strategy planning and detailed design are the areas that have so far been confirmed.

2. **Micro and nanotechnology components and capability development – 'proof-of-concept' (IMRL 2):** Micro and nano-scale technological components and manufacturing capability development and feasibility studies (e.g. material processing capabilities, component technologies' dependencies) are validated. Work steps are confirmed and technology risk is assessed.

3. **Micro and nanotechnology systems and capability demonstration – 'prototypes' (IMRL 3):** Adequacy and integration (scale-down challenges) of micro and nano-scale manufacturing systems and strategy are demonstrated (e.g. systems engineering, prototypes and overall production preparation). Prototypes' technical specifications and actual process demonstration are finalized.

4. **Micro and nanotechnology systems tests and pilot-production (IMRL 4):** Combined systems test, verification and evaluation are conducted including equipment inspection (e.g. reliability, maintainability, availability) and pilot-line production.

Figure 2. The proposed IMRLs readiness matrix

Innovative (micro and nano) Manufacturing Readiness Levels (IMRLs)

Reconfiguration

Key Performance Measures	IMRLs Phases / Key Issues	MNTs Concepts Development and Refinement **IMRL 1**	MNTs Proof-of-Concept Development **IMRL 2**	MNTs Prototypes and Systems Demonstration **IMRL 3**	MNTs Systems Tests and Pilot-line Production **IMRL 4**	MNTs Systems Operation and Market Audit **IMRL 5**
	Science and Technology Performances					
	Product and Process Performances					
	Market and Business Performances					
	Firm and Industry Performances					

Figure 3. A basic comparison of readiness scales

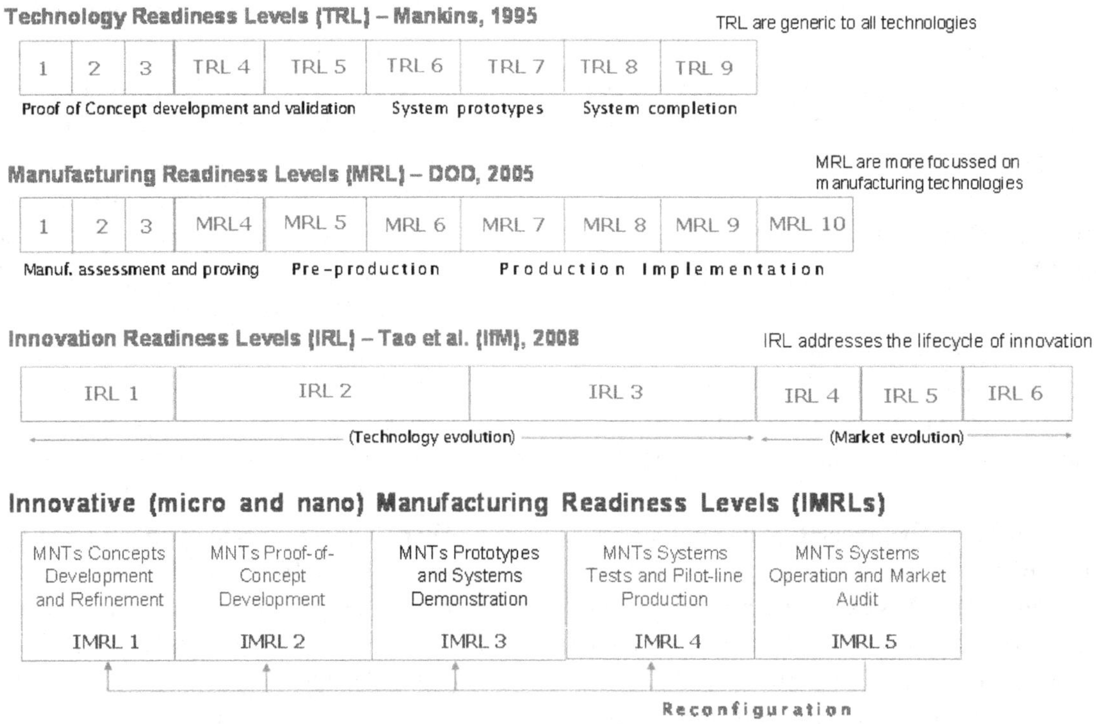

Technology Readiness Levels (TRL) – Mankins, 1995   TRL are generic to all technologies

1	2	3	TRL 4	TRL 5	TRL 6	TRL 7	TRL 8	TRL 9

Proof of Concept development and validation   System prototypes   System completion

Manufacturing Readiness Levels (MRL) – DOD, 2005   MRL are more focussed on manufacturing technologies

1	2	3	MRL4	MRL 5	MRL 6	MRL 7	MRL 8	MRL 9	MRL 10

Manuf. assessment and proving   Pre-production   Production Implementation

Innovation Readiness Levels (IRL) – Tao et al. (IfM), 2008   IRL addresses the lifecycle of innovation

IRL 1	IRL 2	IRL 3	IRL 4	IRL 5	IRL 6

———————————— (Technology evolution) ————————————→ ←———— (Market evolution) ————→

Innovative (micro and nano) Manufacturing Readiness Levels (IMRLs)

MNTs Concepts Development and Refinement	MNTs Proof-of-Concept Development	MNTs Prototypes and Systems Demonstration	MNTs Systems Tests and Pilot-line Production	MNTs Systems Operation and Market Audit
IMRL 1	IMRL 2	IMRL 3	IMRL 4	IMRL 5

Reconfiguration

5. **Micro and nanotechnology systems operation, quality and market audit (IMRL 5):** Overall micro and nanomanufacturing systems are in operation. Quality measurement and initial market audit is conducted. Reconfiguration (for example, re-structuring, re-innovating) of any level is started if necessary.

Defining the Key Performances or Aspects of IMRLs Matrix

Performance measures are the criteria that are used to give support to management in the achievement of goals for guiding and improving the decision making process in companies (Dumond, 1994). Companies use performance measures to evaluate and improve manufacturing processes in order to reach their goals. Managing performance measurement factors in a manufacturing company often increases profitability in the system and provides many competitive advantages. In the management of micro and nanotechnology capacity, the IMRLs matrix includes the following four performance measures:

1. **Science and technology performances:** Science performance delineates overall scientific opportunities captured by micro and nanotechnologies, exploring whether it reconfigures sciences, collaborates sciences, or makes it evolutionary or revolutionary. Technology performance means capturing micro and nanotechnological opportunities and manufacturing capabilities. It could be an indicator of whether the emergent micro and nanotechnologies will be sustainable or disruptive.

2. **Product and process performances:** These type of performance parameters capture product and process development trajectories, exploring whether the micro and nano-products or processes lead to discontinuous (radical) or continuous (incremental) innovation.

3. **Market and business performances:** This measure attempts to identify the scope of businesses and the size of the market (for example, niche and mass markets) showing whether it creates new markets or businesses or strengthens existing markets or businesses.

4. **Firm and industry performances:** This category helps to identify industry or firms' structures development trajectories, delineates core competencies, industry and firms' capabilities and resources, and overall competition among the competitors.

MICRO AND NANOTECHNOLOGY MATURITY AND PERFORMANCE ASSESSMENT: A CASE STUDY

In this study a specific manufacturing company in the laser processing sector in Oxford was chosen and the top management was interviewed. The main aim of the study is to determine the appropriate performance criteria, considering the conditions and the structure of the micro and nanomanufacturing system specific to laser processing, and to determine the relative importance of those criteria by using the IMRLs matrix. In this section, the case study is discussed focusing on the stages of MNTs maturity, key issues and challenges that contributed most to the development of the integrated IMRLs framework.

Building the IMRLs Framework with a Laser Processing Manufacturer

The studied Company is an established UK-based contractor and a commercial laser processing equipment manufacturer. It was formed in 2000 and is situated in one of the largest planned locations of the UK industry and provides fully integrated systems and sub contract services for laser processes. They have headquarters in the UK and are establishing a presence in North America through a laser processing facility in

South Carolina and a sales office in the North East. The business undertaken in the Company is split into

1. Design and manufacture of machine tools that process materials on the micro and nano-scales using lasers, and
2. Sub-contracting work that is actively and steadily being increased by building up the operation in the United States.

The business success of the Company has been a result of developing automated laser processing equipment that offers reproducible, scalable automated solutions to traditional laser processing technologies. There are two main product groups that the Company offers:

1. **Fiber processing equipment:** The business began with its own patented fiber processing technology and is capable of developing products using its own intellectual property. These bespoke machine tools are in service around the world producing high quality and reproducible laser-processed fiber ends that are extensively used in telecommunications, bio-medical, aerospace, photovoltaic, automotive and defense applications.
2. **Non-fiber processing equipment**: When the business began it had to meet its customer's particular requirements and/or needs. The products in this category are standard machine tools offering a range of micro machining processes, for example, cutting, milling, and scribing, welding and lithography.

Key Phases and Levels of IMRLs Observed

Lasers have become popular in micro and nano product development, for rapid prototyping. The case study explores the path of obtaining a first proof-of-concept part from the initial idea

to reaching prototypes and production of laser processing equipment. Based on a series of meetings and face-to-face interviews conducted with the relevant personnel in the Company, a detailed analysis was made exploring the key levels of IMRLs relevant to laser-based micro and nano-manufacturing technologies. Table 3 demonstrates the IMRLs maturity assessment chart.

- **Level 1 (Idea generation):** In this stage, a new and/or improved idea is generated either for the Company itself or for its potential customer (for example, HP, DuPont) and is checked to discover whether it works or not. Technology feasibility is thus confirmed. Once it has been finalized, it is possible to at least make an informed decision for proof-of-concept. This stage usually takes a few hours to a few days and can cost anything from a few hundred to a few thousand pounds. This phase includes tests of basic materials to establish how effectively the customer samples can be machined (for example, choice of laser), whether the material can withstand the feature resolutions and/or depths to be produced, and what the process speeds might be (for example, what the practicalities are going to be). Customer needs and demands, including target market, are confirmed.
- **Level 2 (Proof-of-concept):** Having some basic information from the initial phase allows for a small number of demonstration parts to be produced (e.g. components and capability development). During this stage, new laser-based technological components and capabilities are achieved by replacing conventional laser processing. Work steps are confirmed and IP is protected. Technological risk is assessed and alternative solutions are considered. This again, typically, takes a few days and costs are in line with the first phase, depending on the extent of process development work

Table 3. Maturity and performance assessment of laser-based micro and nano manufacturing technologies using IMRLs matrix

IMRLs phases or levels Key issues assessed	IMRL 1 Concept development and refinement	IMRL 2 Proof-of-concept and capability development	IMRL 3 Prototypes and systems demonstration	IMRL 4 Systems verification and Pilot-line production	IMRL 5 Systems operation and market audit
Science and technology performances	-Basic scientific principles/technologies observed -In-house idea generated (e.g. patents) along with limited innovative idea from outside -Studies on technology feasibility confirmed	-New fibre laser-based technological components and capabilities achieved by replacing conventional laser processing -Success criteria understood and agreed -Work steps confirmed and IP protected -Technological risk assessed and an alternative solution considered	-Components and capabilities tested -Prototypes and actual processes demonstrated -IP protected	-Technology documented -Number of operations reduced -Expertise formed	-Technology operation and maintenance enabled -Technological service provided -After sales support -Re-innovate considered
Product and process performances	-Product and/or process development research conducted -New product planned (e.g. fibre laser processing tools that is new) -Incremental improvement of conventional tools confirmed (e.g. non-fibre laser processing equipments)	-Access to appropriate lasers confirmed -Detailed product and/or process launch plan issued -Customers' required specification considered -Quality in relation to competitors achieved	-Levels of reliability demonstrated -Prototypes' technical specifications issued (e.g. scalability, reproducibility) -Products over-engineered (i.e. better quality of design and prototypes offered)	-Systems Verified and product launched -Equipments evaluated and inspected	-Products, service and solutions provided -Periodical review on product performance conducted (e.g. ten times better than guarantee) -General availability confirmed to the market via channels
Market and business performances	-Either stayed with existing customers or go to the competitors -Customer needs and demands observed -Fibre laser processing market confirmed	-Variation of and relationship with customers and suppliers set up -Business opportunities analysed and plan issued -Market risk assessed -Innovative and end customer identified -Detail market launch plan issued	-Specific needs and requirements of customers known -Market segment, size and share predicted -Pricing & Launching issued -Partnership established	-Positioning in the market -New business model established -Customer-intimate marketing (feedback) -Competitors identified (e.g. no competitors identified in the fibre laser processing)	-Business model refined -Initial market research conducted -Necessary restructure made and re-innovate considered -Formulated direct sell and through sub-contracting
Firm and industry performances	-Considered for: 1) increasing sub contract work, 2) risk reduction exercise	-Key personnel and/or individuals involved -Organisational risk considered (e.g. investment plan initiated and investment started)	-Formalised organisation -Organisational risk assessed (e.g. profit predicted; investment issued) -Strategy formulated (e.g. 'insidovation' for fibre laser processing and 'open innovation' for non-fibre laser processing)	-Formal organisation established -Organisational risk periodically assessed (e.g. financial indicator: cost reduced)	-Organisational risk periodically assessed -Increased sub-contracting to increase self control -Re-innovate or exit

required. The customer's required specification and detailed product and/or process, including market launch plan, are issued. Business opportunities are analyzed and the end customer is identified.

- **Level 3 (Prototypes and systems demonstration):** Once principles have been proven, some parts, maybe with varying designs, will be produced in the second phase. Any prototype's technical specifications are issued and its levels of reliability are demonstrated. Prototypes and actual processes are demonstrated during this stage. Laser-based machine tools are over-engineered to ensure better quality of design and prototypes for the Company's products.

- **Level 4 (Pilot-line production):** In this phase, a particular chosen design is produced so that proper testing of the part can be carried out and process issues, such as repeatability and yield, can be assessed. The number of operations is reduced and equipment is evaluated and inspected. Overall systems are verified and pilot scale production is launched.

- **Level 5 (Overall systems operation):** When all design and process issues have been agreed, much larger volumes of parts can be produced. At this stage, the process is fixed and this phase may lead to significant decisions being made regarding larger volume production options. Technological operation and maintenance are enabled, including review of product performance. Service and technical solutions are provided. Organizational risk is periodically assessed and initial market research is conducted.

There are a number of important factors to keep in mind for the case of laser-based micro and nanomanufacturing technologies. Firstly, laser machining and/or processing is a direct and very flexible technique. Hence, it is unrivalled in the proof-of-concept and prototyping phases when compared with traditional machining, such as mechanical, chemical, molding or etching methods. The speed with which product design ideas can be assessed and/or parts modified using lasers is very quick, which is crucial in the fast-moving micro and nano technology sector. Secondly, the use of laser-based micro and nanomanufacturing allows the high-volume stage to be reached more quickly and allows the client to choose the right time to commit to the full production stage. It is worth noting that laser micro and nano machining and/or processing are becoming important in product development plans across industries.

Key Issues and Performances of IMRLs Confirmed

Science and Technology Performance

The Company has developed original fiber laser processing technologies that have never been seen before. It owns these technologies as it holds the patents and expertise in this area, and is able to offer product repositioning to other sectors. In-house ideas have been utilized for innovation both for themselves and for their customers. The Company has greater knowledge than their customers and better understands the limitations and future potential of the technologies. The non-fiber laser processing technologies (for example, general micro laser equipment) are typically customer driven, and the Company follows an '*open innovation*' strategy through which it welcomes new and/or improved ideas from outside, particularly for group 2 products. The company can offer their expertise when either approached by customers or a need is identified that the customer cannot achieve. The laser-based fiber processing technologies are developed based on their internal expertise and capability. This is termed '*insidovation*', which means that the innovation processes are developed within the Company. Such processes are different from those observed in the non-fiber laser processing technologies for products group 1.

Product and Process Performance

The conventional product and process development research has been conducted prior to new products being planned and incremental improvement of conventional tools confirmed. The machine tools developed in-house are guaranteed for 5 000 hours, but typically give service for 50 000 hours; ten times better than the guarantee. They confess that they have an obsession with product reliability. Quality and customer satisfaction are more important to the Company than cost and time. The Company attaches more importance to quality and customer satisfaction because they see it as giving them a competitive edge. Superior product quality in relation to competitors (for example, no manufacturing competitors exist for fiber laser processing, few exist for non-fiber laser processing) is achieved. Products are over-engineered and the levels of reliability, scalability and/or reproducibility have been maintained.

Market and Business Performance

The Company is delivering products to a wider market that they did initially. It has formulated direct sales and/or sub-contracting. The initial targeting of large Companies (for example, HP, DuPont) accounted for a significant proportion of their business, however two other groups have emerged: (1) Venture capitalist funded Companies – start-up Companies who offer technical solutions to predicted problems that do not as yet exist. (2) Companies that provide services to an established industry who differentiate themselves from competitors by having capability beyond the norm (for example, the Company sells some products to small companies who use the equipment as a show-piece, which becomes a part of its strategic marketing). Flexibility in the relation-

ship with customers and suppliers is successfully maintained. It provides samples with quick lead times and works closely with customers to develop optimum solutions. Potential market risk is assessed, any innovation necessary is worked on and the end customer is identified. Even in an extreme case it has developed products for customers for free over an eighteen month period without any revenue other than the promise of a potentially lucrative contract at the end. A future business strategy would include the development of cheaper machine tools with China as a target market, as well as increase in sub-contracting work.

Firm and Industry Performance

The studied Company can be seen as a niche manufacturer due to its high-end products requiring in-depth technical expertise, along with expensive unit cost. For other firms and/or competitors (for example, no manufacturing competitors exist for fiber laser processing, few exist for non-fiber laser processing), it would be difficult to replicate its R&D and product portfolio. The Company usually uses new designs and sophisticated optics with established lasers instead of using leading edge lasers unless there is a particular demand, in which case it offers a sub-contracting service. This strategy distinguishes it from its competitors. Trust and close relationship building with customers are one of the driving forces for the Company's performance. The senior management work closely together and are in direct contact with customers as part of the day-to-day running of the business. Organizational risk is periodically assessed and finally organizational strategy is considered for increasing sub-contract works and for risk reduction exercise (i.e. the more risk, the more revenue and self-control). Among the Company's strengths are that its performance is better than its competitors in respect to quality, machine servicing hours, expertise and as originators of patents. Start-ups use it as the show-piece kit at the front of their operation. (It has quick turn around on demonstrators I am quite sure this is not right, but I really don't know how to correct it), e.g. good proof of principle. Risks from R&D are offset against the increase in sub-contracting work.

CONCLUSION

Laser-based micro and nanomanufacturing processes are being adopted in increasing numbers of production plants and are now the method of choice for the displays and lighting, printers, bio-medical, telecommunications and micro electronics industries. The studied Company has been routinely involved in delivering precision laser processing to production applications (for example, the work involves improved processes, replacement or displacement of conventional processes). This study presents the performance measures of a laser-based micro and nanomanufacturing Company and analyses their relative importance. The importance of final product quality and delivery reliability can be seen increasingly as the important factors for the sector. In relation to laser lensing, their process is an alternative to a number of conventional techniques including hot drawing, grinding and polishing. The case study has confirmed the practicability and feasibility of the IMRLs matrix. The research findings suggest the following:

- Partition of the maturity stages of laser-based micro and nanomanufacturing technologies.
- Four key issues and performance indicators have been observed in this case that are key to managing the life cycle of laser-based MNTs; science and technology, product and process, market and business, firm and industry performances.
- Key activities and criteria within each phase of the IMRLs have been identified from the case study.

The limitations and gaps of existing theories, particularly on the management of micro and nano technologies led to the initial idea of developing a classification tool to assess maturity throughout its lifecycle, which has hopefully been accomplished by this research. The novelty of the research could be that organizations that develop and use MNTs have an opportunity to apply a specific assessment scale to quantify the technology readiness of unreleased micro and nanotechnologies. The IMRLs assessment chart identifies maturity criteria for laser-based micro and nano manufacturing technologies and describes the key performance activities along with four different criteria. The IMRLs is intended to be used as a descriptive assessment tool for managing the emergent MNTs. Based on detailed discussion with the key persons involved, when all the key activities in one phase are accomplished, implementation proceeds to the next phase. In this way, the development of technologies is always under control and risk is monitored. However, there are some limitations in the findings which provide ground for future research. It would be beneficial to apply the matrix in different environments or in different sectors, and then compare the results to determine the differences and improve the research. Further testing of the IMRLs tool is necessary in order to increase its robustness and to better understand its applicability and feasibility in assessing emerging and disruptive technologies maturity.

ACKNOWLEDGMENT

The author gratefully acknowledges and gives thanks to the C-Steps' (a project on 'Technology management for emerging and disruptive manufacturing technologies') partner company – a leading laser processing company where the case study was carried out. Valuable discussions and comments with the C-Steps colleagues at Cardiff University Innovative Manufacturing Research Centre (CUIMRC) are also greatly appreciated.

REFERENCES

Alting, L., Kimura, F., Hansen, H. N., & Bissacco, G. (2003). Micro engineering. *Annals of the CIRP*, *52*(2), 635–657. doi:10.1016/S0007-8506(07)60208-X

Beacham, J. (2006). *Succeeding through innovation: 60 minute guide to innovation, turning ideas into profit*. TSO.

Brinksmeier, E., Riemer, O., & Stern, R. (2001). Machining of precision parts and microstructures. In *Proceedings of the 10th International Conference on Precision Engineering*, Yokohama, Japan (pp. 3-11).

Cooper, R. G. (2001). *Winning at New Products*. Cambridge, MA: Perseus Books.

Dimov, S. S., Matthews, C. W., Glanfield, A., & Dorrington, P. (2006). A roadmapping study in multi-material micro manufacture. In *Proceedings of the Second International Conference on Multi-Material Micro Manufacture*, Grenoble, France (pp. 11-25).

Dumond, J. E. (1994). Making best use of performance measures and information. *International Journal of Operations & Production Management*, *14*(9), 16–31. doi:10.1108/01443579410066712

Foster, R. (1986). *Innovation: The attacker's advantage*. London: McMillan.

Genaidy, A., & Waldemar, K. (2008). A roadmap for a methodology to assess, improve and sustain intra-and inter-enterprise system performance with respect to technology-product life cycle in small and medium manufacturers. *Human Factors and Ergonomics in Manufacturing*, *18*(1), 70–84. doi:10.1002/hfm.20097

Igami, M., & Okazaki, T. (2007). *Capturing Nanotechnology's Current State of Development via Analysis of Patents*. STI Working Paper 2007/4, OECD Directorate for Science, Technology and Industry.

Islam, N. (2009). Innovative manufacturing readiness levels (IMRLs): A new readiness matrix. *International Journal of Nanomanufacturing*.

Islam, N., & Miyazaki, K. (2009). Nanotechnology innovation system: Understanding hidden dynamics of nanoscience fusion trajectories. *Technological Forecasting and Social Change*, *76*(1), 128–140. doi:10.1016/j.techfore.2008.03.021

Madou, M. J. (2002). *Fundamentals of microfabrication: The Science of miniaturization* (2nd ed.). Boca Raton, FL: CRC press.

Mankins, J. C. (1995). *Technology Readiness Levels. Advanced Concepts Office, Office of Space Access and Technology*. NASA.

Masuzawa, T. (2000). State of the art of micromachining. *Annals of the CIRP*, *49*(2), 473–488. doi:10.1016/S0007-8506(07)63451-9

Moore, G. A. (1999). *Crossing the Chasm*. New York: Harper Business.

Niemeyer, J. K., & Whitney, D. E. (2002). Risk reduction of jet engine product development using technology readiness metrics. In . *Proceedings of the ASME Design Engineering Technical Conference*, *4*, 3–13.

Rogers, E. (1995). *Diffusion of Innovations*. New York: The Free Press.

Rooks, B. (2004)... *Assembly Automation*, *24*(4), 352–356. doi:10.1108/01445150410562534

Tao, L., Probert, D., & Phaal, R. (2008). Towards an integrated framework for managing the process of innovation. In *Proceedings of the R&D Management Conference 2008*, Ottawa, Canada.

Technology Readiness Assessment Desk Book DUSD(S&T). (2005). Department of Defense (DOD) report, United States, May.

The Royal Society & the Royal Academy of Engineering. (2004). *Nanoscience and nanotechnologies: opportunities and uncertainties.* London, UK. Retrieved from http://www.royalsoc.ac.uk/policy

Zerfass, A. (2005). Innovation readiness: A framework for enhancing corporations and regions by innovation communications. *Innovation Journalism, 2*(8), 1–27.

APPENDIX

Table 3. Maturity and performance assessment of laser-based micro and nano manufacturing technologies using IMRLs matrix

IMRLs phases or levels / Key issues assessed	IMRL 1 **Concept development and refinement**	IMRL 2 **Proof-of-concept and capability development**	IMRL 3 **Prototypes and systems demonstration**	IMRL 4 **Systems verification and Pilot-line production**	IMRL 5 **Systems operation and market audit**
Science and technology performances	- Basic scientific principles/technologies observed - In-house idea generated (e.g. patents) along with limited innovative idea from outside - Studies on technology feasibility confirmed	- New fibre laser-based technological components and capabilities achieved by replacing conventional laser processing - Success criteria understood and agreed - Work steps confirmed and IP protected - Technological risk assessed and an alternative solution considered	- Components and capabilities tested - Prototypes and actual processes demonstrated - IP protected	- Technology documented - Number of operations reduced - Expertise formed	- Technology operation and maintenance enabled - Technological service provided - After sales support - Re-innovate considered
Product and process performances	- Product and/or process development research conducted - New product planned (e.g. fibre laser processing tools that is new) - Incremental improvement of conventional tools confirmed (e.g. non-fibre laser processing equipments)	- Access to appropriate lasers confirmed - Detailed product and/or process launch plan issued - Customers' required specification considered - Quality in relation to competitors achieved	- Levels of reliability demonstrated - Prototypes' technical specifications issued (e.g. scalability, reproducibility) - Products over-engineered (i.e. better quality of design and prototypes offered)	- Systems Verified and product launched - Equipments evaluated and inspected	- Products, service and solutions provided - Periodical review on product performance conducted (e.g. ten times better than guarantee) - General availability confirmed to the market via channels
Market and business performances	- Either stayed with existing customers or go to the competitors - Customer needs and demands observed - Fibre laser processing market confirmed	- Variation of and relationship with customers and suppliers set up - Business opportunities analysed and plan issued - Market risk assessed - Innovative and end customer identified - Detail market launch plan issued	- Specific needs and requirements of customers known - Market segment, size and share predicted - Pricing & Launching issued - Partnership established	- Positioning in the market - New business model established - Customer-intimate marketing (feedback) - Competitors identified (e.g. no competitors identified in the fibre laser processing)	- Business model refined - Initial market research conducted - Necessary re-structure made and re-innovate considered - Formulated direct sell and through sub-contracting
Firm and industry performances	- Considered for: 1) increasing sub contract work, 2) risk reduction exercise	- Key personnel and/or individuals involved - Organisational risk considered (e.g. investment plan initiated and investment started)	- Formalised organisation - Organisational risk assessed (e.g. profit predicted; investment issued) - Strategy formulated (e.g. 'insidovation' for fibre laser processing and 'open innovation' for non-fibre laser processing)	- Formal organisation established - Organisational risk periodically assessed (e.g. financial indicator: cost reduced)	- Organisational risk periodically assessed - Increased sub-contracting to increase self control - Re-innovate or exit

Section 4
Ethics, Regulation and Environment

The fields of nanotechnology and microelectronics are involved in broad ethical complications owing to their diversity and relative infancy. With these burdens come the need for regulations towards ensuring safety in their products, processes and tools as well as protecting the environment and human. In this section, ethics, regulation and the climate control are examined. The climate economics and finance, the intellectual property rights challenges of microelectronics and nanotechnology along with information and communication technologies are discussed.

Chapter 12
Diffusion of the Clean Development Mechanism

Shaikh M. Rahman
Texas Tech University, USA

Ariel Dinar
University of California, Riverside, USA

Donald F. Larson
World Bank, USA

ABSTRACT

The Clean Development Mechanism (CDM) of the Kyoto Protocol is an innovation that combines greenhouse gas abatement targets with sustainable development objectives. This chapter provides an estimate of the overall growth pattern of the CDM and makes projections about CDM activity during and beyond the first commitment period of the Kyoto Protocol commitments under current rules. The results imply that if the emission reduction targets remain unchanged beyond the first commitment period, further expansion of the CDM pipeline is unlikely.

INTRODUCTION

Under the Kyoto Protocol, the Clean Development Mechanism (CDM) is the only formal way for the 39 countries that have pledged to reduce greenhouse gas emissions, known as Annex B countries, to tap potential sources of mitigation in countries that have not. For the most part developing countries comprise the second group and are known, in Protocol parlance, as non-Annex B countries. There are two project-based mechanisms under the Protocol, both in terms of the scale of current investments under the program and in terms of its mitigation potential. The CDM is by far the larger of the two.[1]

Briefly, a CDM project is an investment hosted by a non-Annex B country that is intended to reduce greenhouse gas emissions or speed the removal of greenhouse gases from the atmosphere relative to a business-as-usual baseline.[2] The projects are reviewed individually by a CDM Board prior to implementation and are subject to continuous monitoring and a verification process. If successful, the projects generate offsets, known as Certified Emission Reductions (CERs) that Annex B countries can use to meet their Kyoto obliga-

DOI: 10.4018/978-1-61692-006-7.ch012

tions. Overall, the CDM is expected to lower the cost of meeting the environmental goals of the Kyoto Protocol by encouraging investments in low-cost abatement efforts wherever they can be found. Another stated objective of the CDM is to assist the host developing countries achieve sustainable development through the mobilization of direct private foreign investment and technology transfer.[3]

With its dual objectives, the CDM attracts both Annex B and non-Annex B parties to the convention. Since its inception in 2003, Greenhouse Gases (GHG) abatement activity under the CDM has increased rapidly. By December 2007, 2,966 CDM projects were submitted to the UNFCCC for validation that are expected to generate 441 million CERs annually from 2008-2012, the first commitment period of the Kyoto Protocol (UNEP Risoe, 2008). Moreover, many investors expect the CDM or some similar mechanism to continue beyond the first commitment period and many CDM projects currently underway will generate emission reductions well beyond 2012.

Nevertheless, the scope for additional CDM projects is limited by the fundamental components of demand and supply, which are in turn, determined by the rate and composition of global economic growth; current Kyoto targets and expectations about future regulations; domestic mitigation efforts in Annex B countries; and JI efforts among Annex B countries.

As is discussed later, there are a variety of predictions about the size of the eventual CDM market that take these fundamentals into account. In this chapter, we look to see if these predictions are consistent with the historic pattern of growth in CDM projects and conceptual models of technology diffusion. In particular, we test whether the predicted size of the CDM market will be exceeded, based on a sigmoid expansion path that is often associated with the diffusion of new ideas and technologies.

The remainder of the chapter is organized as follows. The next section describes the incidence and extent of participation in the CDM process–which we refer to as adoption–globally and in individual countries. Section 3 discusses how the conventional logistic (epidemic) model can be applied to analyze global CDM adoption. Using data on observed CDM activity during 2003-07 and considering several alternative scenarios, the estimated parameters of the global CDM adoption model are presented in Section 4 along with projections of CDM activity during and beyond the first commitment period of the Kyoto Protocol. The last section discusses the policy implications of the empirical results, indicates areas of future research, and concludes.

GROWTH OF CDM IN THE WORLD AND INDIVIDUAL COUNTRIES

The CDM/JI Pipeline Analysis and Database of the United Nations Environment Programme (UNEP) Risoe Center constructs and maintains an up-to-date dataset consisting of all CDM projects that have been sent to the CDM Board for validation. The dataset includes information about each individual CDM project, such as project name, type, registration/validation status, baseline and monitoring methodologies, involved host country and credit buyers, expected annual and total CERs to be generated in each year during the life of the project, potential power generation capacity, etc. In order to analyze the CDM adoption process, information about all CDM projects that have been sent to UNFCCC for validation up until December 2007 are extracted from that dataset.

The dataset shows that, the CDM portfolio has been growing very rapidly since its inception in 2003. As of 31 December 2007, 2,966 CDM projects have been submitted to UNFCCC for validation. Only 805 of these projects have been registered, 36 rejected, 6 withdrawn, and the rest are in the process of registration or validation (UNEP Risoe, 2008). The 2,924 CDM projects (excluding the 36 rejected and 6 withdrawn projects) in the pipeline are expected to generate 436.8 Million CERs (i.e., reducing approximately 436.8 Million tons of CO_2 equivalent greenhouse gases) each year during the first commitment period of the Kyoto Protocol, and 3,462.3 Million CERs over the duration (life-time) of the projects. Annual and total estimated investments (capital costs) in these projects amount to 10.9 and 86.6 billion US dollars, respectively.

Table 1 presents the number of CDM projects submitted for validation in each year during 2003-07.[4] Annual and total CERs expected to be generated from these projects and estimated capital investments in these projects are also reported in the table. It is evident from the table that CDM activity, measured in terms of number of projects, expected CERs, and investments, has been increasing over time at an increasing rate. The exponential growth of CDM activity is also evident from Figure 1, which depicts the cumulative distribution of expected CERs from the projects in the pipeline. While there has been a substantial increase in CDM activity, from 5 CDM projects in 2003 to more than 1,592 projects in 2007, the rate of global CDM adoption is likely to decline as the market for CERs becomes saturated (i.e., as CERs from these projects meet the aggregate demand of the countries listed in Annex B of the Kyoto Protocol).

Based on the CDM pipeline dataset for 2003-07, the projects are (to be) located in only 66 of the developing countries (hosts hereafter). On the other hand, only 19 of the 39 Annex B countries (investors hereafter) are directly involved in CDM projects.[5] Note that 126 of the projects involve credit buyers from two or more Annex B countries. Also, there are 1,244 unilateral CDM projects initiated by 46 developing countries (with the highest number of 665 unilateral projects in India).

While the global CDM growth is impressive, the extent and growth of CDM activity, widely varies across the host as well as investor countries. Table 2 provides a summary statistics of the CDM projects of the host and investor countries by years. The scenario depicted in the table can be described as follows. First, the numbers of participating host and investor countries have rapidly increased over the five-year period, as have the total number of CDM projects and expected CERs. Second, the total and average numbers of projects for the host and investor countries and the respective average CERs have been consistently increasing as the first commitment period approaches its termination.[6] Finally, the ranges of CDM activities of the host and investor countries (given by the minimum and maximum number of projects and expected CERs to be generated from the projects) have widen over time.

The adoption of CDM in selected host and investor countries are depicted in Figures 2a and 2b. Figure 2a shows the cumulative expected CERs from the projects that are hosted by selected

Table 1. Number of CDM projects submitted for validation during 2003-07, respective CERs expected to be generated, and estimated (pledged) investments

Year[1]	Number of CDM Projects	Expected CERs per Year (Million)	Estimated Annual Investments (Million US$)	Expected Total CERs (Millions)	Estimated Total Investments (Million US$)
2003	5	5.53	9.79	38.75	68.76
2004	53	4.78	124.28	36.03	948.49
2005	434	91.82	956.81	722.35	7, 947.39
2006	840	125.45	3, 102.03	1, 012.72	25, 568.32
2007	1, 592	209.18	6, 678.43	1, 652.43	52, 064.67
Total	**2, 924**	**436.76**	**10, 871.34**	**3, 462.28**	**86, 597.63**

Figure 1. Expected CERs from CDM projects submitted to UNFCCC (2003-07)

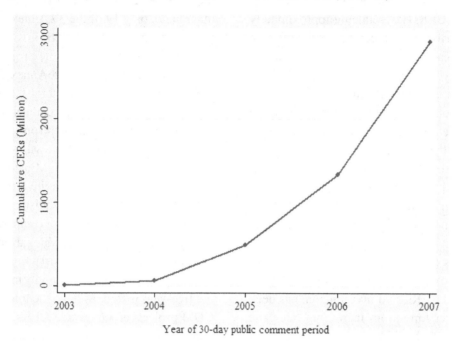

developing countries. Brazil, China, India, and Mexico appear to be the major host countries for CDM projects. More than 75 percent of the projects (2,231 of 2,924) submitted for validation during 2003-07 are located in these countries, which account for about 78 percent of the total CERs (339.6 of 436.8 million) expected to be generated every year. Brazil and India are among

Table 2. Summary statistics of the CDM projects submitted for validation, by year and participating Annex B and developing countries

Year[1]	Obs. (Countries)	No. of CSM Projects				Expected CERs (Mill.) per Year			
		Total		Min.	Max.	Total		Min.	Max.
Annex B Countries									
2003	3	9	3.00	2	4	4.66	1.55	1.082	1.797
2004	8	52	6.50	1	15	4.15	0.52	0.035	1.733
2005	14	341	24.36	1	149	72.49	5.18	0.092	22.959
2006	13	439	33.77	3	194	87.00	6.69	0.220	36.867
2007	19	1011	53.21	1	362	162.76	8.57	0.017	65.189
Developing Countries									
2003	5	5	1.00	1	1	5.53	1.11	0.015	3.393
2004	15	53	3.53	1	16	4.78	0.32	0.001	2.882
2005	38	434	11.42	1	184	91.82	2.42	0.006	37.284
2006	41	840	20.49	1	292	125.45	3.06	0.003	67.790
2007	53	1592	30.04	1	735	209.18	3.95	0.006	136.421

the countries which adopted CDM projects first in 2003 (other countries are Chile, Guatemala, and South Korea). China appears to be a relatively late adopter of CDM (with no projects in 2003 and only 2 projects in 2004), but immediately surpassed Brazil and India both in terms of number of projects and expected CERs. As of December 2007, 961 (33 percent of the total) CDM projects in the pipeline are (to be) located in China, which are expected to generate 241.6 million (more than 55 percent) of the total CERs per year. With 836 (29 Percent) projects generating 59.8 million (14 Percent) CERs per year, India is the second largest adopter of CDM among the host countries. As shown in Figure 2a, other major host countries include Argentina, Indonesia, Malaysia, South Africa, Sri Lanka, and Viet Nam. On the other hand, Lao PDR is the smallest host country with only 1 CDM project that is expected to generate 3.3 thousand CERs per year. Other small CDM host countries include Fiji, Guyana, Malta, Mozambique, Singapore, Tajikistan, and several others.[7]

In terms of expected CERs per year, the United Kingdom, Japan, Italy, the Netherlands, and Canada are the five largest investor (credit buyer) countries, respectively (Figure 2b). Credit buyers from the United Kingdom are involved in 723 (25 Percent) of the CDM projects, and are expected to acquire 128.1 million CERs (29 Percent) per year. Credit buyers from Japan, the second largest investor country, are involved in 252 (9 Percent) of the CDM projects and are expected to acquire 51.0 million CERs (12 Percent) per year. On the other hand, Czech Republic is the smallest investor country involved in only 1 CDM project that accounts for 128 thousand CERs per year.

As can be seen from Figures 2a and 2b, CDM adoption by each of the selected host and investor countries has been increasing over time at an increasing rate.[8] It is also evident that the rate of CDM adoption varies widely among the host and investor countries. Factors that are responsible for the variation in CDM adoption across countries could be empirically determined and are the

subject of a separate study. However, the next section provides an investigation of the underlying theory for analyzing the diffusion process of new innovation, with an application to CDM adoption.

THE LOGISTIC MODEL FOR CDM ADOPTION

Alternative models that have been used to describe the diffusion process of new technological and economic innovations can be categorized in three major groups: the epidemic or logistic growth model (Griliches, 1957, 1980; Mansfield, 1961; Gore and Lavaraj, 1987; Doessel and Strong, 1991; Knudson, 1991; Dinar and Yaron, 1992); the exponential growth model (Gregg et al., 1964; Dixon, 1980; Metcalfe, 1981); and the new product growth model (Bass, 1969; Mahajan and Schoeman, 1977). All of these models are founded upon the epidemic or logistic model that views the diffusion process to be similar to the spread of an infectious disease, with the analogy that contact with other adopters (i.e., learning from the experience of others) and exposure to information on the innovation (i.e., demonstration effect) leads to adoption. The model is based on the assumption that members of a homogeneous population have an equal probability of coming into contact with each other and that the flow of new adopters of the technology in a given point in time is a function of the stock of existing adopters. When the stock of existing adopters is small, there is little risk of "contagion." The risk of "contagion" increases as the stock of existing adopters increases (potential adopters decreases), and the flow of new adopters rises exponentially. However, as the stock comes closer to the total number of potential adopters, the flow of new adopters gradually decreases and eventually becomes zero. Then, new innovations replace the old one and we may see an abandonment of the old innovation and adoption of the new one. The diffusion of the innovation thus follows a symmetric S-shaped function over time.

Figure 2. (a) CER generation capacity of the CDM projects in selected developing countries; (b) CER generation capacity of the CDM projects of selected Annex B countries

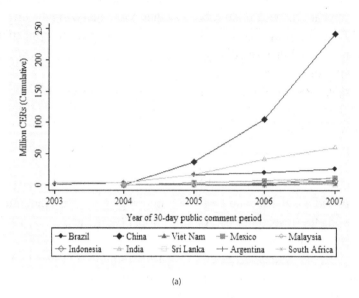

(a)

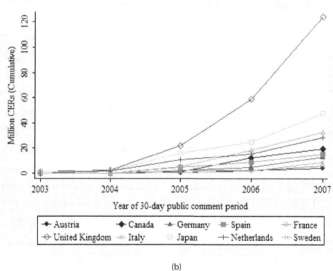

(b)

The CDM of the Kyoto Protocol can be viewed as a new innovation that combines carbon abatement and development objectives, and the epidemic or logistic model can be applied to understand the adoption of CDM by individual countries as well as the world. In the case of CDM, the analogy to the epidemic model is that exposure to the opportunity and learning from the experience of other countries lead countries to adopt the mechanism. Also, the ability of investor countries to invest in several host countries leads to increased learning and adoption. Following Feder and Umali (1993), the general form of the logistic model for CDM adoption is

$$\frac{dq_t}{dt} = \beta \frac{q_t}{q^*}(q^* - q_t), \qquad (1)$$

where q_t denotes the accumulated amount of CO_2 to be abated through the CDM projects at time t (e.g., year), q^* denotes the country-specific ceiling of carbon abatement (e.g., AAUs or total CO_2 emission), and β is a parameter measuring the speed of adoption (the rate of adoption).[9] Given the value of β, the additional amount of carbon abatement through CDM at any point in time is the product of two opposing forces; $q^* - q_t$ declines as q_t / q^* increases. This results in a bell-shaped frequency distribution of CO_2 abatement through CDM and a sigmoid cumulative density function of the logistic frequency distribution over time. The logistic model thus imposes a symmetric S-shaped CDM adoption trend which attains a maximum adoption rate when half of the cumulative potential CO_2 abatement are accomplished by the CDM projects.

Solving the logistic differential equation (1) for q_t yields the standard logistic growth function

$$q_t = \frac{q^*}{1 + \exp[-\alpha - \beta \cdot t]} \qquad (2)$$

where α is the coefficient of integration, which shifts the location (intercept) of the function without affecting its shape, and β_i is the coefficient representing the speed of adoption (rate of adoption over time). Three important elements determine the S-shape of the function in Equation (2): the ceiling of CO_2 abatement (q^*), the timing of adoption (t), and the rate of adoption (β). For a given t, q_t and q^*, α and β (i.e., the location parameter and rate of adoption, respectively) can be estimated from equation (2). The estimates can be further used to forecast adoption of CDM in successive years.

Based on the logistic growth function described above, the following section estimates the parameters of the global CDM adoption curve using empirical data for the period of 2003-07. The parameter estimates are then used to project the extent of global CDM adoption in each year during 2008-2020.

THE GLOBAL CDM ADOPTION CURVE

The global adoption of CDM can be measured in terms of aggregate expected CERs, number of projects, and capital investments over time. From the annual data presented in Table 1, cumulative expected CERs, number of projects, and investments in each year during 2003-2007 can be calculated and used to represent the dependent variable q_t in the logistic growth function specified in equation (2).

While accurate measures of the aggregate demand for CERs are not available, several studies have estimated the potential demand for CERs during the first commitment period, 2008- 2012. Table 3 presents Annex B countries' potential demand for total emissions reduction and the potential size of the CDM market as estimated by different studies.[10] Based on the available estimates, aggregate demand for CERs ranges from 0 to 520 million per year. In the empirical model, we use the upper limit of the range to represent q^* while estimating the CDM adoption function in terms of expected CERs. However, estimates for the upper limits (ceilings) of the number of CDM projects or investments are not available from existing studies reported in the literature.

When the ceiling is unknown and a reasonable estimate is not available, the logistic growth function can still be estimated using an alternative technique that considers q^* as an additional parameter. This technique imposes a logistic curvature on the available observations of q_t, and then simultaneously estimates $\alpha, \beta,$ and q^*.[11]

Using cumulative amount of expected CERs in each year during 2003-2007, the logistic CDM growth function is estimated both with and without a given upper bound (ceiling). The simple nonlinear least-squares estimation procedure is employed to estimate the parameters. We apply define two specifications for the CDM growth function. The first specification relates to a case where the annual CER ceiling is not known and the second specification relates to a case where the annual CER ceiling is known and set based on previous analysis as reflected in Table 3. The estimates of the location parameter (α), the rate of adoption (β) and q^* in the case where an upper limit on CER is unknown are presented in the left column of Table 4. The estimates of α and β in the case where an upper limit on CER is known ($q^* =520$ million CERs per year is used to represent the ceiling) are presented in the right column of Table 4. All coefficients are significant at a 10 percent and better. Goodness of fit is very high (R-square>.99).

Using these parameter estimates, expected CERs to be generated by the CDM projects in each year of the first commitment period of the Kyoto Protocol and beyond are projected. Table 4 presents the results of the capped and un-capped models.

Based on the parameter estimates in Table 4, Figure 3a depicts the predicted logistic growth functions (both with and without a given ceiling) along with the actual plots of cumulative expected CERs over the period of 2003-07. The fitted logistic growth curves show that, with a given maximum aggregate demand of 520 million CERs, the growth rate of CDM adoption starts to decline after 2006, approaches zero by 2010, and remains at zero beyond that point. On the other hand, with an estimated ceiling of 520million CERs, the growth rate starts to decrease after 2007, reaches zero by 2012, and remains constant in the successive years. It is important to note that, given the growth rate of expected CERs during 2003-2007 (Figure 1), the adoption function is yet to reach an inflection point as greenhouse gas abatement through CDM is still increasing at an increasing rate. However, the fitted adoption curve with an exogenous ceiling of 520 million CERs induces an inflection point in year 2006.

Table 3. Estimates of the potential demand for emissions reduction and size of the CDM market

Study	Annex I countries' demand for Kyoto units under the Protocol MtCO$_2$ equivalent/year	Potential size of the CSM market MtCO$_2$ equivalent/year
Blanchard, Criqui, and Kitous (2002)[a]	688-862	0-174
Eyckman et al (2001)[a]	1414-1713	261-499
Grutter (2001)[a]	1000-1500	0-500
Haites (2004)[b]	600-1150	50-500
Halsnaes (2000) [b]	600-1300	400-520
Holtsmark (2003)[b]	1246-1404	0-379
Jotjo and Michaeloa (2002)[a]	1040	0-465
Vrolijk (2000)[b]	640-1484	300-500
Zhang (1999)[b]	621	132-358
Range	**600-1713**	**0-520**

Note: [a] Assumes that only the United States does not ratify the Kyoto Protocol

[b] Assumes that Australia and the United States do not ratify the Kyoto Protocol

Sources: Haites (2004); Zhang (1999)

Table 4. Two and three parameter growth logistic model estimates for CDM diffusion

Variable	Upper limit on CER is unknown	Upper limit on CER is known
A	-2536.65** (-4.74)	-3162.61*** (-9.42)
B	1.26** (4.74)	1.57*** (9.42)
q^*	664.18* (3.80)	Imposed at 520
Number of observations	5	5
F-test	317.54*** (0.003)	360.59*** (0.000)
Adjusted R-squared	0.995	0.993

*, **, *** indicate significant level of the t-statistic of the estimated coefficient of .10, .05. and .01, respectively.

When the ceiling (i.e., q^*) is simultaneously estimated with the location and rate of adoption parameters (i.e., α and β, respectively), the growth rate starts declining after 2007. This is because of the symmetric structure of the underlying logistic growth function.[12] The symmetric structure also implies that the higher the ceiling the later the point of inflection occurs. As depicted in Figure 3b, fitted logistic adoption curves with exogenous ceilings of 1000 and 2000 million CERs induce points of inflection in years 2008 and 2009, respectively.

Given the growth rate of CDM adoption during 2003-2007 and potential size of the market for CERs, further growth of CDM beyond the first commitment period appears to be unrealistic (see Figure 3a). A positive growth of CERs beyond 2012 can be expected only if expectations about the size of the market increases–that is, only if the overall demand for tradable Kyoto "units" increases, or the supply of competing ERUs or AAUs diminishes. As shown in Figure 3b, with larger sizes of potential market for CERs (e.g., with higher ceilings of 1000 and 2000 million CERs), CDM activity grows at a positive rate even beyond the first commitment period. On

the contrary, CDM activity is likely to decline if the expected size of the market for CERs shrinks.

Importantly, it is worth noting that the simple logistic model is inconsistent with the low-range predictions from structural models of the CDM market (Table 3). In general, the low-end predictions were associated with "hot air" scenarios in which large supplies of AAUs from Russia and the Ukraine displaced the CER market as well as a continued absence of the US from the market for CERs and other Kyoto units. An important implication of our results is that market participants have apparently discounted such scenarios.

The logistic growth parameters for CDM adoption in terms of number of projects and investments are also estimated. Since estimates of the upper limits of the number of CDM projects and investment are not available, the standard three-parameter estimation procedure is employed. Using the parameter estimates, accumulated number of projects and amount of capital investments in each year during the first commitment period and beyond are projected. The estimated logistic growth curves for CDM adoption in terms of number of projects and investments are depicted in Figures 4 and 5, respectively.

Figure 3. (a)CDM adoption during and beyond 2008-12–logistic growth with implied and estimated ceilings on CERs (Mill.); (b) Effect of alternative CER ceilings on CDM adoption during and beyond 2008-12 (Mill. CERs)

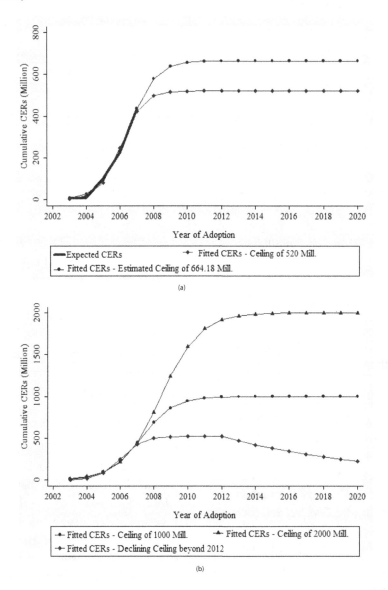

Comparing the structures of the CDM adoption curves presented in Figures 3, 4, and 5, it becomes clear that CDM adoption in terms of expected CERs, number of projects, and investments follow a similar trend. However, actual adoption of CDM is yet to reach an inflection point and the projections are likely to be updated with new data on additional CDM projects in the coming years.

SUMMARY, CONCLUSION AND AREAS FOR FUTURE RESEARCH

This study provides an estimate of the growth pattern of the CDM and, based on the estimated growth parameters, projects CDM activity during and beyond the first commitment period of the Kyoto Protocol. Fitting a logistic growth curve to

Figure 4. Projected CDM adoption in terms of number of projects

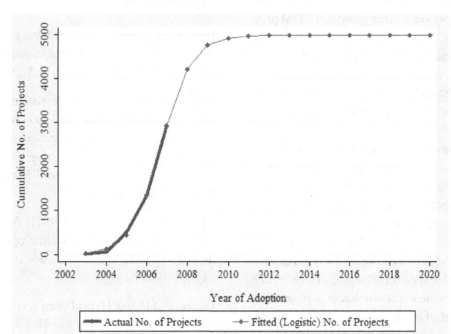

Figure 5. Projected CDM adoption in terms of investments (Mill. US$)

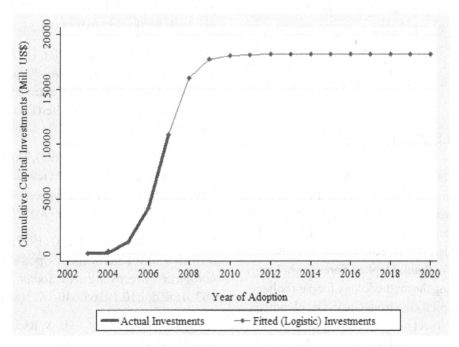

the CDM adoption data over time, it is predicted that the growth rate of the CDM starts to decrease after 2007, becomes zero by 2012 reaching an estimated ceiling of 664.19 million CERS per year, and remains constant in the successive years. This result implies that if the overall demand for

CER obligations remains constant beyond the first commitment period, further growth of CDM projects is unlikely. Alternatively, events that lower the overall demand for offsets or increase the supply of competing Kyoto units could cause the CER market to contract below current levels, although such scenarios are inconsistent with current rates of investment.

Looking ahead, the model presented in this chapter is largely descriptive and absent the set of determinants that can be linked to policy. This is especially important since early experience indicates that individual country experiences among host countries varied greatly. Potentially, the model could be extended to incorporate the types of variables expected to speed or slow project investment, such as the abatement cost structure of host countries and the strength of institutions backing investments generally and CDM projects in particular.

While the estimated CDM growth functions depicted in this chapter help understanding the global trend, the incidence and speed of CDM adoption widely vary between countries as shown in Figure 2a and 2b. Studying the determinants of cross-country CDM adoption, would be the natural next step, and will be performed in a separate study.

ACKNOWLEDGMENT

This chapter is one of the products of the study "Local Actions for Global Impact: Inference of the Role of the Flexible Mechanisms in Reducing GHGs" conducted in the World Bank DECRG-RU that was funded by the Knowledge for Change Trust Fund. We thank Stephen Seres for helping with developing the methodology for data collection on the capital cost from Clean Development Mechanism Project activity. During the work on the project leading to this chapter, Rahman was a consultant to the World Bank and Dinar was a Lead Economist with the World Bank; both in the Development Economics Research Group.

REFERENCES

Bass, F. M. (1969). A new product growth model for consumer durables. *Management Science, 15,* 215–217. doi:10.1287/mnsc.15.5.215

Blanchard, O., Criqui, P., & Kitous, A. (2002). *After The Hague, Bonn and Marrakech: The Future International Market for Emissions Permits and the Issue of Hot Air. Cahier de Recherche no. 27 bis.* Grenoble: Institut d'Economie et de Politique de l'Energie.

Dinar, A., & Yaron, D. (1992). Adoption and abandonment of irrigation technologies. *Agricultural Economics, 6,* 315–332. doi:10.1016/0169-5150(92)90008-M

Dixon, R. (1980). Hybrid corn revisited. *Econometrica, 48,* 1451–1461. doi:10.2307/1912817

Doessel, D. P., & Strong, S. M. (1991). A neglected problem in the analysis of the diffusion process. *Applied Economics, 23,* 1335–1340. doi:10.1080/00036849100000054

Eyckmans, J., van Regemorter, D., & van Steenberghe, V. (2001). *Is Kyoto Fatally Flawed?* Working Paper 2001-18, Center for Economic Studies, Katholieke Universiteit Leuven, Leuven, November.

Feder, G., & Umali, D. L. (1993). The adoption of agricultural innovations: A review. *Technological Forecasting and Social Change, 43,* 215–239. doi:10.1016/0040-1625(93)90053-A

Gore, A. P., & Lavaraj, U. A. (1987). Innovation diffusion in a heterogeneous population. *Technological Forecasting and Social Change, 32,* 163–168. doi:10.1016/0040-1625(87)90037-0

Gregg, J. V., Hassell, C. H., & Richardson, J. T. (1964). *Mathematical Trend Curves: An Aid to Forecasting.* Edinburgh, UK: Oliver and Boyd.

Griliches, Z. (1957). Hybrid corn: An exploration in the economics of technological change. *Econometrica, 25,* 501–522. doi:10.2307/1905380

Griliches, Z. (1980). Hybrid corn revisited: A reply. *Econometrica, 48,* 1451–1461. doi:10.2307/1912818

Grütter, J. (2001). The GHG market after Bonn. *Grütter Consulting and Joint Implementation Quarterly, 7*(3), 9.

Haites, E. (2004). Estimating the Market Potential for the Clean Development Mechanism: Review of Models and Lessons Learned. PCFplus Report 19, The World Bank Carbon Finance Business PCFplus Research Program, Washington, DC.

Halsnæs, K. (2000). Estimation of the Global Market Potential for Cooperative Implementation Mechanisms under the Kyoto Protocol. In Ghosh, P. (Ed.), *Implementation of the Kyoto Protocol: Opportunities and Pitfalls for Developing Countries.* Manila, Philippines: Asian Development Bank.

Holtsmark, B. (2003). Russian behaviour in the market for permits under the Kyoto Protocol. *Climate Policy, 3*(4), 399–415. doi:10.1016/j.clipol.2003.08.004

Jotzo, F., & Michaelowa, A. (2002). Estimating the CDM market under the Marrakech Accords. *Climate Policy, 2*(1), 179–196.

Knudson, M. K. (1991). Incorporating technological change in diffusion models. *American Journal of Agricultural Economics, 73,* 724–733. doi:10.2307/1242824

Larson, D. F., Ambrosi, P., Dinar, A., Rahman, S. M., & Entler, R. (2008). A review of carbon market policies and research. *International Review of Environmental and Resource Economics, 2*(3), 177–236. doi:10.1561/101.00000016

Larson, D. F., & Breustedt, G. (in press). Will markets direct investments under the Kyoto Protocol? Lessons from the activities implemented jointly pilots. *Environmental and Resource Economics.*

Lecocq, F., & Ambrosi, P. (2007). The clean development mechanism: history, status and prospects. *Review of Environmental Economics and Policy, 1*(1), 134–151. doi:10.1093/reep/rem004

Mahajan, V., & Schoeman, M. E. F. (1977). Generalized model for the time pattern of the diffusion process. *IEEE Transactions on Engineering Management, 24,* 12–18.

Mansfield, E. (1961). Technical change and the rate of imitation. *Econometrica, 29,* 741–765. doi:10.2307/1911817

Metcalfe, J. S. (1981). Impulse and diffusion in the study of technical change. *Futures, 13,* 347–359. doi:10.1016/0016-3287(81)90120-8

United Nations Environment Programme (UNEP) Risoe Center. (2008). *CDM/JI Pipeline Analysis and Dataset.* Retrieved from http://cdmpipeline.org

Vrolijk, C. (2000). Quantifying the Kyoto Commitments. *Review of European Community & International Environmental Law, 9*(3), 285–295. doi:10.1111/1467-9388.t01-1-00277

Zhang, Z. (1999). *Estimating the Size of the Potential Market for the Kyoto Flexibility Mechanisms.* FEEM Working Paper No. 8.2000. Milano, Italy: Fondazione Eni Enrico Mattei.

ENDNOTES

[1] The second project-based mechanism, Joint Implementation (JI), covers joint projects among Annex B countries. Offsets from this mechanism are called Emission Reduction Units (ERUs). Both ERUs and CERs are tradable. Another source of flexibility in the

Protocol is a provision that allows Annex B countries to trade assigned amount units (AAUs), national units that correspond to emission levels permitted under pledged caps.

2 For a discussion of pilot programs preceding the CDM see Larson and Breustedt (forthcoming).

3 A detailed description of the CDM and analysis of the issues related to this provision can be found in Larson et al. (2008) and Lecocq and Ambrosi (2007).

4 The starting year of the 30-day public comment period under validation is considered because the date of submission for validation is not available.

5 The United States is yet to ratify the Kyoto Protocol while Russia and Ukraine have emissions less than their AAUs. Excluding Russia, Ukraine, and the USA, there are only 36 potential investor countries.

6 The number of unilateral projects initiated by host countries has also increased considerably over these years.

7 Note that there are many developing countries which do not participate in CDM at all. This point will be addressed in section 5.

8 While CDM adoption is depicted in terms of the cumulative expected CERs in Figures 2A and 2B, similar trends are observed when CDM adoption is represented in terms of cumulative number of projects and capital investments.

9 Similarly, the adoption can also be represented by the number or amount of total investment in CDM projects at time t.

10 Article 6.1 d of the Kyoto Protocol prevent industrialized countries from making unlimited use of CDM by the provision that use of CDM be 'supplemental' to domestic actions to reduce emissions. Thus, the estimates of potential demand for CERs are less than the estimates of potential demand for all Kyoto units.

11 Note that this approach may result in inconsistent estimates of the parameters when using a small set of data from the early stages of growth (especially when the inflection point is unknown).

12 If the actual ceiling is below the estimated one (i.e., less than 658.50 $MtCO_2e$) and the inflection point is observed beyond 2007, then an asymmetric growth function (e.g., the Gompertz function) would better describe the CDM adoption process.

Chapter 13

Challenges to Intellectual Property Rights from Information and Communication Technologies, Nanotechnologies and Microelectronics

Ahmed Driouchi
Al Akhawayn University, Morocco

Molk Kadiri
Al Akhawayn University, Morocco

ABSTRACT

Information and communication technologies, nanotechnologies and microelectronics are progressively challenging the current state of intellectual property rights. This is related to the economic features underlying these technologies. The directions of changes in intellectual property rights are found to require further coping with the overall chain of innovation and with the uncertainty that can be embedded in the new trends of technological development.

INTRODUCTION

This chapter addresses the challenges of advanced technologies to Intellectual Property Rights (IPRs). These institutional arrangements have been changing with the new trends and complexities implied by scientific and technological transformations. These changes include the scale of

operations at which research is pursued, the large number of partners and the diversified interests at different stages of production, diffusion and valuation of outputs. These transformations increase the likelihood of potential conflicts around IPRs. Besides that, the IPRs system is compelled to cope with the enlargement of the number of stakeholders and the entry of new countries into the race. Developing economies urge the IPRs to adapt to

DOI: 10.4018/978-1-61692-006-7.ch013

their needs. Furthermore, the uncertainties related to advanced technologies at both production and use have been also expanded and shared through global and specific vehicles. Overall, IPRs are now facing further challenges imposed by the new trends and advances in Information and Communication Technologies (ICTs), nanotechnologies and microelectronics.

The above issues are addressed through three sections. The first one looks at the current trends in innovation and technologies (ICTs, nanotechnologies and microelectronics) while the second focuses on the economic foundations of the challenges to IPRs. The last section addresses the challenging features linked with each set of new technologies.

The approach used in this chapter is mainly based on the analysis of series of publications from the social science literature with a focus on economics. The trends in advanced technologies were estimated for major developed countries through regression analysis.

CURRENT TRENDS IN INNOVATION AND NEW ADVANCED TECHNOLOGIES

This section contributes to exploring the new trends in innovation, and the factors behind its increasing complexity and pace.

The Enlargement of the Number of Stakeholders

In general, the growth in patents can be perceived as a proxy for R&D (Research and Development) activities and innovation. The number of applications in OECD (Organization of Economic Cooperation and Development) patent offices increased by 40% between 1992 and 2002. This is a doubling of the number of applications at the European and U.S. Patent Offices (EPO and USPTO) with a 15% increase at the Japanese

Patent Office (JPO) (OECD, 2004). These figures point clearly to the accelerating pace of contemporary innovation activity. This accelerating pace is clearly depicted in Figure 1[1] as well since the total number of patents displays an increasing trend between 1999 and 2005.

The innovation activity is not only confined within the academic sphere nor dependent only on individual efforts. A multitude of new actors are reshaping the innovation activity and accelerating its rate. Innovation has become the most important driver of competitiveness as it grants the lead time advantages necessary to cope with diminishing product lifecycle (Williams, 2005). For instance, between 1990 and 2001 industry-financed R&D in the OECD region rose 51% in real terms or from 1.31% to 1.48% of GDP (Mairesse & Mohnen, 2003).

New entrants to the innovation activity consist of small and medium-sized enterprises (SMEs) that are playing an increasingly role and reducing the traditional supremacy of large firms on R&D activities. This trend has been supported by the increasing flow of venture capital funding towards new technology-based firms (Gans et al., 2002).

The Internet has undeniably facilitated the information flow with respect to new technologies. It has reduced communication costs and made it difficult to opt for secrecy as the dominant corporate strategy. Such a change has contributed to forging more collaborative innovation processes that now involve a larger number of actors (public/private) with inter-linkages among them (OECD, 2004). Additional factors are forcing innovation agents to work in greater collaboration mainly, the growing technological complexity of products and processes, rapid technological change, intensified competition, and higher costs and risks associated with innovation. While the largest share of R&D expenditure is devoted to the firm's core competencies, complementary technologies and processes are acquired from other firms, universities and public laboratories. The result has been the rapid surge in virtually all

Figure 1.

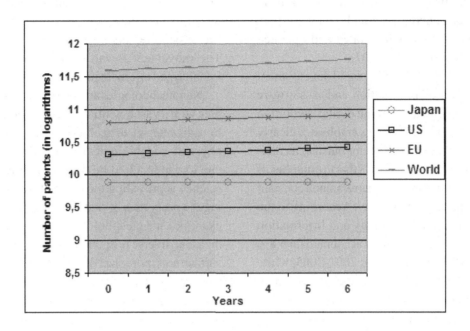

forms of collaboration including sponsored and collaborative research, strategic alliances, mergers and acquisitions, and licensing. Accordingly, the intensification of collaborative research has quickly brought about market-based transaction and diffusion of knowledge schemes.

The emergence of academic patenting was one of the direct consequences of the growth of the market-based approach of knowledge exchange (via patent licenses). It now allows inventors to control returns from their research investments. Academic patenting concerns the inventions resulting from university and public research, whether supported fully or in part by public funds. This is perceived as a way to encourage innovation via the possibility of commercializing research results, with significant private and social benefits. Academic patenting is in fact part of a broader policy framework aimed at amplifying the impact of public research on the economy through public/private partnerships. Notably, in 1980, the United States passed milestone legislation in this regard, the Bayh-Dole Act, which granted recipients of federal R&D funds the right

to patent inventions and license them to firms. The Bayh-Dole Act was introduced to facilitate the valuation of government-funded research results by transferring ownership from the government to universities and other contractors (Schacht, 2008).

Accordingly, the entries of new actors into the innovation arena, the globalization of the innovation process and the establishment of market-based diffusion of knowledge, have triggered an increase in the pace of the innovation activity. But, this increase is not the only aspect of the transformation. The emergence of a new wave of technologies inspired from the recent ability to manipulate materials at an ever-decreasing scale constitutes an additional transformation of the innovation activity.

Increasingly Integrated Technologies

Between 1994 and 2001, the share of ICTs in European patent filings increased from 28% to 35%. During the same period, the share of biotechnology climbed from 4.3% in 1994 to 5.5%. Figures from the U.S. display similar trends. While nanotechnologies, microelectronics, and ICTs were often considered separately, they are now increasingly viewed as integrated and operated at very small scale (Falk, 2007).

ICTs have been generating rapid development in transmission of information and in software through the continuation of the digital revolution (Kaushik, 2000). The statistics emphasize clearly the current strength of this sector but also its relative maturity as Figure 2 shows that the number of patents granted in the sector has remained constant throughout the current decade[2]. According to the World Information Technology and Information Services Alliance (2006), ICTs accounted for 6.8% of global GDP over the period 2001-2005, while worldwide spending on ICTs in 2006 amounted to US$3 trillion Trend.

Microelectronics refer to technologies associated with the design and fabrication of electronic devices, systems or subsystems using extremely small components. They are multidisciplinary technologies that encompass Electrical, Computer, and Optics, Material Science, Physics, Chemistry, Statistics, Experimental Design, Computer Aided Design, and Manufacturing. The origins of microelectronics go back to the invention of the transistor in 1947 at AT&T Bell Labs and since then the technology progressed to enable the development of PCs, the Internet and cellular phones which have quickly become an integrated part of daily life (Rochester Institute of Technology, 2008). The movement towards nano-scaling has also transformed microelectronics to nanoelectronics.

Nanotechnologies are relatively new technologies with a vast potential of applications. It refers to a diverse set of techniques for manipulating physical and biological materials at scales of one to one hundred billionths of a meter. Nanoscience is providing the optimal way of examining the properties of materials at a tiny scale. For example, nanotechnology equipment can now measure diverse properties of materials including surface area, pore size, density, catalytic activities, and water absorption. The exponential growth of patents related to nanotechnology starting from the 1990s reflects their wide and multidisciplinary applications. As a matter of fact, the evolution of nanotechnology patents has been characterized since the late 1970s by a general upward trend in the US, Japan, and the European Union. Nevertheless, the US has initially taken and is still taking the lead in the nanotechnology wave while the EU and Japan have gradually improved their positions in an attempt to emulate the US (Figure 3[3]). Existing products of nanotechnology include particles tailored to have specific proper-

Figure 2.

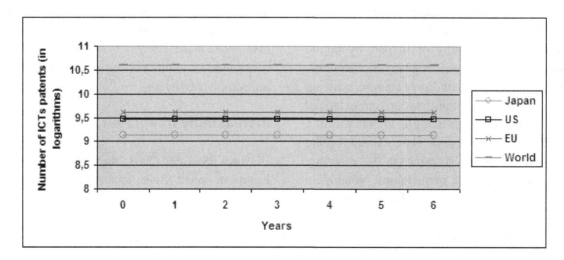

ties, biosensors designed to and view atoms and electrons and nano-crystallography capable of depicting the structure of proteins in three dimensions (Falk, 2007). It allows materials with new chemical, optical or electrical properties, as well as quantum computers, and microscopic engines and other devices, to be constructed. These should have applications in medicine, sensing and information storing and processing in addition to manufacturing.

The recent advances in microelectronics, in nanotechnology, and in ICTs are increasingly accompanied with signs of convergence. As a matter of fact, the microelectronics field on which advances in ICTs were based, is currently evolving into nanoelectronics. For instance, computer chips are manufactured following a process through which numerous thin layers of materials are deposited to assemble the chip. These layers of materials have gradually become nano-scale and have subsequently acquired more sophisticated electric properties (Bainbridge & Roco, 2005). This convergence covers all sciences and technologies. The medical field provides as well strong evidence on scientific disciplines integration and interplay. Also, biotechnology as a variant of nanotechnology deals with complex molecules. The main drivers of the increasing convergence of sciences and technologies were undeniably the groundbreaking capabilities developed by nanosciences to picture and manipulate matter at the nano-scale (Bainbridge & Roco, 2005). Another driver of the convergence is the integration of information systems across all scientific disciplines to the extent that computer scientists are now compelled to build a multidisciplinary background in order to be able to develop solutions for different fields. One aspect of the convergence is the entry of the social sciences into the ongoing integration dynamics (Bainbridge & Roco, 2005).

The characteristics and trends described above provided the background to the following section where the economic foundations of the new challenges facing IPRs are introduced.

The Economic Sources of Challenges to Intellectual Property Rights

IPRs include patents, copyrights, trademarks, geographic indications, industrial designs and other rights related to the ownership and control

Figure 3.

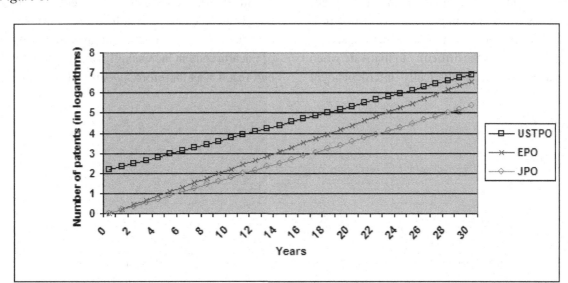

of the industrial, scientific, literary or artistic innovations. IPRs confer to the creator a temporary exclusive right over the financial benefits related to the innovation. Besides the ethical dimension of IPRs, their economic rationale lies on the assumption that their enforcement preserves the incentive to innovate as it guarantees innovators both credit over their inventions and access to the financial rewards related to their creations. IPRs are commonly viewed as a trade-off between encouraging innovation and the consequences of monopoly (Figure 4[4]). While it is necessary to grant a temporary monopoly to the innovator over the production and the marketing of the invention (Nordhaus, 1969), society bears monopoly costs. But, series of contributions have identified sources of failures and traps related to IPRs. These include the tragedies of commons and anti-commons besides the patent-thickets as related to the level of segmentation and competition.

Commons, Anti-Commons, New Technologies and IPRs

Cumulative innovation implies that a number of separate innovations constitute the building blocks for a subsequent invention. Different pieces of software codes are necessary to come up with a new one. Similarly, cumulative innovation arises when it is a matter of improving an existing process. With respect to that, the concern raised by IPRs profusion is the difficulty to innovate when inventions are technologically interdependent and when each 'building block' is considered in isolation.

Such parceling of intellectual property hinders innovation and results in what is termed the "tragedy of the anti-commons" (Figure 5[5]). This refers to a situation where several individuals hold rights of exclusion (i.e. patents) related to several 'building block' technologies. As an additional individual wants to build a new technology that relies on each of these previously existing rights of exclusion, separate negotiations with each of the IPRs holders are necessary. Given that such multiple negotiations are held separately with each patentee, the total price of the technology ends up being higher than if the innovator dealt with a single patentee. Building the new technology in this case can become financially unfeasible. The total impact on innovation is negative due to the fact that the available intellectual resources are underused (Lévêque & Ménière 2004).

The media have reported lately the demise of Encarta (Chronicle of Higher Education, 2009), the encyclopedia produced by Microsoft. The open source Wikipedia has imposed itself now as providing alternatives at no charge with quicker knowledge updates and adjustments. As in the tragedy of anti-commons (Heller, 1998), the too many discrete property rights needed to gather the different pieces of knowledge feeding Encarta have been the cause of its failure. The multitude of copyrights involved made the aggregate collection of Encarta content increasingly costly. The natural end result was therefore the shift of users to an

Figure 4.

Figure 5.

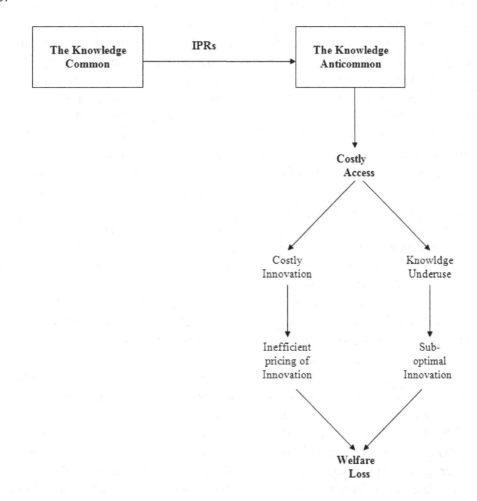

open access solution where additional quantities of content are accessible at a zero price.

The tragedy of "anti-commons" as explained by Heller (Heller et al, 1998) is a mirror of "commons", introduced by Hardin (1968). The tragedy of commons is defined by the existence of several owners with no one having the right to exclude the others. This implies an excess use of common resources due to the existence of multiple users. The result of this non-exclusion is the overuse of the given resource that leads to a "tragedy of the commons" (Heller, 1998). In the situation of anti-commons, owners have the right to exclude others from a scarce resource while no one is allowed to use it effectively. This situation leads to the under use of the scarce resource and thus, to

the "tragedy of the anti-commons" (Heller, 1998). This phenomenon was first described by the latter author in relation to the reforms that took place in ex-socialist economies. This was related to the "storefronts that remained empty while small kiosks were full of goods and mushroomed on the streets" (Osterloh & Rota, 2004). Heller (1998) suggested then, that the root of the problem was the way property rights were distributed and not only the ambiguity of their definition or scope, corruption or defiance of the rule of law.

The issue with the way intellectual property rights are distributed is closely related to how intellectual property rights, over a given idea, impacts the possibility to build upon that piece of knowledge for future research and development activities. In practice, because of high levels of IPRs protection, the domain of ideas is largely segmented with high levels of exclusion of others leading to higher probability of social welfare loss. A large number of cases have been reported in the literature where segmented intellectual property rights, in biomedical research and in the software industry have been serious constraints to the generation of further positive effects on the economy (Heller et al, 1998). Therefore, intellectual property rights or patents, if largely distributed, may decrease social welfare by limiting future inventions, creating anti-commons and encouraging non-competitive behavior (Heller et al, 1998; Jensen et al, 2004). However, if they are granted carefully, this may reduce enterprise willingness to invest in imitated and differentiated inventions. The issue is to find a balance between social and inventors' interests (Jensen et al, 2004).

The practice of "open science" has been considered as a more adequate alternative as it corrects for the inefficient use of resources for the production and dissemination of information caused initially by the intellectual property system (David, 2003). The open science approach can lead to better results in the research system since privatizing data and information, under strong intellectual property rights, has become a cause of decay in the whole system of research. This resembles the lack of cooperation behavior between researchers regarding shared access to raw data and information (David, 2003). The open source science is one of the "virtual communities" innovations related to the "open science" approach. This alternative was also induced by the nature of the technologies at hand, but mainly given that the private ownership of intellectual property rights doesn't encourage innovation. For instance, the open source software code is considered as a classical public good given that these software licenses allow the user to read the software's source code, change and publish the changes to the code but doesn't allow any fees-raising for the initial source code (Osterloh & Rota, 2007).

Other segments of the ICTs field are still suffering from the anti-common tragedy. This is the case of the telecommunications sector. Hazlett (2003) considered America's TV band as an empty storefront in reference to Heller's (1998) study. However, cable TV systems succeed, since "the vertically integrated networks owned by the cable company, which has no obligation to share, prevent tragedy of the anti-commons" (Hazlett, 2003).

It is expected though, that the negative effects related to the existence of anti-commons, have larger impacts on the access to ICTs in developing economies, given the large transactions and social costs involved and that are expected to reduce incentives for innovation in these economies.

The Patent Race, Patent Thickets and Competition in New Technologies

The patent race is another typical example of the additional inefficiencies caused by IPRs. Basically, IPRs are designed to confer a monopoly advantage to the first innovator to introduce the technology. He would temporarily collect all the profits stemming from the innovation. As this creates a "winner takes all" situation, a race between different competitors within the same industry is triggered (Lévêque & Ménière, 2004). An inefficient allocation of resources arises when each agent devotes similar levels of research and development expenditures to reach the same final innovation. Obviously, at the end of the race only one competitor obtains the patent and the collective investment has been exuberantly beyond the optimal required level to put the innovation in place.

Unfortunately, the repercussions of the patent race are far from being limited to the inefficiencies caused when the aggregate expenditures of multiple innovators exceed by far what is needed to come up with the innovation. The patent race has more complex ramifications that are further accentuated by the radically changing context of innovation. These ramifications are also more specific to the new wave of technologies. Conventionally IPRs grant the inventor the temporary advantage of exclusively producing and marketing his invention. The new transformations in the innovation activity however are such that innovation has increased in pace, has become cumulative, and technologically interdependent. Accordingly, the time horizon granted to innovators to appropriate the returns of their innovations has become inconsistent with the recent intensification of the innovation activity leading to its obstruction and slow-down. Very often, the resulting situation is a patent thicket (Figure 6[6]). A patent thicket is "a dense web of overlapping IPRs that a company must hack its way through in order to actually commercialize new technology" (Shapiro, 2001). It occurs when developing a single product requires the purchase of multiple related licenses to avoid the risk of patent infringement. The transaction costs in this case might amount to the degree where the returns from the new products are well below its costs and risks of development leading to the abandoning of the project. An additional risk of

'hold up' is also added. Hold up is "the danger that new products will inadvertently infringe on patents issued after these products have been designed" (Shapiro, 2001).

In order to overcome the patent thicket and hold up problems companies in the ICTs, nanotechnology, or semiconductors industries have naturally resorted to the purely defensive strategy of inflating their patents portfolio and accumulating as many blocking patents as possible with the hope that the interdependence of innovation is such that in case they face infringement allegations they are in a position to threaten the competitor by analogous infringement allegations. With time, this purely defensive behaviour evolved into attempts of co-operation in the form of cross-licensing (Shapiro, 2001). A derived solution consists of the creation of patent pools formed when a consortium of at least two companies agrees to cross-license patents relating to a particular technology. Nevertheless, cross-licensing and patent pools are not absolute remedies to the patent thicket problem as shown in Bessen (2003). This latter author argues that

Figure 6.

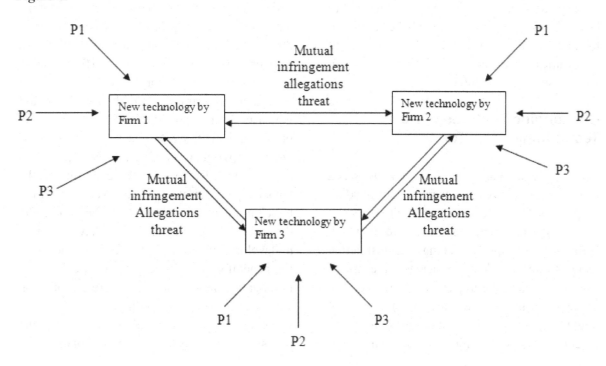

"although cross-licensing and patent pools may resolve some problems of transaction cost and vertical monopoly, these institutions do not correct all problems associated with patent thickets". His model shows how patents are responsible for more inefficiency when technologies are complex and the standards for granting patents are low.

In the absence of patents, innovators can still reap the benefits of their innovations and realize rents thanks to lead time or first-mover advantage. It is the lead time advantage that provides the initial incentive for innovation and guarantees optimal levels of R&D. As simply put by Boldrin (2008), "The fact that you are the first and know how to do it better than the other people--it may be a huge protection". However, in the case of cross-licensing, Bessen's model reveals that R&D incentives are sub-optimal. He explains this finding by the fact that contrary to the "winner-take-all" patent race logic, cross-patenting implies that the rewards of innovation are systematically shared with the other patent holders. Cross-licensing thus impacts negatively the incentive to innovate by canceling off the lead time advantages as the winner's profits are already part of the bargain. They also turn out to be relatively negligible due to both the obligation to share the innovation rents and the disbursement of royalties to the other cross-licensing parties in some cases (Bessen, 2003).

Challenging Features of New Technologies to IPRs

Applying IPRs to any emerging technology is a challenging task given the uncertainty surrounding the process. The outcomes are uncertain since each technology is characterized by features to which IPRs have to adapt. Not taking such distinctive features into consideration increases the risk of IPRs obstructing the trajectory of new technology. Failing to take these features into account might even be detrimental to innovation and its outcomes. This section illustrates the challenging

features of new technologies to IPRs and the effects of uncertainty.

Information and Communication Technologies

As far as ICTs are concerned, the controversy surrounding the application of patents or copyrights protection to software illustrates the difficulty of adapting IPRs to ICTs. Prior to the 1980's, software was unanimously considered non-patentable since it consists basically of a collection of mathematical algorithms. However, prominent court decisions in the US during the last 20 years ruled for the patentability of software establishing fully the practice of patenting software programs in the US and inducing other developed countries to reconsider the patentability of software programs (Evans, Layne-Farrar, 2008). Software entails the double feature of being the creative expression of the programmer used to perform a technical task. Hence, while the logic behind the use of copyrights was to protect the creative expression (the code), advocates of the legitimacy of using patents contended that the technical task embedded in the software and thus its real commercial value ought to be protected as well (Owens, 2004). For instance, a computer programmer can develop software to perform a certain functionality (i.e. controlling machinery). Later on, a second computer programmer can develop software which performs exactly the same functionality but which is written differently. In this example, under a copyright protection the second programmer is compelled to write the code differently. Under a patent protection however, in addition to being prevented from copying the same code, the second programmer does not have the right to develop software with the same functionality. So, generally speaking, copyrights are a protection against piracy and patents are a protection against competition. Accordingly, Put differently, while patenting software grants the inventor the necessary monopoly to own the fruits of his work;

it seems at the same time unfair to confer such monopoly while there can be a thousand different ways of performing this same task.

In this sense, the emergence of the open source alternative is evidence on the difficulty experienced by IPRs to gain consensus in the ICTs sector. The fact that patenting prevents subsequent improvement is the reason behind the attractiveness of the open source software (OSS) model. Basically, OSS is an alternative way of distributing software. In an OSS, the user has access to the underlying code of the software. The user can read it, customize it, or build completely different versions. Nevertheless, OSS is not to be perceived as symmetrically opposed to the IPRs paradigm. In fact, OSS is not fully in the public domain as it constitutes a particular case of copyrights. OSS licenses are nothing but copyright licenses which does not mean the abandoning by the OSS first originator of his intellectual rights. What the latter does actually is organize and regulate the way in which subsequent users can copy, modify and distribute their versions of the software (Owens, 2004). Since 2008 in the US, OSS licenses are enforceable under copyrights law, and violation of licensing terms represents an infringement (Thinking Open, 2008).

Other characteristics of the ICTs industry amplify the debate by suggesting the obsolescence of IPRs in the ICTs sector. One of the indicators of the incompatibility of IPRs with ICTs relates to the nature of the Internet. Hunter (2002) argues that conventional legal constructs applied to human behavior in real life fail to hold in the context of the Internet as it is by definition an endless and open space where human interaction follows different patterns. The Internet defies regulation by challenging the power of the nation-State and even its legitimacy as a regulator. According to Hunter (2002), the assumption that the Internet is a place, has led to stretching real world property rights concepts to the virtual world. The "Internet common" has consequently become fragmented

into virtual private holdings. Of course, the benefit of property rights is to organize the use of resources and avoid the "tragedy of the commons". But, the Internet is not subject to the tragedy of the commons given that it is an infinite space. Also, the nonrival nature of intellectual goods is such that their depletion is inconceivable.

The argument of the incompatibility of IPRs with ICTs is also grounded on the fact that ICTs and the Internet have in fact flourished in a period when IPRs enforcement was very weak (Hunter, 2002). The ICTs industry managed to innovate and managed to have the benefits from innovation outweigh the losses incurred by piracy (Bessen & Maskin, 2004). Further, given that the objective of IPRs is to foster more innovation, one should expect the increase of innovation in ICTs following the strengthened enforcement of IPRs. With respect to this issue, Bessen and Hunt (2004) actually demonstrate the opposite by establishing empirically that starting from the 1990s; firms that increased their patenting focus have at the same time decreased the intensity of their R&D activities. Also, increases in software patent share were associated with decreases in R&D intensity.

Accordingly, during the 1990s patents and R&D were no longer complements but substitutes. In other words, the decrease in the cost of obtaining patents combined with the reliance of firms on similar and overlapping technologies has caused firms to compete in patents rather than in R&D. Interestingly, capital-intensive hardware firms have a much higher software patent propensity compared to software firms. For instance, manufacturers of machinery, electronics, and instruments employed only 6% of all computer programmers and yet they obtained 2 out of 3 software patents. Further, software firms and start-ups have relatively less propensity to patent as they do not view patents as an incentive for innovation (Owens, 2008).

IPRs advocates argue that tighter enforcement should take place as the development of ICTs has tremendously decreased the cost of copying and piracy and as the Internet has become nothing but a big "copy shop" that ought to be tightly regulated (Hunter, 2002). Open access advocates contend however by refuting the traditional view according to which IPRs in the ICTs context equate imitation and copying. They argue that in the specific case of ICTs, innovation is rarely the result of a sole isolated inventor. It is rather the output of a whole community in constant interaction. Such interaction provides the opportunity for innovators to publish their works and make them available for incremental improvement by other users (Bessen & Maskin, 2004). Denying open access would mean that an endless number of opportunities would remain untapped, as no one will have the right to do so. As a matter of fact, IPRs fail to recognize the added value brought by additional parties and go beyond their basic function of limiting direct copying by preventing value-adding imitation to occur (Bessen & Maskin, 2004).

The emergence of the knowledge-based economy has been synonym of the emergence of a new form of competition and growth (Driouchi A. El. Azelmad and G. Anders, 2006). To illustrate that, in the past, innovations and thus IPRs were embedded in tangible products which needed some tangible and thus costly input. With ICTs, the cost of duplication has become very close to zero. So, while in the past it was a matter of diminishing returns to scale given the capacity constraint, the new intangible economy is characterized by increasing returns to scale. Once the initial innovation investment is recovered, every additional copy sold is pure profit. This means that IPRs in the context of ICTs, imply the extraction of unjustified rents (Andersen, 2002). Competition in the ICTs sector is also characterized by many potential sources of market domination by a given firm independently of whether or not the firm benefits from IPRs protection. These are related to customer bonding factors which allow ICT firms to lock in their customer base. Essentially, users once accustomed to a certain interface or software, find it difficult to switch to competitors' platforms.

ICTs are also characterized by network externalities from which a user benefits once becoming part of a larger network that uses the same technology. The existing advantage of the firm that owns the technology is strengthened by such network effect. Further, when the customer opts for the product of a specific technology provider, he is compelled to use in the future only the complementary products that are produced by the same firm due to compatibility constraints. All these monopoly factors are typical to the ICTs sector making it difficult for IPRs to play their role without exacerbating this pre-existing monopoly propensity and thus are likely to create unfair competition issues in the sector (Andersen, 2002).

Nanotechnologies

Nanotechnologies are most likely to trigger deep transformations through their endless applications to all aspects of human life and commercial activity ranging from health, to food, computing and electronics, energy, or transportation. The common perception today is that the very nature of nanotechnologies undermines the assumptions on which the principles of patenting inventions are based (Vaidhyanathan, 2006). Basically, laws and conventional patenting criteria that used to be applicable to inventions during the industrial era are not applicable to nanotechnologies as these laws simply were not intended to examine patent applications for atomic or molecular structures (D'Silva, 2007). Applying conventional IPRs to nanotechnologies is problematic given that it relies on an industrial process that is radically distinct from what prior technologies have relied on. Thus, while prior technologies are based on a top-down approach which aims at deriving new products from large blocks of raw material; nanotechnologies follow a bottom-up approach whereby nano elements (molecules, atoms, or cells) are carefully selected and assembled to form the final product. Such reversed industrial process is obviously different from what IPRs systems have historically dealt with (Siddharth, 2006). Moreover, patenting nanotechnologies often contradicts the generally agreed upon principle that properties of matter and other fundamental scientific discoveries are not "patentable". In fact, nanotechnologies challenge conventional patenting by blurring the boundaries between discoveries and inventions (Bowman, 2007).

In addition, the propensity of IPRs enforcement to stifle innovation through the segmentation of knowledge resources necessary for a particular technology is confirmed in the case of nanotechnologies (D'Silva, 2008). In fact, nanotechnologies constitute an interdisciplinary field that integrates principles and theories inherent to computer science, biology, physics, chemistry and mathematics. It has transformed scientific research in such a way that separate fields have converged under a single engineering paradigm umbrella (Nordmann, 2004). The multidisciplinary nature of nanotechnologies is such that the risk of overlapping patents is exacerbated, as the pre-existing patent classifications are obsolete (Kamra, 2009). For instance, a broad patent granted to a nanotechnologies process applied in robotics can block its subsequent improvement and use in the medical field. Further, patenting nanotechnologies often implies patenting abstract ideas given the embryonic stage of the discipline currently. So, most nanotechnologies patents cover basic inventions and patents are applied for and obtained too early in the research process before an actual relevant final product is obtained (D'Silva, 2007). The issue here is that patents on basic inventions can cover a larger scope (sometimes more than one discipline) than what the potential or targeted final product actually requires. IPRs protection in this case impedes future research as well as the entry of new researchers in the field (i.e. patent thickets) as, in emerging technologies, the breadth and the scope of a patent is determined based on prior state of art (what has been granted before) and thus pioneering research leading to basic inventions end up being granted patents of

a scope that is too broad (Nicholas Godici cited in Kamra, 2009).

Microelectronics

The transition from electronics to microelectronics represents one of the aspects of the shift towards an increasingly knowledge-centric economy. It is as well characteristic of a parallel shift from discrete to complex technologies (Williams, 2005). In discrete technologies, the end product is made of fewer interrelated components and rarely includes a component which can be used in other products or industries. In fact, discrete inventions offer little room for improvement or for the development derivative products. However, complex technologies products do integrate a multitude of components and each of these components is the result of a distinct innovation process. Also, the innovation process and progress in complex technologies is incremental, cumulative, and sequential. This distinction between discrete and complex technologies affects directly IPRs in the sense that the patent system was designed to tackle property rights issues for discrete inventions (Allison & Lemley, 2002). In other words, an invention was covered by no more than one patent. Yet, in complex technologies, a new invention is a combination of components and each of these components can be subject to a separate patent. For instance, assembling a microelectronics-based product such as a 3rd Generation cellular phone involves thousands of patents (Lemley & Shapiro, 2007).

As a matter of fact, microelectronics as a complex technology directly challenges the tenets of traditional IPRs theory as the latter no longer allow the original inventor to appropriate all the monopoly profits derived from the invention. What happens instead is that a firm holding a patent on an essential microelectronics element can use the related license to prevent other firms from incorporating this element in their new products unless they agree to pay licensing fees or accept to grant access to their in-house technology. As a result, the propensity of IPRs to act as an incentive for innovation is weakened within a context where the expected profit form innovation erodes significantly once license royalties' expenses for accessing all the related technology inputs are accounted for.

Instead, there is a growing belief in the industry that competitive advantage is more a function of well-guarded trade secrets, lead-time advantages and complementary capabilities (Williams, 2005). IPRs and patent protection are said to play only a secondary role as they act as speed bumps with respect to the flow of in-house innovation to competitors, rather than a definite protection. Moreover, IPRs in complex technologies in general and in microelectronics is particular, tend to be obsolete with respect to granting a competitive advantage given a set of industry specific factors such as short product life cycles, the practice of reverse engineering, more transparency and increased employee mobility (Williams, 2005). As a consequence, IPRs rarely grant firms the kind of complete proprietary advantage they were designed to grant at the first place.

Thus, if IPRs fail to achieve their original task of encouraging innovation, one might inquire about the reasons behind their profusion in areas such as microelectronics. The surge in patenting is explained in fact by the strategic value of patents as a defensive mechanism against litigation and not as an innovation incentive. Thus, while there is a general consensus in the industry that innovation is driven above all by competition and the necessity to take the lead, the surge for patents is explained by the need of firms to strengthen their position in cross-licensing negotiations. The more the patent portfolio is designed to overlap with competitors' IPRs, the more the company strengthens its position in these negotiations (Williams, 2005). Cross-licensing negotiations are based on a "mutually assured destruction" situation where IPRs overlap allows two companies to threaten each other of filling a lawsuit for infringement.

The other reason behind the surge of IPRs in industries such as microelectronics is the fact that firms are increasingly viewing patents as a new separate lucrative activity regardless of whether or not these patents are embedded in an actual product marketed by the company. As a matter of fact, the phenomenon of royalties stacking emerges easily when a single product requires licenses for a considerable number of overlapping patents, owned by different parties. The firm willing to market the product accumulates or stacks the different royalties related to each necessary license in order to conduct its cost/benefit analysis for the new product, based on the total royalties burden implied (Lemley & Shapiro, 2007). What happens often is that when the different patent holders know that the new product is profitable enough, they become in position to negotiate royalties that are in excess of the true value of their patents contribution to the product. A related phenomenon is the emergence of patents trolls. These are business entities specialized in acquiring patents from bankrupt firms or firms under financial distress. Patent trolls later on screen the market for identifying firms that might be infringing these patents, fill infringement lawsuits, and make profit out of the damages paid by the defendant firms. Patent trolls are referred to as "patent terrorists" because they cannot be countered by a reciprocal litigation threat since they never make use of the patents they acquire for producing any product (Williams, 2005).

Finally, the interaction of IPRs with complex technologies such as microelectronics has led to the emergence of patent portfolio theory. Wagner and Parchomovsky (2005) have demonstrated that over the current decade, patents intensity, measured as the number of patents obtained per R&D dollar, has increased dramatically while the expected value of individual patents has dropped over time. The authors suggest that the answer to this patenting paradox is that in the new IPRs landscape, patenting decisions are independent of the expected value of individual patents. Rather, it is the expected value of the aggregation of individual patents into a patent portfolio that is the new decision parameter. In other words "the whole is greater than the sum of the parts" (Wagner & Parchomovsky, 2005). Accordingly, individual patents are to be understood as necessary means to portfolio construction, as the goals themselves. Basically, patenting occurs whenever the marginal benefit of building a patent portfolio exceeds the marginal cost of acquisition, this marginal benefit being the enhanced bargaining power of a firm in cross-licensing negotiations described earlier.

Uncertainty as Additional Source of Challenges

While new products of new technologies are developing and penetrating the daily lives of individuals, households and communities, important uncertainties have been identified. No direct evidence and causality has been established all the time between the new product of advanced technologies and its likely effects. Releases of new products are not often accounting for the requirements and standards for human safety (Roth, 2007).

Mobile phone, screens (computers, TV), the economic lamp, scanners, kitchen and restaurant new equipments, and others new machines are now suspected to be sources of magnetism, radiations and other harmful physical and chemical products as already recognized to some extent by producers of these new products. Health issues with cancer and new diseases are progressively better identified also through the development of new advanced technologies (Roth, 2007). But, to what extent, direct or indirect causality can be established as linking the intensive use of new products and new health issues? These uncertainties are enlarging the level of challenges to IPRs and consequently the debate that has to account for the contributions of all the stakeholders that intervene in the innovation chain.

CONCLUSION

The technologies analyzed in this chapter appear to be exhibiting important challenges to the IPRs systems worldwide and in every economy. These challenges are expected to be increasing as new technological developments are pursued. These challenges are multidimensional and are related to the main features underlying the new technologies and the processes through which they are produced and used. Among these features, the lowering of the scale of operations and outputs, the enlargement of the number of stakeholders besides the pace of production appear to be critical. The entry of developing countries as users but also as potential producers is also among the major trends that characterize these new technologies and thus the challenges faced by the IPRs systems.

Important adjustments are progressively introduced in the IPRs either through legal changes in some areas in developed countries or through the TRIPs and World Trade Organization (WTO). But some areas remain under continuous debate with agriculture, environment, pharmaceutical and health innovations being dominant. These latter areas are also concerned with potential applications of ICTs, nanotechnologies and microelectronics.

While important benefits are expected to occur worldwide and in each economy, IPRs can be also challenged when looking at the types of uncertainties suspected under each technology. While risks have always accompanied new technologies, the advances induced in different areas including health, keep showing new sources of risks with suspicious causality between new innovations and the welfare of humans.

These ongoing trends that continuously challenge IPRs can be captured with the inclusion of all the interests in the chain of innovation and use while accounting for uncertainties and risks. These challenges can be further captured under new institutional schemes and mechanisms that allow the pursuit of creativity, the development of research and development networks but also the realization of the benefits needed by individuals and human societies. Further networking and cooperation between developed and developing economies can be major engines for ensuring smooth and flexible IPRs systems where different rights are continuously internationally and nationally monitored and adjusted.

Among the consequences of these changes, IPRs are likely to become the appropriate frameworks for enhancing the levels and the types of innovations and directions of creativity as new flexible incentives are promoted throughout all the economic and social channels.

ACKNOWLEDGMENT

This work is part of a research that looks at the Economics of Intellectual Property Rights. This is financially supported by the "Hassan II Academy of Sciences & Technology" in Morocco. The Academy is acknowledged for its support but is not responsible of the content and views expressed in the current paper. The views and contents of the paper are the sole responsibility of the authors.

REFERENCES

Allision, J. R., & Lemley, M. A. (2002). The growing complexity of the United States patent system. *Boston University Law Review. Boston University. School of Law, 82*(77).

Andersen, B. (2002). *The performance of the IPR system in the new economy: implications for digital inventions and business methods*. Copenhagen, Denmark: The Standing Committees of Research and Industrial Development of the Danish Parliament. Retrieved June 17, 2009, from http://www.druid.dk/conferences/summer2002/Papers/ANDERSEN.pdf

Bessen, J. (2003). Patent Thickets: Strategic Patenting of Complex Technologies (ROI Working Paper). *Social Science Research Network Database*. Retrieved July 4, 2009, from http://papers.ssrn.com/sol3/papers.cfm?abstract_id=327760

Bessen, J., & Maskin, E. S. (2005). Geistiges Eigentum im Internet: Ist alte weishit ewig gültig? [Intellectual property on the Internet: What's wrong with the conventional wisdom?] In Bernd, L., Gehring, R. A., & Bärwolff, M. (Eds.), *Open Source Jahrbuch 2005: Zwischen Softwareentwicklung und Gesellschaftsmodell*. Berlin: Lehmanns Media.

Boldrin, M., & Levine, D. K. (2008). Perfectly competitive innovation. *Journal of Monetary Economics, 55*(3), 435–453. doi:10.1016/j.jmoneco.2008.01.008

Bowman, D. M., & Hodge, G. A. (2007). A small matter of regulation: An international review of nanotechnology regulation. *Columbia Science & Technology Law Review, 8*, 1–32.

Chen, H., Roco, M. C., Li, X., & Lin, Y. (2008). Trends in nanotechnology patents. *Nature Nanotechnology, 3*. Retrieved June 20, 2009 from http://mis.eller.arizona.edu/docs/news/2008/nature_nano_42008.pdf

D'Silva, J. (2008). Pools, Thickets and Open Source Nanotechnology. *Social Science Research Network Database*. Retrieved June 4, 2009, from http://ssrn.com/abstract=1368389

David, P. A. (2003). The economic logic of 'open science' and the balance between private property rights and the public domain in scientific data and information: a primer. In J. Esanu & P. F. Uhlir (Eds.), *The Role of the Public Domain in Scientific and Technical Data and Information: Proceedings of a Symposium*. Washington, DC: National Academies Press.

Driouchi, A., Azelmad, E., & Anders, G. (2006). An econometric analysis of the role of knowledge in economic performance. *The Journal of Technology Transfer, 31*(2), 241–255. doi:10.1007/s10961-005-6109-9

Falk, J. (2007). Transitioning to new technologies: Challenges and choices in a changing world. *Journal of Futures Studies, 12*(2), 69–90.

Förster, W. (2007). Rate, direction, and lifecycle of patenting investments. In *Proceedings from the HEC Paris Symposium Intellectual Property: Les Entretiens de Paris 2007.*

Gans, J., Hsu, D. H., & Stern, S. (2002). When does start-up innovation spur the gale of creative destruction? *The Rand Journal of Economics, 33*(4), 571–586. doi:10.2307/3087475

Hardin, G. (1968). The tragedy of the commons. *Science, 162*(3859), 1243–1248. doi:10.1126/science.162.3859.1243

Hazlett, T. (2003, April 18). Tragedies of the telecommons. *Financial Times.*

Heller, M. A. (1998). The tragedy of the anticommons: property in the transition from Marx to markets. *Harvard Law Review, 111*(3), 621–688. doi:10.2307/1342203

Hunter, D. (2003). Cyberspace as place, and the tragedy of the digital anticommons. *California Law Review, 91*(2), 442–519. doi:10.2307/3481336

Jensen, P. H., & Webster, E. (2004). Achieving the optimal power of patent rights. *The Australian Economic Review, 37*(4), 419–426. doi:10.1111/j.1467-8462.2004.00343.x

Kamra, A. (2009). Patenting Nanotechnology: In Pursuit of a Proactive Approach. *Social Science Research Network Database.* Retrieved June 12, 2009, from http://ssrn.com/abstract=1399332

Kaushik, P. D. (2000). *TRIPS: IPR regime for the digital medium.* Retrieved July 6, 2009, from http://unpan1.un.org/intradoc/groups/public/documents/APCITY/UNPAN006305.pdf

Khanijou, S. (2006). Patent inequality?: Rethinking the application of strict liability to patent law in the nanotechnology era. *Journal of Technology . Law & Policy, 12*, 179–181.

Layne-Farrar, A., & Evans, D. S. (2004) Software patents and open source: The battle over intellectual property rights. *Virginia Journal of Law and Technology, 9*(10).

Lemley, M. A., & Shapiro, C. (2007). Patent holdup and royalty stacking. *Texas Law Review*, 85.

Lévêque, F., & Ménière, Y. (2004). *The economics of patents and copyright.* Berkeley, CA: Berkeley Electronic Press.

Mairesse, J., & Mohnen, P. (2003). Intellectual Property in Services. What do we learn from innovation surveys? In *Proceedings from the OECD conference.* IPR, Innovation and Economic Performance.

Nordhaus, W. D. (1962). *Invention, growth, and welfare: A theoretical treatment of technological change.* Cambridge, MA: MIT Press.

Nordmann, A. (2004). *Converging Technologies: Shaping the Future of European Societies (Report of the High Level Expert Group Foresighting the New Technology Wave).* Brussels, Belgium: European Commission Research.

OECD. (2004). *Patents and innovation: trends and policy challenges.* Retrieved May 16, 2009, from http://www.oecd.org/dataoecd/48/12/24508541.pdf

Osterloh, M., & Rota, S. (2007). Open source software development- Just another case of collective invention? *Research Policy, 36*(2), 157–171. doi:10.1016/j.respol.2006.10.004

Owens, R. (2008). Intellectual Property and Software: Challenges and Prospects. *WIPO publications*. Retrieved June 3, 2009, from http://www.tecpar.br/appi/SeminarioTI/Richard%20Owens.pdf

Parchmovsky, G., & Wagner, R. P. (2005). *Patent portfolios* (Public Law Working Paper 56). Philadelphia, PA: University of Pennsylvania Law School Rochester Institute of Technology. (2008). *About MicroE: What is Microelectronic Engineering?* Retrieved June 30, 2009, from http://www.rit.edu/kgcoe/ue/about.php

Roth, C. (2007). New technologies and the ergonomic risk to users: Are new technologies part of a technology nirvana or newly identified ergonomics risk factors? *EHS Today, The Magazine of Environment, Health, and Safety Leaders*. Retrieved June 3, 2009, from http://ehstoday.com/health/ergonomics/ehs_imp_70688/

Schacht, W. H. (2008). *The Bayh-Dole Act: Selected Issues in Patent Policy and the Commercialization of Technology* (*CRS Report RL32076*). Retrieved July 7, 2009, from http://www.usembassy.it/pdf/other/RL32076.pdf

Shapiro, C. (2000). Navigating the patent thicket: Cross licenses, patent pools, and standard setting. *Innovation Policy and the Economy, National Bureau of Economic Research*, 1. Retrieved June 12, 2009, from http://faculty.haas.berkeley.edu/shapiro/thicket.pdf

The Chronicle of Higher Education. (2009). *Microsoft's Encarta, rendered obsolete by Wikipedia, will shut down*. Retrieved May 27, 2009, from http://chronicle.com/wiredcampus/article/3715/microsofts-encarta-rendered-obsolete-by-wikipedia-will-shut-down

Thinking Open. (2008). *The Decision All Of Open Source Has Been Waiting For*. Retrieved June 3, 2009, from http://thinkingopen.wordpress.com/2008/08/15/the-decision-all-of-open-source-has-been-waiting-for/

Vaidhyanathan, S. (2006). Nanotechnologies and the law of patents: a collision course . In Hunt, G., & Mehta, M. (Eds.), *Nanotechnology: Risk, Ethics and Law*. London: Earthscan.

Williams, A. (2005). *The patent explosion* (Working paper). New Paradigm Learning Corporation. Retrieved June 25, 2009, from http://anthonydwilliams.com/wp-content/uploads/2006/08/The_Patent_Explosion.pdf

ENDNOTES

[1] This graphical representation is based on regression analyzes run by the authors where the logarithm of the number of patents is the dependent variable. The authors used annual data (1999-2005) on the number of patents retrieved from the OECD Stat. Extracts database.

[2] This graphical representation is based on regression analyzes run by the authors where the logarithm of the number of ICTs patents is the dependent variable. The authors used annual data (1999-2005) on the number of patents retrieved from the OECD Stat. Extracts database.

[3] This graphical representation is based on regression analyzes run by the authors where the logarithm of the number of nanotechnology patents is the dependent variable. The authors used annual data (1967-2006) on the number of nanotechnology patents from Chen et al. (2008).

[4] Under the IPRs approach, a nonrival good is made excludable by attributing temporary monopoly rights to its inventor. The inventor

requires thus a fee for its use while the real cost of duplication of the nonrival good is zero. On the one hand, charging a fee for the innovation allows the inventor to appropriate a reward for the innovation and thus acts as an incentive for further innovation, preserving the optimal level of innovation necessary for a welfare gain in society. On the other hand, charging a fee for the use of a good of which the production cost is zero results in an efficiency loss and thus in a welfare loss as not all society can benefit from an originally free innovation. Under the open science approach, a free access to innovation is guaranteed to all society resulting in an optimal level of exploitation of the innovation and thus in a welfare gain. Yet, by the same token, innovation ends up unrewarded which decreases the incentive for subsequent innovation resulting in a welfare loss. In sum, both the IPRs and the open science approach lead to concurrent welfare gains and losses and thus to trade-off situations. The magnitude of the welfare losses/gains under both approaches is hardly measurable which feeds the on-going debate in economics over which approach really maximizes social welfare.

5   IPRs enforcement over the basic pieces of innovation needed for further innovation leads to the parcelling of intellectual property among the many originators of innovation.

The knowledge common becomes a knowledge anticommon as access is subject to the unanimous consent of all the originators of innovation. Negotiating access individually with each intellectual property rights holder results in a costly access to innovation. Starting from this point, two outcomes are possible. Either the acquisition of rights to the parcelled intellectual rights becomes so costly that innovation is radically discouraged. Or, the cost of innovation is passed along to customers via artificially high prices of the end product. Both situations lead to a welfare loss in society.

6   In a patent thicket situation, two or more firms hold overlapping patents. Each of Firm 1, Firm 2, and Firm 3 wants to develop a new technology, yet the patent overlap is such that each of the three firms needs the other two firms' patents. For instance, Firm 1 needs, in addition to its in-house patent P1, a license for P2 and P3. In this situation, Firm 1 is subject to patent infringement allegations threat if it is to develop its new technology without negotiating a license from Firm 2 and Firm 3. At the same time, it is a source of patent infringement allegations threat to Firm 2 and Firm 3 if they are to develop their new technologies without negotiating a license from Firm 1 for P1.

Chapter 14
Taking the Lead:
How the Global South Could Benefit from Climate Finance, Technology Transfer, and from Adopting Stringent Climate Policies

Adrian Muller
University of Zürich, Switzerland

ABSTRACT

In this chapter, the author argues that the countries in the Global South can gain from stringent own climate policies. This is so, as in the current situation, the south tends to be dominated by the climate policies of northern countries and climate finance largely supports single projects and technology transfer that are not embedded in a broader policy framework in southern countries. Adopting own stringent policies could help to counteract this and to channel these financial means to their most beneficial use. This could help southern countries to follow an agenda that is different from the fossil fuel based development path of the north. Such a "green new deal" could be a promising economic and technological development strategy. Stringent climate policies would strengthen the southern countries in the international climate negotiations and southern countries could take the lead in the climate change mitigation debate. Technology transfer and the carbon finance sector would play a crucial role for this. Climate policy and climate finance could thus be used to set a new stage, where the south is not at a disadvantage with respect to the north.

INTRODUCTION

Technology transfer plays a crucial role in the context of climate change mitigation and adaptation and the corresponding financial programs: Article 4.5 of the United Nations Framework Convention on Climate Change (UNFCCC 1992)

states that developed country Parties "shall take all practicable steps to promote, facilitate and finance, as appropriate, the transfer of, or access to, environmentally sound technologies and know-how to other Parties, particularly developing country Parties, to enable them to implement the provisions of the Convention." Similarly, Article 4.8 of the convention calls developed country Parties to support technology transfer to developing

DOI: 10.4018/978-1-61692-006-7.ch014

countries for adaptation to the adverse effects of climate change. This emphasis on the importance of technology transfer is renewed in the Bali Action Plan from 2007, which calls for enhanced action on this (UNFCCC 2007), and in several UNFCCC documents and workshops that support project developers in this topic (e.g. UNFCCC 2006). In 2001, a UNFCCC Expert Group on Technology Transfer was established and the UNFCCC sites provide ample information on technology transfer. The convention thereby acknowledges that developing countries have more important general development goals than mitigation and adaptation (Art. 4.7): "[…] economic and social development and poverty eradication are the first and overriding priorities of the developing country Parties." Technology transfer is also expected to be crucial for reaching those general development goals (e.g. UN 2004 on technology transfer and the Millennium Development Goals). The Intergovernmental Panel on Climate Change (IPCC) defines technology transfer "as a broad set of processes covering the flows of know-how, experience and equipment […] amongst different stakeholders such as governments, private sector entities, financial institutions, non-governmental organizations (NGOs) and research/education institutions." (IPCC 2000, p3). This is a very broad definition, but it captures how this term is used in the climate change context, in which this chapter is located.

Technology transfer needs money. Given enough financial means become available in the context of climate change mitigation and adaptation, the big challenge is to channel those to beneficial and promising projects. Technology transfer clearly can improve livelihoods in a society and it clearly can support development. But this need not necessarily be the case. Without duly accounting for the relevant country, society, cultural and regional context, technology transfer may fail (this is also acknowledged by the IPCC 2000, p3). For a concrete example of the importance to account for this context, cf. e.g. Bhat et al. (2001) on biogas plant dissemination in India. Technology transfer may fail to lead to positive consequences in various ways. (a) It can be without effect on development–then it is a missed opportunity, but no worse. (b) Or the transferred technology is not used because of not meeting consumer needs (as happened e.g. to new cooking stoves to replace inefficient fire places, cf. e.g. Barnes et al. 1993). In both these cases, there is however no further harm done. (c) Technology transfer can also have crowding out effects that may have adverse impacts (e.g. engineered crops that crowd out locally adapted varieties). (d) It can also lead to incoherent technical systems in target countries, if a specific technology transfer activity is not aligned with existing policies and infrastructure (e.g. large-scale power plant development in a policy context for a decentralized electricity sector). (e) Or it may lead to a lock-in situation (e.g. biomass energy power development, which can hinder development of sustainable agriculture heavily relying on biomass as a fertilizer, cf. e.g. Muller 2009a). Technology transfer is thus not necessarily positive or, at worst, only without any effect. It can, as in these three last cases, have adverse impacts that could also prevail on a longer time-scale.

These potential drawbacks of technology transfer are of particular relevance in climate policy. There, the industrialized countries that committed to emissions reduction under the UNFCCC (the "Annex I" countries, as they are listed in Annex I of the UNFCCC 1992) have the possibility to realize part of their reduction goals in countries not having reduction targets themselves (i.e. mainly in developing countries, the "non-Annex I" countries). Currently, the specifics of such reduction projects and corresponding technology transfer are largely determined by the needs of the Annex I countries. Furthermore, the goals of donor (i.e. Annex I) and target (i.e. non-Annex I) countries are not aligned, as the former mainly need projects for cheap emission reductions while the latter pursue general development goals such as eradication of extreme poverty and hunger. Thus, the relevant context may be neglected and the problems to be solved with technology transfer and potential solutions may be defined and framed wrongly.

The goal of this chapter is to illustrate how non-Annex I countries may make best use of the potential of carbon finance and related technology transfer, without being dominated by the climate change mitigation agenda of Annex I countries. This chapter first addresses the development goals of non-Annex I countries (section 2). Carbon finance and the goals of the Annex I countries are discussed in section 3. In section 4 the chapter then discusses potential strategies of non-Annex I countries in the light of the aspects mentioned before. Section 5 concludes.

Development Goals

Whether technology transfer is supporting or even crucial for non-Annex I countries' development goals depends on the specific goal and technology. Furthermore, it depends on the specific context technology transfer would take place in (e.g. country, region, culture, sector) and how this context is accounted for. This section thus collects some of the main development goals of non-Annex I countries and which role technology transfer may play for those.

Millennium Development Goals

A first set of goals is provided by the Millennium Development Goals (MDG), see Table 1. These state eight basic goals which are further differentiated into several "targets". The MDG's apply globally i.e. also to industrialized nations, and those do by far not fare perfect on all targets, e.g. regarding gender equality (goal 3). Technology transfer is seen to play a crucial role for meeting the MDGs (e.g. UNCTAD 2004), but it is clearly of greater importance for some goals (e.g. access to information and communications technologies, goal 8.5) than for others (e.g. to comprehensively deal with debt, goal 8.3). It is noteworthy that some

topics such as increased rural electrification and electricity supply security are not listed, although those aspects are crucial to some MDGs, such as to reach universal access to information and communications technologies (goal 8.5).

Climate Change Adaptation and Mitigation Necessities

While the MDGs are general development goals to be pursued largely independent of other global developments, climate change necessitates planning for some further goals. As some climate change takes place irrespective of how fast and how far-reaching global mitigation measures will be implemented, adaptation is unavoidable. This is especially so in many non-Annex I countries that will be hit particularly hard (IPCC 2007). Most extreme is the example of the island states, such as the Maldives, that plan for the resettlement of their entire population, as their home countries will very likely become submerged due to rising sea levels (Times of India 2008).

Non-Annex I countries will have to adapt to increased water stress, decreasing crop yields, and increasing burden from malnutrition and infectious diseases. Regionally, sea level raise and extreme weather events will also necessitate due adaptation measures (IPCC 2007). Some key impacts that necessitate adaptation are collected in figure 1. There is a close link between climate change mitigation, adaptation and development topics. The IPCC states in its Fourth Assessment Report that "Sustainable development can reduce vulnerability to climate change, and climate change could impede nations' abilities to achieve sustainable development pathways." (IPCC 2007, Summary for Policy Makers, p20).

If the goal to keep global average temperature increase below 2°C is to be met with some higher probability such as 75% (this goal is generally seen as an upper limit to avoid "dangerous" climate change, e.g. IARU 2009, p12), there is the necessity of significant mitigation also in non-Annex

Table 1. The Millennium Development Goals

The Millennium Development Goals
1 Eradicate extreme poverty and hunger
1 Halve, between 1990 and 2015, the proportion of people whose income is less than $1 a day 2 Achieve full and productive employment and decent work for all, including women and young people 3 Halve, between 1990 and 2015, the proportion of people who suffer from hunger
2. Achieve universal primary education
1 Achieve Ensure that, by 2015, children everywhere, boys and girls alike, will be able to complete a full course of primary schooling
3. Promote gender equality and empower women
1 Eliminate gender disparity in primary and secondary education preferably by 2005, and in all levels of education no later than 2015
4. Reduce child mortality
1 Reduce by two thirds, between 1990 and 2015, the under-five mortality rate
5. Improve maternal health
1 Reduce by three quarters, between 1990 and 2015, the maternal mortality ratio 2 Achieve, by 2015, universal access to reproductive health
6. Combat HIV/AIDS, malaria and other diseases
1 Have halted by 2015 and begun to reverse the spread of HIV/AIDS 2 Achieve, by 2010, universal access to treatment for HIV/AIDS for all those who need it 3 Have halted by 2015 and begun to reverse the incidence of malaria and other major diseases
7. Ensure environmental sustainability
1 Integrate principles of sustainable development into country policies and programmes and reverse the loss of environmental resources 2 Reduce biodiversity loss, achieving, by 2010, a significant reduction in the rate of loss 3 Halve, by 2015, the proportion of the population without sustainable access to safe drinking water and basic sanitation 4 By 2020, to have achieved a significant improvement in the lives of at least 100 million slum dwellers
8. Develop a global partnership for development
1 Address special needs of the least developed countries, landlocked countries and small island developing States 2 Develop further an open, rule-based, predictable, non-discriminatory trading and financial system 3 Deal comprehensively with developing countries' debt 4 In cooperation with pharmaceutical companies, provide access to affordable essential drugs in developing countries 5 In cooperation with the private sector, make available the benefits of new technologies, especially information and communications technologies

Source: http://www.un.org/millenniumgoals

I countries, even if Annex I countries would cut emissions substantially (Hare and Meinshausen 2006, p127 ff and Gupta et al. 2007, p775ff). The necessity to curb emissions is accepted by the large emitters China, Brazil and India, and many non-Annex I countries have set targets and plans for mitigation policies (Rogelj et al. 2009), but they oppose binding reduction targets, pointing out the responsibility of Annex I countries to first undertake strong mitigation measures (Financial Times 2009).

As with the MDGs, technology transfer plays a role for some of these goals and not for others.

Technology transfer is crucial for own mitigation measures in non-Annex I countries (e.g. UNFCCC 2007). Examples are new energy technologies and efficiency improvements. Also for adaptation, technology transfer is important (IPCC 2007, chapter 20). Technology transfer could play a role, for example, for improved cheaper desalinization technologies as a measure against increased water stress or in the context of drug development against malaria and other diseases whose pressure will increase. Given the broad definition of technology transfer used here, this comprises also transfer of knowledge on improved agricultural practices,

Figure 1. Key impacts as a function of increasing global average temperature change (Impacts will vary by extent of adaptation, rate of temperature change, and socio-economic pathway; Simplified from IPCC 2007, Summary for Policy Makers)

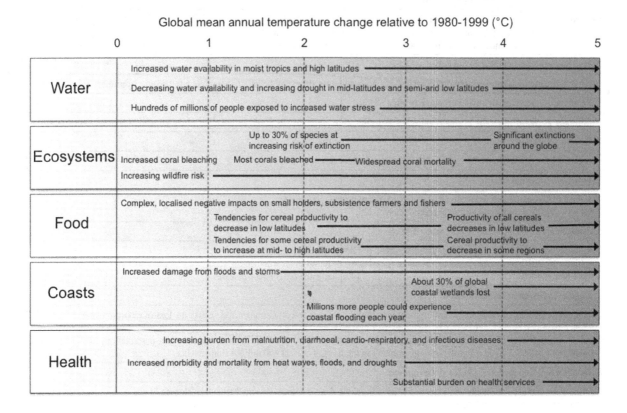

for example (such as organic agriculture with a proven potential for adaptation to and mitigation of climate change, Muller 2009b).

Further Goals

The development goals described above mainly target the poor and vulnerable. This is most important, but in the course of development, other goals may be defined, which target parts of the population, where poverty and hunger is no problem. MDG 8.5, access to information and communications technologies, is such a goal and not a subsistence necessity. Neither does technology transfer related to generation of skilled jobs target the poorest part of the population, for example.

In addition, critical development theory can be employed to assess development goals. As mentioned above, it is currently the Annex I countries that define the goals to be realized under carbon finance. This is in line with observations from critical development theory, that industrialized countries dominate the discourse not mainly by economic or technological power but by their power to define, i.e. to chose and frame the problems and corresponding sets of acceptable solutions (e.g. Sardar 1999). Thus, development goals of non-Annex I countries may look very different from what the north might expect, e.g. if they would like to not emulate the consumption oriented lifestyle of Annex I countries or if they would decide to question the GDP-growth paradigm. On the other hand, it can also be a naïve view of some in the Annex I countries who would like the people in non-Annex I countries to be different from (i.e. "morally better than") the people in the Annex I countries. In fact, the self-chosen development goal of many people in non-Annex I countries might very well exactly be to become as the people in Annex I countries, including their consumption levels. Cf. Table 2 for a large-scale technology example that can have different interpretations in the context of this criticism.

We do not further pursue this fundamental critique of the current development discourse and correspondingly different development goals. Even when adopting a mainstream view on development, it can be stated that there is a danger of Annex I countries dominating the debate in carbon finance at the expense of non-Annex I countries.

Carbon Finance

Several mechanisms for project finance have become implemented in climate change policy. They mainly aim at financing mitigation measures, although provision of finance for adaptation increasingly gains importance as well. Mitigation measures are financed via the regulated and the voluntary carbon market (cf. Table 3; Hamilton et al. 2009).

The first covers the EU and several regional emissions trading schemes, the Clean Development Mechanism (CDM) that allows Annex I countries to invest in reductions in non-Annex I countries, which subsequently can be accounted for the own reduction target under the Kyoto Protocol, and Joint Implementation that is similar, only that both partners come from Annex I countries. The voluntary carbon market has two sections. One is the Chicago Carbon Exchange,

Table 2. DESERTEC

DESERTEC
A) DESERTEC is a recently launched project by 13 industrial investors from Europe and the DESERTEC Foundation to install huge capacity of thermal solar power in the desert areas of North Africa to produce power for Europe and the North African region (DESERTEC 2009). Thermal solar power produced in desert areas and delivered to consumers by High-Voltage Direct Current transmission lines (with comparably low losses over thousands of kilometers) has a huge potential to significantly contribute to carbon-free electricity production. DESERTEC delivers a considerable amount of power to North-African states, and there are claims of technology transfer to Africa and potential side-benefits such as desalination of water with waste heat. On the other hand, DESERTEC has many features of a project planned in the North with relatively few genuine development benefits for the South, as it is primarily a project to assure carbon-free supply for electricity needs in Europe.
B) Such a project would be more in line with development goals from African countries if it would deliver power to Sub-Saharian Africa instead of Europe. Financial requirements are huge and getting it funded poses a main problem – which is currently solved with European Companies being the key players in DESERTEC. But climate finance could provide means to tap other financial sources to implement such a project independently. Clearly, such a project would involve development of international electrification strategies way beyond the level of single power plants or even national strategies. A hindrance for such a large-scale project is the focus of current climate finance on single plant level (cf. section 3). A DESERTEC project that would increase renewable power and electrification in North and Sub-Saharian Africa, funded by climate finance sources could be an example of successful technology transfer in the climate context, where the goals to be achieved would be formulated by the project host countries themselves, embedded in some broader regional cross-country energy strategy, and not by the Annex I countries in the north. Clearly, the African countries do not follow one agenda and aligning their interests would by far not be simple task. It also has to be assured that such a project does not only provide power for already connected parts of the population but increases access and supply security for all.
C) Some critique goes even further, claiming that for a sustainable energy future, one should rely on decentralized solutions for energy supply rather than on large-scale technology projects, which tend to be dominated by the industry investors and their profit motive. Technology transfer is crucial for both strategies; if several non-Annex I countries would however want to follow a decentralized electrification strategy, well-organized concerted strategy development and action to raise funds would be necessary to counter the dominance by large-technology development goals of Annex I country industries. Carbon finance could however be the funding source for decentralized strategies as well.

Table 3. Transaction volumes and values, Global Carbon Market, 2007 and 2008; primary CDM relates to credits directly bought from the project, secondary CDM to credits bought on the market; (a) assume a CA$ 10 price for Alberta offsets and Emission Performance Credits based on interviews with market participants (Source: Hamilton et al. 2009, data: Ecosystem Marketplace and New Carbon Finance)

Markets	Volume(MtCO2eq)		Value (US$ million)	
	2007	2008	2007	2008
Voluntary OTC	43.1	54.0	262.9	396.7
CCX	22.9	69.2	72.4	306.7
Other Exchanges	0	0.2	0	1.3
Total Voluntary Markets	**66.0**	**123.4**	**335.3**	**704.8**
EU ETS	2061.0	2982.0	50097.0	94971.7
Primary CDM	551.0	400.3	7426.0	6118.2
Secondary CDM	240.0	622.4	5451.0	15584.5
Joint Implementation	41.0	8.0	499.0	2339.8
Kyoto (AAU)	0	16.0	0	177.1
New South Wales	25.0	30.6	224.0	151.9
RGGI	-	27.4	-	108.9
Alberta's SGER[a]	1.5	3.3	13.7	31.3
Total Regulated Markets	**2919.5**	**4090.5**	**63710.7**	**119483.4**
Total Global Markets	**2985.5**	**4213.5**	**64046.0**	**120188.2**

a voluntary Cap-And-Trade System, the other is the so-called "over the counter" market, comprising in bilateral deals and mainly operating outside exchanges. Relevant for this chapter are the mechanisms that allow for project funding in non-Annex I countries. Most prominent there is the Clean Development Mechanism. In principle, the CDM ties sustainability, development and climate change mitigation together as it has a twofold goal to reduce greenhouse gas emissions while assisting the host countries in achieving sustainable development (UNFCCC 1998, Article 12). Thereby, it is the prerogative of the hosting non-Annex I countries to define what sustainability means (UNFCCC 2001, Decision 17, p20). For an overview of the current status of the CDM, see IETA 2009.

Sustainability explicitly plays a role also in other parts of the carbon market besides the CDM. The UNDP, for example, established the Millennium Development Goal Carbon Facility that aims at directing financial flows for emission offsets towards projects that contribute to the achievement of the MDGs (UNDP 2007).

Whether emission reduction activities support development or achieve sustainability goals is however contested. Many claim that the sustainability goal of the CDM, emissions reductions and the profit motive of investors are likely to not be in line. Studies analyzing CDM projects find poor sustainability performance. (Huq 2002, CDMWatch 2005, Kuzma and Dobrovolny 2005, Cosbey et al. 2005, UNEP 2005b, Michaelowa and Michaelowa 2007a, 2007b, Borges da Cunha et al. 2007, Sutter and Parreno 2007, Olsen 2007). The CDM, left to market forces, does not significantly contribute to sustainable development.

As an example, we report the results from Sirohi (2007). Based on an assessment of the situation in India in August 2006, she concludes that the CDM does not notably contribute to rural poverty alleviation although such is at the core of India's development policy. This is, on the one hand, due to a business orientation of most projects and the corresponding unimportance of the projects' performance regarding rural development. On the other hand, it is because chronic poverty has deeper structural, institutional and long-term roots beyond the reach of the CDM, such as the distribution of land holdings, productivity of land and quality of labor force. Borges da Cunha et al. (2007) undertook a similar assessment of the CDM in Brazil (September 2006). They also find an only limited potential of the CDM to contribute to achievement of the MDGs while operating under the market logic. Similarly, Michaelowa and Michaelowa (2007b) conclude in their comparison of investments under the official development assistance and in the context of emission reduction projects that the latter will contribute only to a limited degree to poverty reduction as framed in the MDGs and that the CDM in particular, as a market mechanism, will not contribute to poverty reduction.

One reason for the poor sustainability performance of the CDM is that the Kyoto Protocol and subsequent amendments do not provide any operational guidelines on how to analyze sustainable development and its implementation in this context. It is only stated that climate change mitigation projects should be in line with the host country's regulations and legislation on sustainable development and the environment. As it is the prerogative of the host country to decide if a CDM project assists in achieving sustainable development or not, there is a danger that sustainability requirements become very low in order to attract more projects. This is a classical situation of a "race to the bottom"(cf. also Sutter 2003, p 67ff).

A further reason for the poor sustainability performance of the CDM is the fact that many CDM projects are realized "unilaterally" (currently around 30%, UNEP 2009). This is against the original idea that projects should be realized bilaterally, with finance from Annex I countries helping realizing projects in non-Annex I countries. In unilateral projects, a project developer in the non-Annex I country fully bears the risk of the project and sells the emission reductions generated on the market. Thus, no institutional context for technology transfer exists and technology transfer is lacking. The share of unilateral projects also partly explains the skewed country distribution of projects where least developed countries and the poorest communities tend to be bypassed. Currently, most projects are realized in China, India and Brazil (Michaelowa and Müller 2009). Domestic capital is lacking in the least developed countries and poor communities. Few investors from Annex I countries are interested in bilaterally developing projects in these contexts. Other barriers for CDM project implementation in least developed countries are the political risk and lack of information and expertise (de Gouvello et al. 2008).

There are however some international Carbon Funds with specific development goals or targeting poorest countries, namely the World Bank (WB) Community Development Carbon Fund, the WB BioCarbon Fund and partly the WB Prototype Carbon Fund. The volume of these funds ($450 million) is however minor compared to the total volume of the carbon market (cf. Table 3; World Bank 2009), in particular if accounting for the fact that by far not all the funding in these special funds is used for projects in poorest countries and communities. Additional funding comes from the Global Environmental Facility, which supports the transfer of environmentally sound technologies (viz. low carbon in particular) since 1990. Total funds used for this over the 17 years till 2008 are $2.5 billion. This is five times the current size of the above mentioned carbon funds, but it is still little in relation to the volume of the carbon market. It is less than half of the annual CDM volume, for example. In addition, it is over a period of almost 20 years. On the other hand, it led to an additional $15 billion of leveraged co-financing (GEF 2008).

All finance currently available is low in relation to what is needed, though. As suggested by the G77 (the largest intergovernmental group of developing countries in the UN, currently 130 countries) and China and supported by many more countries in the Climate Negotiations 2008, a global fund for technology transfer and capacity building should be installed with funding by 0.5 to 1% of GNP of developed countries (UNFCCC 2008, p6, 37). This would amount to some $200-500 billion annually. Total needs for mitigation are estimated at some $ 200 billion annually in 2030 (UNFCCC 2007b).

In contrast to mitigation, the importance of adaptation gained prominence only recently, and institutionalized funding for this becomes available now as well. In 2007, the Kyoto Protocol Adaptation Fund was established to help developing countries with adaptation. Currently, it is funded from a 2% levy on all credits generated from the CDM. Estimated annual financial needs

for adaptation in developing countries amount to several tens of billions US dollars (UNFCCC 2007b, UNDP 2007, p15; see also Parry et al. 2009). Several suggestions exist on how to increase available funding, but no agreement has been reached so far (Brown et al. 2009).

For this chapter, the most important aspect of carbon finance targeted at non-Annex I countries is however the fact that all its forms largely finance single installations only. This bears the danger that they are not embedded in any national strategy of the host country. CDM projects thus do not support sectoral transformation, improvements in structural development aspects (such as chronic poverty) or ambitious concerted policies that reach way beyond the single installation level, such a smart grids or intelligent urban mobility systems (cf. Sterk and Wittneben 2006 for some collection of the criticisms against the current CDM). Table

Table 4.

Problems of single mitigation projects when a larger framework is lacking: biomass power
In some regions, many biomass energy projects are planned, but their interaction is not systematically assessed. This is strikingly evident, when consulting the project development documents (PDD) from several biomass energy projects in India (similarly in Brasil).
The PDDs generally exhibit a lack of or only very biased and partial economic analysis of the viability and wider effects of the project activity. This is reflected in a general trend that aspects favorable for the projects viability, sustainability impact and barriers tend to be assessed in detail while more critical issues are addressed less detailed. Thus, most projects either state abundant availability of biomass in the project region or they are based on own biomass that is available from their core business in any case (e.g. bagasse-based cogeneration projects). A significant part does not mention the availability at all. This is a predominantly static assessment, as the abundance of biomass makes the projects viable. Lacking is a discussion of potential shortages in biomass supply due to an increase of similar project activities, resp. an increase in opportunity costs of using own biomass as a fuel due to similar developments. Exactly this dynamic is often assessed, however, in the discussion of biomass prices, usually undertaken in the barrier analysis. There, many projects refer to increasing prices (in the past and projected for the future) as a major investment barrier that makes the additional income from carbon finance a necessity for the project activity. This thus draws a contradictory picture of the economic environment regarding biomass availability within which these projects are operating.
The inconsistent consideration of dynamic effects regarding prices and quantities also pervades the assessment of leakage in most projects. There, it is often an issue whether other users of biomass may be forced to switch to fossil fuels due to the project activity. It is then usually argued that this is not the case due to availability of abundant quantities of biomass in the region. The observed and predicted parallel price increases for biomass waste, however, are not mentioned, although they change the situation for existing projects as well and may well prompt a decision to switch to other fuels there.
Project-wise planning without coordination among projects and information on the wider context thus may result in too many projects that mutually hinder each other and reduce their competitiveness as agricultural-waste biomass may become scarce in this region, thus driving up prices for the biomass, i.e. fuel costs. Besides mutual adverse effects between single, uncoordinated projects, large-scale support of some type of projects may hinder other development options. Large-scale agricultural-waste bioenergy development in some region, for example, can hinder other strategies, such as the development of sustainable agriculture in the same region. The latter would be based on composting of the agricultural-waste biomass, which is used in the bioenergy plants and is thus not available for composting anymore (Muller 2009).
A region may decide to support a biomass energy strategy and less so sustainable agriculture. Such a decision should be based on an open debate and should then be taken explicitly. It should not emerge as a lock-in result due to uncoordinated single project activities.
Similar conclusions on the potential shortage of biomass due to biomass power CDM projects are drawn in Sutter (2003, 141ff) based on detailed analysis of biomass power projects in Andhra Pradesh, India.

(Source: Assessment by the author of all Indian Biomass CDM projects registered until 4.9.2009, http://cdm.unfccc.int/index.html)

4 gives an example of potential adverse effects of this single installation focus. The single installation focus also sets incentives to not reform energy policies, as this allows attracting more profitable CDM projects due to the presence of a high emissions baseline against which the reduction achieved by these projects is measured.

In principle, finance for "programs" beyond the single installation level are possible under the CDM (via the so-called "programmatic CDM") but existing rules for this have serious drawbacks and hardly any projects are planned under this category (Michaelowa and Müller 2009). Improvement on this is planned, also via introduction of a "sectoral" CDM, which achieves reductions not by single installations but by a variety of activities in the context of an encompassing broader governmental or private sectoral strategy (Baron et al. 2009, Sterk and Wittneben 2006, Figueres 2006, UNFCCC 2009, Michaelowa and Müller 2009, Schneider and Cames 2009). Finance for actions beyond the single project level should also become available in the context of the "Nationally Appropriate Mitigation Actions" (NAMA) that continuously gain momentum, in particular as they are included in the Appendix to the Copenhagen Accord (UNFCCC 2010). These refer to concerted national strategies for mitigation projects. What their exact contents shall be is kept open, but they should at least foster concerted planning of different projects and policies.

SYNERGIES AND TAKING THE LEAD IN THE DEBATE

The previous sections have shown that there is a mismatch between the host countries' broader development policies and the results achieved by carbon finance projects. This section presents some possible remedies against this.

Synergies

Annex I countries need cheap certificates for emissions reductions incurred abroad. Reducing emissions for other countries is however not the goal of the non-Annex I countries as their various development goals need not correlate with emissions reductions for others.

Although synergies are small in the current CDM portfolio, there are nevertheless situations where synergies between the goals of Annex I and non-Annex I countries can be identified. As the current situation shows, realization of this potential however needs extra effort. Some of the general development goals are actually related to the reduction goals of the Annex I countries. Examples are CDM projects realized under the Gold Standard, a strict label that aims at realizing projects with strong additional sustainability benefits. Second, non-Annex I countries will need to undertake adaptation measures, which are likely to be financed by some international fund sourced mainly from Annex I countries. This money would be directly targeted at some development goals without further aspects such as mitigation linked to and potentially interfering with it. Third, as climate change is a global problem, Annex I countries have an interest that non-Annex I countries undertake mitigation measures for themselves as well.

Taking the Lead

A second strategy to assure beneficial effects of technology transfer consists in non-Annex I countries taking the lead in the debate. This would avoid Annex I countries dominating carbon finance goals, actions and related definitions, problem identification and implementation. Of particular importance is that the non-Annex I countries have strategies on what they want and need with regard to technology transfer—e.g. in the context of broader development strategies. Critical aspects of such strategies are a) a discussion of the underlying technology transfer paradigm–it is not only climate change mitigation that counts; b) detailed formulation of certain development goals on an operational level and how they may fruitfully combine with carbon finance. Many non-Annex I countries have mitigation plans (Rogelj et al. 2009, see also UNFCCC 2010), but the strategies for implementation need to be very concrete to take the lead. This is of importance as carbon finance currently mainly works on a single installation level without embedding the single projects in more encompassing development strategies. An example of related adverse effects is given in Table 4.

Such a leader role should be taken when it comes to mitigation measures for (at least some) non-Annex I countries that need to be implemented in the longer run to reach the 2°C reduction target (Hare and Meinshausen 2006, p127 ff, Gupta et al. 2007, p775 ff, PBL 2009). Adopting binding targets would strengthen the position of these countries in the international debate. Several proposals exist on how such targets could be defined (Gutpa et al. 2007). They could be indexed with GDP growth to not curb economic growth (e.g. Jotzo and Pezzey 2007). Such targets for non-Annex I countries clearly would need to be combined with binding commitments of Annex I countries for financial support, though. Carbon finance could then truly complement development strategies in the non-Annex I-countries and both the Annex I and the non-Annex I countries could profit.

Mitigation measures and binding reduction targets in particular offer the opportunity that the respective countries have full control of the whole process, given finance is available from own sources or from the international community. Such targets would also avoid a potential sell-out of cheap emission reduction possibilities as it now may happen with the CDM (Muller 2007, Narain and van t'Veld 2008) and less opportunities for reductions abroad would be available for Annex I countries. Implementing a global emissions trading scheme should however be avoided, as

players (Annex I and non-Annex I countries) would be too unequal, potentially leading to strong market failure.

Taking the lead would thus help to make best use of carbon finance for the non-Annex I countries. Technology and knowledge transfer, education, and development of own technologies could thus be supported, avoiding, for example, that mainstream technology from the Annex I is implemented project-wise without further benefits besides certified emission reductions for the Annex I countries. There is growing awareness for the potential of such leadership. See e.g. LEAD (2009) on the importance and some aspects of leadership on climate change for the example of Africa. Advocating the lead of non-Annex I countries does not mean that we would advocate any goal a non-Annex I country may choose or that we would judge as problematic any goal an Annex I country may propose. Consultancy from outside can be an important support for development and corrupt structures and rent-seeking behavior of elites can hinder development. Just as corporations do not act to the best of all their stakeholders, it is a fact that decision makers in any country may not decide for the best of their population. It is, for example, just this situation that the non-Annex I countries define what sustainable development means for them (and that Annex I corporations act as profit maximisers), which counteracts implementation of the sustainability goals of the CDM (cf. above).

Promising Technology Transfer

Finally, it has to be acknowledged that there are examples of potentially promising technology transfer in the climate context also on the level of single projects (e.g. technology for water capture from dew or for cleaning or desalinization of water). For those, a broader perspective seems most promising, though: infrastructure developments relying on information and communications technologies, such as smart grids or intelligent mobility structures in urban areas. Improved grid-less or grid based information and communications technologies would be the basis for this (cf. the recent development of a large capacity glass-fiber connection on the West African coast). Based on these technologies and specific measurement applications, improved weather forecasts could, for example, result in optimized irrigation schemes, better local adaptation capacity and crop planning (cf. improved seasonal forecasting used to offset the effects of climate change on hydropower generation, IPCC WG2, chp. 3, p 198; see also IPCC 2000, GEF 2008 for examples).

As different non-Annex I countries face similar problems, technology transfer between non-Annex I countries should also be supported. The special kind of technology transfer comprised in knowledge and science transfer may be particularly supported, e.g. by programs where students become educated in Annex I countries and then return to their home-country (many such programs already exist at different universities). This would support own development of new technologies–given funding can be organized. Organizing such is also likely to be easier if technology development is embedded in some broader development strategy. Knowledge and science transfer is thus not confined to engineering but needs also to cover know-how on policies and strategies to develop and support such broader development strategies. A program focusing on this in the context of environmental policy instruments and development is established in Gothenburg, Sweden, for example, combined with a set of newly established environmental economics research institutions in developing countries (www.efdinitiative.org). See also Cannady (2009) for an example of a strategy how climate change technology transfer could be organized based on developing country players and strengthening developing country R&D.

Knowledge and science transfer is also needed, as Annex I countries develop many technologies of limited use for the broad development goals of non-Annex I countries, at least in their current situation. Based on knowledge and science transfer, non-Annex I countries could themselves develop cheap drugs against the most important diseases (e.g. malaria), which would be crucial for adaptation. Or some of the non-Annex I countries may have a need for self-mending appliances, due to large and very remote areas. Such could be used, for example, in desert solar power plants or information and communications technology infrastructure. Particularly promising for non-Annex I countries will be niche technologies of direct relevance for themselves, where Annex I countries currently have low interests (e.g. for adaptation in coastal areas in the tropics). Thus, a reverse process than currently pursued in carbon finance should be strengthened: potential receiver countries should identify problems to be solved and then look for the technology available to be transferred. If this technology is not available, they should look for possibilities to develop this technology themselves, based on specific transfer of knowledge.

CONCLUSION

The previous sections have illustrated that (a) the potential future volume of carbon finance and corresponding finance of technology transfer is huge. Currently available volumes are still small, though. (b) There is a danger that the Annex I countries dominate the debate on how this money is to be used in non-Annex I countries. Currently, carbon finance mainly finances mitigation projects without much further sustainability benefits for the hosting non-Annex I countries. (c) The goals of non-Annex I countries are different from the goals of Annex I countries and current funding is mainly in line with the goals of the latter. (d) Technology transfer can have positive, no and also negative effects. The outcome depends on the specific context and goals to be achieved. (e) The structure of carbon finance projects is currently focused on single installations without taking into account the broader development context.

To make beneficial use of carbon finance and related technology transfer, synergies between the goals of Annex I and non-Annex I countries have to be identified and, in particular, non-Annex I countries need to take the lead in the debate. It has to be avoided that Annex I countries get imported energy policy and energy systems. Non-Annex I countries need to determine their development, mitigation and adaptation goals on an operational level in such a way as to channel carbon finance to most beneficial uses. It is crucial to assess technology transfer in a broad context of other development policies and strategies in a specific country or region and not only in relation to specific single projects, as is currently the case in carbon finance.

"Taking the lead" should be understood as a fundamental transformation. Climate change is a big threat to humanity but it also offers opportunities. Mitigation is costly for industrial societies–but it may be much less so for developing societies, which are at a different stage on the development path. Climate policy offers the opportunity to fundamentally question this development and to pursue alternatives to the western growth-based model. But even within this paradigm, developing nations could directly engage in a "green new deal", while the inertness of the grown western economies with their fossil-fuel dependence and industrialized agriculture has much more problems with such a change. Developing countries could gain from not postponing mitigation commitments and from not following industrialized countries on their path. Developing their energy supply and transportation infrastructure and their agriculture, they could directly aim at low carbon solutions, avoiding increasing dependence on increasingly scarce and expensive fossil fuels. They could take the lead in mitigation now, as the industrialized countries are not contributing their part. Admittedly, this is not just, given the responsibility of

industrialized countries for past emissions. But claiming justice without action does not help, as drastic action now is necessary in any case. Such a strategy would greatly strengthen developing nations. Questions of justice could be addressed in parallel or later–and used to put increased pressure on industrialized countries to take own actions and to provide finance for the developing nations' actions.

Based on knowledge transfer and built-up expertise in the context of own mitigation targets, non-Annex I countries could even go a step further and start giving advice to Annex I countries on how to best achieve their reductions. Currently, Annex I countries dominate the debate but do not take the lead in reducing emissions and they struggle to meet their Kyoto targets. This would further strengthen the position of Non-Annex I countries.

A main objection against such a strategy is its political infeasibility due to the presence of strong interests against stringent reduction targets in any country. Admittedly, this paper is visionary in this respect–but visions are needed and a second thought beyond common-sense perception of what climate policy and mitigation may offer for development is necessary. A second objection states that mitigation in developing countries is problematic, as it curbs their development perspectives. This is however only true in the frame of replicating a western, fossil-fuel based development path. If alternatives are developed, this needs not be true. This is the challenge for developing countries: to use climate policy to considerably strengthen their position in relation to industrialized countries. The current global system is not favorable for developing countries–thus the necessity of big changes offered by climate change should be used to initiate truly fundamental changes. The failure of industrialized countries in climate policy offers the opportunity to do better and is no excuse to do nothing.

REFERENCES

Barnes, D., Openshaw, K., Smith, K., & van der Plas, R. (1993). The design and diffusion of improved cooking stoves. *The World Bank Research Observer, 8*, 119–141. doi:10.1093/wbro/8.2.119

Baron, R., Buchner, B., & Ellis, J. (2009). *Sectoral Approaches and the Carbon Market*. Paris: OECD.

Bhat, P. R., Chanakya, H. N., & Ravindranath, N. H. (2001). Biogas plant dissemination: success story of Sirsi, India. *Energy for Sustainable Development, 5*(1), 39–46. doi:10.1016/S0973-0826(09)60019-3

Borges da Cunha, K., Walter, A., & Rei, F. (2007). CDM implementation in Brazil's rural and isolated regions: the Amazonian case. *Climatic Change, 84*(1–2), 111–129. doi:10.1007/s10584-007-9272-1

Brown, J., Vigneri, M., & Sosis, K. (2009). *Innovative Carbon-Based Funding for Adaptation*. Overseas Development Institute. Retrieved September 9, 2009, from http://www.odi.org.uk/ccef/resources/reports/s0198_oecd_adaptation.pdf

Cannady, C. (2009). *Access to Climate Change Technology by Developing Countries: A Practical Strategy. ICTSD's Programme on IPRs and Sustainable Development, Issue Paper No. 25*. Geneva, Switzerland: International Centre for Trade and Sustainable Development.

Cosbey, A., Parry, J., & Browne, J. Babu, Y., Bhandari, P., Drexhage, J., & Murphy, D. (2005). *Realising the Development Dividend: Making the CDM Work for Developing Countries*. Report of the International Institute for Sustainable Development IISD, Canada. Retrieved September 3, 2009, from http://www.iisd.org/ publications/pub.aspx? id=694

de Gouvello, C., Dayo, F., & Thioye, M. (2008). *Low carbon energy projects for development in Sub-Saharan Africa*. Washington, DC: World Bank. Retrieved September 9, 2009, from http://wbcarbonfinance.org/docs/Main_Report_Low_Carbon_Energy_projects_for_Development_of_Sub_Saharan_Africa_8-18-08.pdf

DESERTEC. (2009). *Clean Power from Deserts, The DESERTEC Concept for Energy, Water and Climate Security* (4th Ed.). Retrieved from http://www.desertec.org

Ellis, J. (2006). *Issues Related to a Programme of Activities Under the CDM*. Paris: OECD. Retrieved September 9, 2009, from http://www.olis.oecd.org/olis/2006doc.nsf/FREDIRCORP-LOOK/NT00000A2E/$FILE/JT03208489.PDF

Figueres, C. (2006). Sectoral CDM: Opening the CDM to the yet unrealized goal of sustainable development. *International Journal of Sustainable Development Law and Policy, 2*(1).

Financial Times. (2009, April 29). Climate talks: What India, China and Brazil want. Retrieved September 9, 2009, from http://blogs.ft.com/energy-source/2009/04/29/developing-countries-and-the-climate-negotiations/

GEF. (2008). Transfer of Environmentally Sound Technologies: The GEF Experience. *Global Environmental Facility*. Retrieved September 9, 2009, from http://thegef.org/uploadedFiles/Publications/GEF_TTbrochure_final-lores.pdf

Gupta, S., Tirpak, D., Burger, N., Gupta, J., Höhne, N., Boncheva, A., et al. (2007). Policies, Instruments and Co-operative Arrangements. In *Climate Change 2007: Mitigation*. Cambridge, UK: Cambridge University Press. Retrieved September 9, 2009, from http://www.ipcc.ch/pdf/assessment-report/ar4/wg3/ar4-wg3-chapter13.pdf

Hamilton, K., Sjardin, M., Shapiro, A., & Marcello, T. (2009). Fortifying the Foundation: State of the Voluntary Carbon Markets. *Ecosystem Marketplace and New Carbon Finance*. Retrieved September 9, 2009, from http://ecosystem-marketplace.com/documents/cms_documents/StateOfTheVoluntaryCarbonMarkcts_2009.pdf

Hare, B., & Meinshausen, M. (2006). How much warming are we committed to and how much can be avoided? *Climatic Change, 75*, 111–149. doi:10.1007/s10584-005-9027-9

Humphrey, J. (2004). The Clean Development Mechanism: How to Increase Benefits for Developing Countries. *IDS Bulletin 35.3: Climate Change and Development*, 88.

Huq, S. (2002). Applying Sustainable Development Criteria to CDM Projects: PCF Experience. PCFplus Report 10, Washington DC, April 2002.

IARU. (2009).*Climate Change–Global Risks, Challenges and Decisions*. Scientific Conference, Copenhagen 2009, 10-12- March, Synthesis Report, International Alliance of Research Universities IARU. Retrieved September 9, 2009, from http://www.climatecongress.ku.dk

IETA. (2009). State of the CDM 2009: Reforming for the Present and Preparing for the Future. *International Emissions Trading Association (IETA)*. Retrieved April 12, 2010, from http://www.ieta.org/ieta/www/pages/getfile.php?docID=3363

IPCC. 2000. Methodological and Technological Issues in Technology Transfer. Intergovernmental Panel on Climate Change (IPCC). Cambridge: B. Metz, O.R. Davidson, J-W. Martens, S.N.M. van Rooijen and L Van Wie McGrory, eds. Cambridge University Press.

IPCC. (2007). *Contribution of Working Group II to the Fourth Assessment Report of the Intergovernmental Panel on Climate Change*. (M. L. Parry, O. F. Canziani, J. P. Palutikof, P. J. van der Linden & C. E. Hanson, Eds.). Cambridge, UK: Cambridge University Press. Retrieved September 9, 2009, from http://www.ipcc.ch/publications_and_data/publications_ipcc_fourth_assessment_report_wg2_report_impacts_adaptation_and_vulnerability.htm

Jotzo, F., & Pezzey, J. C. V. (2007). Optimal intensity targets for greenhouse gas emissions trading under uncertainty. *Environmental and Resource Economics, 38*, 259–284. doi:10.1007/s10640-006-9078-z

LEAD. (2009). Leading the way: A role for regional institutions. African leadership on climate change in Africa. Based on the workshop 'African Leadership on Climate Change: Challenges and Opportunities', Tunis, January 2009. LEAD Africa, Enda Energy, LEAD International, African Development Bank.

Michaelowa, A., & Michaelowa, K. (2007a). Does climate policy promote development? *Climatic Change, 84*(1-2), 1–4. doi:10.1007/s10584-007-9266-z

Michaelowa, A., & Michaelowa, K. (2007b). Climate or development: is ODA diverted from its original purpose? *Climatic Change, 84*(1–2), 5–21. doi:10.1007/s10584-007-9270-3

Michaelowa, A., & Müller, B. (2009). The Clean Development Mechanism in the Future Climate Change Regime, Summary for Policy Makers. *Climate Strategies*. Retrieved September 9, 2009, from http://www.climatestrategies.org/our-reports/category/39/149.html

Muller, A. (2007). How to make the CDM more sustainable: The potential of rent extraction. *Energy Policy, 35*(6), 3203–3212. doi:10.1016/j.enpol.2006.11.016

Muller, A. (2009a). Sustainable agriculture and the production of biomass for energy use. *Climatic Change, 94*(3-4), 319–331. doi:10.1007/s10584-008-9501-2

Muller, A. (2009b). *Benefits of Organic Agriculture as a Climate Change Adaptation and Mitigation Strategy in Developing Countries.* Scandinavian Working Papers in Economics 343 and EfD Discussion paper 09-09, Environment for Development Initiative and Resources for the Future, Washington DC.

Narain, U., & van t'Veld, K. (2008). The clean development mechanism's low-hanging fruit problem: When might it arise, and how might it be solved? *Environmental and Resource Economics, 40*(3), 445–465. doi:10.1007/s10640-007-9164-x

Olsen, K. H. (2007). The clean development mechanism's contribution to sustainable development: a review of the literature. *Climatic Change, 84*(1-2), 59–73. doi:10.1007/s10584-007-9267-y

Parry, M., Arnell, N., Berry, P., Dodman, D., Fankhauser, S., & Hope, C. (2009). *Assessing the Costs of Adaptation to Climate Change: A Review of the UNFCCC and Other Recent Estimates.* London: International Institute for Environment and Development and Grantham Institute for Climate Change.

PBL. (2009). *Meeting the 2 °C target. From climate objective to emission reduction measures.* Netherlands Environmental Assessment Agency (PBL). Retrieved April 12, 2010, from http://www.rivm.nl/bibliotheek/rapporten/500114012.pdf.

Rogelj, J., Hare, B., Nabel, J., Macey, K., Schaeffer, M., Markmann, K., & Meinshausen, M. (2009). Halfway to Copenhagen, no way to 2°C. *Nature Reports Climate Change.* Retrieved September 9, 2009, from http://sites.google.com/a/climateanalytics.org/test/welcome/briefing-papers

Sardar, Z. (1999). Development and the Locations of Eurocentrism . In Munck, R., & O'Hearn, D. (Eds.), *Critical Development Theory.* New York: Zed Books.

Schneider, L., & Cames, M. (2009). *A framework for a sectoral crediting mechanism in a post-2012 climate regime.* Report for the Global Wind Energy Council Berlin, May 2009.

Sirohi, S. (2007). CDM: is it a win–win strategy for rural poverty alleviation in India? *Climatic Change, 84*(1–2), 91–110. doi:10.1007/s10584-007-9271-2

Sterk, W., & Wittneben, B. (2006). Enhancing the clean development mechanism through sectoral approaches: Definitions, applications and ways forward. *International Environmental Agreement: Politics, Law and Economics, 6*(3), 271–287. doi:10.1007/s10784-006-9009-z

Sutter, C. (2003). *Sustainability Check-Up for CDM Projects. How to assess the sustainability under the Kyoto Protocol.* Berlin: Wissenschaftlicher Verlag.

Sutter, C., & Parreno, J. C. (2007). Does the current Clean Development Mechanism (CDM) deliver its sustainable development claim? An analysis of officially registered CDM projects. *Climatic Change, 84*(1-2), 75–90. doi:10.1007/s10584-007-9269-9

The Times of India. (2008, November 11). *Maldives plans to buy 'new homeland.'* Retrieved from http://timesofindia.indiatimes.com/World/Maldives_plans_to_buy_new_homeland/articleshow/3696018.cms

UN. (2004). *The role of science and technology in the achievement of the MDGs.* Note by the UNCTAD secretariat, United Nations Conference on Trade and Development (UNCTAD). Retrieved September 9, 2009, from http://www.unctad.org/en/docs/tdxibpd4_en.pdf

UNCTAD. (2004). Round Table on Harnessing Emerging Technologies to Meet the Development Goals Contained in the Millennium Declaration. In *United Nations Conference on Trade and Development*. Retrieved September 9, 2009, from http://www.unctad.org/en/docs/tdl383_en.pdf

UNDP. (2007). *Human Development Report 2007/2008 - Fighting Climate Change: Human Solidarity in a Divided World. United Nations Development Programme*. New York: Palgrave Macmillan.

UNEP. (2009, September 1). *UNEP Risoe CDM/ JI Pipeline Analysis and Database*. Retrieved September 9, 2009, from http://www.CDMpipeline.org/

UNFCCC. (1992). *United Nations Framework Convention on Climate Change*. Retrieved September 9, 2009, from http://unfccc.int/resource/docs/convkp/conveng.pdf

UNFCCC. (1998). *Kyoto Protocol to the United Nations Framework Convention on Climate Change*. Retrieved September 9, 2009, from http://unfccc.int/resource/docs/convkp/kpeng.pdf

UNFCCC. (2001). *Report of the COP 7 - Marrakesh-Accords*. Retrieved September 9, 2009, from http://unfccc.int/resource/docs/cop7/13a01.pdf and .../13a02.pdf

UNFCCC. (2006). *Preparing and presenting proposals - A guidebook on preparing technology transfer projects for financing*. Retrieved September 9, 2009, from http://unfccc.int/ttclear/pdf/PG/EN/unfccc_guidebook.pdf

UNFCCC. (2007). *The Bali Action Plan*. Retrieved September 9, 2009, from http://unfccc.int/files/meetings/cop_13/application/pdf/cp_bali_action.pdf

UNFCCC. (2007). *Investment and financial flows relevant to the development of an effective and appropriate international response to Climate Change*. Retrieved September 9, 2009, from http://unfccc.int/cooperation_and_support/financial_mechanism/items/4053.php

UNFCCC. (2008). *Ad Hoc Working Group On Long-Term Cooperative Action Under The Convention, Fourth Session*. Retrieved September 9, 2009, from http://unfccc.int/resource/docs/2008/awglca4/eng/misc05.pdf

UNFCCC. (2009). Further elaboration of possible improvements to emissions trading and the project-based mechanisms under the Kyoto Protocol. *Ad Hoc Working Group On Further Commitments For Annex I Parties Under The Kyoto Protocol, Seventh Session*. Retrieved September 9, 2009, from http://unfccc.int/resource/docs/2009/awg7/eng/inf02.pdf

UNFCCC. (2010). *The Copenhagen Accord, Appendix II - Nationally appropriate mitigation actions of developing country Parties*. Retrieved April 12, 2010, from http://unfccc.int/home/items/5265.php

World Bank. (2009). *Annual Report 2008: Carbon Finance for Sustainable Development 2008*. The World Bank Carbon Finance Unit. Retrieved September 9, 2009, from http://wbcarbonfinance.org/Router.cfm?Page=DocLib&CatalogID=47061

Chapter 15
Emissions Distribution in Post–Kyoto International Negotiations:
A Policy Perspective

Nicola Cantore
Overseas Development Institute, UK & Università Cattolica del Sacro Cuore, Italy

Emilio Padilla
Univ. Autónoma de Barcelona, Spain

ABSTRACT

An abundant scientific literature about climate change economics points out that the future participation of developing countries in international environmental policies will depend on their amount of pay offs inside and outside specific agreements. Though these contributions represent a corner stone in the research field investigating future plausible international coalitions and the reasons behind the difficulties incurred over time to implement emissions stabilizing actions, they cannot disentangle satisfactorily the role that equality plays in inducing poor regions to tackle global warming. Scholars recently outline that a perceived fairness in the distribution of emissions would facilitate a wide spread participation in international agreements. In this chapter the authors overview the literature about distributional aspects of emissions by focusing on those contributions investigating past trends of emissions distribution through empirical data and future trajectories through simulations obtained by integrated assessment models. They will explain methodologies used to elaborate data and the link between "real data" and those coming from simulations. A particular attention will be devoted to the role that technological change will play in affecting the distribution of emissions over time and to how spillovers and experience diffusion could influence equality issues and future outcomes of policy negotiations.

DOI: 10.4018/978-1-61692-006-7.ch015

INTRODUCTION

The Conference of Parties in Copenhagen at the end of 2009 represents a crucial step for future negotiations about emissions stabilizing policies. The world is facing one of the biggest challenges to development that has never experienced in the past: the strong environmental and socioeconomic problems deriving from global warming caused by economic activity. A wide majority of scientists and policy makers agree on the fact that if appropriate policies will not be implemented within a reasonable lapse of time, the human kind could experience disasters that will strongly affect standards of life of future generations. If we consider the Brundtland's Report definition (WCED, 1987; p. 43) of sustainability according to which sustainable development is intended as a form of development that satisfies "the needs of the present generation without compromising the ability of future generations to meet their own needs" we can understand why weak and fragmented actions against global warming can lead to undesirable growth paths.

Within this perspective the priority is to set up the most effective policies to curb the increasing trend of emissions over time that until now does not appear to stabilize yet. A strand of literature in environmental economics refers to the well known Environmental Kuznets Curve (EKC) hypothesis. The prior idea behind this concept (Grossman and Krueger, 1991) is that if we are able to identify a bell shaped relationship between the level of income and pollution and a turning point beyond which the level of emissions begins to decrease the best way to deal with environmental problems is to foster growth (see Figure 1). Though some EKC evidence has been found for many pollutants, evidence is very weak for pollutants generating climate change (Figure 2).

One of the main reasons would mainly lie in the public good nature of clean air. In other words, all countries face an incentive to "free ride" by enjoying positive externalities deriving from emissions

reduction policies without bearing the relative costs (Ansuategi and Escapa, 2002). If we focus on the Stern Review findings stressing that climate change will generate heavy damages and policy actions will be costly in a finite time horizon, we understand why there is a great incentive to "free ride" in order to exploit benefits from emissions reduction efforts of others. The reluctance of poor countries in joining international agreements is mainly supported by historical responsibility of rich regions in generating atmospheric carbon concentration, whereas rich countries claim that emissions stabilizing policies will be effective only when developing countries will join them.

The main finding of the EKC in a context of increasing emissions is that growth is not the best tool to deal with global warming, but appropriate policies are needed to reach a turning point in the relationship between income and emissions.

As mentioned by the scientific literature (Cantore and Canavari, in press), the main problem concerning climate change policies is that to reduce effectively the level of emissions a very strong condition is needed: the involvement and a high reduction burden for developing countries. This finding seems to meet the main concern raised by the past Bush American administration claiming that a USA participation in climate change international policies would depend on a strong commitment of emerging areas in meeting pollution mitigation constraints. The involvement of Annex B[1] countries in emissions stabilizing policies that created the conditions for the signature of the well known Kyoto Protocol was frustrated by the refusal of the main polluter country, namely USA, to ratify the agreement. The stop to a full implementation of the Kyoto agreement generated many doubts on its effectiveness to tackle global warming in the short run and uncertainty in the negotiations to set up post Kyoto international emissions constraint agreements.

The argument provided by the past USA administration to refuse the commitment to climate policies does not take into account the main

Figure 1. The Environmental Kuznets Curve hypothesis

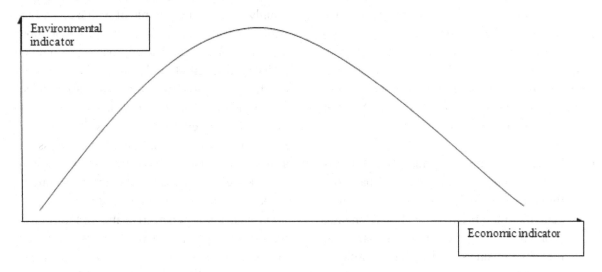

Figure 2. CO$_2$ world carbon emissions from the consumption and flaring of fossil fuels (million metric tons of carbon dioxide) Source: IEA (2008)

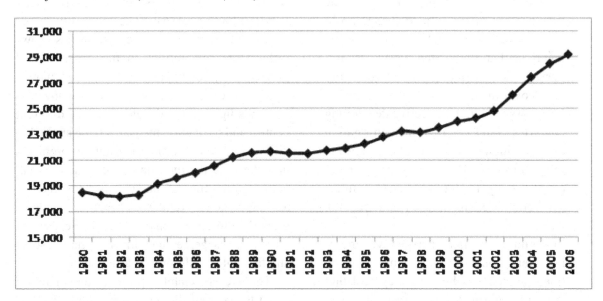

argument provided by the poorest regions: the involvement of developing countries in emissions stabilizing policies could strongly reduce their growth rates and welfare levels (Cantore and Canavari, in press). In other words it is very unlikely that emerging countries will be able to reduce greenhouse gas emissions by bearing reasonable welfare costs. This situation mainly creates two problems. From one side, as outlined by the game theoretical literature, if countries take policy choices on the basis of the pay off generated by a specific strategy it is very unlikely that stable world coalitions aimed at reducing emissions can arise (Altamirano Cabrera *et al.*, 2008), though a wider participation could be favoured by specific money transfers from rich countries to poorest regions in order to create incentives to join the club of pollution stabilizing countries (Carraro *et al.*, 2006). From the other side even if developing countries should gain or pay a low cost from an emissions reduction commitment an equity issue arises. Developing countries could not be still available to join emissions stabilizing policies for many reasons:

- In an intertemporal perspective poorest regions could not be available to pay now for mitigation policies in order to enjoy benefits in the future. The climate change economics literature widely stresses that it exists a mismatch between the timing of paying for policies and that for enjoying environmental benefits (Tol *et al.*, 2004). Electoral and political conditions often impose short term policy views and the time horizon of the climate change problem is well beyond that of many governments settled around the world. An important issue of intergenerational equity arises over time.
- In a cross country perspective developing countries could perceive their mitigation costs too high (or their benefit too low) in relation to those paid (enjoyed) by richer countries, by raising an issue of intra-generational equity (Roson and Bosello, 2002).
- In a multidimensional perspective emerging areas could refuse to join international agreements because they feel that they were not historically responsible in generating global warming or they could not perceive to be vulnerable to climate change. In other words other variables such as the geographical distribution of emissions (Padilla and Serrano, 2006) and climate change damages (Tol *et al.*, 2004) are relevant in climate change negotiations to set up emissions stabilizing policies.

Previous climate change economics literature mainly focussed on the role of welfare costs and benefits to influence climate change negotiations within a game theoretical approach. Our opinion is that researchers should also consider how they are distributed among countries according to equity principles and other relevant variables that may affect policy considerations. This chapter mainly focuses on the distribution of emissions as important factor to investigate with the aim to extract useful insights for the understanding of global warming political negotiations.

From a policy perspective and for future environmental negotiations two elements are crucial: the perception of developing countries of past responsibility of rich countries and the perception of rich countries towards future responsibility of developing countries in generating atmospheric carbon concentration. As outlined by Duro and Padilla (2008; p. 456): "The inequality in per capita CO_2 emissions between countries shows different responsibilities in the generation of greenhouse gases and the contribution to climate change. Therefore, the analysis of this inequality sheds light on the debate about the different control and mitigation measures to be applied in different regions. In fact, distribution problems have become the most important issue to deal with in global climate change policy negotiations and agreements. Taking distribution problems properly into account in policy design leads to an increase in the perceived fairness of the measures and facilitates widespread participation".

The present chapter presents the literature about the path of the inequality in the distribution of emissions in the past arising from the empirical evidence (Section 2), the literature about the future distribution of emissions from integrated assessment model (Section 3) and the role that technological change and the transfer of technology will play in affecting the distribution of emissions and consequently, political negotiations of emissions stabilizing policies (Section 4). Section 5 will conclude.

EMPIRICAL EVIDENCE ABOUT THE DISTRIBUTION OF EMISSIONS

There are several studies that have applied distributive analysis tools to the analysis of climate change. Alcántara and Duro (2004) and Sun (2002) analysed the inequality in energy intensity across OECD countries, which, according to their results, experienced a strong decrease in the period 1971–1998. Sun (2002) used mean deviation as a dispersion measure and also examined the differences between certain subgroups of countries. Alcántara and Duro (2004) employed the Theil index to measure these inequalities. This index allowed them to weight observations according to their GDP and to apply a consistent decomposition of inequality by subgroups. Hedenus and Azar (2005) employed the Atkinson inequality index to analyse the inequality for different natural resources, including carbon emissions per capita. According to their results, the Atkinson index showed a decrease in the inequality in emissions per capita over time. Heil and Wodon (1997) used the Gini index to investigate the inequality of emission across countries. They decomposed the Gini index to study the contribution of two income groups (poor and rich countries) to this inequality. They found that between group inequality was much more important than within group inequality to explain global emission inequality and its evolution.

Heil and Wodon (2000) used the same methodology to analyse future inequality in carbon emissions using business-as-usual projections to the year 2100. They considered four income groups in their study. They also considered the impact on emission inequality of the Kyoto Protocol and other mitigation measures. They found that emission inequality decreases faster than income inequality and that the reduction between groups of countries is more important than the decrease within groups to explain the evolution of overall emission inequality. Duro and Padilla (2008) employed the EGR index (Esteban et al.,

1999) to analyse the polarisation in CO_2 per capita distribution across countries[2]. They found that polarization strongly decreased during 1971–2001 and that the groups of countries grouped according to polarisation optimisation leads to two groups which broadly coincided with Annex B and non-Annex B of the Kyoto Protocol. Padilla and Serrano (2006) employed the Theil index to study the development of emission inequality over time. They showed the contribution of four income groups to inequality. They showed that between group emission inequality was much more important than within group emission inequality. They used concentration indexes and found that the concentration of emissions in richer countries—inequality when countries are ranked by income per capita—diminished less than "simple" inequality—inequality when countries are ranked by emissions per capita. Duro and Padilla (2006) explained the main driving forces of emission inequality by decomposing emission inequality into the different Kaya factors[3] (see Kaya, 1989; and Yamaji *et al.*, 1991) and two interaction terms. They also decomposed inequality and its sources into between and within group components. They found that income per capita was the main factor explaining emission inequality level and development, although differences in energy intensity, which were strongly reduced during the period, and in carbon intensity of energy were also relevant. A recent work by Coondoo and Dinda (2008) incorporates more sophisticated econometric techniques[4] to outline that in the long run the distribution of emissions per capita is mainly governed by the distribution of Gross Domestic Product (GDP per capita) over time.

We can highlight some results and policy implications for climate policy negotiations from the literature on the distributive analysis of emissions. First, the results of the literature confirm that there are strong inequalities between the per capita emissions of different countries and world regions, although there is a clear trend to reduction of this inequality (Heil and Wodon,

1997; Padilla and Serrano, 2006). These strong differences between countries and regions have complicated the achievement of agreements in the past and will difficult future agreements, as these differences imply different interests in the negotiations. Synthetic indicators of inequality such as Gini and Theil indexes provide useful indicators of the evolution of these differences. The differences between countries are much greater if cumulative emissions are considered (Heil and Wodon, 1997). These results might support the arguments of the countries with less responsibility in causing the problem, which have been often reluctant to participate in agreements involving costs to them.

Second, there is a strong correlation between income and emission inequalities (Padilla and Serrano, 2006). Both inequalities have strongly decreased during past decades. The analysis of regions and groups of countries show that the differences between different income groups (between group component) explain most of emission inequality, while the inequality within these groups of countries (within group component) is much smaller (Heil and Wodon, 1997; Padilla and Serrano, 2006). Rich countries, which are the main responsible of climate change, are still the major contributors to the problem. The strong inequality in emission between rich and poor countries indicates that aggressive short-term measures focused on reducing emissions in rich countries might have some impact in the control of global emissions. However, some growing economies, such as China and India, have experienced a strong income and emission growth, which explain the strong reduction of emission inequality, and their participation in future agreements is essential in order to achieve an effective mitigation policy. These results reinforce the need to take into account the distribution consequences of different policy alternatives in order to facilitate the participation of the different parties in the measures.

Third, income inequality is the main driving force of emission inequality (Duro and Padilla, 2006). Moreover, if we decompose inequality in the distribution of emissions according to the Kaya identity factors, most of the decrease of emission inequality is due to the decrease of income inequality, although the strong decrease of energy intensity inequality has also been significant. The strong correlation between income and emissions per capita inequalities, and the importance of income inequality in explaining emission inequality, indicate that global policies oriented by the perspective of approaching to a fair share of atmosphere—an equal per capita emissions rights criteria—and so oriented to reduce emission inequality would be more feasible if income inequality were reduced. Policies reducing global income inequality would also lead to a reduction in emission inequality.

Fourth, the concentration of emission in richer counties has decreased less than simple emission inequality (Padilla and Serrano, 2006). This reinforces the relevance of taking into account the situations of richer and poorer countries rather than taking only into account simple emission inequality in negotiations and agreements.

Fifth, polarisation analysis shows that, although polarisation has decreased since 1971 it has not been reduced in the last years analysed (Duro and Padilla, 2008). Polarisation in 2001 was not lower than in 1997, so a distribution which leads to the formation of groups with confronting interests might still be one of the factors hampering the achievement of new agreements. Moreover, the groups endogenously obtained by polarisation analysis broadly coincide with Annex B and non-Annex B groups of countries of the Kyoto Protocol. This result also shows that polarisation in emissions might be a relevant indicator of groups formation in policy negotiations and of the good or bad environment to form coalitions and achieve agreements.

Sixth, there are strong international differences in energy intensity. This inequality has strongly decreased in last decades, especially in some regions (Duro and Padilla, 2006). Energy efficiency gains in some developing economies have been very important and have contributed to reduce energy intensity differences between countries. Moreover, the reduction in energy intensity inequality, achieving similar levels of efficiency in countries with different income, has contributed to attenuate emissions growth. Clearly, energy efficiency gains and diffusion and convergence in energy intensity might play an important role in the future to help mitigate emissions. Technological transfers and diffusion should be important points to be taken into account in negotiations and future agreements.

Even within developed economies there has been a strong decrease of this inequality in last decades (Sun, 2002; Alcántara and Duro, 2004). According to Duro et al. (2009), in the case of OECD countries, the reduction in energy intensity differences is the main factor explaining the reduction of emissions per capita. The reduction of final energy intensity for a set of OECD countries considered in their analysis for the period 1995–2005 is mainly explained by a trend to the convergence of energy efficiency between different countries sector by sector, while an increasing sector specialisation has played an opposing role. Energy-saving strategies and technology diffusion have importantly contributed to reduce and equalise the energy intensity levels sector by sector of the different countries and have lead to the reduction of energy intensity inequality observed. These results also reinforce the relevance of taking technological policies into account in future negotiations and agreements. However, a policy to this effect would not eliminate global disparities in energy intensity due to the existence of different sectoral specialisation patterns. In fact, the results show that specialisation has increasingly contributed to the energy intensity inequality between countries in the last decade.

Finally, there are also important inequalities in carbon intensity of energy. They are quite relevant for explaining differences within some regions (Duro and Padilla, 2006). This result might indicate that countries with similar income show different

efforts in the change from fossil fuels to renewable energy sources. This points out the strong potential for controlling emissions by increasing renewable energy sources and converging to lower levels of carbon intensity of energy. These differences should also be taken into account by future policies.

Among the wide amount of results that have been outlined by the previous literature three findings should be particularly considered in terms of policy relevance and for their suitability to be compared with other analysis tools (in particular Integrated Assessment Models (IAMs) as we will see more accurately in the next section):

1. Inequality in the distribution of emissions tended to decrease in the past.
2. If we create groups that are differentiated on the basis of different levels of income per capita criteria inequality in the distribution of emissions per capita is mainly composed of a between group than a within group inequality.
3. Inequality in the distribution of emissions between rich and poor countries is mainly explained by differences in the levels of GDP per capita rather than energy intensity or carbon intensity in a Kaya identity perspective.

From a policy perspective they are very important because they all provide evidence that there is a strong correlation between emissions and income distribution and that the policy agenda of equity in income distribution is strongly related to that of equity in emissions per capita distribution. An interesting research question that has been raised by the literature at a certain point is if these trends that have been identified in the past will hold in the future even if different scenarios of international environmental agreements will be implemented. The next paragraph explains what the literature about IAMs focussing on simulations says about previous results found in a context of empirical data.

THE DISTRIBUTION OF EMISSIONS AND INTEGRATED ASSESSMENT MODELS

A recent literature has tried to connect the traditional literature about the drivers of growth with climate change issues. The seminal work by Nordhaus (1994) represented a corner stone in the literature to understand the impact of economic activities on the level of emissions and the feedback from effect from the environment, specifically from the rise of the atmospheric carbon concentration and consequently from the temperature rise to growth. The authors created a powerful tool to search the best climate policies within an uncertainty framework. Nordhaus' work created an interesting strand of research trying to implement effective tools to understand the integration between economy and environment through IAMs. IAMs represent stylized facts describing the mechanisms generating pollution in a dynamic context on the basis of mathematical and computational techniques.

Many studies have focused on some equity implications of different measures with the help of IAMs. Miketa and Shrattenholzer (2006) analyse equity implications of two burden-sharing rules ("equal emissions per capita" and "carbon intensity"). Leimbach (2003) analyses how the equal emissions per capita allocation principle influences future emissions paths and mitigation costs of different regions, taking into account permissions trade. Vaillancourt and Waaub (2004) analyse the consequences on the allocation of emissions to different regions over time of two weight sets of allocation criteria. Vailancourt and Waaub (2006) also analyse the costs for each region in each case. However, none of these works use inequality indexes or other distributive analysis tools to analyse the development of emissions distribution.

A recent contribution from Cantore and Padilla (2007) tries to fill this gap and to verify if the results arising from the empirical literature are robust over time and if more insights can be gained about the impact of future pollution stabilising policies. Any effective climate policy requires limiting global emissions. The authors investigate how the effort to control these global emissions is distributed among different regions and countries. There are several proposals on the distribution of future emission "entitlements": from those based on current emission levels or GDP shares to the distribution of "entitlements" in per capita terms, and many combinations of these. In any case, only if a global policy is perceived as fair could it lead to the necessary agreement and participation of all the relevant parties. Therefore, it becomes essential to analyse income and emission distributions, their relationship, and the consequences of different mitigation policies and scenarios on future emission and income distributions.

Cantore and Padilla (2007) use an optimal growth model to analyse emissions and their distribution under different scenarios assuming the implementation of international agreements about pollution constraints at regional level (only for Annex B countries or for both Annex B and non Annex B countries) or at global level (atmospheric carbon concentration constraint or a temperature increase constraint). Results from Cantore and Padilla substantially confirm many findings arising from the empirical literature: they confirm a strong correlation between income and emissions inequality. In other words this paper largely confirms for future trends what has been remarked for the past. Though inequality in the future decades between rich and developing countries will continue to shrink there is evidence that still emerging countries such as China and India will be far from becoming the main polluters in the near future in terms of emissions per capita and that the current status will continue to hold in the short/medium term. Of course results derive from a model and are affected by assumptions and calibration, but they seem quite robust when relevant parameter vary or (in more recent studies), when other IAMs are considered (Cantore *et al.*, 2008).

The Cantore and Padilla's research leads to other results enriching analyses and findings related to the previous empirical literature. First, the results show that concentration of emission would be smaller than concentration of income for the different scenarios considered. That is, first, there is a "progressive" distribution of emissions. The "progressivity" in emission distribution decreases over time, especially in the scenarios involving a greater abatement effort in developing economies and so a higher redistribution of emissions towards rich countries. This, in the absence of economic compensations to poor countries, leads to a negative redistribution of the assimilative capacity of the atmosphere. Thus, these scenarios leading to a lower "progressivity" might be seen unfair by poor countries and undesirable from a distributive perspective.

Second, the analysis of between and within group inequality for three world regions grouped according to their income per capita shows a decrease in the "simple" emission inequality of both components for most scenarios. The results change significantly in the scenarios requiring a major mitigation effort to poor countries. In these cases the Theil index—and both between and within group components—increase. These policy scenarios seem the less appealing from a distributive point of view.

Third, when equity principles are considered for a given emissions reduction target, these principles lead to a higher abatement effort in rich countries and a redistributive effect. The redistribution is especially strong in the case of the application of the equality principle rule, although it is also significant in some cases. Although the variations of income inequality are not very relevant, it might be noticed that the application of scenarios like the Brazilian proposal[5] or the equalitarian rules[6] are the ones involving the scenarios with greater income inequality decrease. In these cases, climate policies could make a contribution to the reduction of income inequality. Clearly, these criteria would be more

acceptable from a distributive point of view and more appealing for developing economies.

Finally, the impact of emissions trading policies on emission distribution depends on the structure of marginal costs for each country, on the level of global abatement reduction and especially on how the abatement effort is shared among regions. Emissions trading is a crucial mechanism governing the efficiency of policy implementation and compliance costs introduced by the Kyoto Protocol on the basis of a well known result in the literature on environmental economics claiming that the introduction of a carbon market would allow the accomplishment of a policy target with the lowest global cost. As regards its impact on distribution, it has some impact on emission inequality and a much smaller impact on income inequality. In short, this flexibility mechanism allows to alleviate the impact of the scenarios involving a stronger abatement exigency to developing countries leading to a redistribution of emissions to poor countries as regards the scenario without trade and also involves a greater decrease of income inequality. In these cases, emissions trading improves both efficiency and emission distribution and would be preferred by all parties. In contrast to these scenarios, in the scenarios considering the application of equity rules for achieving a given atmospheric concentration—which imply a strong redistribution when no trading is considered—the results indicate that trading increases emission inequality as rich countries are the ones that buy the emission permits in this case. However, it has to be taken into account that, despite its effect on inequality indexes, emissions trade would have a positive global impact for the income of both rich and poor countries.

Summing up, we find an interesting consistency between results coming from empirical evidence and those derived from simulations data. The advantage of IAMs if compared to empirical analyses is that they are useful to implement scenario analyses based on plausible future evolutions of international environmental negotiations for policy agreements and they allow to understand the drivers of emissions distribution because of the formal mathematical equations fed by numbers extracted by reality. The great trouble concerning analyses driven by IAMs is that very often models are very different in terms of assumptions and calibration. A recent work implemented by Cantore *et al.* (2008) interestingly show that many results such as the decreasing path of the emissions distribution over time, the predominance of the between group component and the identification of GDP per capita as main driver of emissions distribution rather than carbon intensity or emissions intensity can be extended to other IAMs. Of course much research is needed to confirm further these results.

EMISSIONS DISTRIBUTION AND THE ROLE OF TECHNOLOGY AND TECHNOLOGY DIFFUSION

In the previous paragraphs we presented an overview of the literature on emissions distribution in terms of empirical evidence and data coming from simulations run with IAMs and we outlined the policy implications that we may extract from this abundant set of contributions. We stressed that inequality in the distribution of GDP per capita is the main determinant to explain inequality in emissions per capita. Especially IAMs can let the researcher make a step forward and understand what are the forces driving levels of GDP and its distributions between rich and poor regions and the other factors that can influence emissions distribution over time even if they are not so crucial as GDP. What emerges from IAMs is that a primary role will be played by the evolution of technology over time.

IAMs generally distinguish between two different forms of technology: industrial technology intended as Total Factor Productivity (TFP) affecting growth rates over time and environmental friendly technology affecting energy intensity

through the structural change of the economies and carbon intensities reductions. The scientific literature generally deals these two kinds technology formation mechanisms as relatively equivalent. The literature about industrial technology developed earlier. After that the first contributions explaining industrial technology generated the typical Ramsey–Solow–Koopmans frameworks where technology over time evolves exogenously as "manna from heaven", a fruitful strand of research introduced endogenous technological change governed by Research and Development expenditures (R&D) and learning by doing (lbd) processes. In other words, many contributions explained technology as outcome of specific strategies devoted to promote ideas, patents, licences with the aim to increase productivity of the economic system (knowledge formation through R&D) or as outcome of learning experiences over time (knowledge formation through lbd). On the basis of this literature convergence in terms of income over time depends on two factors:

1. the speed of the process by which R&D and lbd mechanisms penetrate over time in each country.
2. the diffusion of technologies across time and across countries in terms of economies openness (imitation and learning processes) through international spillovers or technology transfer through specific programs.

These contributions explained the process of technological change as determined by an accumulation process of knowledge deriving from learning or from research and innovation according to a neoclassical scheme comparable to that describing the accumulation of physical capital. The main difference between R&D knowledge and learning by doing knowledge is represented by the fact that the latter is usually represented as costless over time.

The literature about environmental friendly technological change followed that dealing strictly with industrial processes. Whereas specifically in the "industrial" literature technological change is typically shaped to affect the long run growth rate over time, in the climate change economics literature it typically decreases carbon intensity or energy intensity (Bosetti *et al.*, 2006). In some cases industrial and environmental friendly technological change are present in the same model framework, in others international spillovers are introduced in order to incorporate effects of technology transfers from rich to developing countries or transboundary learning effects (Buonanno *et al.*, 2003).

There are many points to understand about technological change, technology diffusion and possible effects on emissions distribution. First, it is still not clear how endogenous technological change evolves over time in different areas. Generally models include equations that are identical for rich and poor areas by implying that the mechanisms governing the knowledge formation over time are similar everywhere. Even much more obscure is the process by which technological change spreads across regions over time. Whereas for R&D investments there is wider consensus on the fact that the direction of transfer lies from rich regions to poor regions, transboundary effects of learning can be modelled as symmetric or asymmetric (Gerlagh and Kuik, 2007; Cantore, 2009). Second, for both industrial and environmental friendly technological change a consensus of the scientific evidence is still lacking on the effective capability of developing countries to learn from experience of richer countries or to use appropriately their technology. The literature about the potential of poorer regions to absorb technological change from developed areas is still slowly emerging (Bosetti *et al.*, 2008). Third, it is still not clear the relationship between industrial and environmental friendly technological change. In other words, there is still great ambiguity on the fact that industrial and environmental friendly knowledge are interconnected over time (Buonanno *et al.* 2003) or should be dealt as independent

processes. Our opinion is that there is a degree of correlation between the two forms of endogenous technological change. It is straightforward that the current technologies to capture and store carbon derive from technologies deriving from oil industries especially for what concerns pipelines. The great challenge is to verify how strong this correlation is and the modalities by which it can be expressed in mathematical and logical terms. It is reasonable to think that a higher correlation between industrial and technological change provides more opportunities for the diffusion of technologies across countries.

Fourth, a crucial issue is represented by crowding out effects that can arise in the presence of technological diffusion. As stressed by the mainstream literature, when knowledge becomes a public good that is available for all countries through technology transfers of transboundary effects this phenomenon can crowd domestic investments in technology by generating lower investments in technology for developing countries. This is one of the main interesting results found by Bosetti *et al.* (2008) when they use their IAM including international spillovers in an R&D expenditures environment.

Finally, it is difficult to understand how policies will interact to affect the process of technological diffusion. In the field of environmental technology a wide importance is assumed by flexible mechanisms for the accomplishments of emissions constraints, with a particular focus on the role played by the Clean Development Mechanisms (CDM). CDMs are mechanisms introduced by the Kyoto Protocol by which the implementation of project such as those aimed at transferring technology from Annex B to non Annex B countries can give the right to developed countries to claim a reduction of the emissions abatement effort. As interestingly argued by Millock (2001) CDMs can be viewed by a typical principal–agent scheme in which the rich country is the principal that cannot observe the effort produced by developing countries (agent) to put into force effectively the program. This situation creates a market failure named as information asymmetry that can be overcome by opportune policies aimed at monitoring with effectiveness the effort produced by the recipients of the programs through opportune indirect observable indicators.

In the industrial technology literature and in the environmental economics literature many contributions deal on the economic consequences of international spillovers (Barro and Sala Martin, 1995; Rao *et al.*, 2006) and in general outline a positive effect in terms of policy compliance costs and growth. In some cases spillovers can generate losses when trading effects are included and spillovers generate lower energy market prices for fossil fuel exporters or when, in spatial contexts, openness can make peripheric areas of some regions still more remote or unproductive (respectively Böhringer and Löschel, 2003; Caniëls and Verspagen, 2001).

With a particular focus on the environmental economics literature no previous studies deal with the possible consequences that spillovers can generate on the distribution of emissions[7]. It is very important at this point to outline that the most important policy target in the climate change context is the reduction of emissions. The convergence in terms of emissions per capita cannot be considered per se a policy target because a context of convergence in pollution levels is compatible with scenarios incorporating very negative outcome such as high levels of atmospheric carbon concentration and temperature increase.

As outlined by Table 1, we can argue that the idea of international spillovers and convergence in terms of technological knowledge (industrial and environmental friendly) accumulation is very unlikely to contextually reduce emissions and generate a convergence in the levels of emissions per capita.

As we outlined in the previous sections, inequality in the distribution of GDP per capita governs the path of inequality in terms of emissions per capita. From this finding we can extract some interesting claims. In a scenario involving convergence in the levels of income per capita we should also expect a lower gap between developed and developing countries in terms of emissions per capita, but also a higher level of emissions because of the higher levels GDP per capita for developing countries. On the other side, a technological convergence in abatement technologies should reduce emissions, but, in a context of higher levels of income per capita for rich regions, this would not reduce the gap in terms of emissions per capita. In other words, a scenario presenting an effective reduction of greenhouse gas emissions and equality in the distribution of emissions per capita is a desirable situation that is feasible only in scenarios where convergence in industrial and environmental friendly technological change will be jointly pursued. This is a challenging issue to investigate on which the right coordination of growth, equity and environmental policies could depend.

Table 1. Effects of technological convergence policies on the reduction of emissions and on the convergence in terms of emissions per capita

	Reduction of emissions target	**Convergence in emissions per capita**
Industrial technological convergence	An industrial technological diffusion promotes growth in developing countries and more emissions	Convergence in terms of output per capita and emissions per capita
Environmental friendly technological convergence	An environmental friendly technological diffusion promotes an emissions reduction driven by developing countries	The gap in terms of output per capita and consequently in terms of emissions per capita does not shrink as the abatement technologies of rich and developing areas are similar

Moreover, we point out that a scenario presenting convergence in income per capita and emissions per capita and convergence in abatement technologies could not generate a satisfactory emissions reduction to tackle global warming as, even in a convergence path of industrial and environmental friendly technology, the industrial technological change could enhance growth more than what environmental friendly technological change reduces emissions in each country. Reconciliation between emissions reductions policies that stabilize or decrease the path of pollution and emissions per capita convergence options is compatible with a scenario in which income and emissions levels are delinked. The EKC literature stresses that the relationship between levels of income per capita and emissions per capita is governed by three factors: the scale effect governed by the increase of economic activities that rises emissions; the technical effect that reduces emissions through technologies reducing carbon intensity; and finally the structural change effect reducing emissions through the decarbonisation of the economy driven by the transition from fossil fuel to fossil free economies.

Many contributions point out that for CO_2 the scale effect increasing emissions dominates the technological and the structural change effect by generating an increasing path of the income-emissions relationship. As we have seen before, this finding is confirmed by the inequality literature stressing that differences in the levels of income per capita govern differences in emissions per capita.

Our opinion is that the pursue of the emissions reducing and convergence in emissions per capita targets is compatible with future scenarios in which the technological and the structural change effect will dominate the scale effect and convergence in terms of GDP per capita for developing countries will be reached by harmonized green paths of growth among countries. For this purpose, as outlined before, it will be interesting to address the research to directions aimed at investigating synergies between technological change developed for "industrial" purposes and environmental friendly technology purposes.

Until now we have emphasized that technological convergence will affect emissions reduction, but the opposite causal relationship is also relevant. Whereas some interesting contributions emphasize that energy efficiency can be a win-win strategy for firms as it increases output and reduces pollution (Taylor *et al.*, 2008), others outline that environmental friendly technological change is essentially a process "induced" by policy (Gerlagh, 2006). Within an induced technological change perspective the setting up of environmental policies by public authorities to encourage the introduction of green technologies is crucial and represents the first step for the spreading of knowledge diffusion mechanisms. An important second step is the identification and removal of barriers to technological transfers. As outlined by a recent SBSTA (Subsidiary Body for the Scientific and Technological Advice) report (2006) a wide set of factors can represent obstacles to the diffusion of technological change in developing countries (see Table 2).

Though a wide literature focussed on the interaction between technological convergence processes and environmental policies, this issue is still controversial for many aspects and further research is needed to understand in depth their relationship.

Table 2. % representing the share (in a sample of 23 countries) of developing countries that identify a technological transfer barrier as crucial in the Technology Needs Assessment document submitted to the UNFCC (United Nations Framework Convention on Climate Change)

Economic and market barriers	83%
Policy barriers	78%
Technical barriers	74%
Human capacity	70%
Institutional barriers	65%

Source: SBSTA (2006)

CONCLUSION

In this book chapter we explained a new strand of research that is arising from climate change economics literature about the study of the international distribution of emissions per capita and its determinants. We stressed that whereas there are already some contributions dealing with this topic by investigating evidence about past trends, a new field of analysis concerns the study of the future evolution of emissions distribution over time.

Interestingly we argued that results concerning the past distribution of pollution generation at international level arising from the empirical evidence and those coming from simulations obtained by IAMs according to scenario analyses are consistent for many reasons. In fact past and "future" evidence both agree about a decreasing path of emissions per capita inequality over time, about a predominance of the "between group" component of inequality in the distribution of emissions when we analyse groups according to different income per capita levels and about the fact that GDP per capita rather than carbon intensity or emissions intensity is the main determinant to explain differences in emissions per capita between rich and poor regions. In particular, IAMs and empirical evidence agree on the fact that inequality in the distribution of income governs inequality in the distribution of emissions. From a policy point of view this finding is particularly interesting as it implies that the policy agendas about income and emissions distribution are strictly connected.

Moreover we explained the role that technological change will play in relation to targets concerning income and emissions distribution and specifically we made a distinction between technological change aimed at industrial development or at environmental friendly methods of production. We outlined that targets concerning income distribution, emissions distribution and emissions reduction will be likely to be in conflict as scenarios involving a convergence in income could heavily increase emissions and scenarios spreading abatement technologies would fail to reach an equal distribution of emissions per capita if income convergence were not possible.

The "first best" scenario would be that in which convergence in income per capita and diffusion of industrial technology are jointly reached and in which levels of emissions are delinked to the levels of income. In other words from a policy perspective the best scenario is the one in which there is wide international diffusion of industrial and environmental friendly technology and in which industrial technological change progressively generates opportunities to set up environmental friendly technological change. This is a very optimistic scenario that will very unlikely occur. Rather, policy makers should be prepared to tackle conflicting targets, but often they are not appropriately supported by research. The policy relevance of the EKC hypothesis is quite weak, because it is a concept dealing only with two policy dimensions: economy and environment. As mere example it does not say anything about the welfare levels that are associated to each income-emissions possible path. For this reason the EKC hypothesis represents an inappropriate tool to tackle sophisticated and articulated policy targets. Policy makers should be supported by scientific tools that take into account a wider set of policy targets. Multicriteria analysis in this context is an interesting tool (Janssen and Munda, 1999), but also the inequality literature referring to Atkinson's (1970) and Sen (1976) represents an interesting strand of research to deal trade off and complexities[8]. The way towards the reconciliation between science and policy is still ongoing, but fruitful directions exist and should be pursued with more effectiveness.

ACKNOWLEDGMENT

Emilio Padilla acknowledges support from projects SEJ2006-04444 (Ministerio de Ciencia e Innovación), 2005SGR-177 and XREPP (DGR).

Nicola Cantore acknowledges the support from the project "Modelli matematici per le decisioni economiche-finanziarie ed attuariali–Anno 2008".

REFERENCES

Alcántara, V., & Duro, J. A. (2004). Inequality of energy intensities across OECD countries. *Energy Policy*, *32*, 1257–1260. doi:10.1016/S0301-4215(03)00095-8

Altamirano Cabrera, J., Finus, M., & Dellink, R. (2008). Do abatement quotas lead to more successful climate coalitions? *The Manchester School*, *76*(104).

Ansuategi, A., & Escapa, M. (2002). Economic growth and greenhouse gas emissions. *Ecological Economics*, *40*, 23–37. doi:10.1016/S0921-8009(01)00272-5

Atkinson, A. B. (1970). On the measurement of inequality. *Journal of Economic Theory*, *2*, 244–263. doi:10.1016/0022-0531(70)90039-6

Barro, R., & Sala Martin, X. (1995). *Economic growth*. New York: McGraw-Hill.

Böhringer, C., & Löschel, A. (2003). Climate policy beyond Kyoto: Quo vadis? A computable general equilibrium analysis based on expert judgements. *Kyklos*, *58*, 467–493. doi:10.1111/j.0023-5962.2005.00298.x

Bosetti, V., Carraro, C., & Galeotti, M. (2006). The Dynamics of Carbon and Energy Intensity in a Model of Endogenous Technical Change. The Energy Journal, Special Ed.: Endogenous Technological Change and the Economics of Atmospheric Stabilisation.

Bosetti, V., Carraro, C., Massetti, E., & Tavoni, M. (2007). International energy R&D spillovers and the economics of greenhouse gas atmospheric stabilisation. *Energy Economics*, 2912–2929.

Buonanno, P., Carraro, C., & Galeotti, M. (2003). Endogenous induced technical change and the costs of Kyoto. *Resource and Energy Economics*, *25*, 11–34. doi:10.1016/S0928-7655(02)00015-5

Caniëls, M., & Verspagen, B. (2001). Barriers to knowledge spillovers and regional convergence in an evolutionary model. *Journal of Evolutionary Economics*, *11*, 307–329. doi:10.1007/s001910100085

Cantore, N. (2009). International spillovers and learning by doing in a regionalised model of climate change: a post-Kyoto analysis. In Marques, H., Soukiazis, E., & Cerqueira, P. (Eds.), *Integration and globalisation: challenges for developed and developing countries* (pp. 125–142). Cheltenham, UK: Edward Elgar Publishing.

Cantore, N., & Canavari, M. (in press). Reconsidering the Environmental Kuznets Curve Hypothesis: the trade off between environment and welfare . In Montini, A., & Mazzanti, M. (Eds.), *Environmental efficiency, innovation and economic performance*. New York: Routledge.

Cantore, N., Canavari, M., & Pignatti, E. (2008). *International distribution of CO_2 emissions according to the climate change integrated assessment models*. Paper presented at the AISSA (Congress of the Italian Societies of the Scientific Agricultural Associations), 26–28 November, Imola, Italy.

Cantore, N., & Padilla, E. (2007). *Equity and emissions concentration in climate change integrated assessment modelling*. DeiAgra WP-03-2007 and Working Paper 07-05, Department of Applied Economics, Univ. Autónoma de Barcelona.

Carraro, C., Eyckmans, J., & Finus, M. (2006). Optimal transfers and participation decisions in international environmental agreements. *The Review of International Organizations*, *1*, 379–396. doi:10.1007/s11558-006-0162-5

Coondoo, D., & Dinda, S. (2007). Carbon dioxide emission and income: a temporal analysis of cross-sectional distributional patterns. *Ecological Economics*, *65*, 375–385. doi:10.1016/j.ecolecon.2007.07.001

Duro, J. A., Alcántara, V., & Padilla, E. (2009) *La desigualdad en las intensidades energéticas y la composición de la producción*. Un análisis para los países de la OCDE, Working Paper 09.05, Departamento de Economía Aplicada, Universidad Autónoma de Barcelona.

Duro, J. A., & Padilla, E. (2006). International inequalities in per capita CO_2 emissions: a decomposition methodology by Kaya factors. *Energy Economics*, *28*, 170–187. doi:10.1016/j.eneco.2005.12.004

Duro, J. A., & Padilla, E. (2008). Analysis of the international distribution of per capita CO_2 emissions using the polarisation concept. *Energy Policy*, *36*, 456–466. doi:10.1016/j.enpol.2007.10.002

Engle, R., & Granger, C. (1987). Co-integration and error-correction: Representation, estimation and testing. *Econometrica*, *55*, 251–276. doi:10.2307/1913236

Esteban, J., Gradin, C., & Ray, D. (1999). *Extension of a Measure of Polarization, with an Application to the Income Distribution of Five OECD Countries. Papers 24*. El Instituto de Estudios Economicos de Galicia Pedro Barrie de la Maza.

Gerlagh, R. (2006). ITC in a Global Growth-Climate Model with CCS. The Value of Induced Technical Change for Climate Stabilization. *The Energy Journal, Special issue on Induced Technological Change and Climate Change*, 55-72.

Gerlagh, R., & Kuik, O. (2007). *Carbon leakage with international technology spillovers*. Nota di Lavoro FEEM 33.2007.

Grossman, G. M., & Krueger, A. B. (1991). *Environmental impacts of the North American Free Trade Agreement*. NBER working paper 3914.

Hedenus, F., & Azar, C. (2005). Estimates of trends in global income and resource inequalities. *Ecological Economics*, *55*(3), 351–364. doi:10.1016/j.ecolecon.2004.10.004

Heil, M., & Wodon, Q. (1997). Inequality in CO2 emissions between poor and rich countries. *Journal of Environment & Development*, *6*, 426–452. doi:10.1177/107049659700600404

Heil, M., & Wodon, Q. (2000). Future inequality in CO2 emissions and the impact of abatement proposals. *Environmental and Resource Economics*, *17*, 163–181. doi:10.1023/A:1008326515058

IEA. (2008). *International Energy Annual 2006*. Retrieved from http://www.eia.doe.gov/iea/

Janssen, R., & Munda, G. (1999). Multi-criteria methods for quantitative, qualitative and fuzzy evaluation problems . In Van den Bergh, J. (Ed.), *Handbook of environmental and resource economics*. Cheltenham, UK: Edward Elgar.

Kaya, Y. (1989). *Impact of Carbon Dioxide Emission Control on GNP Growth: Interpretation of Proposed Scenarios*. Paper presented to the Energy and Industry Subgroup, Response Strategies Working Group, Intergovernmental Panel on Climate Change, Paris, France.

Leimbach, M. (2003). Equity and carbon emissions trading: a model analysis. *Energy Policy*, *31*, 1033–1044. doi:10.1016/S0301-4215(02)00180-5

Miketa, A., & Schrattenholzer, L. (2006). Equity implications of two burden-sharing rules for stabilizing greenhouse-gas concentrations. *Energy Policy*, *34*, 877–891. doi:10.1016/j.enpol.2004.08.050

Millock, K. (2001). Technology transfers in the clean development mechanism: an incentive issue. *Environment and Development Economics, 7,* 449–466.

Nordhaus, W. (1994). *Managing the global commons, The Economics of Climate Change.* Cambridge, MA: MIT Press.

Padilla, E., & Serrano, A. (2006). Inequality in CO_2 emissions across countries and its relationship with income inequality: a distributive approach. *Energy Policy, 34,* 1762–1772. doi:10.1016/j.enpol.2004.12.014

Rao, S., Keppo, I., & Rihai, K. (2006). Importance of technological change and spillovers in long term climate policy. The Energy Journal, Special Ed.: Endogenous Technological Change and the Economics of Atmospheric Stabilisation.

Roson, R., & Bosello, F. (2002). Carbon emissions trading and equity in international agreements. *Environmental Modeling and Assessment, 7,* 29–37. doi:10.1023/A:1015218031905

SBSTA. (2006). *Synthesis report on technology needs identified by Parties not included in Annex I to the Convention, FCCC/SBSTA/2006/INF.1. Subsidiary Body for Scientific and Technological Advice to the United Nations Framework Convention on Climate Change.* New York: United Nations.

Sen, A. (1974). Information bases of alternative welfare approaches: Aggregation and income distribution. *Journal of Public Economics, 3,* 387–403. doi:10.1016/0047-2727(74)90006-1

Sun, J. W. (2002). The decrease in the difference of energy intensities between OECD countries from 1971 to 1998. *Energy Policy, 30,* 631–635. doi:10.1016/S0301-4215(02)00026-5

Taylor, R., Govindarajalu, C., Levin, J., Meyer, A. S., & Ward, W. A. (2008). *Financing Energy Efficiency: Lessons from Brazil, China, India and Beyond.* Washington, DC: The World Bank Group.

Tol, R., Downing, T., Kuik, O., & Smith, J. (2004). Distributional aspects of climate change impacts. *Global Environmental Change, 14,* 259–272. doi:10.1016/j.gloenvcha.2004.04.007

Vaillancourt, K., & Waaub, J.-P. (2004). Equity in international greenhouse gases abatement scenarios: A multicriteria approach. *European Journal of Operational Research, 153,* 489–505. doi:10.1016/S0377-2217(03)00170-X

Vaillancourt, K., & Waaub, J.-P. (2006). A decision aid tool for equity issues analysis in emission permit allocations. *Climate Policy, 5,* 487–501. doi:10.3763/cpol.2005.0538

WCED. (1987). *Our common future.* New York: United Nations.

Yamaji, K., Matsuhashi, R., Nagata, Y., & Kaya, Y. (1991). An Integrated Systems for CO2/Energy/GNP Analysis: Case Studies on Economic Measures for CO_2 Reduction in Japan. In *Workshop on CO_2 Reduction and Removal: Measures for the Next Century,* 19–21 March 1991. Laxenburg, Austria: International Institute for Applied Systems Analysis.

ENDNOTES

[1] The Kyoto Protocol designs as Annex B countries those which committed themselves as a group to reduce their emissions of greenhouse gases. Annex B includes developed countries and those in transition to market economies.

[2] The concept of polarization consists in examining the degree to which the observations of a distribution are allocated around different poles and therefore form significantly

homogeneous groups which are different between them. The EGR is a synthetic index of polarization developed by Esteban *et al.* (1999).

[3]   The Kaya identity decomposition is a well known finding in the environmental economics literature by which the levels of emissions per capita is expressed as the product of emission intensity of energy (quantity of carbon per unit of energy), energy intensity (quantity of consumed energy per unit of GDP) and GDP per capita.

[4]   In particular they use cointegration techniques (see Engel and Granger, 1987).

[5]   The Brazilian proposal is a scenario widely investigated in the climate change economics literature where the sharing of the global abatement burden according to the historical responsibility in generating the temperature increase.

[6]   The equalitarian rule is a scenario in which a global abatement burden is shared among regions by equalizing the levels of emissions per capita.

[7]   A work in progress paper by Cantore *et al.* (2008) outlines that in a model including endogenous technological change, international spillovers do not widely affect the distribution of emissions. Even if these results are already quite interesting because they signal that spillover could not be drivers of convergence across regions, more research is needed to confirm the results with different models assumptions and calibrations.

[8]   In particular they create indices dealing with scenarios presenting tradeoffs between income and income distributions, but they could interestingly be extended to incorporate other policy dimensions.

Chapter 16
Potential Ethical Concerns in Nanotechnology

Chi Anyansi-Archibong
North Carolina A&T State University, USA

Silvanus J. Udoka
North Carolina A&T State University, USA

ABSTRACT

Nanotechnology is science at the size of individual atoms and molecules. At that size scale, materials have different chemical and physical properties than those of the same materials in bulk. Research has shown that nanotechnology offers opportunities to create revolutionary advances in product development. It also has the potential to improve assessment, management, and prevention of environmental risks. There are however, unanswered questions about the impacts of nanomaterials and nanoproducts on human health and the environment. This chapter describes state-of the-science review, exposure assessment and mitigation, and potential macro ethical issues that must be considered to mitigate risk implications of emerging technologies such as nanotechnology.

INTRODUCTION

Nanotechnology is the science of manipulating extremely small particles in materials. It involves the manipulation of matter at the atomic and molecular level, and has the potential to make groundbreaking advances in technology, medicine, and green environmental initiatives (Balbus, 2005). 'The novel properties that emerge as materials reach the nanoscale (changes in surface chemistry, reactivity, electrical conductivity, and other properties) open the door to innovations in cleaner energy production, energy efficiency, water treatment, environmental remediation, and "lightweighting" of materials, among other applications, that provide direct environmental improvements" (Baum, 2003).

Although numerous studies have documented that nanotechnology can provide astounding benefits to humankind, there are others that suggest we must be aptly cautious before allowing nanotechnology based products to become part of our environment. There is concern that the novel properties that make nanotechnology so attractive

DOI: 10.4018/978-1-61692-006-7.ch016

may pose yet undiscovered risks to consumers of nanotechnology based products, occupational safety and health of the nanotechnology workplace, and the environment. This chapter seeks to highlight a variety of ethical concerns that arises from and must be adequately addressed to ensure that exposure of humans and our environment to nanotechnology is rigorously assessed and mitigated.

Effective discussion of any potential ethical issues relating to Nanotechnology must be preceded by definition and understanding of this phenomenon. There are several versions of definitions as well as claims on what nanotechnology can do and cannot do. This paper examines some select definitions and uses the identified characteristics to examine the potential ethical issues.

BACKGROUND

What is Nanotechnology? Frequently used in science and electronics, the prefix nano means one-billionth of a measure, such as a second or a meter. Nanoscience and nanotechnology generally refer to the world as it works on the nanometer scale, ranging from approximately one to one hundred nanometers (Bell, 2007). Nanotechnology is science at the size of individual atoms and molecules, with objects and devices measuring in billionths of a meter. At that size scale, materials have different chemical and physical properties than those of the same materials in bulk. Nanotechnology has potential applications across various sectors of the global economy, including consumer products, health care, transportation, energy and agriculture. The technology also promises new opportunities to improve how we measure, monitor, manage, and minimize contaminants in the environment (Bello, 2009).

One of the first to articulate a future with possibilities of nanotechnology was Richard Feynman, a Nobel laureate who died in 1988. He presented a lecture entitled "There is Plenty of Room at the Bottom," in December 29, 1959 at the California Institute of Technology. In this lecture, he was talking about nanotechnology before the word existed. Feynman discussed the problem of manipulating and controlling things on a small scale, a staggeringly small world, a technological vision of extreme miniaturization (IWGN, 1999, p. 4). Extrapolating from known physical laws, Feynman argued it was possible to write all 24 volumes 1959 edition of the Encyclopedia Britannica in an area the size of a pin head. He calculated that a million such pinheads would amount to an area of about a 35 page pamphlet. According to Feynman: "All of the information which mankind has ever recorded in books can be carried in a pamphlet in your hand, not written in code, but a simple reproduction of the original pictures, engravings and everything else on a small scale without loss of resolution" (Feynman, 1960).

Feynman discussed "[a] biological system [that] can be exceedingly small. Many of the cells are very tiny, but they are active; they manufacture substances; they walk around; they wiggle; and they do all kind of marvelous things–all on a very small scale. Also, they store information. Consider the possibility that we too can make a thing very small which does what we want—that we can manufacture an object that maneuvers at that level!" (Feynman, 1960, p. 6).

Nanotechnology holds exciting promise in medical applications, with possibilities such as targeted cancer therapies where cancer can be eradicated without making the rest of the body sick (Bell, 2007). A great example of such therapy is described in the work of Samuel Wickline, M.D., principal investigator of The Siteman Center of Cancer Nanotechnology Excellence (SCCNE) at Washington University, and his colleagues who developed nanobees to deliver toxic peptides such as melittin specifically to cancer cells while sparing healthy cells from the otherwise nonselective havoc these molecules cause. "The nanobees fly in, land on the surface of cells, and deposit their cargo of melittin, which rapidly merges with the target cells," said Dr. Wickline. "We've shown that the bee toxin gets taken into the cells where it pokes holes in their internal structures" (NCI, 2009).

"Nanoparticles are already widely used in certain commercial consumer products, such as suntan lotions, "age-defying" make-up, and self-cleaning windows that shed dirt when it rains. One company manufactures a nanocrystal wound dressing with built-in antibiotic and anti-inflammatory properties. On the horizon is toothpaste that coats, protects and repairs damaged enamel, as well as self-cleaning shoes that never need polishing. Nanoparticles are also used as additives in building materials to strengthen the walls of any given structure, create tough and durable, yet lightweight fabrics" [5]. Given these changes in chemical and physical properties of nano scale materials with effects yet to be fully understood, it becomes very imperative to protect human health and safeguard the environment through better management of potential risks arising from the exposure to nanoscale materials and their products.

As reported in the Environmental Protection Agency Nanotechnology White Paper,." EPA 100/B-07/001 of February 2007, researchers are also finding vital environmental applications of nanomaterials and are developing approaches that promote pollution prevention, sustainable resource use, and good product stewardship in the production, use and end of life management of nanomaterials (EPASPC, 2007). The white paper also recommends that the "Agency (US EPA) should draw on new, "next generation" nanotechnologies to identify ways to support environmentally beneficial approaches such as green energy, green design, green chemistry, and green manufacturing" (EPASPC, 2007, p. 2).

There is agreement among nanotechnology researchers that the concept is interdisciplinary and spans several disciplines across sciences and engineering. Among the many individual researchers and authors on the subject is the group – Center

for Responsible Nanotechnology (CRN) which efforts and goals are to create awareness and encourage responsible research relating to potential benefits and detriments of a "transformational power of manufacturing through nano - science and engineering". The focus of many of these studies is on Molecular Manufacturing.

The CRN recommended a series of thirty studies that it defined as very essential. It indicated that since there is a large unexpected transformational power of manufacturing, it was urgent to understand the issues raised by the concept. The thirty recommended studies were organized into five sections – Fundamental theory of Nanotechnology, Possible capabilities of the technology, bootstrapping potentials, product capabilities, and policy questions. CRN believed that it was urgent to understand several issues related to molecular manufacturing (MM) in order to prepare for its possible development in the near future. The group believed that the technology will be more transformative than expected, and could develop too rapidly for reactive policy to have an impact. They believe that MM is the result of convergence of many technologies and will benefit from synergies among them.

Several specific study titles under the five sections include: 1) A study of whether mechanically guided chemistry is a viable basis for a manufacturing technology; 2) What is the performance and potential of biological programmable manufacturing and products; 3) What types of applicable sensing, manipulation, and fabrication tools exist?; 4) What is the probable capability of the manufacturing system; and 5) What effect will molecular manufacturing have on military and government capability and planning, considering the implications of arms races and unbalanced development, as well as many other topics of study?

In his book, Engines of Creation, (1986) Eric Drexler used the term "Nanotechnology" to describe a new technology manufacturing at the molecular scale (Baum et al; 2003; Drexler, 1986). The author believed that tiny factories could be created and could assemble anything at the atomic scale. He believed that these "assemblers" would have molecular computers with blueprints for anything that is naturally possible and from which they would then construct things from raw materials (atoms). Assemblers, according to Drexler will be able to make anything from common materials without labor, replacing smoking factories with clean system. They will transform technology and the economy, opening a whole new world of possibilities. In addition to solving environmental problems, Drexler proclaimed that these "engines of abundance" could also extend human life, solve energy problems, and enable the colonization of space, just to name a few of the potential benefits (Drexler, 2001).

However, the question remains on how long it will take for these assemblers to be constructed. Some believe it will not require new science but a large engineering project in which all the basic knowledge is already available. (Baum et al, 2003). Nanotechnology advocates link it to artificial intelligence – where humans will merge with machines, uploading consciousness into computers with super computational capacity, with plans that these micro- and nano machines will lead to smart environments that can change instantly, a type of "Brave new world" (Kurzwell, 1999).

There was another group, associated with Drexler who opposed the above view of nanotechnology. They believe that such a view threatens humanity with extinction. In an essay, " Why the Future Doesn't need us", published in "Wired" magazine, Bill Joy, former head of Sun Microsystems, warns against the potential catastrophe that could result from the convergence of nanotechnology, genetics, and information science (Joy, 2000). Joy claimed that the potential risk is that the assemblers will run wild and replicate themselves uncontrollably, using biomass as raw materials. They will ultimately destroy the environment, including human life. Other researchers (Fukuyam, 2002; Mckibben, 2003) have joined Joy in calling for a moratorium on developing and using this new technology.

The more scientists learn about the subject, the more they suspect nanoscience and nanoengineering will become as socially transforming as the development of running water, electricity, antibiotics, and microelectronics. Many scientists, including physicist and Nobel laureate Horst Stormer of Lucent Technologies and Columbia University, are themselves amazed that the emerging nanotechnology may provide humanity with unprecedented control over the material world. Says Stormer: "Nanotechnology has given us the tools...to play with the ultimate toy box of nature—atoms and molecules. Everything is made from it...The possibilities to create new things appear limitless" (IWGN, 1999, p. 1)

Nanotechnology: Issues, Controversies and Problems

The vision for nanoscale science and technology as described in the National science Foundation (NSF) report – Nanotechnology: Shaping the World Atom by Atom (NSTC, 1999), involves understanding and manipulating matter at the atomic scale – "The emerging fields of nano-science and nano-engineering are leading to unprecedented understanding and control over the fundamental building blocks of all physical things. This is likely to change the way almost everything, from vaccine to computers to automobile tires to objects not yet imagined, is designed and made" (IWGN, 1999, p. 1).

Nanotechnology is as an interdisciplinary movement encompassing such disciplines as chemistry, physics, biology, engineering, toxicology, and environmental sciences. Although enriched by the confluence of these crucial disciplines, the field is still challenged by ability to accurately predict physico-chemical properties and risk/benefit boundaries (Tomalia, 2009). Predictive capabilities for risk/benefit assessment remain an urgent challenge for nanotechnology (Bell, 2007). This gap has far reaching implications and raises a host of ethical and social issues.

Occupational Safety and Health in Nanotechnology

Murashov et al (2009) discuss a variety of activities covering occupational safety and health of nanotechnology and the nanotechnology workplace, along with exposure assessment and mitigation carried out by The Organisation for Economic Cooperation and Development (OECD), an intergovernmental organization, based in Paris, France. OECD is playing a critical global role in ensuring that emerging technologies, such as nanotechnology, are developed responsibly, as part of its Environment, Health, and Safety program. One of the various committees set up by OECD was Working Party on Manufactured Nanomaterials (WPMN), established in 2006 to, among other things, proactively address issues such as risk assessment, regulatory programs and schemes to examine regulatory policies and how they address (or might address in the future) the risk assessment and risk management of nanomaterials. One of OECD's steering groups, U.S. National Institute for Occupational Safety and Health, is in the process of prioritizing and developing specific projects aimed at measuring and mitigating exposures in workplaces involving nanotechnology.

Although some nanoscale materials have been in existence in industry for many years, with a variety of uses, there is growing concern that engineered nanomaterials can be appreciably more toxic, with concerns that with their size, they have more propensity to penetrate into the human body than more conventionally sized materials (Tran, et al, 2009). Those currently receiving the highest levels of exposures to nanomaterials are people that are occupationally exposed in industry. There is thus an urgent need to understand the toxicity and intensity of workers' exposure to nanoparticles and fibers during manufacture of nano-enabled products.

Exposure to Nanoparticles during Manufacturing

As the production of nanomaterials increase with the explosion in the number of nano-enabled products, the number of workers directly in contact with manufactured nanomaterials correspondingly increases. In their study of exposure to manufactured nanostructured particles in an industrial plant, Demou, Peter and Hellweg (2008) sought to find association between exposure and corresponding risks, such as pulmonary inflammation and oxidative stress. These authors cited a number of studies that point to the fact that: inhalation is the main exposure pathway for particulate matter and the most critical for nanoparticle uptake; tendency for increased toxic effects per mass dose of ultrafine metal particles when compared to their larger counterparts; translocation capabilities of nanoparticles to target organs in the body such as the brain, spleen and liver. The results of their study demonstrated elevated concentrations of manoparticles during production that were orders of magnitude higher than control environments, indicating real-time worker exposure. Since toxicological properties of manufactured nanomaterials are not fully understood, there is a corresponding lack of information on the magnitude of the risks of worker exposure to nanoparticles during production.

There is a significant concern for worker safety during processing of nanocomposites (nanostructures in a matrix) and hybrid composites due to nanoparticle toxicity. Possible release of carbon nanotubes from composites that contain carbon nanotube, along with other nanoparticles and respirable fibers during their processing is the central theme of a study by Belo, et al. (2009). Their study investigated airborne exposures to nanoscale particles and fibers generated during dry and wet abrasive machining of advanced composite systems containing carbon nanotubes. This study found that dry cutting of base and carbon nanotube containing composites without emission controls resulted in high airborne exposures to nanoscale and fine particles as well as submicron respirable fiber. They concluded that since the toxicological profile of airborne nanoscale and fine particles and fibers generated during machining of carbon nanotube composites is not well known, care must be exercised to control such exposures.

In a study that monitored and analyzed the exposure characteristics of silver nanoparticles during a liquid-phase process in a commercial production facility, Park, et al (2009) they concluded that silver nanoparticles were released from the reactor during the liquid phase production process and silver nanoparticles were released into the atmosphere of the workplace. Silver is classified as an environmental hazard by US Environmental Protection Agency (EPA) because it is more toxic to aquatic plants and animals than any metal except mercury. According to the Nanotech Project (NTP, 2008) even if a nanparticle is not especially toxic, silver nanoparticles increase the effectiveness of delivering toxic silver ions to locations where they can cause toxicity.

According to Ostrowski- "Understanding the toxicity and toxicological consequences of nanomaterials and nano-enabled products is important for human and environmental health, safety, and public acceptance" (Ostrowski, et. al, 2009, p. 251). The authors of this review point to the fact that the scientific literature, the primary source of information on nanomaterial toxicology, by a large measure, agree that there is a lack of standard and systematic approaches to assessing the toxicology of nanomaterials, and that there is a need for standard methodologies to examine human health and environmental implications of novel nanomaterial characteristics. According to Dr. Samuel N. Luoma, the author of the Project on Emerging Nanotechnologies (PEN) report *Silver Nanotechnologies and The Environment: Old Problems or New Challenges?* "We need not assume that because nano is new, we have no scientific basis for managing risks," (NTP, 2008, p.

2) further highlighting the need for more research on the potential risks posed by nanomaterials.

Regulatory Issues with Applications of Nanotechnology

The novelty of the properties that make nanoscale materials so attractive in a wide field of applications may present unconventional risks to consumers, workers, and the environment. There is therefore an urgent need for robust oversight mechanisms to support safety in the manufacture and use of nanotechnology-enabled products. Without a systematic framework to regulate the activities in this burgeoning field, the full benefits of nanotechnology may never be realized.

The scientific literature is rife with evidence of a lack of clarity in approaches that are currently available to accurately assess risks and benefits of nanotechnology. This introduces an unacceptable level of uncertainty that affects efforts to develop regulations governing the nanotechnology industry. Regulatory agencies and policy makers globally are grappling with policies aimed at protecting workers and consumers from potential harmful effects of nanotechnology. Corley, Scheufele, and Hu (2009) in their study determined that the nature of nanotechnology makes it difficult for one agency to regulate all of nanotechnology research and applications. Thus multiple federal agencies regulate products that employ nanotechnology or nanomaterials, but there is no comprehensive framework to govern how they are regulated. They argue that with such a piecemeal approach, it is likely that certain technologies or products will not be fully regulated by any agency.

Since its creation in 1972, the U.S. Consumer Product Safety Commission (CPSC) hailed at that time as "the most powerful federal regulatory agency ever created." It has never lived up to these expectations, struggling since its inception to carry out its mandate: *to protect Americans from unreasonable risks associated with consumer products*. CSPC has been stretched too thin from years of neglect, underfunding and the challenges posed by an increasingly global manufacturing system "CPSC's inability to carry out its mandate with respect to simple, low-tech products such as Thomas the Tank Engine toy trains, Barbie dolls and Easy-Bake Ovens bodes poorly for its ability to oversee the safety of complex, high-tech products made using nanotechnology. The agency lacks the budget, the statutory authority and the scientific expertise to ensure that the hundreds of nanoproducts now on the market, among them baby bottle nipples, infant teething rings, teddy bears, paints, waxes, kitchenware and appliances, are safe. This problem will only worsen as more sophisticated nanotechnology-based products begin to enter the consumer market" (Fletcher, 2008, p3).

Ethical Issues and Responsible Nanotechnology

Two major areas of potential ethical concerns are discussed- micro- and macro-ethical issues as they relate to molecular manufacturing (MM). For the Micro or researcher-oriented integrity, we defer to the recent pronouncement from a major funding agency, the National Science Foundation (NSF). The requirements will be examined to identify the potential ethical issues that may relate to nanotechnology.

In August 2009, the NSF issued statements on the implementation of Section 7009 of the America COMPETES Act – America Creating opportunities to Meaningfully Promote Excellence In Technology, Education, and Science (COMPETES). This section of the Act requires that "each institution or group that applies for financial assistance from the Foundation for science and engineering research or education describe in its grant proposal a plan to provide appropriate training and oversight in the responsible and ethical conduct of research". It explains that the responsible and ethical conduct of research (RCR) is critical for excellence, as well as public trust, in

science and engineering. But the issue is not just in the training, it is an issue of personal integrity which goes beyond funding criteria.

The integrity and moral behavior of the researcher to collect and report accurate data is critical in nanotechnology. There is an explicit link between fact and value in nanoscience, thus ethics with its reflections on values goes at the root of techno sciences.

Because of the interdisciplinary nature of nanotechnology and the potential impact on society as a whole, ethical reflections must accompany its research every step of the way. According to Khushf, (2006) ethical reflection should be a defining future of nanotechnology, not just a statement of how ethical issues should be addressed. Potential micro ethical issues include but not limited to various research misconduct, data falsification, fabrication, and possible misinterpretation of data for personal gains.

The potential ethical concerns of Molecular Manufacturing for the society flows from the benefits that such a powerful "general purpose technology" brings to the society. With the benefits come some risks. Molecular manufacturing has been referred to as the industrial revolution compressed into a few years. It has potential to impact society and politics. It has the power to create a conflict among nations including a disruptive and unstable arms race. Weapons and surveillance devices could be made small, cheap, powerful, and very numerous. Reverse engineering, copying and inexpensive manufacturing could lead to proliferation of such devices.

Despite the reported vast potential of nanotechnology to develop and improve products and production processes, Simons, et al (2009) contends that the public at large has little knowledge of what nanotechnology is. This lack of knowledge is exemplified in the responses from respondents to a call for signatures for a petition by the Environmental Defense Fund (EDF), which posed the question: Unregulated Nanotechnology - Are Your Products Safe? The comments left by the respondents clearly documented the information void between nanotechnology researchers and the public. Comments such as "companies are so concerned with making money that they market products before we have all of the scientific data on their long-term effects. … put science and research ahead of profits and seriously investigate the effects of nanotechnology;" "I would like a list of all products using Nanotechnology so I can make the decision whether to use them or not. I don't want that decision made for me" formed a predominant thread in the respondents' reaction to the question.

Over 1,000 nanotechnology-enabled products are now available to consumers around the world, according to the Project on Emerging Nanotechnologies (PEN). Ethical dilemmas arise when there is a short-term profit motive driving nanomaterial research and subsequent infusion in consumer products. Some of the early applications of nano-materials are in the realm of cosmetics, where to date, there is no clear regulatory authority by the Food and Drug Administration (FDA) to review these products prior to market introduction. In such instances, consumers are unaware of the presence of nanomaterials in the product, and even if they are aware, they rely solely on the manufacturers of these products, without proper regulatory oversight to ensure their health and safety and that of the environment. In this construct, consumer protection from risk relies at present on ethical behavior by those who are rapidly exploiting nanotechnology for profit-making and an assumption that such particles are unlikely to be harmful to the user and his/her environment.

Given the above potential risks, it is imperative that regulatory bodies such as the Food and Drug Agency (FDA) as well as the funding agencies, NSF (National Science Foundation) and other private agencies, create a taskforce to assess and address the issue of potential risks. Identification of the major potential risks, especially in consumer products will create a platform for appropriate regulations and policies designed to control the risks. Consideration should also be given to the speed for possible proliferation of products with nanoparticles.

REFERENCES

Balbus, J., Denison, R., Florini, K., & Walsh, S. (2005). Getting nanotechnology right the first time. *Issues in Science and Technology*, 65–71.

Baum, R., Drexler, K. E., & Smalley, R. (2003). Nanotechnology: Drexler and Smalley make the case for and against molecular assemblers. *Chemical and Engineering News*, *81*(48), 37–42.

Bell, T. E. (2007). *Understanding Risk Assessment of Nanotechnology*. Retrieved August 12, 2009, from http://www.nano.gov/Understanding_Risk_Assessment.pdf

Bello, D., Wardle, B. L., Yamamoto, N., deVilloria, R. G., Garcia, E. J., & Hart, A. J. (2009). Exposure to nanoscale particles and fibers during machining of hybrid advanced composites containing carbon nanotubes. *Journal of Nanoparticle Research*, *11*, 231–249. doi:10.1007/s11051-008-9499-4

Corley, E. A., Scheufele, D. A., & Hu, Q. (2009). Of Risks and Regulations: How Leading U.S. Nanoscientists Form Policy Stances about Nanotechnology. *Journal of Nanoparticle Research*. Retrieved from DOI 10.1007/s11051-009-9671-5

Daily, S. (2006). *Nanotechnology? What's That?! Engineers Create Exhibits on Achievements, Promise*. Retrieved August 20, 2009, from http://www.sciencedaily.com/videos/2006/0611-nanotechnology_whats_that.htm

Demou, E., Peter, P., & Hellweg, S. (2008). Exposure to manufactured nanostructured particles in an industrial pilot plant. *The Annals of Occupational Hygiene*, *52*(8), 695–706. doi:10.1093/annhyg/men058

Drexler, K. E. (1986). *Engines of Creation: The Coming Era of Nanotecnology.* New York: Anchor Books, Doubleday.

Drexler, K. E. (2001). Machine-phase nanotechnology. *Scientific American, 285*(3), 74–75. doi:10.1038/scientificamerican0901-74

Environmental Defense Fund (EDF). (2009). *Nanotechnology and Health: Getting nanotechnology Right.* Unregulated Nanotechnology - Are Your Products Safe.

EPA's Science Policy Council (EPASPC). (2007 February). *Nanotechnology Workgroup, US Environmental Protection Agency Nanotechnology White Paper.* EPA 100/B-07/001.

Felcher, E. M. (2008). *The Consumer Product Safety Commission and Nanotechnology. Project on Emerging Nanotechnologies (PEN) Report.* Retrieved July 10, 2009, from http://www.nanotechproject.org/process/assets/files/7033/pen14.pdf

Feynman, R. P. (1960). *There's Plenty of Room at the Bottom: An Invitation to Enter a New Field of Physics.* Retrieved August 20, 2009, from http://www.zyvex.com/nanotech/feynman.html

Fukuyama, F. (2002). *Our Post Human Future: Consequences of Biotechnology Revolution.* New York: Farrar, Straus, and Giroux.

Joy, B. (2000). Why the future doesn't need us. *Wired, 8*(4), 238–262.

Kurzweil, R. (1999). *The Age of Spiritual Machines.* New York: Penguin Books.

McKibben, B. (2003). *Enough: Staying Human in an Engineered Age.* New York: Henry Holt & Co.

Murashov, V., Engel, S., Savolainen, K., Fullam, B., Lee, M., & Kearns, P. (2009). Occupational safety and health in nanotechnology and Organisation for Economic Cooperation and Development. *Journal of Nanoparticle Research - Special Issue: Environmental and Human Exposure of Nanomaterials.* Retrieved April 29, 2009 from DOI: 10.1007/s11051-009-9637-7

National Cancer Institute (NCI) Alliance for Nanotechnology in Cancer. (2009). *Nanotech News: Tumors Feel the Deadly Sting of Nanobees.* Retrieved August 17, 2009, from http://nano.cancer.gov/news_center/2009/aug/nanotech_news_2009-08-27a.asp

NSTC (National Science and Technology Council). (1999). *Nanotechnology: Shaping the World Atom by Atom.* Washington, DC: National Science and Technology Council. (NTP) Nanotech Project. (2008). *Nanoscale Silver: No Silver Lining?* Retrieved July 10, 2009, from http://www.nanotechproject.org/news/archive/silver/

Ostrowski, A. D., Martin, T., Conti, J., Hurt, I., & Harthorn, B. H. (2009). Nanotoxicology: characterizing the scientific literature, 2000–2007. *Journal of Nanoparticle Research, 11*(2), 251-257. Retrieved from DOI 10.1007/s11051-008-9579-5.

Park, J., Kwak, B. K., Bae, E., Lee, J., Kim, Y., Choi, K., & Yi, J. (2009). Characterization of exposure to silver nanoparticles in a manufacturing facility. *Journal of Nanoparticle Research.* Retrieved August 2, 2009, from DOI 10.1007/s11051-009-9725-8.

Seaton, A. L. T., Aitken, R., & Donaldson, K. (2009). Nanoparticles, human health hazard and regulation . *Journal of the Royal Society, Interface.* Published online 2 September 2009. doi:. doi:10.1098/rsif.2009.0252.focus

Simons, J., Zimmer, R., Vierboom, C., Harlen, I., Hertel, R., & Bol, G-F. (2009). The slings and arrows of communication on nanotechnology. *Journal of Nanoparticle Research.* Retrieved May 20, 2009, from DOI 10.1007/s11051-009-9653-7.

The Interagency Working Group on Nanoscience (IWGN). (1999). *Nanotechnology: Shaping the World Atom by Atom.* Washington, DC: National Science and Technology Council Committee on Technology. Retrieved from http://www.wtec.org/loyola/nano/IWGN.Public.Brochure/IWGN.Nanotechnology.Brochure.pdf

Tomalia, D. A. (2009). In Quest of a Systematic Framework for Unifying and Defining Nanoscience. Journal of Nanoparticle Research, 11, 1251–1310. Retrieved from DOI 10.1007/s11051-009-9632-z

Tran, L., Aitken, R., Ayres, J., Donaldson, K., & Hurley, F. (2009). Human Effects of Nanoparticle Exposure . In Hester, R. E., & Harrison, R. M. (Eds.), *Nanotechnology: Consequences for Human Health and the Environment* (pp. 102–113). Cambridge, UK: The Royal Society of Chemistry.

Section 5
Lessons from Agricultural Technology

This section uses agriculture and agricultural technology to help to understand technology adoption, diffusion and penetration trajectory as they pertain to developing nations. Agriculture being one of the most common industries offers lots of insights on the struggles many developing nations have faced in adopting new technologies. Where these nations have failed and succeeded in agricultural technologies could provide indicators on how well they can do on nanotechnology and microelectronics adoption. The lessons could provide the needed directions in developing plans for global transfer of emerging technologies like nanotechnology and microelectronics.

Chapter 17

Technological Change and the Transformation of Global Agriculture:
From Biotechnology and Gene Revolution to Nano Revolution?

Alejandro Nin-Pratt
International Food Policy Research Institute, USA

ABSTRACT

This chapter discusses the economic impact of science-based research in agriculture. Global agriculture was transformed in the 20th century by the Green Revolution that resulted from applying Mendelian genetics to crop and animal breeding. Developments of biotechnology in the last 20 years marked the dawn of a gene revolution that is thought to replace Mendelian genetics as the driver of technical change in agriculture. In recent years and still far from reaching the full potential impact of biotechnology in agriculture, developments in nanotechnology promise to further push the research and innovation frontier in agriculture. In this new environment, the private sector emerges as the main actor in agricultural R&D displacing the public sector, which played a central role during the Green Revolution period. However, more public investment in R&D rather than less and new institutions will be needed in developing countries if they are to benefit from the new technologies.

INTRODUCTION

The need for agricultural growth during the early stages of development has been at the center of debates in the development field. In early classical economic theory, agriculture was characterized as a sector with low productivity, traditional technology and decreasing returns, a sector that only passively contributed to development by providing food and employment. Because of being dependent on a fixed amount of land it was assumed that agricultural output couldn't increase proportionally with increases in labor supply under a given technology. On the demand side, and because agriculture was seen as the sector providing food for the satisfaction of basic needs, it was assumed that it was not possible for agriculture to grow faster than population in order to avoid stagnation

DOI: 10.4018/978-1-61692-006-7.ch017

in a Malthusian trap. In contrast with this view of agriculture, manufacturing appeared under the eyes of the early growth theorists as the sector with high productivity and increasing returns.

In part, this view of agriculture was based on the constraints that some of the specificities of agricultural production imposed on growth. Firstly, production in agriculture deals with live organisms and unlike any other sector in the economy it involves land as a key factor of production. The importance of land in agriculture introduces rigidities into the production process and strong interactions with the environment. For example, some of the production activities are seasonal, which determines that the demand for labor concentrates at a certain time of the year, with other periods where resources stand idle. The biological nature of agricultural production and its interaction with the environment also introduces a high degree of uncertainty on the final results of the production process. Secondly, and because land and live organisms are an integral part of agricultural production, the interaction with the environment makes this production process dependent on agro-ecological conditions, which are specific to a particular geographic location. Thirdly, the importance of land also introduces space-related economic factors affecting production decisions like the location of land with respect to markets, transportation costs for inputs and outputs, and infrastructure. All these factors affect the spatial allocation of production of different crops and livestock activities. Lastly, land as a factor of production brings the space dimension into the production process at the farm level: machinery, unlike in manufacturing, needs to move to the production site introducing new costs and technical challenges.

The view on the role of agriculture in the economy started changing with the transformation of the agricultural production process through the introduction of science-based technology. Early advances in mechanical and biological technologies showed that constraints in land or labor endowments could actually be overcome by fostering the links between science, agriculture and manufacturing. It was also found that agricultural growth could be transmitted to the rest of the economy through linkages with other sectors, for example, through a growing and cheaper supply of inputs to manufacturing, or by providing more and cheaper food to urban workers. Recent studies go beyond these linkages showing positive relationships between nutrition and economic growth, between enhanced food security and growth and between agricultural growth and poverty reduction.

As mentioned above, the potential impact of agriculture on economic growth and development was recognized when technological advances were incorporated in the production process. In fact, it was the development of science-based technologies that allowed the transformation of agriculture overcoming constraints like the fixed supply of land that were thought to impose rigid limits to agricultural growth. At present, the mainstream agricultural production methods and technologies that are being applied are the result of Mendelian genetics. The growing number of applications and developments of biotechnology applied to agriculture in the last 20 years marked the beginning of a new era in agricultural research. In recent years and still far from seeing the full potential impact of biotechnology in agricultural production, developments in nanotechnology offer to introduce changes in agricultural research and innovations that are as large as those promised by biotechnology 20 years ago, if not larger.

TECHNICAL CHANGE IN AGRICULTURE

There is an extensive literature that explains and analyses technical change in agriculture as the result of changes in factor prices, thus, looking at the process of technical change as endogenous to the economy. The theory of "induced technological innovation" states that "…a rise in the price of one factor relative to that of other factors induces a sequence of technical changes that reduces the use of the more expensive factor relative to the use of other factor inputs. As a result, the constraints on economic growth imposed by resource scarcity, are released by technical advances that facilitate the substitution of relatively scarce factors" (Hayami and Ruttan 1985). If for example, the endowment of labor becomes more abundant relative to land as a result of population growth and limited land availability, land becomes more expensive relative to labor. In this particular case, the theory of induced innovation predicts a change in technology towards using labor more intensively while saving land. This technical change that increases the use of one factor relative to the use of other factors is called "biased technical change," in opposition to "neutral" technical change where the proportion in the use of different factors by the new technology does not change. This bias change is the result of a demand of agricultural producers for new technologies and of the efforts of profit-seeking entrepreneurs that facing the demand for technical change, find incentives to invest in the development of new technologies that reduce production costs by substituting relatively more abundant resources for scarce resources (Hayami and Godo, 2005).

Even though the previous characterization of the theory of induced innovation explains technical change as the exclusive result of market forces and the supply and demand of new technology, Hayami and Ruttan (1985) have extended this framework to include the participation of public research institutions to explain agricultural development. In this case, changes in relative prices are still the major forces motivating producers and generating the demand for new technology. What changes is the relationship between the demand and supply of new technologies which does not necessarily occurs through the market, but results from interactions between agricultural producers, agro-industries, the government and public research institutions. Under this framework, the interaction between different interest groups and public institutes result in new institutions allowing society to internalize the benefits of innovative activities (de Janvry, 1973). According to Hayami and Godo (2005), this theory is also applicable to subsistence-oriented non-market economies if it is assumed that relative resource scarcities are recognized by producers as shadow prices reflecting the social opportunity costs of resources.

The importance of biased technical change and the specification of demand for innovations in agriculture can become more apparent using a classification that allows us to categorize different types of agricultural technologies according to their use of different factors and their effects on factor substitution. According to de Janvry (1973), technologies can be classified very broadly into four categories: 1) mechanical (tractor, harvester, windmill); 2) biological (hybrid seeds and cattle breeds); 3) chemical (fertilizers, insecticides and pesticides); and 4) agronomic (cultural practices and management techniques like crop rotation, changing the time for planting or harvesting, etc.). de Janvry also distinguishes between 'on line' management or the actual direction of farm activities and "staff" management, which deals with financial and fiscal administration, and commercial activities.

The four groups of technologies in de Janvry's classification can be characterized by their own impact on the marginal rates of technical substitution among capital, labor, land and management, and on yield level (the quantity of output per unit of factor of production, e.g. measured as the ratio output/land or output/labor). Table 1 presents this characterization of agricultural technologies and highlights their use of different factors.

Mechanical innovations making intensive use of machinery (tractors) require less labor (on-line management), more capital, and more staff management per unit of land. These technologies raise the yield per unit of labor, but have no major effect on the quantity of output produced per unit of land.

Table 1. Impact of different techniques on factor use

	Land	Labor		Capital	Yield	
		On-line	staff			
Machinery	(+ +)	(- -)	(+)	(+ +)	(0)	
Biological	(0)	(0)	(0)	(+)	(+)	
Chemical	(- -)	(+ +)	(+ +)	(+ +)	(+ +)	
Agronomic	(- -)	(+ +)	(+ +)	(+ +)	(+ +)	

Note: (++) highly intensive use of factor; (+) intensive use of factor; (- -) intensively factor saving; (-) factor saving; (0) no effect in the use of the factor.

Source: Elaborated by the author

Biological innovations are slightly capital using and moderately yield increasing if they are not combined with other inputs, while they are neutral on labor and management requirements. Chemical innovations are strongly yield-increasing, use capital and labor intensively (they require both more on-line and staff management per unit of land) and save land. Agronomic innovations like chemicals are strongly yield-increasing, saving land and using labor and on-line-management intensively. Finally, "technology packages" combining biological, chemical, and agronomic technologies as the one adopted in the so called "Green Revolution" tend to be labor using, land saving and very strongly yield increasing.

The classification of technologies in Table 1 shows that constraints imposed on agricultural development by scarcity of land may be offset by advances in biological technology, while the constraints imposed by an inelastic supply of labor may be offset by advances in mechanical technology. According to Hayami and Ruttan (1985), "the ability of a country to achieve rapid growth in agricultural productivity and output seems to hinge on its ability to make an efficient choice among the alternative paths. Failure to choose a path that effectively loosens the constraints on growth imposed by resource endowments can depress the entire process of agricultural and economic development."

The importance of changes in relative factor abundance and biased technical change is ubiq-

uitous in economics, given that in most cases, technical change is not neutral. In agriculture, technical change is motivated by changes in relative prices of capital and land, capital and labor, and labor and land as in the previous example. But the theory applies to other sectors as well and has been showing a growing importance in the economic literature in recent years. Acemoglu (2002) presents various examples of economic developments that need to be analyzed using a framework that accounts for biased technical change. Some of these examples are the fast increase in the US of the relative supply of skills (skilled labor expanding relative to unskilled labor and capital); the contrasting case of unskilled-labor bias in the nineteenth century where the artisan shops were replaced by the factory; and the high unemployment in continental Europe during the 1980s together with a steep decline in labor share, a change that was interpreted as being the consequence of capital-biased technical change.

THE GREEN REVOLUTION AND THE TRANSFORMATION OF AGRICULTURE

The term 'Green Revolution' usually refers to the development and diffusion of high-yielding cereal varieties, primarily wheat and rice, in the developing countries of the tropics and semi-tropics, beginning in the mid-1960s (Ruttan and

Binswanger, 1978). Other authors extend the term "Green Revolution" to include the release of hybrid maize in the United States in the 1950s (de Janvry et al. 2000). In any case and as pointed by Pingali (2007), the major technological break-throughs that started the Green Revolution in the 1960s were built on the advanced state of research on wheat and rice in the late 1950s (Evenson and Gollin, 2003). The origins of the wheat materials used in the 1960s in developing countries can be traced back to semidwarf wheat varieties devel-oped in Japan by 1873 and brought to the US in 1946. The first new rice varieties released by the International Rice Research Institute were based on genetic materials drawn from China, Taiwan and Indonesia while the model for the efficient rice plant promoted by the Green Revolution in Asia was based on the experience in developing improved rice varieties in Japan and Taiwan (Rut-tan and Binswanger, 1978).

Scientific Basis of Agricultural Technology in the Green Revolution

The technology promoted in developing countries under the Green Revolution was the development of high-yielding varieties of cereals. Conventional plant breeding approaches were responsible for the development of these modern varieties. Cross-ing plants with different genetic backgrounds and selecting among the progeny for individual plants with desirable characteristics, repeated over sev-eral generations, led to varieties with improved characteristics such as shorter stature, higher yields, improved disease resistance and improved nutritional quality (Pingali 2007).

The yield improvement in the new varieties of rice and wheat resulted from a reduction in plant height through the incorporation of genes for short stature. Prior to the Green Revolution, the varieties of wheat and rice available were tall, leafy, with weak stems and a maximum yield potential of 4 tons per hectare. When nitrogenous fertilizer was applied at rates exceeding 40 kilograms per hectare, these varieties yielded less than they would with lower fertilizer inputs because of ex-cessive growth. As a result of their short stature, the improved varieties are responsive to fertilizer inputs, thus, their maximum yield potential is 10 tons per hectare (Khush 1995).

Improved wheat and rice varieties were also developed for wide adaptability, selecting genetic materials insensitive to changes in day length and date of planting while yield stability was increased through the incorporation of genes for disease and insect resistance. Tolerance to some abiotic stresses, such as adverse soils and temperatures, and moisture stress, also contributed to the adapt-ability and yield stability of the new varieties. To exploit the full potential of the new varieties, genetic selection was complemented with research on management practices like formulation of suitable fertilizer recommendations; optimum rates, dates and methods of sowing; irrigation management; information for pest and weed control; and mechanization for land preparation, timely sowing, and harvesting (Khush 1992). In addition to improving yield potential and stability, and adaptability, the new varieties were selected for appropriate cooking quality, taste preferences, and milling recovery in rice, and baking and mill-ing characteristics in wheat.

The Institutional Model of Innovation and Diffusion

The diffusion of the technology promoted during the Green Revolution was the result of institutional innovation. By 1950, national agricultural research systems (NARS) had been established in only a small number of countries, primarily in Western Europe, North America, Oceania and the USSR and Japan. In developing regions, investment in agricultural research was confined mainly to major export crops while research in commodi-ties used by domestic consumers or produced by smallholders was lacking. In this context, the strategy to promote the Green Revolution and the international adaptation and diffusion of new crop technologies was built on the creation of international research centers. The new international agricultural research system (IARS) is funded by a consortium of private foundations, national aid agencies, and international development banks. They are governed by boards of trustees which, in turn, are responsible to the Consultative Group on International Agricultural Research (CGIAR) (see Binswanger and Ruttan 1978 for a thorough discussion of this process).

The first two international centers, the International Rice Research Institute (IRRI) and the International Maize and Wheat Improvement Center (CIMMYT), were established in 1959 and 1964 respectively to develop research on rice and wheat. The success of these first two international institutes had an important "demonstration effect" that led many national governments to accelerate the development of their NARS. Because of the strong interaction effects between national and international institutions the returns to investment in agricultural research were high compared with returns from other investment alternatives and by the mid 1970s a new set of national and international crop and animal research institutes had been established in Latin America, Asia and Africa with a dramatic increase in investment in agricultural research. These newly created international institutions developed a capacity to link the emerging national systems into an international network capable of developing, communicating and diffusing research methodology and its results and coordinate efforts to produce a continuous stream of agricultural technology suited to the environment of the developing countries (Binswanger and Ruttan, 1978).

According to Pingali (2007), one of the major benefits of the creation of an international system of agricultural research was to make available to NARS in developing countries the genetic materials developed over several decades of research in developed countries. Prior to 1960, there was no formal system in place that provided plant breeders access to germplasm[1] available beyond their borders. Since then, international public sector (the CGIAR) has been the predominant source of supply of improved germplasm developed from conventional breeding approaches. Evenson and Gollin (2003) suggest that germplasm contributions from international centers helped national programs to hinder the 'diminishing returns' to breeding that might have been expected to set in had the national programs been forced to work only with the pool of genetic resources that they had available at the beginning of the period.

The adoption of green-revolution technology was facilitated by investment and development of irrigation facilities, the availability of inorganic fertilizers, government policies facilitating the adoption, like investment in infrastructure development, and price support for outputs and inputs. Both CIMMYT and IRRI devoted a great many resources to training scientists and technicians from developing countries and from the agricultural universities established by national governments.

Impact of the Adoption of the New Technology

To better understand the impact and implications of the Green Revolution it is important to be aware of the climate and the discussion on food production and food security in the early 1960s. In his 1999 paper, Khush draws a good picture of this climate: "The 1960s was a decade of despair with regard to the world's ability to cope with the food-population balance, particularly in the tropics. The cultivated land frontier was closing in most Asian countries, while population growth rates were accelerating, owing to rapidly declining mortality rates resulting from advancements in modern medicine and health care. International organizations and concerned professionals were busy organizing seminars and conferences to raise awareness regarding the ensuing food crisis and to mobilize global resources to tackle the problem on an emergency basis." Khush also cites what was a famous book at that time entitled "Time of Famines" (Paddock and Paddock 1967). In this book, the authors predicted that "ten years from now, parts of the underdeveloped world will be suffering from famines. In 15 years, the famines will be catastrophic, and revolutions and social turmoil and economic upheavals will sweep areas of Asia, Africa and Latin America."

In this context, and given the pessimistic prognosis of the future in the 1960s, the impact of the Green Revolution was remarkable in several grounds. First, the new technology was massively adopted, although with geographical differences, adoption being faster and higher in Asia (Evenson and Gollin 2003). Second, this adoption had a dramatic impact on yields and production of wheat and rice. FAO data indicate that for all developing countries, yields rose 224 percent for wheat, 114 percent for rice and 171 percent for maize in the last 45 years (Pingali, 2007). Pingali and Heisey (2001) on the other hand, compiled yield and Total Factor Productivity (TFP) evidence for several countries and crops, and find that TFP trends are similar to the partial productivity trends captured by yield per hectare. According to Pingali (2007), wheat production in developing countries quadrupled since 1960 and rice production trebled, an expansion driven by yield increases with only a 35 percent increase in area for both crops, while in Asia, wheat production doubled and rice production trebled, in both cases with only a 30 percent increase in area.

Third, the Green Revolution had a significant impact in food security of many populated and at that time hunger prone developing regions mainly by affecting three major drivers of food security: output per capita, the price of food, and the incorporation of marginal land to production. The accelerated growth rate of rice and wheat production had surpassed the rise in population, leading to a substantial increase in cereal consumption and calorie intake per capita.

Figure 1 shows the impact of the Green Revolution in China and India reflected in the evolution of the number of hectares harvested. Had 1961 yields still prevailed today, three times more land in China and two times more land in India would be needed to equal 2006 cereal production. In other words, to produce 187 million tons of rice as China did in 2006 at the yield levels of 1961, 60 million extra hectares of land would have been required. According to Khush (1999), if the Asian countries attempted to produce a 1990 harvest at the yield levels of the 1960s, most of the forests, woodlands, pastures, and range lands would disappear, and mountain sides would be eroded, with disastrous consequences for the upper water shed and product lowlands, the extinction of wildlife habitats, and the destruction of biodiversity.

Evenson and Rosegrant (2003) estimate that without the CGIAR and national program crop germplasm improvement efforts, food production in developing countries would have been almost 20 percent lower requiring another 15-20 million hectares of land under cultivation in addition to at least 5 percent higher food imports; world food and feed prices would have been 35 to 65 percent higher, and child malnutrition would have gone up by 6-8 percent - affecting some 30 to 45 million more children than otherwise.

Finally, several studies have shown that growth in the agricultural sector has economy-wide effects. Some of the effects from agricultural growth were positive long term effects from the rapid growth in rice production on land and labor markets and the non-agricultural sector in a village in the Philippines (Hayami et al. 1978); development of backward and forward linkages from increased agricultural productivity growth in India (Hazell and Ramaswamy 1991), and similar evidence for Africa (Delgado et al. 1998).

Pingali (2007) also discusses some of the negative aspects that have been attributed to the Green Revolution in the past. He concludes that while the benefits of agricultural productivity growth have been shared widely, some social

Figure 1. Land areas needed to obtain the rice harvests obtained in 1961-2006 in China and India with 1961 yields compared to land areas actually planted to obtain the same production with current yields (in millions of hectares) (Source: Author based on FAO 2009 data)

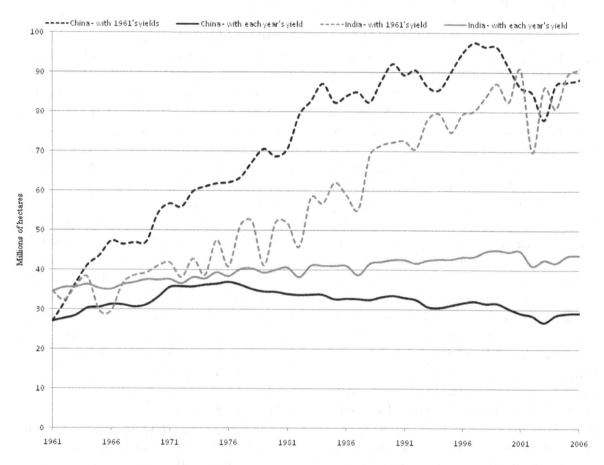

groups have gained relatively less than others, mainly landless labor, female headed households, and farm households in marginal environments. Other negative effects of the Green Revolution are the negative environmental effects through genetic erosion and chemical run-offs (de Janvry et al. 2000).

Despite the evidence showing favorable impacts in developing countries, a long debate has been taken place between advocates and opponents of the Green Revolution on the cost and benefits of the spread of the high yield variety-fertilizer technology. Buttel, Kenney and Kloppenburg (1985) present a good summary of this debate.

Proponents of the Green Revolution have argued that the benefits that resulted from the transfer of Green Revolution practices have far outweighed the costs of introducing this technology in developing countries. Green Revolution detractors, on the other hand, have rejected the assumption that developing nations can or should develop along the same path as the present industrial/high-income countries. These critics have argued that the Green Revolution strategy has exacerbated class inequality leading to "premature" rural emigration and urbanization.

Ruttan and Binswanger (1978) dismissed some of the early criticisms because, they argue, they appeared to be ideologically motivated. They state that hope for the radicalization of the lower peasantry and landless laborers has been viewed as dependent on the continuation of the process of immiserizing growth. The radical critics have welcomed the Green Revolution in socialist economies like those of Cuba and China but in non-socialist economies the introduction of high-yielding varieties was viewed as raising the cost of radical change by channeling new income to the middle and upper peasantry.

In recent years the positions taken by advocates and critics have become somewhat less polarized according to Buttel, Kenney and Kloppenburg (1985). They point out that advocates have recognized the problems that can be caused by rapid market penetration and by the tendencies toward biased development and applications of practices built around high-yielding varieties and petrochemical inputs. This has led to an increased targeting of the smallholder sector for special research and extension attention by development organizations and related agencies. Buttel, Kenney and Kloppenburg (1985) also assert that on the other side of this dispute, critics have come to recognize the importance of stimulating productivity improvements in developing regions of the globe probably disillusioned by the failures of "appropriate technologies" to provide productivity improvements and by the increasingly remote prospects for meaningful land reform. But even as this concession is made, many analysts speculate that the yield-enhancing potential of mechanical and petrochemical inputs may be already exhausted, and there is widespread concern that past levels of agricultural performance may be increasingly difficult to sustain. Attention turns now on recent advances in applied genetics and molecular biology, which appear to contain the potential to undergird a new era of productivity gains in agriculture in both developed and developing nations.

Biotechnology and the Gene Revolution in Agriculture

Until the 1990s, the paradigm[2] of agriculture innovation and technical change was centered on the development and distribution of high-yield seeds, highly productive animals and the inputs to make them grow and produce to their full potential (Raney and Pingali, 2007). The gravity center of agricultural technology innovation was the National Agricultural research System using conventional methods of selective breeding and the crossing of different parental lines to produce varieties and hybrids with desirable characteristics.

The scientific foundations for this paradigm were established late in the 19th century and the early decades of the 20th century, and was the research developed on these foundations what allowed most developed countries to achieve a Green Revolution in the first half of the 20th century. This technology was then extended to developing countries in the second half of the century with the creation of a "public" international research system interacting with national research systems in developing countries.

The science based traditional breeding techniques behind the Green Revolution have been used since the late 19th century to improve crops by transferring genes that code for desirable traits such as disease and insect resistance from one variety to another. These traditional breeding techniques involve crossing two organisms mixing thousands of genes, which in addition to providing the beneficial trait also result in the loss of other traits of that organism considered to be valuable. Significant time and effort, usually several generations or planting seasons, is required to produce a plant or animal carrying the beneficial traits using these traditional breeding techniques.

By the end of the 20th century we witnessed a change in the scientific foundations of agricultural technology as a result of new developments in genetics, a science that was radically changed after

James D. Watson and Francis Crick describe the structure of deoxyribonucleic acid (DNA). Since then, knowledge and advances in biology have increased exponentially and modern biotechnology was born with the creation of the recombinant DNA technology in the late 1970s and 1980s.

In its present usage, the term biotechnology refers to genetic engineering, a cluster of techniques of which the most powerful is DNA recombination. This technique allows to directly alter the structure and characteristics of genes of an organism, including the insertion of genetic material from one organism into the genetic code of another. This means that the new biotechnology techniques can be used to introduce desired traits into an organism as traditional breeding techniques do, however, genetic engineering brings in two main changes into this process that have vast implications for agricultural technology. First, genetic engineering is more precise, allowing food developers to add or enhance useful traits more specifically by transferring only desirable genes (e.g. for weed and pest resistance, enhanced nutrition, or longer shelf life). This means that a plant can be created that is the same in all respects to the original plant except for the addition of the beneficial trait. Second, there is no need to cross two organisms to obtain the desired genes. The recombinant DNA technique has made possible to simply insert a gene in the organism the source of which can be from any living organism thereby causing the 'genetically engineered' organism to exhibit a trait uncharacteristic of natural members of the species. This possibility allows getting around conventional barriers of genetic incompatibility opening extraordinary production possibilities not only in agriculture but also in manufacturing: plant and animal varieties incorporating useful characteristics of other varieties or species, microorganisms programmed to manufacture large quantities of useful chemical substances not easily or economically extractable from their natural sources, or bioengineered bacteria that efficiently convert one organic chemical to another more valuable one.

The importance and implications of these developments for agriculture at first sight appear to be enormous. According to Raney and Pingali (2007), we could be witnessing a nascent "Gene Revolution." Transgenic crops are spreading faster than any other agricultural technology in history, despite continuing controversy about potential risks such as gene flow (the escape of inserted transgenes into related crops or wild plants), the emergence of resistant pests, and fears that eating genetically modified foods might affect the health of consumers. The U.S. and Canada grow the bulk of transgenic crops (56 percent by area cultivated). The Southern Cone of South America is the second major producing area in the world. Argentina, Brazil, Paraguay, Uruguay and Bolivia together plant one-third of total cultivated area in the world, while China and India together contributed 10 percent to total planted area.

Characterization and Scientific Basis of the New Technology

According to Zilberman et al. (1997), most of the experimentation with new biotech products has concentrated on pest control products which are varieties designed to resist insect and viral pest or tolerate broad-spectrum herbicides. These crops account for most of the biotechnology crops available commercially. *Bt* maize, potato, and cotton incorporate select genes from the Bacillus thuringiensis to resist the European corn borer, Colorado potato beetle and pink boll worm, respectively. The Bt genes allow the crop to produce the pesticide within the plant, eliminating the need to spray for these pests. Other commercial plants that have been modified to resist viral infection include squash, cucumber, watermelon, and papaya. These plants resist viruses through a mechanism known as cross-protection, which is somewhat similar to immunization. Farmers growing these plants are able to reduce pesticide applications to control virus-carrying insects.

Soybean, maize, canola and other crop plants have been modified to tolerate safe, broad-spectrum herbicides. Herbicide tolerance allows farmers to use weed controls more selectively. Rather than applying herbicide before planting, farmers can wait until after the crop emerges to apply herbicides only where and in the quantities needed.

The commercial success of this line of products, according to Zilberman et al. (1997), is due to the relative simplicity of the genetic manipulation that they entail and the fact that they seem to cost-effectively meet a need. These products offer new market opportunities to seed companies that generally have a relative advantage in biological processes, with some of the major seed companies expanding their capacity in pest control, and the boundaries between pest control companies and seed companies gradually eroding.

A second group of biotech products is the one incorporating quality modifying innovations (Zilberman et al. 1997). Biotechnology techniques allow modification of agricultural products to enhance desirable characteristics. The most well known example is that of "Golden rice." Using biotechnology techniques, scientists have developed a new strain of rice, called Golden rice, that naturally produces beta-carotene, the precursor to vitamin A. Golden rice can provide enough beta-carotene to make up vitamin-A deficiencies in the diets of poor children, and it can also increase the amount of vitamin A in breast milk, an important source of nutrition for infants. This same strain of rice has also been enriched with additional iron to combat anemia, which affects hundreds of millions of the world's poor. A second variety of Golden rice, named Golden rice 2, was created by Syngenta, a biotechnology company in the private sector.. The production of beta-carotene in golden rice 2 is 23 times as much as that in the original golden rice. This beta-carotene enriched rice, created in 2005, will be available to any who wish to use it, as the private company has given out access to the technologies without cost.

There has been less research and investment in biotechnology to modify quality than to address pest control attributes: pest control biotechnology is being pursued mainly by established companies, while experimentation with quality augmenting biotechnology is often done by start-up companies and university researchers.

The Institutional Model of Innovation and Diffusion

Among the multiple differences between the Green and Gene revolutions, the defining feature of the Gene Revolution is its predominantly private character (Evenson 2002, Buttel, Kenney and Kloppenburk 1985). During the Green Revolution period, 1960s through the 1980s, private sector investment in plant improvement research was limited, particularly in the developing world, due to the lack of effective mechanisms for proprietary protection on the improved products. According to Pingali (2007), this situation changed in the 1990s with the wider availability of hybrids that led to the development of the seed industry in the developing world. Buttel, Kenney and Kloppenburk (1985) highlight the fact that the Green Revolution was spearheaded by the IARCs organized under the auspices of the Consultative Group on International Agricultural Research which institutionally played the role of a an international public sector, but at present, industry finds itself in a superior technological position to develop biotechnical products vis-à-vis the IARCs and the national programs of the development countries. While this has brought a surge of new investment into agricultural research it has also increased the difficulties faced by developing countries in accessing new technology. Access problems are particularly severe for poorer countries with small markets and weak institutions, traditional targets of public sector research on "public goods" (Pingali 2007). This could have a high negative impact for developing countries given that the economic penalties and rewards associated with differential access to Green Revolution technology will also be an issue for the Gene Revolution (Evenson, 2002).

The literature on this issue raises three main factors that explain the predominance of the private sector in developing the new technology. First, the private sector is in a much better position at present than during the years of the Green Revolution to appropriate benefits from agricultural research. One of the main reasons for this is the strengthening of intellectual property rights for artificially constructed genes and genetically modified plants. Proprietary protection together with a declining market for pesticides (Conway, 2000) provided incentives for private sector entry into plant improvement research which resulted in investment of the large multinational agrochemical companies early in the development of transgenic crops by purchasing existing seed companies, first in industrialized countries and then in the developing world. According to Pingali (2007), these changes have transformed the system through which agricultural technologies are developed and disseminated.

A second reason explaining the predominance of the private sector is that the new institutional model of generation and transfer of biotechnology in agriculture is difficult to replicate by the public sector. Pingali and Traxler (2002) describe this process as including three steps, starting with the generation of basic knowledge on useful genes and the insertion of these genes into plants. In the second step the modified genetic material moves downstream to adaptive breeding, through where it is back-crossed into commercial lines, and finally to delivering the seed to farmers. The most sophisticated, human and physical capital intensive is the first stage for which big multinational firms have advantages, with research costs and research sophistication declining in the progression towards downstream activities. This results in a clear division of labor with the multinationals providing the upstream biotechnology research and the local firms providing crop

varieties with commercially desirable agronomic backgrounds. Within this scheme, multinational firms control the new genetic material created in the first stage of the process through patents and are able to transfer these material across borders looking for commercial applications in different countries. The public sector, on the other hand, has limitations for direct country-to-country transfer of technologies because public sector research programs are generally established to conform to state or national political boundaries (Traxler and Pingali, 2002), which severely curtails spillover benefits of technological innovations across similar agro-climatic zones. In the past, the operation of the CGIAR germplasm exchange system has mitigated this problem for conventional breeding of several important crops, but Pingali (2007) fears that it is not clear whether the system will work for biotechnology products and transgenic crops, given the proprietary nature of the technology.

Finally, the growing role of the private sector in agricultural R&D results from an erosion of the national and international public systems. I summarize here part of the discussion on this issue in Pingali (2007). One of the reasons of the erosion of the public sector is the low R&D budgets of the public sector in many developing countries. Few developing countries have highly sophisticated biotech research programs except for Brazil, China and India which have extensive research programs in all areas of biotechnology, including advanced genomics and gene manipulation techniques. These countries followed similar paths to have a jump start in the biotech research, starting with an early investment in biotechnology R&D, heavy investment to develop the regulatory frameworks and capacity for intellectual property rights, food safety and environmental protection that are essential for the successful commercialization of transgenic crops, and cooperation between national and multinational companies to develop technologies (Pingali 2007). The case of China is illustrative of this strategy. China started her biotechnology research programs in the mid-1980s and was the first Asian country to commercialize a transgenic crop – insect resistant cotton. Private joint ventures were developed aimed at spreading biotechnology to poor farmers, mostly between American multinational firms and seed companies in China. Under these joint venture contracts, the multinational firm supplies the *Bt* gene and the cotton varieties while the Chinese partners provide the variety testing, seed multiplication, and seed distribution networks in their respective provinces and beyond (Pingali 2007). In this way, the US companies obtain the local knowledge they needed to have their genetically modified (GM) cotton varieties approved by the national biosafety (food safety and environmental) committee for commercial production, while the provincial seed companies acquire access to the *Bt* gene construct and the improved germplasm held by the US companies. The government of China has also invested heavily in developing the regulatory frameworks and capacity for intellectual property rights and biosafety protection that are essential for the successful commercialization of transgenic crops (Pingali, 2007).

According to IFPRI (2004), in addition to these countries, there are an additional 12 developing countries with significant and growing capacity for bio-technology research, but these are typically more advanced or middle-income countries such as Argentina and Egypt which are developing the capacity to adapt and adopt innovations developed elsewhere. However, most of the least developed countries have no research experience with genetically modified organisms and many have limited capacity for agricultural research of any kind (Pingali, 2007).

The erosion of the public sector related to agricultural R&D started with the publicly funded land-grant universities in the US. One of the reasons for this is that private universities, traditional outside the agricultural research community have programs in molecular and cell biology that are superior to those of the land-grant universities, leading to a growing number of contracts between multinationals and private universities to develop corporate-sponsored research.

A second reason for the erosion of the public sector is the fiscal problem in many States in the US and the austerity policies that came with those problems. National and state governments have limited the ability of the land-grant universities to maintain even conventional breeding programs, much less expand their biotechnology research efforts.

Finally, the passage of the Plant Variety Protection Act in the US and the Supreme Court decision permitting the patenting of genetically modified life forms have greatly increased the attractiveness to private industry of both conventional and biotechnological modes of plant breeding. The land-grant universities, key partners of the CGIAR institutes in the development and diffusion of the Green Revolution technology are under increasing pressure to withdraw from releasing finished plant varieties and to limit their efforts to maintaining, evaluating, and improving germplasm (Buttel, Kenney and Kloppenburg 1985).

The scenario for the international diffusion of the biotech technology has changed dramatically compared to the situation prevailing during the Green Revolution years. At present the IARCs no longer have a strong, unrivaled public agricultural research sector in the United States to rely on for crucial expertise in the new technology. Moreover, the IARCs themselves are facing difficulties similar to those confronted by the land-grant universities with the prospect that a significant part of the new technology being developed will bypass the international centers and national programs and occur under the orbit of private capital and competitive markets. Examples of these are already happening in China, India and other countries as discussed above (Buttel, F.H., M. Kenney and J. Kloppenburg 1985).

Another factor eroding the leading role of the IARCs in the international system of R&D and technology diffusion is the very expensive nature of the instrumentation, facilities, and, above all, personnel required for biotechnology research, development, and production in biotechnology. In particular, scaling up for production is very costly because recombinant DNA facilities entail a large investment and even much higher investment will be required to create the infrastructure for production for the market (Buttel, Kenney and Kloppenburg 1985). According to Byerlee and Fischer (2001), private sector investment in agricultural research at the beginning of the decade exceeded the combined investments of all public sector research institutes in the world dwarfing the investments in public research targeted to developing country agriculture. To show the importance of the differences between investments in the private and public sector, Byerlee and Fischer (2001) present data showing that the world's top ten multinational corporation investing in bioscience had a total annual expenditure on agricultural R&D that was nearly US$3,000 million, of which about half was devoted to biotechnology. In comparison, the CGIAR was spending at that time less than US$300 million annually of which less than 10 percent went to biotechnology. During the same period, the public sector agricultural research programs in Brazil, China, and India, the largest programs in the developing world, had annual budgets of less than $500 million each, with less than 10 percent devoted to biotechnology (see also Pingali 2007).

Impact of the Adoption of Agricultural Biotechnologies

Several issues that have been in the concern of governments in developing and developed countries, development agencies, NGO's and producer and consumer organizations around the world, among others need to be considered in an evaluation of the impacts of the diffusion and adoption of the agricultural biotechnologies. The first issue is the extent to which these technologies are benefiting small agricultural producers around the world and in particular in developing

countries. Equally important has been the concern about how the institutional arrangements that privilege the private sector in the new model of technology generation and diffusion are affecting different economic agents. In particular, it is important to see up to what extent, the central role of private multinationals in development of the new technologies resulted in market power changes significantly shifting the benefits of the new technology towards multinationals. A third concern is related to international trade and the impact of the adoption of the new technology on international agricultural markets. Finally, we look at the impacts of the adoption of agricultural biotech on health and the environment.

Global adoption of transgenic crops reached 125 million hectares in 2008 from 67.7 million in 2003 and 2.8 million in 1996. The number of countries electing to grow biotech crops has increased steadily from 6 in 1996, the first year of commercialization, to 25 in 2008. In 2008, developing countries out-numbered industrial countries by 15 to 10, and this trend is expected to continue in the future with 40 countries, or more, expected to adopt biotech crops by 2015, the final year of the second decade of commercialization (ISAAA 2008).

Traxler (2004), evaluating the first decade of availability of the new technology, finds that delivery of the technology has occurred almost entirely through the private sector and adoption has been rapid in areas where the crops addressed serious production constraints and where farmers had access to the new technologies. By 2003, three countries (USA, Argentina and Canada), three crops (soybean, cotton and maize) and two traits (insect resistance and herbicide tolerance) accounted for the vast majority of global transgenic area. Traxler also finds that while some farmers in some developing countries are benefiting, most do not have access to transgenic crops and traits that address their needs. Although a high degree of temporal and spatial variation was found, Traxler concludes that farmers who adopted the transgenic varieties experienced higher effective yields (due to less pest damage), higher revenue and lower pesticide costs. These factors more than compensated for higher prices paid for transgenic seeds so that net profits increased for adopters. This explains the fact that transgenic crop varieties have delivered large economic benefits to farmers in some areas of the world between 1996 and 2003. In several cases the per hectare savings, particularly from *Bt* cotton, have been very large when compared with nearly any technological innovation introduced over the past few decades (Traxler 2004).

GM technology has had a very positive impact on farm income derived from a combination of enhanced productivity and efficiency gains. According to Traxler (2004), the direct global farm income benefit from GM crops was $4.8 billion. If the additional income arising from second crop soybeans in Argentina is considered, this income gain rises to $6.5 billion. This is equivalent to adding between 3.1 percent and 4.2 percent to the value of global production of the four main crops: soybeans, maize, canola, and cotton, a substantial impact. The largest gains in farm income have arisen in the soybean sector, primarily from cost savings, where the $4.14 billion additional income generated by GM soybeans in 2004 has been equivalent to adding 9.5 percent to the value of the crop in the GM-growing countries, or adding the equivalent of 6.7 percent to the $62 billion value of the global soybean crop. Substantial gains have also arisen in the cotton sector through a combination of higher yields and lower costs. The 2004 income gains are equivalent to adding nearly 12 percent to the value of the cotton crop in these countries, or 5.8 percent to the $28 billion value of total global cotton production (Traxler 2004)

Even more significant than the gains reported above is the fact that the economic evidence available to date does not support the widely held perception that transgenic crops benefit only large-scale farmers. Smale et al. (2009) find that rising adoption rates in developing countries are a fact,

as are the high adoption rates of herbicide-tolerant soybeans across Latin America. Also, *Bt* cotton is found to reduce pest damage and insecticide use and increases yields, particularly in India (Smale et al. 2009).

Having shown the importance of profitability for the adoption of the new technology, Traxler observes (mainly in Raney, 2006; Smale, et al., 2006) that "institutions – national agricultural research capacity, environmental and food safety regulations, intellectual property rights and agricultural input markets – matter at least as much as the technology itself in determining the level and distribution of economic benefits" (Traxler 2004).

One of the lessons derived from these studies is that the institutional context in which the technology is introduced is a key determinant of the adoption of biotech varieties by small-scale farmers. This institutional environment is far more important than the effectiveness of any particular transgenic trait. According to Pehu and Ragasa (2007), the quick and extensive adoption of *Bt* cotton in China was the result mainly of publicly developed *Bt* cotton varieties and to the decentralized breeding system, which could quickly transfer the Bt gene to locally adapted cultivars that could be sold at a low price. These authors report as a contrasting experience that the institutional support in South Africa, especially for credit, failed to complement the technological success. The high profitability experienced by early adopters was superseded by a drastic decline in the number of smallholder *Bt* cotton producers. Pehu and Ragasa (2007) conclude that a certain level of national research and regulatory capacity is a prerequisite for success, along with an effective system for supplying inputs, especially seed. This means that where institutional conditions are weak, as is the case for the poorest farmers in the poorest countries ensuring access will remain a formidable challenge.

Another important conclusion that can be drawn from the analysis of research studies evaluating the impact of the diffusion of biotechnology in agriculture of the past 20 years is that the benefits generated by the adoption of biotech crops are being shared by consumers, technology suppliers and adopting farmers. No support was found for the fear that multinational biotechnology firms are capturing all of the economic value created by transgenic crops (Qaim and Traxler 2005). On the other hand, non-adopting farmers are being penalized, as their competitors achieve the efficiency gains they are denied (Traxler 2004).

One of the problems that a country adopting agriculture biotechnology can face is potential export losses because of the fragmentation of international markets and the ban on imports of these products imposed by many regions, the most notorious of which is the European Union (EU). In a paper using an empirical economy-wide model of the global economy (GTAP) to quantify the effects of selected countries enjoying an assumed degree of productivity growth from adopting GMO maize and soybean, Anderson and Nielsen (2002) conclude that a ban on imports of crops from GMO-adopting countries, would be very costly in terms of economic welfare for the region imposing the ban, the European Union itself. Imposing a ban hinders European consumers and intermediate demanders in gaining from lower import prices, domestic production of corn and soybean is forced to rise at the expense of other production, and hence overall allocative efficiency in the region is worsened. Another conclusion from Anderson and Nielsen's study is that the GM-adopting regions still enjoy welfare gains due to the dominating positive effect of the assumed productivity boost embodied in the GM crops, but those gains are reduced by the import ban as compared with the scenario in which GM crops are traded freely.

With respect to the impact of biotech crops on the environment, studies so far have shown that the effects of *Bt* cotton and genetically modified soybeans have been strongly positive, as insecticide use on *Bt* cotton is significantly lower than on conventional varieties and glyphosate has been substituted for more toxic and persistent herbicides in soybeans (Traxler 2004). Similarly Pehu and Ragasa (2007) report that: "to date, the data available provide no scientific evidence that large-scale cultivation of transgenic crops has harmed the environment. There has been no accumulation of *Bt* toxins after several years of cultivation; neither laboratory nor field studies have shown lethal or sub lethal effects of Bt toxins on non-target organisms; monarch butterflies have not been driven to extinction, and "superweeds" have not invaded agricultural or natural ecosystems." According to Pehu and Ragasa, there is growing evidence of the environmental benefits of transgenics. A consistent reduction in pesticide use has been documented in the major larger users of biotech crops: USA, Australia, Argentina, India, China, South Africa, and Mexico This reduction is equivalent to a 14 percent reduction in the associated environmental impact of pesticide use on these crops, as measured by the Environmental Impact Quotient (EIQ)—a composite measure based on the various factors contributing to the net environmental impact of an individual active ingredient. These positive results obtained so far are not conclusive on the evidence on many potential environmental risks and the conclusion is that continuous research must be conducted on the environmental impacts of transgenic crops (Pehu and Ragasa 2007).

Finally, one of the main food safety concerns associated with biotech products relates to the possibility of increased allergens, toxins, or other harmful compounds and to horizontal gene transfer from transgenic to conventional crops. According to Pehu and Ragasa (2007), there is presently no evidence that transgenic foods cause allergic reactions.

Prospects and Challenges

The conclusions of most papers analyzing the prospects of agricultural biotech in developing countries is that there is large potential for these countries to benefit from the new technology.

This is, for example, the conclusion arrived at by Evenson (2003) who states that there are enormous potential benefits in using transgenic technology to develop pro-poor crops. Despite these positive conclusions, most authors also agree on that the problems and limitations most developing countries face to successfully adopt the new technology are many and important.

One of the main problems developing countries face is building the adequate institutions for technology development and adoption. This is required not only to access to genetic materials controlled by private firms, but also to overcome the large costs of biotech R&D (Evenson 2003). This same author suggests measures that poor countries need to implement if they are to have access to the technology. These measures entail negotiations between private agro biotech suppliers of GM traits and farmers in developing countries to provide specific GM products to farmers in specific countries; or to convert the GM line into a form where "conventional" breeding methods can be utilized to replicate the GM products.

A second limitation that developing countries face to utilize biotechnology is the need to define adequate regulation and incentives to allow and encourage the introduction of improved varieties. Developing countries need to invest in enhancing the capacity of regulatory bodies to assess environmental and food safety risks and approve the release of biotech products given that this capacity is quite limited in most countries. An important part of this regulatory framework is the role played by intellectual property rights, such as patents or plant breeders' rights to provide incentives to both private and publicly funded innovators to develop varieties suited to a developing country's production conditions and to supply these varieties to domestic producers (Evenson 2003).

Finally, a third major problem that developing countries face for the adoption of biotech products is related to the political economy of this adoption and the resistance of consumers. Although there is a growing adoption of GM crops worldwide, in certain countries and for certain crops there is strong resistance to the introduction of GMOs given the many risks or potential risks that have been raised in public debate. The perception of risks may bring great social costs, even when the actual harm is not present, as was discussed previously when analyzing trade and biotech crops.

FUTURE PATHS FOR AGRICULTURAL TECHNOLOGY DEVELOPMENT: A NANO REVOLUTION?

According to Scrinis and Lyons (2007) nanotechnology is expected to become the technology for the next industrial revolution and is set to be the dominant technological form of the early twenty-first century because it will frame the further development, integration and convergence of other techno-sciences. In this context, nanotechnology is also emerging as the technological platform for the next wave of development and transformation of agri-food systems.

At present the US is the leading nation in advances in nanotechnology with a 4 year, 3.7 billion USD investments through the National Nanotechnology Initiative. The US is followed by Japan and the European Union, which have also committed substantial funds (0.75 and 1.2 billion USD respectively). Investment in developing countries is comparatively smaller but countries like China have a large share in the total number of scientific publications related to nanotechnology amounting to 18 percent in 2004, the second country in the world in number of publications (Joseph and Morrison 2006).

Nanotechnology can be defined as an emerging field that combines fundamental science, materials science and engineering at nanometer scales (billionth of a meter) to the design and manufacture of structures with dimensions about the size of a molecule, manipulating individual atoms, molecules, or molecular clusters into structures to create materials and devices with new or vastly different properties (Joseph and Morrison 2006).

More directly related with agriculture, nano-biotechnology refers to the use of nanotechnology to manipulate living organisms, as well as to enable the merging of biological and non-biological materials to facilitate genetic engineering breeding programs, the incorporation of synthetic materials into biological organisms, and ultimately the creation of new life forms.

Characterization and Scientific Basis of the New Technology

We follow here the review on different methods and applications made by Scrinis and Lyons (2007). In the agricultural sector, nanotech research and development is likely to facilitate the next stage of development of genetically modified crops, animal production inputs, chemical pesticides and precision farming techniques. Through the convergence of nano and bio techniques, it may be possible to improve the precision of genetic engineering breeding programs, thereby ensuring greater control in delivering new character traits to plant and crop varieties (ETC Group, 2004).

Techniques at the nano-scale are also being applied to enable the targeted delivery or increased toxicity of pesticide applications including nano-scale active ingredients into pesticides. The advantage of the nano-particle pesticide applications is that they have an improved capacity for absorption into plants compared to larger particles. As such, they may not be washed off as readily, thereby increasing their effectiveness, but also posing a new order of risks to consumers of these products. Pesticides may also be encapsulated via nano-encapsulation techniques to enable greater control over the circumstances in which encapsulated pesticides will be released, depending on need and under specific conditions, such as moisture and heat levels (see for example Zhang et al., 2006).

Nanosensors represent the intersection of nanotechnologies and information technologies. Alongside geographical positioning systems and other information technologies, nanosensors could be scattered across farmers' fields to enable the 'real time' monitoring of crops and soils, and the early detection of potential problems, such as pest attacks and declining soil nutrient levels. Nanosensors could be applied, for example, to precision agriculture which makes use of computers, global satellite positioning systems, and remote sensing devices to measure highly localized environmental conditions thus determining whether crops are growing at maximum efficiency or precisely identifying the nature and location of problems. By using centralized data to determine soil conditions and plant development, seeding, fertilizer, chemical and water use can be fine-tuned to lower production costs and potentially increase production, benefiting the farmer. Precision farming can also help to reduce agricultural waste and thus keep environmental pollution to a minimum. Although not fully implemented yet, tiny sensors and monitoring systems enabled by nanotechnology will have a large impact on future precision farming methodologies (Joseph and Morrison 2006).

Nanotechnology also has the potential to displace traditional food and non-food farm commodities through the development of artificial nanomaterials in factories. For example, the global cotton market and cotton prices could be affected by the development of synthetic fibers such as Nano-Tex, which is reported to be a much stronger material than cotton but with a similar texture. Similarly, nanoparticle alternatives to rubber are already in production, such as silica carbine nanoparticles and carbon nanotubes for use in car tires (Scrinis and Lyons 2007).

Nanotechnology also can be applied in animal production systems, including new tools to aid animal breeding, targeted disease treatment delivery systems, new materials for pathogen detection, and identity preservation systems. Manufacturers are already applying nano techniques with the aim of improving the quality, durability and shelf life of packaged foods. A new range of so-called 'smart packaging' is being developed through the application of nano-sensors able to detect the release of particular chemicals. The packaging may be engineered to change color to warn the consumer if a food is beginning to spoil, or has been contaminated by pathogens. To do this, electronic 'noses' and 'tongues' will be designed to mimic human sensory capacities, enabling them to 'taste' or 'smell' scents and flavors (Scrinis and Lyons 2007).

Nanotechniques are also being used to develop food identifiers that may be able to detect contaminants in food and animal feed. The aim is to increase the security of manufacturing, processing, and the shipment of food, by enabling early detection of contaminants, and the removal of infected products from the food chain. A new range of nano-barcodes and monitoring devices are also being developed. This includes nano-scale radio frequency identification tags (RFid) able to track containers or individual food items. These RFid tags could also transmit information after a product leaves the supermarket, unless the tags are disabled at the check-out register (Scrinis and Lyons 2007).

At the same time, nanotechnology may provide food industries with a new platform to define and regulate the terms of food safety. The types of application include: smart packaging, on demand preservatives, and interactive foods. Building on the concept of "on-demand" food, the idea of interactive food is to allow consumers to modify food depending on their own nutritional needs or tastes. The concept is that thousands of nanocapsules containing flavor or color enhancers, or added nutritional elements (such as vitamins), would remain dormant in the food and only be released when triggered by the consumer. Most of the food giants including Nestle, Kraft, Heinz, and Unilever support specific research programs to capture a share of the nanofood market in the next decade (Scrinis and Lyons 2007).

Working at the scale of microns where the sphere of nanotechnology resides, it is possible to bring together knowledge and techniques used in biotechnology, bioengineering and nanobiology to solve practical problems in agriculture creating staggering possibilities in crop and animal production through genetic manipulation of species. These advances open up tremendous scope for nanofabrication in modern molecular biology or agricultural biotechnology of plants and animals.

The Institutional Model of Innovation and Diffusion

As in the case of genetic engineering, nanotechnology appears to be a *corporate technology*. More than 400 companies around the world today are investing in nanotechnology R&D and this number is expected to increase to more than 1,000 within the next 10 years. In terms of investment, the US leads, followed by Japan, China, and the EU. An estimate by the Business Communications Company, a technical market research and industry analysis company shows that, the market for the nanotechnology was 7.6 billion USD in 2003 and is expected to be 1 trillion USD in 2011 *(Scrinis and Lyons 2007)*.

There are at least four main reasons for this predominance of corporations in the development and expansion of nanotechnology. First, huge investments in budget are needed for ongoing nanotechnology R&D, which make cost and affordability major challenges for the future adoption of nanoagriculture in developing countries.

Second, the majority of the value of nanotechnology is currently embodied in specific knowledge which makes the intensive patenting of nano-scale techniques and materials a key feature of the nanotech industry (Scrinis and Lyons 2007). Given the materially fundamental nature of these patents and their widespread applicability across applications and industries, the control of these broad nano-patents may be a strategy for corporate concentration both within and across industrial sectors.

Third, nanotechnology also introduces new possibilities for the differentiation of production systems and final food products. For example, the development of food crops with modified nutrient and functional traits; the manufacture of processed foods with a wider variety of features and functionalities; and packaging to enable the improved transportation, shelf-life and year-round availability of quality foods such as fresh foods and ready meals (Moraru et al. 2003).

Fourth, nanotechnology will also facilitate the growing demand for the identification or identity preservation of products across the food system, for the purposes of food safety, quality control, segmented supply chain logistics, consumer data gathering, and patent surveillance, all of which is likely to favor the larger agri-food and retailing corporations (Scrinis and Lyons 2007).

Prospects and Challenges

The future development of nanotechnology corporations and the changes that this will bring to the food system could have important implications for agriculture in developing countries. Firstly, nanotechnology can potentially extend industry control into the production chain by using new techniques for integrating seeds and chemical inputs (such as chemically-triggered seeds traits), and new crop or animal traits that address emerging agronomic problems or consumer demands, and that thereby entice farmers to switch to patented seeds that are subject to 'technology fees' and binding contracts, as are many genetically modified crops; and the further undermining of subsistence practices, such as on-farm breeding ((Kloppenburg, 2004; Scrinis and Lyons 2007).

For the food processing industry, techno-commodification may take the form of new proprietary techniques for modifying the nutrient profile of foods and introducing new packaging functionalities that provide new value-adding possibilities.

The economic impacts of nanotech developments are likely to affect farmers differentially, depending on the size and capital-intensity of the production unit. As with earlier technological innovations, it is larger-scale, capital-intensive farming operations that will be more able to reap any early economic advantages from adopting nano-applications. On the other hand, farming communities in developing countries, particularly smaller-scale and local market and subsistence oriented farmers, as well as agricultural laborers, may be adversely affected in a number of ways. This includes the continuation of commodity price depression through productivity increases in high income countries; the displacement or undermining of traditional agricultural commodities through the development of nanotechnological industrial alternatives; and the reduction in farm labor through the increased efficiency, mechanization or automation of farming practices (Scrinis and Lyons 2007).

As mentioned in the case of biotechnology, as the new techno-sciences have come to play an increasingly important role across the agri-food system, they have also become the focus of civil society and social movement contestation. The strong opposition to genetically modified crops has arisen on the basis of a number of concerns, ranging from "the defense of peasant and small farmer interests, to bio-diversity, environment, animal welfare, ethics and consumer health issues" (Wilkinson, 2002). The range of issues raised reflects a growing recognition of the power of these new techno-sciences to not only introduce new health and ecological hazards, but also to increase the power of agri-food corporations over farmers and citizens' interests. This opposition has so far restricted the development and commercialization of genetically modified crops, and has raised concerns about the emergence of a similar level of public resistance to nano-foods (Scrinis and Lyons 2007).

This has led to repeated calls for the nanotechnology and food industries to "learn the lessons" of biotechnology and GM foods (Grove-White *et al*, 2004). If a significant level of public and consumer resistance to the introduction of nano-foods does emerge, an important issue will be whether the now dominant retail sector will take a position of responding to consumer concerns — as supermarket chains in some countries have to GM foods, thus pitting themselves against the interests of corporations at the other end of the food chain — or whether the broad scope of nano-food applications, and supermarkets' own adoption of nano-applications, compromises their ability to respond in similar ways.

Finally, the novel characteristics of nanoparticles and nanomaterials may also be the source of new forms of hazards to environments and people, such as new forms of toxicity and new forms of pollution (Colvin, 2003; ETC Group, 2004; Scrinis, 2006). There is little known about the health effects of eating foods that contain nano-particles, or of workers handling nano-materials.

These developments have been identified as the basis of a new technological paradigm, and as framing the restructuring of contemporary agri-food systems. However, the nano-corporate food paradigm does not represent a major break with the biotech paradigm in agricultural production. In the case of agricultural biotechnologies, for example, there are strong similarities and continuities between genetic engineering and nano-technology in regard to the types of agricultural practices, farming styles, patenting regimes, and corporate structures these technologies are being used to support and transform. Nanotechnology will in fact serve as an enabling technology for genetic and cellular technologies, as well as for the information technologies which have played an increasingly important role across food sectors for managing and coordinating production and distribution systems.

CONCLUSION

Modern developments in agriculture have been associated to advances in science and technology. The mainstream agricultural production methods and technologies that are being applied at present in high income countries and in commercial agriculture of many developing and middle income countries are the result of Mendelian genetics and the incorporation of the power engine from the industrial revolution. We are a long way from the initial developments and applications of the first improved varieties in Japan and of the appearance of the tractor that allowed the substitution of the power engine for the horse. However, the high yields and harvests obtained today in industrial and post-industrial economies using modern technology are still the result of the same scientific principles that brought about the agricultural revolution in the late 19th century.

By the end of the 20th century, with the impressive advances made in biology and genetics we witnessed the dawn of a new paradigm in agricultural research that was thought to replace the Mendelian genetics as the driver of technical change in agriculture. After the first overly optimistic forecasts about the speed at which the new paradigm would replace the old, and the resistance that applications of biotechnology to agriculture found among some groups of environmentalists and consumers in high income countries, the coming years will see a growing number of applications and developments of biotechnology applied to agriculture. At least, this is what is to be expected according to the theory of induced innovation discussed in this chapter. The potential of biotechnology to reduce costs of production, rationalize the use of chemical inputs, and improve the quality and properties of final products is enormous. If this is the case, this potential will be realized through markets, changing relative input and output prices, changes that will incentive further investment in R&D and the generation of new technologies.

The transition between the old and the new paradigms in agricultural research that we are already witnessing will not occur without problems. One of these problems is the institutional changes that the new paradigm is bringing. The displacement of the center of gravity of agricultural R&D from the government and non-profit research institutions to the private sector will generate many benefits but will also create some new problems for which new solutions need to be created.

One of the benefits is already happening, as large and growing amounts of private investment are being directed to agricultural R&D. This fact shows that the private sector has clear expectations of benefiting from the potential offered by biotechnology applied to agriculture and will assure a flow of innovations from the private sector in the coming years.

Some of the benefits from these new developments are already being experienced by, for example, the countries that embraced the new technology for soy production. The US, Brazil, and Argentina are today the dominant forces in soybean production and privileged suppliers of the Chinese market. And this is not only because of the natural comparative advantage of these countries for agricultural production, but also because the new technology brought benefits to commercial producers as the evidence reviewed in this chapter shows.

However, potential conflicts between private and public interests could result from the dominant position of the corporations in agricultural R&D. One of these possible conflicts is the priority that private R&D is giving to a second generation of biotech products that puts emphasis on quality traits while no consideration is given to the expansion of yields. Vernon Ruttan in a 1999 article explains his concerns about research priorities in the private sector. According to Ruttan, by the early 1990's there was growing evidence that yields of a number of important cereal crops, such as maize and rice, might be approaching yield ceilings determined by physiological constraints as information coming from rice yield trials at the International Rice Research Institute and other sources apparently show. For Ruttan, "If present yield ceilings are to be broken, it seems apparent that improvements in photosynthetic efficiency, particularly the capture of solar radiation and reduction of water loss through transpiration, will be required. Even researchers working at the frontiers of plant physiology are not optimistic about the rate of progress that will be realized in enhancing crop metabolism." Given the time lag between investment in R&D and the moment when these technologies are producing results in the field, Ruttan is concerned of the fact that more intensive research efforts are not being devoted to attempts to reduce the constraints that will limit future increases in yields.

A second issue that results from the changing nature of institutions for innovation in agriculture is the future role to be played by public sector oriented agricultural research. As discussed in this chapter, the public sector has been displaced from the gravity center of agricultural innovation and its role in the future is by now challenged. This is a result not only of changes in science and the advantages for the private sector to develop new technologies under the new innovation paradigm, but also because governments in high income and developing countries have been giving less priority to investment in agricultural R&D. Ruttan (1999) asserts that this is a mistake and points out that even though the viability of private R&D depends on the possibility of developing proprietary products, the technical and scientific viability of private sector research depends on the capacity of public sector institutions to conduct basic and generic research.

Public research institutions should also play a key role under the new research paradigm in relation to economic growth and development in poor countries. There are two main reasons for this. First, most developing countries still need to realize the benefits of the first Green Revolution. This means that they have still the possibility of experience large leaps in productivity and yield growth using traditional crop breeding, higher levels of technical inputs, better crop management and strategic use of biotechnology crop protection which were the foundation of the changes during the Green Revolution. Given the key role played in the first Green Revolution by the international research system represented in the IARC and national research institution in developing countries, more efforts not less must be put on developing national and regional research centers in places that are lagging behind the technological frontier as is notoriously the case of Sub-Saharan Africa. Closing the technology gap in these regions will depend mainly on these efforts and not on those of private firms investing in agricultural R&D whose interests are not and could not possibly be in the underdeveloped markets of low income countries.

A second reason to develop a strong system of national and international public research centers, results from the need to grant access to these countries to the benefits of the biotech products of the Gene Revolution. As discussed in this chapter, the problems and limitations most developing countries face to successfully adopt the new technology are many and important and because of this, to build the adequate institutions to have access to the new technology becomes a central issue for these countries. This is required not only to access genetic materials controlled by private firms, but also to overcome the large costs of biotech R&D, and to define appropriate regulations and incentives to allow and encourage the introduction of the crop varieties needed in these countries.

Lastly, the promises and possibilities that nanotechnology offers to introduce innovations and new technologies in agricultural production are as large as those promised by biotechnology 20 years ago, if not larger because these new developments can work as a platform integrating

several disciplines from biology to information technology and from production to packing in the food industry. One of the main contributions that nanotechnology can bring for the future of agriculture is making possible what Evenson (2003) calls "quantitative enhancement" through the use of genomics, and proteonomics research, a mechanism that entails "quantitative" trait breeding, or what Ruttan refers as "raising the ceiling of crop yields." At present, biotech products are basically "qualitative trait" products meaning that they endow plants with specific cost advantages that vary from environment to environment, but are "static" in nature. Quantitative or dynamic gains are those that allow for gains in photosynthetic efficiency in plants, as Evenson puts it, an area of research that is very demanding of skills and creativity. The importance of opening this new avenue for agricultural research and innovation is on itself the new scientific revolution in agriculture. As stated by Evenson (2003) "It is sometimes said that the Gene revolution will replace the Green Revolution. But this will not happen until and unless breeders are able to produce "dynamic" gains in generations of varieties. Until such time the Gene Revolution GM products can only complement conventional Green Revolution breeding." The Nano Revolution in agriculture could bring the opening of this last frontier.

REFERENCES

Acemoglu, D. (2002). Directed technical change. *The Review of Economic Studies, 69*(4), 781–809. doi:10.1111/1467-937X.00226

Anderson, K., & Nielsen, C. (2002). *Economic effects of agricultural biotechnology research in the presence of price-distorting policies.* Discussion Paper No. 0232, Centre for International Economic Studies, University of Adelaide.

Binswanger, H. P., & Ruttan, V. W. (1978). *Induced innovation. Technology, institutions and development*. Baltimore: The Johns Hopkins University Press.

Buttel, F. H., Kenney, M., & Kloppenburg, J. Jr. (1985). From green revolution to biorevolution: some observations on the changing technological bases of economic transformation in the Third World. *Economic Development and Cultural Change, 34*(1), 31–55. doi:10.1086/451508

Byerlee, D., & Fischer, T. (2001). Accessing modern science: policy and institutional options for agricultural biotechnology in developing countries. *IP Strategy Today,* 1. Retrieved June 12, 2009, from http://www.biodevelopments.org

Colvin, V. (2003). The potential environmental impact of engineered nanomaterials. *Nature Biotechnology, 21*(10), 1166–1170. doi:10.1038/nbt875

Conway, G. (2000). *Crop biotechnology: benefits, risks and ownership*. Rockefeller Foundation, New York (. *Foundation News*, 03062000.

De Janvry, A. (1973). A socioeconomic model of induced innovations for Argentine agricultural development. *The Quarterly Journal of Economics, 87*(3), 410–435. doi:10.2307/1882013

de Janvry, A., Graff, G., Sadoulet, E., & Zilberman, D. (2000). *Technological Change in Agriculture and Poverty Reduction*. Concept paper for the World Development Report on Poverty and Development 2000/01 University of California at Berkeley.

Delgado, L. C., Hopkins, J., & Kelly, V. A. (1998). *Agricultural growth linkages in Sub-Saharan Africa*. IFPRI Research Report No. 107. Washington, DC: International Food Policy Research Institute.

Evenson, R. E. (2002). From the Green Revolution to the Gene Revolution . In Evenson, R. E., Santaniello, V., & Zilberman, D. (Eds.), *Economic and Social Issues in Agricultural Biotechnology* (pp. 1–16). London: CAB International. doi:10.1079/9780851996189.0001

Evenson, R. E. (2003). *GMOs: Prospects for increased crop productivity in developing countries*. Center Discussion Paper No. 878, Economic Growth Center, Yale University. Retrieved June 12, 2009, from http://ssrn.com/abstract=487503

Evenson, R. E., & Gollin, D. (2003). Assessing the impact of the Green Revolution: 1960-1980. *Science, 300*, 758–762. doi:10.1126/science.1078710

Evenson, R. E., & Rosegrant, M. (2003). The economic consequences of crop genetic improvement programs . In Evenson, R. E., & Gollin, D. (Eds.), *Crop Variety Improvement and Its Effect on Productivity: The Impact of International Agricultural Research* (pp. 473–498). Wallingford, UK: CAB International. doi:10.1079/9780851995496.0473

ETC Group. (2004 November). *Down on the Farm. The Impacts of Nano-Scale Technologies on Food and Agriculture.*

Grove-White, R., Keraned, M., Miller, P., Machnaghten, P., Wilsdon, J., & Wynne, B. (2004). *Bio-to-Nano? Learning the Lessons, Interrogating the Comparison*. Working Paper. Institute for Environment, Philosophy and Public Policy and Demos.

Hayami, Y., et al. (1978). Anatomy of a peasant economy: a rice village in the Philippines. Los Baños, The Philippines: International Rice Research Institute (IRRI).

Hayami, Y., & Godo, Y. (2005). *Development Economics: From the Poverty to the Wealth of Nations* (3rd ed.). Oxford, UK: Oxford University Press.

Hayami, Y., & Ruttan, V. M. (1985). *Agricultural Development: An International Perspective*. Baltimore, MD: Johns Hopkins University Press.

Hazell, P., & Ramaswamy, G. (Eds.). (1991). *The Green Revolution reconsidered. The impact of high-yielding rice varieties in South India*. Baltimore, MD: Johns Hopkins University Press.

ISAAA. (2008). *Global Status of Commercialized Biotech/GM Crops: 2008. The First Thirteen Years, 1996 to 2008*. ISAAA Brief 39-2008.

Joseph, T., & Morrison, M. (2006 May). *Nanotechnology in agriculture and food*. Institute of Nanotechnology. Retrieved June 12, 2009, from http://www.nanoforum.org

Khush, G. S. (1992). Selecting rice for simple inherited resistances . In Stalker, H. T., & Murphy, J. P. (Eds.), *Plant breeding in the 1990s* (pp. 303–322). Wallingford, UK: Commonwealth Agricultural Bureaux International.

Khush, G. S. (1995). Modern varieties: Their real contribution to food supply and equity. *GeoJournal*, *35*, 275–284. doi:10.1007/BF00989135

Khush, G. S. (1999). Green Revolution: preparing for the 21st century. *Genome*, *42*, 646–655. doi:10.1139/gen-42-4-646

Kloppenburg, J. R. (2004). *First the Seed: The Political Economy of Plant Biotechnology 1492-2000* (2nd ed.). Madison, WI: University of Wisconsin Press.

Moraru, C., Panchapakesan, C., Huang, Q., Takhistov, P., Liu, S., & Kokini, J. (2003). Nanotechnology: A new frontier in food science. *Food Technology*, *57*(12), 24–29.

Paddock, W., & Paddock, P. (1967). *Time of Famines*. Boston: Little Brown & Company.

Pehu, E., & Ragasa, C. (2007). *Agricultural Biotechnology Transgenics in Agriculture and their Implications for Developing Countries*. Background paper for the World Development Report, 2008.

Pingali, P. (2007). *Will the Gene Revolution reach the poor? Lessons from the Green Revolution*. Mansholt Lecture, Wageningen University, 26 January 2007.

Pingali, P. L., & Heisey, P. W. (2001). Cereal-crop productivity in developing countries. Past trends and future prospects . In Alston, J. M., Pardey, P. G., & Taylor, M. (Eds.), *Agricultural science policy*. Washington, DC: IFPRI & Johns Hopkins University Press.

Pingali, P. L., & Traxler, G. (2002). Changing the locus of agricultural research: will the poor benefit from biotechnology and privatization trends? *Food Policy*, *27*, 223–238. doi:10.1016/S0306-9192(02)00012-X

Qaim, M., & Traxler, G. (2005). Roundup ready soybeans in Argentina: Farm level and aggregate welfare effects. *Agricultural Economics*, *32*(1), 73–86.

Raney, T. (2006). Economic impact of transgenic crops in developing countries. *Current Opinion in Biotechnology*, *17*, 1–5.

Raney, T., & Pingali, P. (2007 September). Sowing a gene revolution. *Scientific American*.

Ruttan, V. W. (1999). Biotechnology and agriculture: a skeptical perspective. *AgBioForum*, *2*(1), 54–60.

Ruttan, V. W., & Binswanger, H. P. (1978). Induced innovation and the Green Revolution . In Binswanger, H. P., & Ruttan, V. W. (Eds.), *Induced innovation: Technology, institutions and development* (pp. 358–408). Baltimore, MD: Johns Hopkins University Press.

Scrinis, G. (2006). Nanotechnology and the Environment: The Nano-atomic reconstruction of nature. *Chain Reaction, 97*, 23–26.

Scrinis, G., & Lyons, K. (2007). The emerging nano-corporate paradigm: nanotechnology and the transformation of Nature, food and agri-food systems. *International Journal of Sociology of Food and Agriculture, 15*(2), 22–44.

Smale, M., & Zambrano, P. Falck-Zepeda, J., & Gruère, G. (2006). *Parables: Applied economics literature about the impact of genetically engineered crop varieties in developing economies.* EPTD Discussion Paper 158, International Food Policy Research Institute, Washington, DC.

Smale, M., Zambrano, P., Gruère, G., & Falck-Zepeda, J. Matuschke, I., Horna, D., Nagarajan, L., Yerramareddy, I., & Jones, H. (2009). Measuring the Economic Impacts of Transgenic Crops in Developing Agriculture during the First Decade: Approaches, Findings, and Future Directions. Food Policy Review 10. Washington, DC: International Food Policy Research Institute.

Traxler, G. (2004). *The Economic Impacts of Biotechnology-Based Technological Innovations.* ESA Working Paper No. 04-08 Agricultural and Development Economics Division. The Food and Agriculture Organization of the United Nations. Retrieved June 12, 2009 from http://www.fao.org/es/esa

Wilkinson, J. (2002). Genetically modified organisms, organics and the contested construction of demand in the agrofood system. *International Journal of Sociology of Agriculture and Food, 10*(2), 3–11.

Zhang, F., Wang, R., Xiao, Q., Wang, Y., & Zhang, J. (2006). Effects of slow/controlled release fertilizer cemented and coated by nano-materials on biology. *Nanoscience, 11*(1), 18–26.

Zilberman, D., Yarkin, C., & Heiman, A. (1997 August). *Agricultural biotechnology: economic and international implications.* Invited Paper Presented at the meeting of the International Agricultural Economics Association. Sacramento, CA.

ENDNOTES

[1] Germplasm is a collection of genetic resources, which in the case of crops it may be stored as a seed collection

[2] As will be used in this chapter, paradigm refers here to "pattern or model" as define in the Oxford English Dictionary, and is not related to the concept of "scientific paradigm" as developed by Khun in Kuhn, Thomas (1970). *The Structure of Scientific Revolution*, 1962; 2d. ed., Chicago.

Chapter 18

Technology Adoption and Economic Development:
Trajectories within the African Agricultural Industry

Taiwo E. Mafimisebi
Federal University of Technology, Nigeria

ABSTRACT

Africa's economic development will result from conscious efforts directed towards diversification and increased productivity in its low-performing agricultural sector. Technology development, transfer and uptake, which are low for now, are indispensible necessities in this respect. The purpose of this chapter is to review the characteristics, importance, constraints and technology adoption process of African agriculture to identify factors that enhance or hinder technology uptake. This is with a view to isolating lessons for developers or packagers of new agricultural or other technologies for Africa, especially nanotechnology and microelectronics which are evolving and transformational. The attributes of technologies that have made desired impact in African agriculture included cheapness, simplicity, observability, visibility of results, usefulness, compatibility with existing technologies and farm- or farmer-specific socio-economic or socio-cultural conditions. Case studies of the welfare-enhancing impacts of adopted agricultural technologies were examined under use of fertilizers, improved varieties and biotechnology. Useful lessons for development and transfer of nanotechnology and micro-electronics to Africa were highlighted.

INTRODUCTION

Agriculture is one of the most important and possibly the oldest economic activity developed by humans and it is becoming increasingly dependent on development of technologies (Alexan-

DOI: 10.4018/978-1-61692-006-7.ch018

dre, 2009). Agriculture in Africa is a subsistent, low technology and low-performing one which requires reduced drudgery and increased commercialization. For crop production, farmlands are cleared using cutlass and/or fire after which land preparation is done using the hoe; hence the term "slash and burn" agriculture. Animal husbandry is based on extensive or semi-intensive systems in which feeds, water and shelter are not usually provided for farm animals. The growth of farm animals thus follows a staircase pattern (Williamson and Payne, 1978) in line with the trend in feed availability.

The dependence of African agriculture on human labour and crippling land tenure problems are the main reasons for small and uneconomic farm sizes unsuitable for mechanization. Furthermore, seeds, propagules and breeding stocks accessible to farmers and highly adaptable to African agriculture are the hardy, unimproved and low-producing types. The generic problem of low productivity of the existing seed-stock is one of the factors responsible for Africa meeting just about 50.0% of her per *capita* demand for food (Olalokun, 1998; Okunmadewa, 1999, Okunmadewa *et al.*, 2002). The problems of pests and diseases are particularly serious because a larger proportion of African land area falls in the tropical region which is described as "a paradise of parasites." Poor animal health is the main reason for losses in animal production. Direct and indirect losses of meat, milk and work output are estimated at about US$40 billion a year in Sub-Saharan Africa (SSA) alone (World Bank, 2005). There is also soil fertility and other environmental management problems which negatively impinge on agricultural production and productivity (World Bank, 2006).

To assure increased food supply and attain sustainable economic development, there is a compelling need to increase agricultural production. This is necessary from the point of view of food, income, employment, poverty reduction and economic stability. In spite of the fact that a considerable proportion of Africa's annual budget goes for food import, under-nourishment and malnutrition are still prevalent (World Bank, 1990; UNICEF, 1990; Okojie, 1990, Mafimisebi, 2007). There are problems inherent in further food import which is projected to expand considerably in future. The most practical and sustainable way for Africa to achieve a reliable food supply is to give a powerful boost to her own agricultural system.

In order to turn the tables in the race between the growing population and dropping food levels, more countries of the continent must strive to achieve the recommended 4% agricultural production growth rate per year (World Bank, 1989; 1993) necessary to achieve overall economic development. Only very few countries, less than one-fifth of African countries, shown in Table 1, had achieved or came close to achieving this agricultural growth rate in the late 1980s or early 1990s.

Some of these countries; Botswana, Chad and Comoros are classified as resource-poor and this is heart-warming news to African countries in that category. Most of these countries were all poor performers in the 1960s and 1970s and the change in scorecard holds important lessons on what can be done to reverse agriculture's fortune in other African countries.

The objective of this chapter is to give a review of the present state of agriculture in Africa and the role of technologies in it. The chapter also touches on the process of adoption of technologies and factors that enhance or hinder it in Africa. A review of attributes and suitability of models used in technology transfer is also carried out. The overall

Table 1. African countries with high agricultural growth rates

Country	Year		
	1986-1989	**1990**	**1991**
Chad	6.1	8.9	20.0
Cape Verde	12.0	-3.8	9.3
Nigeria	4.3	4.1	5.0
Botswana	19.5	3.7	2.7
Guinea-Bissau	6.4	2.5	5.7
Uganda	6.0	3.4	2.9
Benin	5.0	1.4	4.5
Kenya	4.3	3.5	-0.7
Tanzania	4.5	2.9	-
Comoros	4.5	2.8	3.9

Source: World Bank, 1993.

objective of the chapter is to distill lessons from the African agricultural industry for successful introduction and uptake of nanotechnology and microelectronics in Africa.

The remaining part of the chapter is organized as follows. Section I presents the background to the chapter while in section II is contained a review of the process of technology adoption in African agriculture. Section III examines the impacts of already adopted technologies on Africa's agricultural and economic development. In section IV, lessons for transfer of nanotechnology and micro-electronics to Africa and other developing economies are presented. Section V contains the conclusion of the chapter.

BACKGROUND

Agriculture in the African Economy

Agriculture constitutes a very significant sector of the economies of African countries. Apart from its traditional role as food and fiber provider, it ensures supply of agricultural raw materials to industries and employs considerable proportion of labour (up to 78% of the economically active population) in some cases. It is a generator of income to farmers and a guaranteed foreign exchange earner from the export of the surplus. The continued fulfillment of these traditional roles requires the formulation of strategies for modernizing agricultural practices. Technology development, transfer and uptake are indispensable necessities in achieving continued performance of these traditional functions (World Bank, 1993; 2005; Mafimisebi *et al.*, 2006).

According to IFPRI (2001), there are two main reasons why agriculture is of great significance in Africa. First, population growth at an average of 3.2% far outpace growth in food production at an average of 1.8% resulting in a terrific supply-demand gap for food. Second, agriculture plays a crucial role in the economy of the various countries with the exception of South Africa and very few

others that depend on solid minerals. An average of 70 percent of the population lives by farming, and 40 percent of all exports are earned from agricultural products. One third of the income of the continent is generated by agriculture. The highest incidence, depth and severity of poverty are found among the rural, agriculture-dependent members of the society (World Bank, 1993; 2002). With most of the rural population living below the poverty line, spending deficit ratio is high and worse still, productivity increases do not automatically translate to more money for efficient farmers. Conditions which compel farmers to sell their produce in local markets such as bad roads, inefficient transportation and a shortage of buyers often make the farmer accept lower prices for more merchandise (IFPRI, 2001).

It is rewarding in spite of these limitations to proffer a lasting solution to Africa's food and economic problems for obvious reasons. Studies

Table 2. Agriculture's share in the GDP of selected African countries

Country	Agriculture Share in GDP	
Sub-Saharan Africa	1965	1990
Benin	59	37
Botswana	34	03
Cameroon	33	27
Chad	42	38
Ethiopia	58	41
Gabon	26	09
Ghana	44	48
Malawi	50	33
Mali	65	46
Niger	68	36
Nigeria	55	36
Rwanda	75	38
Somalia	71	65
Tanzania	46	59
Zimbabwe	18	03

Source: World Bank, 1993

carried out in many African countries revealed that the multiplier effect of farmers' extra dollar income is three because they buy what they do not produce, purchase simple farm tools, use more hired labour and embark on new capital projects or they diversify into other ventures or crops/commodities attracting higher prices with possibility for boosted growth (IFPRI, 2001).

Concept and Indicators of Agricultural Development in Africa

There is no doubting the fact that it is difficult defining clearly what is meant by agricultural development. Despite this, many people have an idea of what things should be expected when agricultural development is being attained. The question that arises then is that how can one show that there has been agricultural development in a country or region? One way is by considering the importance of agricultural development for the country or region in question. It is recognized that this approach is capable of generating considerable controversies because the importance of agricultural development is tied to the stage of economic development and therefore, it is subject to variation even in the same geographical location as argued by Adegeye and Dittoh (1985).

Thus, indices of agricultural development differ according as an economy is a purely subsistent economy, a fully monetized economy, an industrializing economy or an agricultural economy. In an agricultural economy, the situation that approximates that of Africa, agricultural development will imply the provision of other facilities and/or services necessary for enhancing agricultural sector performance. Apart from increasing quantity and quality of food, there will be improvement in infrastructures such as good roads, clean water, electricity, healthcare and telecommunication facilities and schools within easy reach of villages. Reducing dependence on urban people for their important needs should be one of the hallmarks of agricultural development

in such a scenario (Adegeye and Dittoh, 1985). From the fore-going, it is clear that a confirmatory test for occurrence of agricultural development is when its benefits accrue to the rural people. This is because substantial changes can take place within the rural settings without an appreciable effect on the people concerned. This is the case of agricultural growth. Therefore, in agricultural development, the majority of the farmers must experience significant improvement in their incomes and standard of living. This must be accompanied by programmes to sustain this standard of living and in fact increase it substantially. From the consideration above, it is certain that agricultural development cannot take place without the support of the other sectors of the economy and without certain essential requirements as listed by Adegeye and Dittoh (1985).

Concept and Indicators of Economic Development in Africa

For a long time, development economics was dominated by the ideas of the orthodox, traditional school who borrowed their major tools of analysis from neo-classical economics. In most of the underdeveloped countries (UDCs) especially in Africa, which adopted the ideas of the traditional school, failure of prescription, which led to perpetuation of mass poverty was the result. In fact, rather than helping to improve the situation of the UDCs, some of the strategies prescribed by orthodox development economics have complicated the problems of attaining significant and sustainable economic development. It was obvious from the failure that there was improper comprehension of the problems of the UDCs.

Development implies a progression from a lower and often undesirable state to a high and preferred one. A *sine-qua-non* for such a progression is availability of a correct characterization of the higher state. It is only then that one can arrive at the best strategies of attaining the desired higher state. What traditional economics used as bases for defining such a higher state are the characteristics of the industrialized countries of Western Europe and North America without taking cognizance of the fact that the process of economic development in peasant agrarian economies with little or no modern economic structure must be different from that of advanced societies. It is obvious that prescriptions based on such a one-sided analysis and mistaken diagnosis of the problems of underdevelopment cannot succeed much in removing mass poverty (Olopoenia, 1983).

Development can be defined in terms of the major socio-economic characteristics of the state which the UDCs desire to attain. This characterization is best done from the perspective of the majority of the peoples in these countries. Mainstream orthodox development economists assumed that the benefits of growth would automatically spread and trickle down once 'take off'- the most critical component in the economic development process- has been achieved. What had happened, however, was that the accessories of growth either have failed to function or have been systematically offset by disequilibrium forces, leaving development to trickle up rather than down. Therefore the phenomena of unemployment and inequality became the logical consequences of the pattern of development advocated by the orthodox school.

The most fundamental questions relating to the meaning of development were however posed by Seers (1972) when he argued that the question to ask about a country's development is: What has been happening to poverty, unemployment and inequality? If there has been a decline in all three, then it is incontrovertible that development is being witnessed for the country or region in question. If one or two of these central problems have been growing worse, especially if all three have, it would be unreasonable to say there has been development even if national or per *capita* income had expanded greatly.

While it is agreeable that development is a normative concept in implying progress from a less desirable state to more preferred one, it would

be difficult for anyone to argue rationally that reduction or elimination of poverty, inequality and unemployment for the largest majority of the population are not laudable goals in any society. In current development literature therefore, the emphasis is on meeting basic needs and redistributing the benefits of growth. The 'basic needs' approach emphasizes the need for development to provide the necessities for a decent and sustainable livelihood to the majority of the population.

Constraints to Agricultural Development in Africa

In determining what is needed to achieve the objectives established for agriculture in Africa, there is a need for critical appraisal of what has gone wrong in the past. Technology-based agriculture has not come to Africa on a significant scale. Expansion of cultivated area and not productivity increases is responsible for agricultural growth. Many reasons can be adduced for this. First, investment in irrigation, a major input in Asia's Green Revolution, had been negligible in Africa. Then again, compared with Asia, there has been little demand by farmers for yield-enhancing technologies most often because of lack or insufficiency of funds to invest in such (World Bank, 1993*)*. This situation is exacerbated in many African countries by many other factors which discourage risk-averting farmers from using market-dependent technologies.

There is thus a need for more effective approach to technology creation, packaging and transfer. Current research findings report an extra-ordinary deterioration of agricultural research in Africa. A new approach to technology generation and dissemination is required which greatly expands investment in and quality of agricultural research, extension, livestock services, agricultural education and training. Irrigation development must be given greater attention in this technology strategy owing to climatic change that now impinges negatively on agriculture.

Neglect of roads connecting town to country, and the prevailing focus of government infrastructures investment in the mega-cities has, in many countries, cut the agricultural sector off from urban and export markets. It has also cut farmers off from the source of improved inputs and equipment, which is in the cities. The implications of this on agriculture's performance are obvious. Also, there are other constraints that have been well discussed in the literature such as deterioration of the natural resource base in many African countries, non-recognition of the role of women in African agriculture (World Bank 1993, Mafimisebi and Fasina, 2009), weak or non-existent supportive institutions and neglect of development of an African capacity to manage the key areas of focus for agricultural development.

The Nexus between Agricultural and Economic Development in Africa

The promise for the overall development of African economies lies in the agricultural sector for reasons that have been explained earlier in this chapter. Analysis undertaken for SSA found that agricultural development is the most important contributor to the development of industrial, manufacturing and services sectors. This is because agriculture is the major source of raw materials for industry, is a main purchaser of simple farm tools (farm implements), is a purchaser of services (farm mechanics, transport, hired labour) and farmers are the main consumers of consumption goods produced locally (World Bank, 1993).

Enhanced food security and poverty reduction are critical indices of economic development in African countries. Agricultural production will remain the most important element for addressing food insecurity and poverty since most of the poor and the food-insecure are rural people. Being the largest private sector in the continent, stimulating the private sector is tantamount to stimulating agriculture and agriculture-related industries. Improving the economic activities of women, whose principal economic activities is farming, means in large part helping them to become better farmers (World Bank, 1993). For rural people to increase income in most cases, they will have to increase agricultural production. In addition, since local transport and distribution constraints in many African countries restrict many rural people from accessing food markets in far places, the most secure supply is their own production.

Agriculture holds the promise for Africa's economic growth also because the industry has a lower base from which to grow. A 1% growth in agriculture in SSA has been found to cause economic growth of 1.5 times this magnitude owing to the stimulus to industry, transport and services (Haggblade *et al.*, 1989). Crop yields are ridiculously low while fertilizer use is a small fraction of quantity used in other countries (World Bank, 2006). Irrigated area is about 20% of potential. Infrastructures development lags far behind those of other developing countries while private investment is negligibly low. All this provides large scope for rapid growth which will translate to overall economic growth.

TECHNOLOGY ADOPTION PROCESS IN AFRICAN AGRICULTURE

Innovation and Technology

An understanding of these twin concepts is essential especially because it will quicken our understanding of some other concepts which indirectly provide a picture of the activities involved in the adoption process.

Innovation

The term 'innovation', ordinarily implies something new. This somewhat simplistic definition leads to a set of implications which often change the meaning of the concept and its application (Adekoya and Tologbonse, 2005). It is obvious that the newness indicates, to some extent, a strange-ness of idea. The degree to which something is really new will be a function of time and space because these two factors affect the presence and movement of innovation as it traverses space; communities, countries and continents.

All technologies, ideas and practices have points of origins and will be considered as in-novations in an environment until it has become overwhelmingly popular. The concept is therefore a description of newness which declines with time and across space until it becomes common and loses the name. An innovation is thus an idea, practice or product that is perceived as new by the potential users or adopters. Improved variet-ies, agrochemicals and fertilizers are examples of agricultural innovations (Adekoya and Tolog-bonse, 2005).

Technology

In Africa, development could be perceived to include the sum of efforts directed towards the growth and diversification of production and to-wards increases in productivity. This is assumed to be effectively realized through technology transfer, a process of facilitating the relationship between science and agriculture, allowing ideas to flow or be transferred both ways and resulting in the development of technologies. Technology according to Swanson (1996) is the application of knowledge for practical purpose which is gener-ally used to improve the condition of human and natural environment, and carry out some other socio-economic activities. It is also conceptual-ized as a complex blend of materials, processes and knowledge (Roger, 1995; Van den Ban and Hawkins, 1996). From these definitions, the dif-ference between 'innovation' and 'technology' in agricultural extension literature is as imper-ceptible as that between 'late in the afternoon' and 'early in the evening'. In the context of this chapter, "technology" is not restricted to machines and equipment, but includes the skills, abilities,

knowledge, systems and processes necessary to make things happen in the desired direction. Thus, technologies are meant to be total systems that include know-how, procedures, goods and services, as well as organizational and operational measures.

Technology, therefore, is an essential input in production, and as such, it is bought and sold in the markets as a commodity in one of the fol-lowing forms: (a) capital goods, and sometimes intermediary goods, (b) human labour, usually skilled and sometimes highly skilled and special-ized manpower and (c) information, whether of a technical or of a commercial nature. Ways in which technology can be acquired from external sources, according to Torimiro and Kolawole (2005) include published information, migration of people, education and training, employment of foreign experts, technical co-operation, imports of machinery and equipment and license agreement for production. It should, however, be noted that it is not so much the form or level of technology that is of great importance but its functional relevance in relation to five development issues viz: human, resource, societal, cultural and environmental development.

Types of Innovation/Technology

There are two categories of technology/innova-tion namely:

1. Material technology also known as 'hard-ware' component e.g. improved varieties.
2. Knowledge-based technology also known as 'software' component e.g. planting date (Swanson, 1996; Roger, 1995; Van den Ban and Hawkins, 1996).

Material technology is said to be knowledge that is embodied into a technological product such as tools, equipment, agrochemicals, improved plant varieties, improved breeds of animals, and vaccines. Knowledge-based technology is regarded as technical knowledge and management skills such as planting dates and other information that will help the farmers to increase production. It is worthy of note that crop and livestock technologies of different types, have both hardware and software components that are very complementary (Swanson, 1996; Roger, 1995; Van den Ban and Hawkins, 1996).

Diffusion and Adoption of Innovation/Technology

In discussing the innovation diffusion process, it is appropriate to first briefly mention two other related concepts which are 'change' and 'communication'.

Change and Communication

It is often said that the only thing that is permanent is change because it is always taking place as everyone adjusts to one thing or the other every time. However, change can be viewed as a state of dynamism which preclude stagnation and if well managed and directed, always implies progress and development. Change is as desirable as it is feasible and sustainable, which only ensures that the target is made better off. This is perhaps why it is more desirable to introduce changes through mutually acceptable designs, plans and participatory approaches. Only in such circumstance does change become permanent, longed-for and sustainable (Adekoya and Tologbonse, 2005).

Communication has been variously described as exchange of ideas, skills or simply information. Thus terms like source, channel, content, target and feedback, are usually involved in conceptualizing communication. The similarity between communication and diffusion rests on the transfer of information with the expectation of reaction/response. It should however, be emphasized that diffusion requires greater activity because it is goal-driven and usually appraised using only the expected outcome, all other alternatives being con-

sidered undesirable. The above concepts present preliminary pictures which can be merged to conceptualize the process of diffusion and adoption.

Technology Diffusion Process

The process of diffusion can easily be seen to be a precursor to adoption but not necessarily always ending up positively. Diffusion begins with the actual entry of an innovation into a target system. However, this can be passive or active. The passive one usually takes the form of innocuous information exchange often as a result of human movement from one system to another and the resulting casual interaction which subtly impinges on individual characters and influences the level of modernization at individual level. Subsequently, folks observe this modernization, sometimes emulate it and development of the system results. It should be noted here that no effort or activity has been expended by the 'translocated' individual towards influencing the folks.

However, the active diffusion requires a more technical approach because it is consciously done with a purpose in mind. As mentioned earlier, communication techniques are usually employed since diffusion itself is similar to communication. The extension agent, in this wise, represents the source of information and this behooves of him/her to have a complete knowledge of the innovation/technology from the onset. Bringing his/her professional experience to bear, the extension agent understudies the target and thereby determines the appropriate entry point.

Process of Technology Adoption

Achieving an acceptable level of assimilation and adoption of new and available technologies at the farm level is a function of science, economics and human behaviour. Adoption begins as a mental process often reinforced by other emotions or circumstances and it is termed the innovation-decision process. Adoption is regarded as a decision to make full use of an innovation or technology as the best course of action available (Rogers, 1995). Most farmers are said to go through a logical, problem-solving process known as adoption process when considering any new technology or innovation. A farmer's decision about whether or not to adopt a recommended agricultural practice is recognized to occur over a period of time in stages rather than instantaneously. The adoption process involves an interrelated series of personal, cultural, social and situational factors including the five distinguishable stages of awareness, further information and knowledge, evaluation, trial and adoption (Barao, 2005; Polson and Spencer, 1991; Van den Ban and Hawkins, 1996, Bonny and Vijayaragavan, 2001). These stages are code-named AIETA.

For the purpose of this chapter, the adoption stage is emphasized. The adoption stage is the final stage when an individual applies the innovation on a large scale and continues to use it in preference to other old methods. The adoption process may therefore be defined as the acquisition and processing of information about an innovation followed by behavioural change. The adoption process stated above does not always follow the sequence in practice (Van den Ban and Hawkins, 1996) as it depends on the technology and individual in question.

There are a number of deficiencies inherent in the AIETA adoption process. The process is viewed as always resulting in adoption whereas in reality, rejection of the technology is also possible. The process also presupposes that the stated steps always occur in sequential order while some may be skipped and some, merged. There are characteristics of the innovation which the prospective adopter considers and serve to bolster the decision to adopt. These are

1. Relative advantage: this is a crucial issue especially when the innovation is about an existing practice. It can be in terms of cost, ease of operation and rate of return or associated values.

2. Compatibility: the extent to which a technology fits into an existing situation of values and practices of the target. In Zaire, farmers discontinued with adoption of improved cassava variety simply because they could not cope with the strain of processing using the technology available (Adekoya and Tologbonse, 2005).

3. Divisibility: since innovation holds a degree of uncertainty, adopters usually consider the risk factor. Thus, innovations are considered with a measure of uncertainty, implying the need to try it out either in bits or on a small area of land.

4. Observability: this is a reinforcement issue in diffusion which helps adoption decision. Demonstration plots like Small Plot Adoption Techniques (SPAT) and On-Farm Adaptive Research (OFAR) are quite useful in technology uptake.

5. Complexity: this is the degree to which the adopter finds the innovation easy to manipulate. When innovation builds upon existing practices and skills, the adopters finds it easier.

These characteristics should be considered when trying to transfer any technology. From the above issues, it becomes very clear, that farmers have demand for innovations and therefore, the extent to which the farmers can shift to accommodate the demand determines the level of adoption. The capacity of the farmers to cope in this regard is further influenced by personal characteristics like age, education, social-economic status, farm size etc.

Non Adoption of Technology

Lack of improved technologies to use or non-adoption of extended ones by farmers has been given as the major reasons for low productivity of small scale farmers. Roling and Pretty (1996) opined that one major reason for non-adoption of technologies is because they are finalized before farmers get to see them. Such technologies that do not fit farmers' conditions or needs or that farmers are unable to change are usually rejected. Understanding the farmers' goal and decision-making criteria increases the likelihood of addressing the right problems and valuing innovations correctly. Adhikarya (1996) opined that non-adoption of recommended technology is often related to or caused by non-technological factors such as social, psychological, cultural and economic problems. Farmers were reported to reject available technologies not because they are conservative or ignorant but because they rationally weigh the changes in incomes and risks associated with given technologies under natural and economic circumstances before they take any decisions (Nagy and Sanders, 1990). Unfavorable agricultural policies discourage farmers from adopting market-dependent improved technologies.

An important component with regards to non-adoption of technologies by farmers may be connected with the communication process. One of the reasons for taking the audience, the farmers, for granted may be connected to the inability to make communication effective by using a two-way (rather than a one-way) mechanism in which the farmer is given an opportunity to function as a sender (Fliegel, 1984). This is what is termed 'feedback'.

According to Rogers (1995), current human behavioural research suggests that technology and change will most likely be assimilated and implemented when the benefit of implementation will be quickly realized (usually within 12-18 months), the tools for implementation are readily available and accessible in the local marketplace, the risk of implementation can be diminished and when the change or new technology can be comfortably integrated into other basic aspects of daily life.

TECHNOLOGY ADOPTION AND ECONOMIC DEVELOPMENT IN AFRICA

Models of Technology Transfer

The extension agent represents the source of information and this behooves of him/her to have a complete knowledge of the technology from the onset. With professionalism, the extension agent understudies the target system to determine the appropriate entry point and the model to use. The entry point may be an individual, a group or any other medium. The appropriateness of a model is determined by the following criteria.

1. The popularity and thus representativeness among the group thereby providing legitimization for the information.
2. Availability and readiness to serve in the required way.
3. Personal disposition towards change in general and the technology specifically.

It is not to be expected that individuals that will fit into these criteria will abound in the target system, therefore, it requires the extension or change agent to ferret them out. It should never be assumed that political office holders, institutionalized or formal leaders will automatically fit, but no matter the situation, someone, somewhere will wield more control and claim more allegiance from the generality of the people than any others in the system. Such personality can be used to anchor the information by providing the entry point. Any of the models below can then be applied depending on the relationship between the masses and the recognized leadership.

Hypodermic Needle Model

The model assumes a lack of effective or influential leadership and considers the target system as mass audience among whom there is free flow of information. It derives its name from using a single entry point and subjecting the information to a natural flow depending on the level of interaction in the system. In this model, the technology/innovation is simply introduced into the system through any individual chosen by the extension agent. No further intervention is necessary not even to reinforce the technology.

One- Step Flow Model

The model is designed to disseminate information in a way that every member of the system gets it at the same time. This, expectedly, will be reinforced by interactions among the people as the information is further discussed for clarification depending on its salience and incisiveness. The change agent just ensures that an appropriate communication means is employed to ensure an even reach among the people.

Two-Step Flow Model

This model bases its principles on the assumption that there is no equality in the target system. Emphasis is on the existence of at least two hierarchies; leaders and followers, and that information cannot reach both at the same time. It assumes a virile class system that does not limit interaction but determines capacity and order, hence, the leaders are vantagely endowed to access and examine new technologies as against others. So, information is directed at leaders and, expectedly, will eventually flow to the others. This is the approach being used in the Agricultural Development Programme system in Nigeria in which the extension staff first concentrates attention on contact farmers in teaching and trial of a new technology. This is to enable contact farmers to effectively teach followers when the followers are ready to fall in.

Multi- Step Flow Model

This involves the use of several channels directed at several units in the target system. It aims to create information redundancy and thus force generation of reactions on the issue among dyadic groups. The planning for this, however, depends on several factors prominent among which are availability of channels, ability to use the channels and nature of message.

Diffusion can therefore be explained as a process of information exchange or flow between specific groups or units among a people. The speed of flow will depend on the topical nature or essence of the information, model employed and management of the process by extension professionals. It should be noted that the understanding of the target system by the professional will guide the choice of model and direction of flow. With the information technology revolution, the choice of model assumes reduced relevance as information boundary is broken. All sources of information are now accessible by everyone including the target systems. The issue at stake then reduces to information management (Adekoya and Tologbonse, 2005).

Technology Uptake, Agricultural and Economic Development: Case Studies

A number of cases of increase in general welfare from adoption of agricultural technologies abound in Africa despite the lag in technology adoption between Africa and other developing countries. The popularly mentioned ones are:

Fertilizer Use in Kenya and other Countries

Inorganic fertilizer use is one major component of the broader goal of achieving healthier soils for increased agricultural productivity and food security in Africa. The low use of inorganic fertilizer in SSA compared with other developing regions led to a joint study by the World Bank and UK Department for International Development (DFID) to undertake an Africa Fertilizer Study Assessment (World Bank, 2006). Kenya was used as focus in the e-forum background papers as one African country that has a moderate level of fertilizer use with appreciable growth of national fertilizer consumption in the last decade. The study revealed need for specific interventions that can promote efficient and sustainable use of fertilizer by African farmers especially smallholders. The interventions include enhancing affordability achievable through credit availability, small packs and subsidies. One or a combination of these had been used to enhance fertilizer uptake in Togo, Malawi or Zimbabwe with attendant increases in yield and farmers' income from farming. Also, there is a need to increase incentives for farmers' use. Land leasing and sharecropping arrangements discourage use as landowners tend to retrieve fertilized land with improved fertility status. Also suggested was training of fertilizer stockists in both marketing and business technical skills to enable them provide informed advisory services to farmers. The stockists need also to be assisted in building linkages within the supply chain to qualify them for suppliers' credit. It was found that for there to be considerable increased use of fertilizer, national governments and donor community must note these lessons for inclusion in their agricultural development policy. Therefore, apart from fertilizer technology being extended to smallholders in Africa, a much more proactive public role is required in stimulating fertilizer demand for Africa as a whole. The same strategies above are applicable to appreciably increasing use of other agro-chemicals such as herbicides, insecticides, hormones, etc that even record poorer levels of adoption compared with fertilizer in African agriculture.

Use of Biotechnology Techniques

Agricultural biotechnology refers to a wide range of technologies and products that can improve productivity or quality of crops, livestock, fisheries or forest. Modern biotechnological tools have the potential to significantly raise agricultural productivity in a more environmentally- friendly manner, supply cheaper and more nutritious food, ensure more stable yield (due to increased tolerance to diseases and pests) and contribute to poverty alleviation. Techniques of biotechnology such as plant tissue culture, micro-propagation, molecular diagnostics of plant and animal diseases and embryo transfer in livestock have already been adopted in many African countries. These are simple to use, often inexpensive and relatively free of regulatory requirements and public controversy. Disease free planting materials for plantain was developed by tissue culture and plantlets have been effectively distributed to farmers in Kenya and other East African countries. Net income of participating farmers has increased by 35% with the project being scaled up (Persley and George, 1999, Wambugu and Romano, 2001). Other examples are virus-resistant sweet potatoes in Kenya, insect- resistant maize in East Africa and marker- assisted selection for sleeping sickness in African cattle. Molecular approaches require advanced skills, research laboratories and the capacity to manage intellectual property rights. These requirements pose a constraint for developing countries thus preventing the uptake of these advanced tools.

Use of High-Yielding Crop Varieties

In a large number of African countries, high-yielding crop varieties are widely adopted for crops such as cassava, yam, sugarcane, maize, rice, cowpea, plantain and a host of others. Fantastic yield responses of between 35-300% have been the results with reports that project farmers are better off than non-project farmers. Farmers adopting improved crop varieties and using complementary practices correctly are reported better off than farmers adopting traditional varieties despite selling farm products at lower farm-gate prices compared with non-adopting farmers. Thus, there is a welfare-enhancing effect of agricultural technologies adoption in Africa.

Agricultural Projects with Adoptable Technology Components

There are several success stories in African agriculture of projects with technology adoption components funded by donors, national governments and private companies using own or public extension units etc. These include private sector development efforts in Kenya's horticulture sector; in Cote d Ivoire's tree crop sector and small-scale private agricultural trade and processing in Ghana, Nigeria, Tanzania and Uganda. Other examples are cotton sector development in West Africa, soil conservation projects in Burkina Faso, Rwanda, and Kenya; technology development and uptake in dairy production, fruits and vegetable, small-scale irrigation in Nigeria and Niger and park management in Kenya and Zimbabwe.

LESSONS FOR TRANSFER OF EMERGING TECHNOLOGIES TO AFRICA

Having touched on the process of technology adoption and the impact of technologies on African agriculture, this section distills the information provided to be able to isolate lessons for effectiveness of transfer of nanotechnology and micro-electronics to Africa in particular and developing economies in general. These lessons are:

1.   The fact that incomes from agriculture are low and poverty level very high in Africa must be remembered and this requires that any emerging technology to boost produc-

tion and productivity in any sector must be cheap and affordable to be widely adopted. Any emerging technology that requires considerable cash investment will not be adopted to any reasonable extent in Africa.

2.   Involvement of beneficiaries of any emerging technology in the research for generating such technologies is an absolute necessity if they are to better meet the needs of target beneficiaries. Lessons from agriculture shows that technologies that are finalized before end-users get to see them are poorly adopted. If modifications can be made to the technology on request by users, there is hope for wider uptake.

3.   It cannot be assumed that the agricultural or other technologies developed and used in the industrialized countries will work in Africa and other developing countries. Not all innovations need to be developed locally, but at some point, a technology will be evaluated with respect to whether it meets local needs and conditions. For example findings from African agriculture reveals that (i) the development of vaccines for cattle will need to be tested against regional variants of a pathogen in local breed of cattle and (ii) crop breeding requires the evaluation of phenotypes under local environmental conditions to make improved varieties suitably adapted. Developers of nanotechnology and micro-electronics should take cognizance of this.

4.   The successful implementation of an agricultural technology requires that farmers be convinced of its benefits and understand how it works. The question to ask at this point is what is the magnitude of the expected benefit from adoption of nanotechnology and micro-electronics in Africa? If the target beneficiaries are not significant in number, if it does not address a widespread or severe problem, if it does not provide a complete solution to identified problems, if it does

not empower adopters by directly increasing their income, then it is not worthwhile transferring the technologies to Africa.

5. A long term commitment to the development of human resources in Africa is an urgent necessity at the level of technicians, extension agents, agricultural engineers, and research professionals. Despite having many agricultural extension agents paid by government, a good number of African countries do not have the required resources to effectively serve the farmers. Developers of emerging technologies will then need to give assistance to public extension institutions in Africa or use privately funded ones to push their technologies to target beneficiaries.

6. There is now wider access to the internet and cell phones and this can be explored to introduce an emerging technology. In terms of nanotechnology and micro-electronics, beneficiary-to-beneficiary or peer network could use new tools of information technology as a powerful means of communication. This satisfies the requirement to make the new technology compatible with technologies that are already in used.

7. It is required that any emerging technology in Africa should be a gateway to other innovations in agriculture and other sectors of the economy such that it leverages the development and packaging of other technologies to assist farmers and other beneficiaries.

8. Entry point for nanotechnology and micro-electronics into Africa must be professionally determined. This will involve a critical mass of individuals or groups that will serve as the contact clientele. It should be easy to identify the target having established the relevance of the technology to such groups or individuals. Thereafter, there must be a demonstration of the effectiveness and simplicity of the technology over existing others that achieve the same result. Farmers refuse to adopt apparently attractive and supposedly effective technologies if there is no effective method of technology transfer. Cash incentive had not worked as farmers abandoned technologies when cash incentive dried up. Technologies that are transferred through authoritarian imposition are reported not to be stable because once the coercion is withdrawn or relaxed; the adoption of the technology is discontinued.

From a personal point of view, given the present low level of agricultural productivity in Africa compared with world average performance, enormous opportunity exists for a remarkable improvement with emerging technologies in nanotechnology and micro-electronics. What is capable of undermining success of outcomes however is the weak and inefficient state of public and private capacity for research, development and extension Africa-wide.

CONCLUSION

The chapter looked at the present state of agriculture in Africa and the roles that technology adoption has been playing and can further play in it. While it is appreciated that technology is not the only factor necessary for the development of the largely agrarian African economy, access to improved inputs, methods and knowledge can make a substantial contribution to better agricultural production. In reviewing the technology adoption process in African agriculture, it was shown that technology development is not merely the diffusion, adoption and application of new products or services of technological nature into an area. The technologies that have made impact in African agriculture were the effective, simple, cheap, compatible and observable ones. These attributes enhance voluntary adoption which depends solely on the effectiveness of the model used. The lessons that can be learned by developers and packagers of new technologies for introduction into Africa were distilled for research or policy attention.

REFERENCES

Adegeye, A. J., & Dittoh, J. S. (1985). *Essentials of Agricultural Economics*. Ibadan, Nigeria: Impact Publishers Nigeria Ltd.

Adekoya, E. A., & Tologbonse, E. B. (2005). Adoption and diffusion of innovations . In Adedoyin, S. F. (Ed.), *Agricultural Extension in Nigeria* (pp. 28–37). Ibadan, Nigeria: Federal Agricultural Coordinating Unit.

Adhikarya, R. (1996). Implementing strategic extension campaigns . In Swanson, B. E., Bentz, R. P., & Sofranko, A. J. (Eds.), *Improving Agricultural Extension: A Reference Manual*. Rome: FAO.

Alexandre, M. da Silva. (2009). Carrying capacity in agriculture: Environmental significance and some related patents. *Recent Patients on Food . Nutrition and Agriculture, 1*(2), 100–103.

Barao, S. M. (2005). Behavioural aspects of technology adoption. *Journal of Extension, 43*(2). Retrieved April 14, 2005, from http://www.joe.org/joe/2005april/comm1.shtml

Bonny, B. P., & Vijayaragavan, K. (2001). Adoption of sustainable agricultural practices by traditional rice growers. *Journal of Tropical Agriculture, 39*, 151–156.

Fliegel, F. C. (1984). Extension communication and the adoption process . In Swanson, B. E., Bentz, R. P., & Sofranko, A. J. (Eds.), *Improving Agricultural Extension: A Reference Manual*. Rome: FAO.

Haggblade, S., Hazell, P., & Brown, J. (1989). Farm-non-farm linkages in rural Sub-Saharan Africa. *World Development, 17*(18), 22–31.

IFPRI. (2001). *Good News from Africa - Farmers, Agricultural Research and Food in the Pantry* (Schioler, E., Ed.). Washington, DC.

Mafimisebi, T. E. (2007). Long-run price integration in the Nigerian fresh fish market: Implications for marketing and development. In S. M. Baker & D. Westbrook (Eds.), *Proceedings of a Joint Conference of The International Society of Marketing and Development and the Macromarketing Society,* Washington, DC (pp.149-158).

Mafimisebi, T. E., & Fasina, O. O. (2009). Rural women's productivity and welfare issues: A cause for concern . In Agbamu, J. U. (Ed.), *Perspectives in Agricultural Extension and Rural Development* (pp. 361–386). Owerri, Nigeria: Springfield Publishers Ltd.

Mafimisebi, T. E., Onyeka, U. P., Ayinde, I. A., & Ashaolu, O. F. (2006b). Analysis of farmer-specific socio-economic determinants of adoption of modern livestock management technologies by farmers in Southwest Nigeria. [Retrieved from http://www.World-foodnet]. *Journal of Food Agriculture and Environment, 4*(1), 183–186.

Nagy, J. G., & Sanders, H. (1990). Agricultural technology development and dissemination within a farming systems perspective. *Agricultural Systems, 32*, 305–320. doi:10.1016/0308-521X(90)90097-A

Okojie, J. A. (1999). The role of government and Universities of Agriculture in improving animal production and consumption in Nigeria. *Tropical Journal of Animal Science, 2*(2), 1–7.

Okunmadewa, F. Y. (1999). Livestock industry as a tool for poverty alleviation. *Tropical Journal of Animal Science, 2*(2), 21–30.

Okunmadewa, F. Y., Mafimisebi, T. E., & Fateru, O. O. (2002). Resource use efficiency of commercial livestock farmers in Oyo State, Nigeria. *Tropical Animal Production Investigations, 5*(1), 47–57.

Olalokun, E. A. (1998). Sustainable animal production for food self-sufficiency in the 21st century. In Animal Science at the University of Ibadan: A Commemorative Brochure of Anniversary Home-coming Celebrations.

Olopoenia, R. A. (1983). On the meaning of economic development. In Osayimwese, I. (Ed.), *Development Economics and Planning: Essays in Honour of Ojetunji Aboyade* (pp. 13–29).

Persley, G. J., & George, P. (1999). *Banana, breeding and biotechnology: Commodity advancement through improvement project research, 1994-1998. Banana Improvement Project Report 2.* Washington, DC: World Bank.

Polson, R. A., & Spencer, D. S. C. (1991). The technology adoption process in subsistence agriculture: The case of cassava in Southwestern Nigeria. *Agricultural Systems, 36*, 65–78. doi:10.1016/0308-521X(91)90108-M

Roling, N., & Pretty, J. N. (1996). Extension's role in sustainable agricultural development. In Swanson, B. E., Bentz, R. P., & Sofranko, A. J. (Eds.), *Improving Agricultural Extension: A Reference Manual*. Rome: FAO.

Seers, D. (1972). What are we trying to measure? *The Journal of Development Studies, 8*(3), 36–47. doi:10.1080/00220387208421410

Swanson, B. E. (1996). Strengthening research-extension-farmer linkages. In Swanson, B. E., Bentz, R. P., & Sofranko, A. J. (Eds.), *Improving Agricultural Extension: A Reference Manual.* Rome: FAO.

Torimiro, D. O., & Kolawole, O. D. (2005). New Partnership for Africa's Development (NEPAD) vision for agricultural technology transfer. In Adedoyin, S. F. (Ed.), *Agricultural Extension in Nigeria* (pp. 170–176).

UNICEF. (1990). Strategy for Improved Nutrition of Children and Women in Developing Countries. UNICEF Policy Review Paper 1990/71.

Van de Ban, A. W., & Hawkins, H. S. (1996). *Agricultural Extension* (2nd ed.). London: Blackwell Science Ltd.

Wambugu, F. M., & Romano, M. K. (2001). The Benefits of Biotechnology for Small-scale Banana Producers in Kenya. *ISAA Briefs,* 22.

Williamson, G., & Payne, W. J. A. (1978). *An Introduction to Animal Husbandry in the Tropics* (3rd ed.). New York: ELBS/Longman.

World Bank. (1989). *Sub-Saharan Africa, From Crisis to Sustainable Growth: A Long-term Perspective Study*. Washington, DC: World Bank.

World Bank. (1990). Household Food Security and the Role of Women. (J. P. Gittinger, Ed.). World Bank Discussion papers.

World Bank. (1993). A strategy to develop agriculture in Sub-Saharan Africa and a focus for the World Bank. *Africa Technical Department Series, 203*, 83–90.

World Bank. (1996). Nigeria, Poverty in the Midst of Plenty: The Challenge of Growth with Inclusion. Report No 13053 UNI Western Africa Department, Country Operations Division, May 13.

World Bank. (2002). *A Sourcebook for Poverty Reduction Strategies*. Washington, DC: World Bank.

World Bank. (2005). *Agriculture Investment Sourcebook*. Washington, DC: Agriculture and Rural Development Department, World Bank.

World Bank. (2006). Increasing Fertilizer Use in Africa: What Have we Learned? In C. Poulton, J. Kydd & A. Doward (Eds.), *Agriculture and Rural Development Discussion Paper 25*. Washington, DC: World Bank.

World Bank. (2008). *Country Assistance Evaluation*. Nigeria: Independent Evaluation Group Approach Paper.

Chapter 19
Technology Development and Transfer:
Lessons from Agriculture

Saikou E. Sanyang
National Pingtung University of Science and Technology, Taiwan

ABSTRACT

Agricultural technology development and transfer is a driving force for national development. The perspective is to reduce poverty and hunger facing people by adopting new measures to raise income and attain household food security. This is attainable through the establishment of research institutions, extension services, farmer organizations, and public private participation. The primary function of extension service is to deliver efficient and effective transfer of agricultural technologies to farmers as a factor for rural development. The importance of technology development and transfer approaches in developing countries has been recognized as a tool for economic development. Technology development and transfer economics create employment opportunities, reduce poverty, and enhance economic growth. Therefore, better agricultural technologies and better policies can serve as catalysts for economic return. This chapter examines the components of agricultural technology development and transfer and offers implied lessons for emerging technologies like nanotechnology and microelectronics transfer and diffusion.

INTRODUCTION

Agriculture has two ways to increase its output, expanding the land area under cultivation and improving the yields on cultivated land. If agricultural growth and performance taken to increase the farming incomes of rural families, then a third way can be added, i.e. shifting the

product composition to higher value products. For decades, it has been commented that, globally the possibilities for expanding the land area under cultivation are diminishing steadily. Pursuing that route increasingly risks environmental degradation in many parts of the world as forests are cut down and soils on slopes are eroded and degraded. Therefore, the only viable options that remain are increasing yields and changing the product composition (Ellis, 2007). However, while shifting

DOI: 10.4018/978-1-61692-006-7.ch019

to higher-valued crops and livestock products is a sound strategy from the viewpoint of farmers, it does not help to increase food supplies in aggregate. For this reason, the only course left is to increase yields.

In addition, for many poor farmers and less privileged, especially women and youths, who do not have adequate access to diversified markets or cannot fulfill other requirements for shifting to higher valued crops, increased yield is the only route to higher incomes. Increasing agricultural productivity is all the more urgent because majority of the developing world poor are found in rural areas and the sector's average productivity is declining in many low-income countries (Umali, 1997).

The significant majority of the poor continue to depend on agriculture for their livelihood. Of the 720 million poor identified by the World Bank, 75 percent live in rural areas. Thus increasing farmers' incomes through improved productivity is an important element in agricultural policy development and poverty reduction strategies. Yields can be increased dramatically by the application of appropriate agricultural technologies like small scale irrigation system. Adopting irrigation technologies requires the training of farmers and the provision of extension services over a sustainable period, but it can achieve substantial increases in yields without new agricultural research.

The paradigm of involving farmers in research and development based on evidence-enhancing farmers' technical skills, research capabilities, and involving them in decision-making in technology, transfer and adoption process is vital. This would result to innovations that are more responsive to their priority needs, and would surmount the constraints. Linking the technology development process to research, extension and farmer has the potentials to promote agricultural production, and adoption of agricultural technologies (Morris and Doss, 2002, Bozeman et al., 1995). Agriculture follows the principles of nature to develop system conducive for crops and livestock production for

sustainability (Wilson, 1986). Sustainable development of agricultural policies and technology is a social value whose success is indistinguishable from vibrant rural communities, for household food security (Tilt, 2006).

Understandably, agencies that sponsor research and development projects have adapted to the rationale of technology development and transfer as one of the benefits. Appropriate use of better agricultural technologies and better policies are sets of commonly accepted practices of farm economy as a perquisite for sound economic development. Off-farm, consumers and grassroots farmers' are working to create local markets and farm policies that would support sustainable practices (Day, 2003).

Technology transfer is a kind of spontaneous process from farmer to farmer and from one field to another with the ultimate aim of promoting human welfare (Bozeman et al, 1995). One of the most effective tools has been the acceleration of technological innovation in the developed and developing world (Bell1,1987). Adesina and Forson, (1995) emphasized that, not effective is as an attribute of technological change or adoption to agricultural research and extension but on innovation and the institutional arrangements that favors the implementation of programmes. Cohen and Leviathan, (1990) indicated that agricultural technologies sometimes adapted without any effort if new technology fits farmers' needs and assuming that farmers are knowledgeable and skillful. The idea of research, extension and farmer linkages in the field of appropriate technology transfer and adoption is to bring different stakeholders together acquiring and sharing of knowledge. The participatory technology development approach can strengthen local research and adaptive capacities by involving a vast number of communities based researchers, farmers and extension agents of various government agencies, and NGOs (Blalock, 2004). Just and Huffman (2004) notes that rural labor shortage, and high wage rates to labor saving cultivation methods are necessary to speed up farm mechanization and promote the use of integrated farming techniques. The ability of farmers to adopt new agricultural technologies is due to economic difficulties, beliefs and culture. All aspect of their production system and economic behavior must be taken into account when policies and agricultural technologies are developed (Douthwaite et al, 2001).

THE POLICY CONTEXT OF INSTITUTIONAL CONSIDERATION AND TECHNOLOGY TRANSFER

A favorable policy framework and adequate funding may be necessary for the success of agricultural research and extension. In order to reverse these trends and put agricultural productivity back on upward course, agricultural research must undergo extensive institutional transformation. One of the main challenges is to find a viable way to involve non-governmental organizations, universities, foundations, producers' associations, and private companies in the research process (Bozeman et al. 1995). A second challenge is to orient research more effectively toward the producers' needs, by more closely involving farmers in decision-making processes regarding research strategies.

Transformations of this nature are to increase access to assets, i.e. land, water, education, and health that needs to be accelerated for policy development and technology transfer processes. If these factors are in place it will be much easier to convince lending institutions to redirect their attention to agricultural research as in the past. The promotion of innovation through science based agricultural technology should be a priority concern to present governments of developing countries. The driven approach by rapidly growing private investment in research and development (R&D), the knowledge divide between the industrial and developing countries is widening. Therefore, including both public and private sources, developing countries invest only a ninth of what

industrial countries put into agriculture R&D as a share of agricultural GDP. The government of developing counties should create an environment friendly to enhance public-private participation for agricultural development. Therefore, to bridge the gap, sharply increased investments in research and development must be at the top of policy agenda (Escobar et al, 2000).

Agriculture is not an island in the economy and ultimate objective is to support national development. In agriculture, as in other areas, economic policy responds to national imperatives and to a social and political vision (Ellis, 2007). Therefore, the basis of a strategy, or set of policies, should be the enunciation of broad social goals for agricultural and the rural sector. Fundamentally, there is need to relate the promotion of *human development.* Specific objectives for the agricultural sector derived from this overarching goal. In most economies, the ways in which agriculture can most effectively support human development are (a) ensuring that *nutrition and other basic material needs* are available in rural areas, and (b) contributing indirectly to the satisfaction of those needs in urban areas. In some transition economies, nutrition levels are high enough that they are no longer a general concern, but meeting other material needs is very much an issue, given the prevalence of poverty in rural areas. In many developing countries, nutrition levels are still deficient among a significant part of the rural population, although it is important to recognize that, the world as a whole, the share of the population in poverty has dropped markedly over the past three decades (Erlebach, 2006). In many parts of the world, it has long been the practice to define the aim of agricultural technology development strategy as increasing production levels (Nassauer, 2002).

However, producing more staple foods is important, a physical target of that nature is not sufficient for promoting the goal of human development, or even the objective of raising levels of material well-being. *Income* is a better indicator, as it takes into account the price farmers receive and their costs of production. *Real income* which adjusts net income levels for the rate of inflation to measure the *purchasing power of rural households*. Therefore, agriculture can make its most effective contribution to nutrition and other basic material needs by generating income that is more real for rural households. This contribution, in turn, depends on three factors, namely *production, real farm gate prices,* and *non-farm employment in rural areas*. Real prices are usually beyond the control of farmers themselves but be influenced by policies. As limits on the availability of cultivated land reached, and sometimes exceeded, in many places in the world, production increases in the future will increasingly depend on policies and technologies to deliver improvements in productivity (Haggable, 2005).

The Appropriateness of Technology for Smallholder Farmers

In the end, various kinds of diagnoses of research systems and their effects underscore one central message: in the words of Umali (1997), it is the *'importance of getting the technology right'*, and the message valid 'whether the technology evolves over time by farmers themselves, borrowed directly from other parts of the world, or borrowed and then locally adapted'. Reversing the decline in research budgets may be an integral part of any reform to the system, but how to ensure the appropriateness of the technologies developed is the single greatest, and most enduring issue faced by agricultural research systems Roling and Pretty, (1998). Risk aversion goals tackle by many means, including reducing the height of stalks of grains, accelerating the crop maturation process, reducing the dependence on purchased inputs to reduce financial risk, increasing resistance to pests, and diseases (Sanyang et al, 2008 and 2009).

Moreover, farmers themselves have developed many traditional modes of reducing risk, including intercropping, crop diversification, and holding widely scattered plots. Hence, the 'appropriateness' of new agricultural technology has to be evaluated not only in the dimension of yield increases or in terms of increases in net income per hectare as well as risk aversion. Other factors that bear on the appropriateness of technology include its environmental implications, gender equality, access to market, and the need for agro industrial process of product quality. These considerations bear directly on the issue of *defining priorities for research programs* and criteria for selection of programs targeting maximum outputs. Improving the productivity, profitability, and sustainability of smallholder farmer farming is the pathway out of poverty using agriculture for development. A broad array of policy instruments, many of which apply differently to commercial smallholders and to those in subsistence farming to achieve the following:

- Improve price incentives and increase the quality and quantity of public investment.
- Make product markets work better.
- Improve access to financial services and reduce exposure to uninsured risks.
- Enhance the performance through science and technology.
- Make agriculture more sustainable and a provider of environmental services.

An agricultural technology that may be appropriate for large-scale farms with fertile, flat topography land and ready access to production finance may not be as appropriate for small farms on hillsides and with no collateral (Montague et al., 1998). Roling and Pretty, (1998) have stated on the urgency of making research more relevant to the situation of smallholders, especially in demanding agro ecological conditions. Demand-driven research should involve the intended beneficiaries' farmers and other stakeholders in its design, implementation, monitoring, and evaluation. The expansion of on-farm adaptive research encourages beneficiary involvement, but this has not always occurred in projects and often in a limited manner. Researchers need to know the circumstances of farmers whether direct interaction with farming communities or their representatives, significant reliance on intermediaries in public or private extension systems, or a combination of these approaches (Herdt, et al, 1984).

Regardless of the methods used, this interaction has to be an integral part of the research process. The logical implications and relevance of this imperative have engendered an approach called *'participatory technology development'*, in which researchers and farmers become full partners in the process of research and technology dissemination. This approach is basing on the recognition that, scientists alone cannot generate site-specific technologies for the wide diversity of conditions of resource-poor farmers throughout the world, or even within one country the knowledge and skills of farmers.

Participatory technology development serves as a tool to strengthen the capacity of farmers and rural communities to analyze ongoing processes and to develop relevant, feasible and useful innovations. The process of technology development closely linked with a process of social change the planning and assessment obliges the participants to take account of their situation and the responsibilities of different people in the community (Walker, 2007).

How to organize such collaboration with farmers and rural communities is a major issue facing national agricultural research systems. The key to dissemination is training of trainers, both farmers and researchers, sensitizing national research and extension institutions to avoid trying to deliver technology messages to farmers in top-down fashion. As opposed to having farmers to participate in developing their capacities, giving true control over key aspects of the research process to farmers, and endowing each with a small fund for financing the inputs into the research.

Government institutions sometimes cannot give cash to particular groups of farmers. NGOs play a critical role in the process in financing small groups of farmers as well as working with farmers on research issues.

In addition, NGOs and donor agencies take grass roots farmers to be most successful in localities where the degree of farmer organization is already cohesive and strong. The ability of farmers to contribute fruitfully to the research process has been well illustrated not only by NGOs but other case studies in Rwanda, and Zimbabwe. This ability was exemplifying by experiments in which women farmers were allowed to make their own varietal selections. In the test, the varieties selected the women made comparison with those of researchers. The women farmers were invited to examine more than 20 bean varieties at the research stations and to take home and grow two or three varieties they thought are most promising. They planted the new varieties using their own methods of experimentation. Although the women's criteria were not confining to yield, the breeders' primary measure for ranking, their selections outperformed those of the bean breeders by 60 to 90 percent.

Van de, Pulley, Gurad, and Venkataraman (1999), pointed out that participatory technology development represents a response to a different *goal of the development process* that is more than the usual goal of increasing incomes of poorest groups in developing countries. It is the goal of helping rural people to gain more control over the direction to improve on living standard. The role of participatory research is develop and enrich human capabilities has been made by (Doss and Morris, 2001). The integration of formal research into the participatory technology development process enabled both farmers and researchers to jointly develop technologies to have benefits in terms of data (researchers and policy makers) and a deeper understanding of processes, farmers and researchers.

This human factor is the starting point for sustainable bottom-up development approach for technology dissemination. Working directly with rural communities to strengthen their capacity to articulate their agricultural knowledge base and carry out adaptive research on their own farms contributes to their lives. Echevarría, (1998) iterated on slow response of agricultural research systems to the new challenges, and the need for better institutional golden hand-cuff for the kinds of research method appropriate for agricultural development. National research organizations need to broaden their research agendas and to give greater attention to surmount poverty alleviation, environmental degradation and resource management. In addition, agricultural technologies are becoming more management intensive, through the substitution of improved information for environmentally harmful chemicals (e.g. integrated pest management) and demands on all sectors of society to reduce costs in order to increase competitiveness. Advances in molecular biology and information technology have opened new avenues for agricultural research that can reduce the costs of developing improved technologies. .

However, government activities in the agricultural sector in most developing countries have shrunk because the private sector is not 'filling the gap'. The move toward a more poverty oriented and environmentally sound research agenda has been very slow Umali, (1997). Institutional structures to break down the separation of research by commodity, discipline, incentive systems and to develop accountability of impacts at the farm level, need demands on research systems.

The Importance of Agricultural Extension in Technology Transfer and Development

Funding and management deficiencies led to observed syndrome of extension agents spending more time in the office than on farms, and linkages between extension and research services generally have been weak. According to, William, (2001), many low-income developing countries, agricultural and rural extension is in disarray, a fact that countries must now accommodate the new paradigm increasingly shaping by global trends toward market-driven and highly competitive agribusiness enterprises. One of the most serious systemic problems with public agricultural extension services is the lack of adequate incentives for the agent to serve the client satisfactorily. The client or customer is the farmer, and most extension services have not had a strong orientation of 'serving the customer'. This has translated into lack of timeliness of the services, lack of responsiveness to the farmer's own problems, which may be different from those envisaged by researchers and in worst cases lack of attention to the majority of the farmers (Bozeman, 1995).

Clearly, this critique is not true of all extension systems or agents in the weaker systems, but relevant to many situations. The opposite side of the accountability is *expecting the beneficiaries of extension to be responsible for some of the support*, even if only a proportion of total costs. First, it gives the beneficiaries ownership and drawing rights on the services. Secondly, it takes some of the financial pressure off of the central government and therefore responds to the issue of financial sustainability. Lastly, if ownership and responsibility rest with clients, the basis for more demand-driven, responsive service is established. *The most important policy initiative that is needed in extension is to shift the primary focus of power and responsibility for extension to the clients. To borrow Robert Chambers' phrase, we need to 'put the farmers first'. There is abundant evidence that the 'normal' incentive system facing government employees, even under the most enlightened circumstances, puts a premium on not making a mistake and on length of service but not necessarily on service to clients, particularly small farmers,* Just and Huffman, (2004).

Agricultural Policy and Technology Transfer for Poverty Alleviation

The extent of the rural poverty problem in the developing world for rural poor will continue to outnumber the urban poor well into this century. Rural development programs as originally conceived have been declined in some international agencies. There is recognition of the decline of importance of rural development in national agendas and decreased lending portfolio of the World Bank for rural development activities. Despite the strategic importance of rural development and its potential to reduced poverty. To some extent, broad rural development programs replaced by a series of specific initiatives to promote agriculture, such as land funds better management of irrigation system and community participation in agricultural research and extension Food and Agriculture Organization and World Bank, (2002).

Rural development can become another integrating dimension of an agricultural strategy and at the same time carry its policy beyond the sector. It is understood that the linkages between agricultural and non-agricultural activities in rural areas are strong, and the latter constitute significant sources of employment and income for rural families and rural development issues must be integrated into agricultural policy (Renkow, 2005).

In many countries, rural development policy is limited to agricultural policy and no country has ever solved the problem of rural poverty exclusively on the farm. Furthermore, new approaches understanding the dynamics of rural households have emerged in recent years. The analysis of single production activities replaced by the study of households as a diversify enterprise, including non-farm activities that plays an important role in rural livelihood strategies. Non-farm activities not play an important role in combating rural poverty; may also have direct effect on agricultural decision-making. In addition, to research on agricultural production, the research agenda for rural development should also consider non-farm activities, institutional arrangements that facilitate rural development and environmental such as water, carbon and biodiversity in developing countries (Roetter, Vankeulen, Hengsdijk and Laar, 2007).

In addition, to the direct benefits of creating non-agricultural jobs in rural areas, taking into account off-farm employment opportunities for smallholders the measure of the opportunity cost of their time can be crucial for the design of agricultural policies and production technologies that are acceptable, workable and pragmatic. In the short run, this opportunity cost can be different for different members of a household. An integrated view of how a rural household would function including the traditional divisions of labor by gender is necessary in order to formulate realistic approaches to rural development (Kuiper et al 2007). One principal initiative or deals with rural social issues and economic policies for the rural poor. A major concern was the very low standard of living of former collective farm members, mainly the elderly, whose only remaining productive assets are tiny household plots.

Accordingly, the strategy recommended titling those plots, with no charge to the families concerned, and widening the net of retirement benefits to cover former collective farm members (Kuiper et al, 2007). It is beyond the scope of this present study to try to synthesize the literature on rural development, or the richness of experience in this field in agencies such as the Food and Agriculture Organization. Observations offered were illustrative of rural development experiences and conceptual issues, and the link between agricultural development policies and rural development explored. The orientations underlying a renewed approach to rural development projects summarized, and a conceptual framework for guiding resource allocation in rural development suggests for sound and sustainable development (Roling et al 1998). Finally, the importance of better education for rural populations strongly needs attention toward devolving responsibility for acquiring new knowledge to the farmers themselves. The receptivity of rural populations to new information and their ability to assimilate and apply it increases markedly with levels of education (Tilt, 2006). Education is the most important determinant of the ability of rural populations to improve their well-being. When there is a choice at the margin between allocating resources, for a given rural population, to agricultural extension and allocating them to basic literacy training, the decision inevitably need in favor of the latter.

Linking Agricultural Policies and Technology Transfer to Rural Economy Development

A central question for poverty alleviation programs is how to devise a set of policies that will help to place the rural poor on a self-sustaining growth pathway. Than continue to provide handouts to meet their immediate urgent needs, these may be programs of food assistance and health care are required for the poorest groups in the population and must continue, to assist them to develop capabilities to satisfy their needs through their own efforts (Sachs, 2003). In addition to programs that respond to the symptoms of poverty such as malnutrition and high incidences of disease, policies are required to diminish the causes of poverty, i.e. enhancing the earning capacity of low-income households. In the agricultural sector, other possibilities arise including, but not limited to improved access to cultivable land, improved access to technology, training, and improved access to production credit Henson, (2006). The question has broader ramifications, it is of concern as *how to design policies that not only promote growth in general but which improve the lot of the poor*, or at least prevent deterioration in their condition while other groups prosper. Traditionally, stabilization and growth viewed as falling in the purview of *policy*, but the alleviation of poverty reduced to the realm of *programs* and *projects*. The question is can *policies* be developed that will simultaneously promote income growth in general and growth for the poor and hence, agricultural policies need to incorporate a special focus on poverty alleviation (Shetty, 2006). Market-oriented is one of the factors that can link agriculture to rural economy. Market-oriented smallholders can be highly successful in food markets and new agricultural and economic development (Shetty, 2006). For many smallholders, agriculture is a way of life that offers security and complement earnings in the labor market and from migration.

Households in prosperous agricultural regions may diversify into non-agricultural activities to take advantage of attractive opportunities. Those in less-favored environment may shift into low-value nonagricultural activities to cope with risks. Households with asset endowments may seize remunerative opportunities in the nonfarm sector. Farmers lacking land or livestock may have driven into low-value non-farm employment. Labor market income may be important where population pressures on limited land resources are high or where seasonal income from farming is insufficient for survival in the off-season, due to chronic rainfall deficits, prices, or diseases (Mosley and Sulieman, 2007). Off-farm income can be important for both poor and rich households. The rich often dominate lucrative business niches. The rural poor lacking access to capital, educations, and infrastructure, are not the main source beneficiaries of the more lucrative sources of non-farm income (Sumner, 2005). In some cases, it may be necessary to develop a system of rural or agrarian tribunals that can act swiftly, and at low cost to in reinforcing the integrity of contracts. These are examples of policies that reduce the economic distance between small farmers and the institutions that serve the sector, providing those with more nearly equal economic opportunities, as compared to those enjoyed by large farmers and their urban counterparts (Rass, 2006).

CONCLUSION

In the developing countries, there is a need to prioritize better policies and better agricultural technologies as it can facilitates quick return of economic growth. The importance of agriculture to development approaches been recognized. It is the economic pillar of any developed or developing country and it contributes to employment opportunities, poverty alleviation, economic growth and foreign exchange earnings. Many of the world's poorest people depend on agriculture for a living, their development success or failure is often an outcome of what happens in the agricultural sector (Addison, 2005). Furthermore, harnessing the best of better agricultural policies, scientific-based knowledge and technological breakthrough is crucial as governments of developing countries should shift and re-direct their blue print programmes to agriculture to face the challenges of an increasingly commercialized and globalized agricultural sector. Appropriate use of agricultural technologies can also provide new impetus for addressing the problem of production variability and food security of rural populations living in marginal production environment.

Participatory approaches to extension are effective and efficient for agriculture development programmes. These are approaches that use local knowledge as much as possible, that use farmers as extension agents and researchers to an extent, that work with groups of farmers rather than individual contact farmers for each area. They involve producers in problem identification, setting priorities among issues to tackle, solving problems through analysis, and making choices.

Besides, the importance of better education for rural populations strongly needs attention especially in light of the trends toward devolving responsibility for acquiring new agricultural knowledge and skills to the farmers. The receptivity of rural population to new information and their ability to assimilate and apply it increases markedly with levels of education. Education is the most important determinant of rural populations to improve their well-being. When there is a choice at the margin between allocating resources, for a given rural population, to agricultural extension and allocating them to basic literacy training, the decision inevitably need to favor of the latter. Basic literacy opens the doors to many kinds of development that are otherwise impossible.

Therefore, in conclusion better policies and access to appropriate agricultural technologies, training, and related improvements in research, extension and farmer linkage system, and marketing institutions are essential for the adoption of better agricultural policies and better technologies for farmers in developed and in developing countries. It should be part of government policy agenda to invest public funds in to research and development and encourage the private sectors to follow suit. This will help the government to boost its economic growth, generate income and attain household food security for its population.

REFERENCES

Addison, T. (2005). *Agricultural Development for Peace*. Research paper No. 2005/07, United Nations University, World Institute for Development Economics Research.

Adesina, A. A., & Forson, J. B. (1995). Farmers perception and adoption of new agricultural technology. Evidence from analysis in Burkina-Faso and Guinea. *West Africa, 13*, 1–9.

Bell, M. (1987). *The acquisition of imported technology for industrial development: Problems of strategies and management in Arab region*. Baghdad: ESCWA, United Nations University.

Blalock, D. J. (2004). Commercializing USDA innovations via public-private partnerships. Presentation to the agricultural Biotechnology Research Advisory Committee, Washington, DC.

Bozeman, B., Papadakis, M., & Coker, K. (1995). *Industry perspectives on commercial Interactions with Federal Laboratories*. Report to the National Science Foundation, Research on Science and Technology Program, Contract No. 9220125.

Cohen, W. M., & Levinthal, D. A. (1990). Absorptive capacity: A new perspective on learning and innovation. *Administrative Science Quarterly, 35*, 128–152. doi:10.2307/2393553

Day, R. K. (2003). Transferring public research: The patent licensing mechanism in agriculture. *The Journal of Technology Transfer, 28*(2), 111–130. doi:10.1023/A:1022934330322

Doss, C. R., & Morris, M. L. (2001). How does gender affect the adoption of agricultural innovations? The case of improved maize technology in Ghana. *Agricultural Economics, 25*(1), 32.

Douthwaite, B., Keatinge, J. D. H., & Park, J. R. (2001). *Learning Selection: An Evolutionary Model for Understanding, Implementing and Evaluating Participatory Technology Development*. UK: Department of Agriculture, University of Reading.

Echevarría, R. (1998). Agricultural research policy issues in Latin America: an overview. *World Development, 26*(6), 107.

Ellis, F. (2007). *Small-Farms, Livelihood diversification, and Rural-Urban Transitions: Strategic Issues in Sub-Saharan Africa*. Paper presented at the future of small farms workshop, June 26, Wye, Kent, UK.

Erlebach, R. W. (2006). *The Importance of Wage Labor in the Struggle to Escape Poverty: Evidence from Rwanda*. UK: University of London.

Escoba, J., Reardon, T., & Agreda, V. (2000). Endogenous institutional innovation and agro-industrialization on the Peruvian coast. *Agricultural Economics, 23*(3), 267–277. doi:10.1111/j.1574-0862.2000.tb00278.x

Haggable, S. (2005). *The Rural Nonfarm Economy: Pathway out of Poverty or Pathway In*. Paper presented at the Future of Small Farms conferences, June 25, Wye, UK.

Herdt, R. W., Castillo, L., & Jayasuriya, S. (1984). The economics of insect control in the Philippines. In *Judicious and Efficient Use of Pesticides on Rice, Proceedings of the FAO/IRRI Workshop, International Rice Research Institute*, Los Baños, Laguna, Philippines.

Just, R. E., & Huffman, W. E. (2004). *The role of patents, royalties, and public-private in university funding. Mimeo*. College Park, MD: University of Maryland.

Kuiper, M., Meijerink, G., & Eaton, D. (2007). Rural Livelihood interplay between farm activities, non-farm activities and the resource base. In *Science for agriculture and rural development in low-income countries* (pp. 77–95). Dordrecht, The Netherlands: Springer. doi:10.1007/978-1-4020-6617-7_5

Montague, Y., Ratta, A., & Nygaard, D. (1998). *Pest Management and Food Production: Looking to the Future*. Food, Agriculture and the Environment Discussion Paper 25. Washington, DC: International Food Policy Research Institute.

Morris, M. L., & Doss, C. R. (2002). *How does gender affect adoption of agricultural innovations? The case of improved maize technology in Ghana*.

Mosley, P., & Sulieman, A. (2007). Aid, agriculture and poverty in developing countries. *Review of Development Economics, 11*(1), 139–158. doi:10.1111/j.1467-9361.2006.00354.x

Nassauer, J. I. (2002). Agricultural Landscapes in Harmony with Nature. In Kimbrell, A. (Ed.), *The Fatal Harvest Reader: The Tragedy of Industrial Agriculture*. Washington, DC: Island Press.

Rass, N. (2006). *Policies and Strategies to Address the Vulnerability of Pastoralist In Sub- Saharan Africa*. Rome: FAOP, Pro-poor Livestock Policy Initiative. Working paper series.

Renkow, M. (2005). Poverty, productivity, and production environment: A review of the evidence. *Food Policy*, *25*(4), 463–478. doi:10.1016/S0306-9192(00)00020-8

Roetter, R. P., Van Keulen, H., Hengsdijk, H., & Laar, H. H. (2007). Editorial sustainable resource management and policy option for rice ecosystem. *Agricultural Systems*, *94*(Special Issue), 763–765. doi:10.1016/j.agsy.2006.11.003

Roling, N., & Pretty, J. (1998). Extension's role in sustainable agricultural development. In B. Swanson, R. Bentz, & A. Sofranko (Eds.), *Improving Agricultural Extension. A Reference Manual*. Rome, Italy: Food and Agriculture Organization of the United Nations (FAO).

Sachs, J. (2003). *The Case for Fertilizer subsidies for Subsistence farmers*. New York: Colombia University.

Sanyang, S. E., Te-Cheng, K., & Wen-Chi, H. (2008). Comparative study of sustainable and non-sustainable interventions in technology and transfer to the women's vegetable gardens in the Gambia. *J. Technology Transfer*, *34*, 59-75. Retrieved from DOI 10.1007/s10691-008-9084-0.

Sanyang, S. E., Te-Cheng, K., & Wen-Chi, H. (2009). The impact of agricultural technology transfer to women vegetable production and marketing groups in the Gambia. *World Journal of Agricultural Sciences*, *5*(2), 169–179.

Shetty, S. (2006). *Water, Food Security and Agricultural Policy in the Middle East and North Africa Region*. World Bank: Middle East and North Africa Working paper 47.

Sumner, J. (2005). *Sustainability and the Civil Commons: Rural Communities in the Age of Globalization*. Toronto, Canada: University of Toronto Press.

Tilt, B. (2006). Perceptions of risk from industrial pollution in China: A comparison of occupational groups. *Human Organization*, *65*, 115–127.

Umali, D. D. (1997). Public and private agricultural extension: partners or rivals? *The World Bank Research Observer*, *12*(2), 203.

Van de Ven. A. H., Polley, D. E., Garud, R., & Venkataraman, S. (1999). Innovation and agricultural technology transfer and adoption. New York: Oxford University Press.

Walker, T. (2007). *Participatory Varietal Selection, Participatory Plant Breeding, and Varietal Change*. Background paper for the WDR, 2008.

William, M. R. (2001). *Agricultural and Rural Extension Worldwide: Options for Institutional Reform in Developing Countries. Paper prepared for the Extension*. Rome: Education and Communication Service, Draft, Food and Agriculture Organization of the United Nations.

Wilson, I. (1986). The strategies management of technology: corporate fad or strategic necessity? *Long Range Planning*, *19*(2), 45–80. doi:10.1016/0024-6301(86)90216-5

Chapter 20
Technology Transfer and Diffusion in Developing Economies:
Perspectives from Agricultural Technology

Edwin. M. Igbokwe
University of Nigeria, Nigeria

Nicholas Ozor
University of Nigeria, Nigeria

ABSTRACT

The early years of the green revolution heralded a new era of technology adoption and increasing productivity in agriculture. This momentum has not been sustained, giving rise to food shortages and widespread poverty in developing countries. This chapter reviews processes and models of technology transfer in agriculture in developing economies and concludes that previous efforts were not demand driven and therefore lacked the ingredients for diffusion. The drivers of technology transfer are discussed. A number of factors responsible for the low rate of technology transfer especially the absence of public policies on technology transfer are identified and linked to the transfer of emerging technologies, mainly biotechnology and nanotechnology. The chapter recommends the development of public policies, development of the private sector, establishment of partnerships between the two sectors and development of infrastructures especially in rural areas.

INTRODUCTION

When the green revolution, characterized by heavy doses of external inputs such as fertilizers and improved seeds, was initiated in the 1960s and 1970s in much of the developing world and especially South Asia, it was hailed as a panacea

DOI: 10.4018/978-1-61692-006-7.ch020

to the ever recurring incidence of hunger, famine, malnutrition and to some extent diseases. Early evaluation reports showed high levels of technology generation and transfer, adoption and diffusion of technologies and substantial increases in yield. Less than half a century later the developing world is characterized by rising food prices and food shortages and the consequent food riots, poverty and disease and conflicts. In much of the region, especially sub-Saharan Africa farmers have reverted to farm 'extensification' rather than intensification resulting to grievous damage to the environment and low productivity. Have the technologies of the green revolution failed or did they ever take root in societies? Can emerging technologies including nanotechnology, micro-processing and biotechnology offer remedies? What lesson can be learnt from earlier processes of transfer of technology?

The chapter examines the models and processes of agricultural technology transfer (adoption/diffusion) and the factors affecting them and draws relevant lessons for emerging technologies. Characteristics of emerging technologies that differ from those of traditional agricultural technologies and strategies that contribute to successful transfer of technology are discussed. Within this context, transfer of technology refers to the process of technology transfer in a social system. This may result to adoption and diffusion in which a technology is selected for use by an individual and the social system. Technology refers to any new knowledge, material or process that improves productivity, and is synonymous with innovation.

AGRICULTURAL TECHNOLOGY

Technology has been a major driver of both the agricultural productivity increases of the past century and the financial success of many farm and agribusiness firms. The challenges of bringing new technology to market in the agricultural industry are changing - it is no longer adequate to conceive a new invention and convince farmers with a strong marketing campaign that they should adopt the technology that results from this invention. Technology, especially in the developing world has, to a large extent, been dominated by multinational corporations looking for large profits. Without the means to buy the technologies they offer, farmers have retained old farming techniques and

the result has been a less productive agriculture sector. And with the cautious approach to use of biotechnology in agriculture, farmers have also been unable to benefit from these technologies. Thus many farming systems face low yields, high crop losses, and high production costs.

Agricultural technology focuses on techno-logical processes used in agriculture. The New World Encyclopedia (2008) defines agricultural technology as technology for the production of machines used on a farm to help with farming. Agricultural machines have been designed for practically every stage of the agricultural process. They include machines for tilling the soil, plant-ing seeds, irrigating the land, cultivating crops, protecting them from pests and weeds, harvesting, threshing grain, livestock feeding, and sorting and packaging the products. The definition is narrow as it focuses on mechanization of agriculture leaving out non mechanical processes such as development of improved breeds of animals and crop species/genetic engineering.

Types of Agricultural Technology

Research investment produces two kinds of technology: production technology and research and development (R&D) technology and the corresponding impacts are respectively produc-tion impact and institutional impact. Production technology refers broadly to all methods which end-users use to cultivate, harvest, store, process, handle, transport and prepare food crops and livestock for consumption. They are basically physical inputs. R&D technology refers to the organizational strategies and methods used by R&D programs in their work. These are mainly services, like extension advice and capacity build-ing. Production impact refers to the physical, social and economic effects of new technology on crops and livestock production, distribution and use, and on social welfare in general. Institutional impact refers to the effects of the R&D technology on the capacity of research and extension programs to generate and disseminate new production technology.

Need for Agricultural Technology

Extension services will need to be radically restructured to make technology dissemination responsive to small farmers. Innovative institutional arrangements would need to be evolved to make the extension system farmer-driven and farmer accountable. There is need to seek technological interventions that will improve production systems, agricultural trade and commerce, and stimulate broader and more equitable economic growth on a sustainable basis. With new, more productive cultivars and innovations, farmers can produce more output at the same cost, or the same level of output at a lower cost. Farmers will benefit unequivocally from lower marginal production costs—they would be able to supply more at a constant price, giving them higher revenue and higher net returns.

TECHNOLOGY TRANSFER AND DIFFUSION IN CONTEXT

Technology transfer is more than just training, but training certainly is an integral part of technology transfer. Technology transfer involves:

- Identification of user needs (needs assessment via questionnaire, focus group, market research, and direct contact, to name a few methods),
- Information exchange (via newsletters, manuals, videos, training courses, demonstrations, direct technical assistance, software, etc.),
- Implementation of research findings (which can include licensing, training, marketing, and more), and
- Feedback (to the developers and manufacturers of the technology concerning prob-

lems identified, suggestions for improvement, etc.).

In the broadest sense, technology transfer is a process of communication that results in putting research findings or new information into practice. Research is implemented as a result of technology transfer activity, whether the process of technology transfer is formally engaged in or not. Implementation of research is more likely to occur, however, when technology transfer is practiced formally and purposefully. To be most successful, technology transfer must engage all those involved in the research and implementation process. Technology transfer should not only be a consideration upon the conclusion of research; instead, it is a process that most effectively is integrated throughout the entire research effort, resulting in greater benefit from the research results.

Technology Diffusion

Everett M. Rogers in his classic work, *Diffusion of Innovations,* describes diffusion as the process in which an innovation is communicated through certain channels over time among the members of a social system (Harder and Benke, 2005). He also states that diffusion is concerned with new ideas and includes social change. Rogers' four main elements are the innovation, communication channels, time and a social system.

Mock, Knenkermath and Janis (1993) discuss the diffusion process as developed by G.W. Hough. Hough's diffusion process includes the following elements: current science and technology (is it possible?), culture (is it allowed?), market needs (economics—will it pay?) and social needs (is it wanted?). From these elements come informing, innovating, and integrating processes. Outcomes of the processes are technical, geopolitical, economic, and social developments. In contrast to the private sector, the public sector may not be availing itself sufficiently of the research and foundational methodologies about technology diffusion and technology transfer developed in other scientific disciplines, such as the social and behavioral sciences.

Models of Technology Transfer

Tenkasi and Mohrman, (1995), discuss four predominate models of technology transfer: the appropriability model, the dissemination model, the knowledge utilization model and the contextual collaboration model. The appropriability model follows the belief that good technology sells itself. Based on this model purposive attempts to transfer technologies are believed to be unnecessary. When the developer of the technology makes it available through common communication channels (e.g. television, newspapers, technical reports, journals, and conference presentations) the need for the technology automatically creates a demand. The disseminative model takes the view that transfer is best accomplished when experts transfer specialized knowledge to a willing receptor (Rogers, 1995). This model suggests that technology flows from source to the end user just as water flows through a pipe as long as restrictions are kept to the minimum. The knowledge utilization model focuses on strategies that put knowledge to effective use in the recipients setting. The contextual collaboration model is more of a diffusion model, building on the constructivist notion that knowledge cannot simply be transmitted, but must be subjectively constructed by the receiver through contextual adaptations (Tenkasi and Mohrman, 1995). If technology is to be transferred successfully, both the knowledge and the technology being transferred must be contextually adapted. This model goes beyond other models that view transfer as information transmission or communication by implying that successful transfer requires learning on the part of both parties and the need to recognize the perspective of others.

Another set of technology transfer models has been proposed by Ruttan and Hayami (1973). Their model distinguishes three phases of inter-national technology transfer: material transfer, design transfer and capacity transfer. Material transfer is characterized by the simple transfer of a new material or equipment such as machinery, seed etc and the technology associated with the use of the materials. In this case, adaptation of the technology to the local environment is not a direct concern. Design transfer is accomplished through the transfer of designs such as blueprints and tooling specifications so that the receiver can use the new technology on site. Capacity transfer is the most comprehensive of all and involves the transfer of knowledge which provides the end user with the capability to design and manufacture a new technology on their own. This type of transfer serves to expand and build upon a technology base while at the same time providing for learning and development of the receiver. Examples are licensing and franchise. While cost constraints may require that only the materials be transferred, the benefits of the technology are sustainable only if, the user population can adapt the technology to meet their cultural and environmental needs (Parayil, 1992).

Structure and Elements

There are three major technology transfer operating styles or approaches as was identified by Harder and Benke (2005) and used in the U.S. highways transportation. The approach can generally be described as research-unit-led, operating-unit-led, and intermediate-center-led. There is overlap in techniques and services; however, each of these three approaches addresses different needs for technology transfer. The two most common approaches are those led by the research unit and the intermediate center led.

The research-unit-led technology transfer is primarily comprised of facilitating the implementation of research results from its own program or successful research venues. For research results produced by its own program, research unit staffs provide or enlist the expertise, identify necessary resources, and work in partnership with operating units to do what is needed to put an innovation into practice. For programs that contract for research, research units have the added role of being a liaison between the external researcher and the operating unit user. For technologies or innovations originating outside the agency, the research unit will perform these same functions, but will also act as a magnet and filter to pull those innovations into the organization. The research unit will then act as a catalyst to get the operating units to adopt the innovation.

The intermediate-center-led approach is based on the mission of the centre: "foster a safe, efficient, environmentally sound system by improving the skill and knowledge of the said organizations through training, technical assistance, and technology transfer." Core services to clients provided by the centers are training programs, new and existing technology dissemination, personalized technical assistance, website information, and newsletters. The vision developed in the strategic plan for the program includes a focus on interactive relationships, information exchange, and the ability to enrich the knowledge base of the stakeholders. The centers are typically very familiar with their constituencies. They pull into their operations the technologies or knowledge (innovations) suitable for transfer. They find the right packaging or develop it for the needs of their customers, and they use a broad array of tools and mechanisms to deliver the innovation. The centers are particularly experienced in communication and outreach activities, such as instructional activities by means of conferences and symposia, training and short courses, demonstrations, technical assistance/communications, print and web-based publications and materials.

The third approach focuses on the technology transfer that is pulled into the organization by operating units or through the influence of senior management who have been exposed to an external technology push (e.g., a colleague or peer recommending adoption of a technology,

an organizational endorsement of an innovation, or being enlisted to support an innovation and to be instrumental in the adoption and deployment decisions.) This technology transfer is more ad hoc; it occurs most frequently with professionals through communications among technical committees, peer person-to-person discussions, and other general word of mouth. The technology transfer happens when a viable innovation is brought to the attention of prospective users within these networks. Generally there is no assigned responsibility or defined position within operating units for managing this type of technology transfer, although awareness of this function is growing. Although this approach is not as formal as the other two, it is very effective because it is uniquely user- and needs-driven.

In the past, those interested in technology transfer were most likely to be located in the research offices. Now, participants come from within operating divisions and regions (Harder, 2003). For the most part, technology transfer is now recognized as an important part of state research programs. However, recognition of the relationship between technology transfer and achieving agency goals is relatively recent. In general, the term "implementation of research results" is used nearly synonymously with the term technology transfer. There is a general acknowledgement that specific resources are required for accomplishing technology transfer and implementation activities and that providing these resources facilitates the adoption and deployment of innovations.

According to Harder and Benke (2005), American states are beginning to budget funds and human resources for technology transfer and implementation of research results. This is very different from past practices of relying on the operational environment to supply all resources for any implementation or technology transfer activity. Moreover, there is an awareness of the research units being the focus for expertise in technology transfer whether the innovation under consideration is a result of the program's research activities or from some other source. Another characteristic of the current environment includes not only the more common practice of pushing technology out to users, but users seeking innovations and existing solutions to problems by pulling technology into the operational setting. Technology transfer no longer is solely the responsibility of the research group trying to get its results put into practice. Increasingly, operational units lead participants in bringing innovations to the practice.

There is growing recognition that technology transfer now is both the practitioner's responsibility and the researcher's responsibility. The collaborative nature of technology transfer is becoming more accepted. In several states, cross-disciplinary teams of practitioners, researchers, and technology transfer agents exist as formally structured mechanisms rather than as a hit-or-miss team-forming, ad hoc process. Currently, technology transfer is a more planned and deliberate process than ever before. The planning of technology transfer activities and tracking and monitoring of performance are becoming necessary components of technology transfer and particularly of research results implementation using a myriad of communication processes.

In most developed nations, there is an advanced level of technological innovations resulting in high productive capability in agriculture as well as in industry. These technologies are not often available to farmers in developing countries. Farmers find such innovations difficult to maintain even when they are appropriate to local condition. This results in low agricultural production levels that cannot sustain the rapidly increasing population and the growing demand of the people for better living standards.

A number of factors all strongly correlate with successful technology transfer or implementation of research results and are discussed in Table 1.

FACTORS AFFECTING TECHNOLOGY TRANSFER IN THE DEVELOPING WORLD

Technology does not stand alone but encompasses political, social, economic and cultural values that can serve as barriers to its transfer. They exist for all innovations, but some transfers are more affected by the barriers than others.

Socio-Cultural Barriers

It is important to recognize that transfer occurs within a social system. The social system defines the boundary or limits within which the innovations will be transferred and diffused. Most barriers assume some sort of societal judgment. An individual will not recommend a technology to neighbors if it is detrimental to him or if it lacks substantial benefit. According to Al-Thawaad (2008), cultural barriers are the greatest challenge to the successful transfer of technology. By analyzing the culture of the host nation, the donor would be able to identify factors to motivate higher efficiency and production from the workforce, thus enhancing the success of technology transfer, culture is one of the most powerful factors affecting technology transfer. Tradition, religion, historical habits and personal aspirations for a new life are important factors facing technology digestion and absorption. Barriers to a successful technology transfer include cultural and language gaps, low technical and other capabilities, inadequate infrastructure in developing countries, and insufficient investment in research and development, particularly relating to technology adaptation. The transfer of technology is not a movement of idle machinery / equipment from place to place, but rather a transfer of knowledge and information. In addition, the recipient must be thoroughly trained to operate, make use of and understand / and absorb the technology transferred. The barrier of culture and language differences would affect the transfer of training. Different countries have

Table 1. Contributions to successful technology transfer

Technology push	One significant factor affecting successful technology transfer is the push that technology exerts on prospective users. This technology push occurs often in the new product development area when vendors seek to sell an innovation to users.
Champions	For research-unit-led technology transfer, the most successful strategy or factor in a technology transfer situation was the presence of a champion. Champions were seen as critical participants in the successful outcome of transfer. Champions are drawn from the practitioners, from management, and from within advisory committees.
Pilot projects and demonstrations	Pilot projects and demonstrations are another factor for success and are considered a valuable addition to the strategies for facilitating technology transfer. Pilot projects and demonstrations are important success factors; workshops, demonstrations, and pilot projects.
Senior management support	The support of senior management is a significant factor for success; investment decisions, mandated use of the innovation and support from the legislature.
Early involvement of users	A tenet of research results implementation success is to involve the user early in the process of the research (Bikson, Law, Markovich and Harder, 1996). Participants in technology transfer include this factor in their practice whether it is transferring the results of research or an existing technology or innovation transfer.
Technology transfer or implementation plan	Having a plan for the conduct of the technology transfer or implementation activities an important factor for success and implementation plans become working documents that are used to guide the implementation process.
Qualified technical personnel in lead roles	Top ranked among the success factors for technology transfer is qualified technical personnel in lead roles. Without technical expertise little transfer of knowledge and understanding of an innovation would occur.
Partnerships	As with qualified technical personnel, the participants in technology transfer are a factor for success. The team or partnership formed must have the right skills and abilities to positively affect the effort. For success to be achieved, the participants with the highest average involvement in the program/operations personnel, local or municipal experts, university educators or researchers, and state field office personnel. In this case, the key to success is the selection of the various participants to form a partnership to facilitate technology transfer. This example also shows the benefits of a qualified person in a lead role, the value of a champion, and the assistance of identifiable benefits to facilitate technology transfer.
Benefits of technology—meeting users' needs	In addition to the techniques and methods used to accomplish technology transfer, there is one essential success factor that should not be overlooked—the benefits of the technology to be transferred. Supplying what the user needs, when the user needs it, in a form that can be used, at a cost that is reasonable is a compelling success factor.
Progress monitoring and committed funding	Progress monitoring and committed funding have influence on success
Focus area for technology transfer effort	Three areas are vital in successful efforts; the most frequently cited for research units being knowledge transfer and training and education.
Marketing and communications	Effective marketing and communications are key success factors of technology transfer. Every successful technology transfer activity in some manner involves the packaging or marketing of the innovation to suit the intended audience or user. Additionally, effective communications techniques are required to convey the knowledge and skills for users to promote change in their respective settings.

different cultural values, environments, work ethics and motivation and the capabilities of these differences have to be considered for successful transfer of technology. Machinery and equipment portions of the technology to be transferred must be scrutinized with great care before they can be transferred. Machinery has to be adapted to the needs and prevailing conditions of the host country; that lack of adaptability may be due to numerous factors, which may include difficulty of repairs, lack of adaptation to climate, lack of equipment / machinery sturdiness and incompatibility with other equipment.

Political Factors

One of the greatest constraints to agricultural and rural development in developing economies is the absence of systems of policy objectives. The absence of agriculture sector objectives aborts the development of agricultural research and extension objectives and prioritization of development programs. The failure has often led to absence of a linkage between state priorities and lower level priorities and arbitrariness in resource allocation (Idachaba, 2006; Igbokwe, 2005 and 2004; Ayoola, 2001). This might explain the failure of many of the World Bank sponsored agricultural development programs and projects in developing countries. Closely associated with this is the high turnover rate of governments and government policies and programs. The tendency of a government to pursue party program or in the days of military governments, to initiate populist programs too often, leaves the implementing institutions confused and frustrated.

The policy component of a technology can enhance or limit its successful transfer. The principal areas of influence are the price signals to the recipients of the technology. The government often makes policies to enhance the transfer of a technology but this is affected so much by the investment it makes on that which influences the pace and scale. The endowment level is always fixed by the government and this influences public expenditure. Government makes policies on consumer and producer commodity prices, subsidies for inputs, credit availability, import substitution, export earnings, sufficiency and natural resource management. These send direct and indirect price signals to the public and influence the adoption decisions. Higher fertilizer prices for example discourage farmers from adopting it. Government policy on commodity prices acts as incentive or disincentive to recipients of the technology. In agriculture, if there is no profit incentive for production of a specific crop, there is little point in developing or transferring improved technology related to that crop. For example, policies that favor the import of cereal grains at concessionary prices in the international market discourage in-country production of those crops. Technology generation and transfer output that focus on such crops are not likely to interest farmers (Swanson, Sands and Peterson, 1990).

The influence of political barrier on transfer of technology was evident in a problem that occurred in India, where a near-famine situation prompted the development of an agricultural research system and the reform of the bureaucracy that had driven the peasants to poverty (Parayil, 1992). Before the development of the new technology, the colonial government was interested solely in increasing the production of exportable cash crops. In this case, the political agenda largely ignored the needs of the citizens between 1947 and 1965. According to Donald (1999), most developing countries with gross national products far below the total sales of many large companies lack the political muscle to set the rules determining international trade, such as TRIPS agreement. Furthermore, individuals who have no effective political representation in government are unlikely to demand the technology of their choice and cannot resist those they do not want. Another reason for people not getting the most appropriate technology is the system of government they have. Suppliers of technology would prefer a country with political stability other than one where political unrest and war are inherent. The presence of these two conditions i.e. war and political unrest will not afford the supplier of the technology the opportunity to recover the cost of transferring the technology.

Environmental Factors

Technology transfer requires a certain environment (physical, economic, industrial) to be of commercial value. There are no assurances that a form of technology well-suited to one culture and/ or physical environment would be equally effective when transplanted to a different culture and

/ or environmental setting. Local environmental factors are important in conditioning technology transfer. When technology is transferred to a foreign country, suitable adaptation to the local environment is usually required. The physical environmental factor can influence the success of the technology transfer from the point of the topography of the country, climate/weather (for example, typhoon season); all this needs to be considered for technology transfer. The process of ideal conditions for the transfer of technology is by providing the ideal physical environmental qualities that include a temperate (natural or artificial), favorable working conditions and sufficient supply of materials. In other words, providing an ideal physical environment will enhance successful technology transfer (Al-Thawaad, 2008).

Geographical Location Factors

Geographic proximity can enhance technology transfer and the cost associated with maintaining the technology in operation will be less if the transacting parties are in close proximity to one another. The effects of geographical location on technology transfer can be seen from two different perspectives. First, if natural resources or new materials needed to produce certain products through the application of the new technology do not exist because of geographic conditions, this will directly affect the applicable technology. Second, if the geographical location lacks one or more key ingredients necessary to make the technology transferable, such as the lack of plentiful water supply vital to the technological process, even though all the other necessary raw materials are available, such a transfer will fail.

Appropriateness of the Technology

Pursell (1993) suggests that the appropriateness of a technology affects its transfer. Appropriate technologies are inexpensive, easily maintained, suitable for small-scale application, compatible

with one's need for creativity and are relatively easy to learn and to use. Appropriate technologies are those that match the needs and want of the individuals or groups receiving the technology. A good example of an appropriately technology occurred during the 'Green Revolution' in India in the 1960s and 1970s. The introduction of new varieties of wheat into Indian agriculture was successful partly because the wheat was appropriate for the setting to which it was transferred. In this case both the agricultural production conditions and the personal taste of the consumers match the characteristics of the wheat (Dewalt, 1978). Efforts were made by the government to provide tractors to peasant farmers to improve their production. This transfer failed because the tractors were too expensive, they were too large for planting seeds on their small plots of land; maintenance facilities were unavailable; and fuel was costly and scarce. Clearly, tractor was not an appropriate technology for these farmers. In an attempt to increase their yields, by reducing the labor costs of planting and to better control the planting process by seed drill that was pulled by animals and could be manufactured by a local blacksmith was introduced. Because this technology could be developed by the indigenous farmers, was simple to fabricate, and easy to use, it was appropriate for this setting. It quickly gained acceptance by farmers and was diffused throughout the region. This is also an example of intermediate technology; a technology that is at a level between the current technologies of the area and the 'high tech' technologies that are available elsewhere (Johnson, Gatz and Hicks, 1987).

Another way to consider the appropriateness of a technology is to examine its characteristics. Rogers (1995), argues that the characteristics of a technology, as perceived by individuals, influence the rate at which an innovation is transferred and diffused into the society or organization. He describes the five characteristics of relative advantage, compatibility, complexity, triability and observability. The first refers to the degree to which an innovation is perceived as better than the idea it supersedes as measured in economic terms, social prestige, convenience and satisfaction. Compatibility is the degree to which an innovation is perceived as being consistent with the enlisting values, past experiences, and needs of potential adopters. An idea that is incompatible with values and norms of a social system will not be adopted as rapidly as an idea that is compatible. Complexity is the degree to which a technology is perceived as difficult to understand and use. New ideas that are simpler to understand are adopted more rapidly than ideas that require new skills and understanding. The degree to which an innovation can be used in small amounts as trials that minimize risk represents less uncertainty to the individual who is considering adopting the technology and therefore is more likely to be accepted. Observability is the degree to which the results of an innovation are visible to others. The easier it is for other individuals to see others using the innovation with positive results, the more likely they are to adopt it. Timing is an important factor in the success or failure of a technology's ability to progress from the technological activity output phase to beneficial use. There are numerous examples of technologies that appeared either ahead or behind their time, that is they are made available either too late or too early to benefit the user. Successful transfer requires that technology be transferred at the optimum time it is needed or wanted by the user. Timing also affects the technologies to overcome barriers. When timing is right the barriers will be more easily overcome during transfer.

Change Agents

Technology transfer is accomplished by change agents not agencies (Burns, 1969). Within the social environment are key players, opinion leaders or change agents who have the power or influence to change people's attitude about an innovation. Change agents have a professional responsibility to be sensitive toward the receiving culture. They need to consider issues and take part in decisions regarding transportation, land use, pollution control, defense, and restricting or encouraging technological activities. Sound decisions demand an understanding of the impacts, relationship and costs of such technological activities (International Technology Education Association, 1996). It is important that discussions between the change agents and the receiving population be two-sided (Pacey, 1986). After all, the potential users of technology are experts in another sense, that of understanding their culture and society. Through this cooperation technological solutions can be developed that adequately address social, cultural, economic and political concerns. Here lies the case for participatory technology development and transfer.

Personal Barriers

An individual's particular concern about a given technology seems to be an influencing factor in the degree of acceptance (Hall and Loucks, 1978). They stress that individuals have different concerns about technology and proceed through various stages before they fully accept the change. Rogers, (1995) also asserts that transfer depends on certain characteristics of the end user. He contends that a very small percentage of the population called the innovators, constantly seek out new technologies. This group is followed by a larger group called early adopters who and constantly or often sough for advice. This is a key group for change agents working to transfer a technology to identify because they can have a strong impact on their peers.

Intellectual Property

It is legal property rights over creations of the mind, both artistic and commercial. Here the intellectual owners are granted certain rights to a variety of intangible assets such as musical, literary, and artistic work; ideas, discoveries and inventions; and words, phrases, symbols and designs. The owner is covered by the copyright laws that protect his works from reproduction or adaptation without his authority. These rights allow the owner of the inventions to reap monopoly profits. The time and money invested on an invention might be wasted if others could copy it. Competitors could charge a lower price because they did not incur the start-up costs. The purpose of intellectual property right is to encourage innovation by giving creators time to profit from their new ideas and to recover the development costs. The implication of IPR in technology transfer is that the inventors of a certain technology can disallow its use by others. This limits the transfer of technology as the inventor would want to monopolize the market.

Infrastructure

Technology transfer is greatly influenced by what has been called national systems on innovation, which includes the networks of institutions that initiate, modify, import and diffuse new technologies. National systems on innovation are defined as the set of institutions to create, store and transfer the knowledge, skills and artifacts which define technological opportunities. Technology infrastructure consists of science, engineering and technical knowledge available to private industry. Such knowledge can be embodied in human, institutional and facility forms. More specifically, technology infrastructure includes generic technologies, intra-technologies, and technical information and research and test facilities. Technology infrastructure may lack direct economic value to any one firm and thus individual firms may lack adequate incentives to build technology infrastructure on their own. Thus policies to develop public technology infrastructure can build the capabilities that exist elsewhere but need to be imported, adapted and absorbed in the local economy. Odigie and Li-Hua (2008) pointed out that technology transfer can never be successful

in developing countries unless there exist basic infrastructure such as good communication network, transportation electrically and market. There may be areas that cannot be readily reached by vehicles. In this case the recipients of the technology may not be reached.

Linkages between Institutions

The lack of government participation and partnership with universities and research and development project (R & D) has been a great draw-back to progress in knowledge transfer. Universities and R & D projects are not well funded by government of the developing countries. There are no close relationships/linkages between companies and universities (Odigie and Li-Hua, 2008). Sometimes the universities do not produce the required number of personnel for the transfer process.

Local Economic Environment

While the usefulness of technology for socio-economic development has been demonstrated, the sustainability of such development has been demonstrated, the sustainability of such efforts in developing countries has proven challenging. At the end of the day, the local economic environment determines the extent and frequency of technology use in the long-run. So technology transfer should put into consideration the local economic conditions. If people cannot use a technology now, they cannot also afford to use in future if nothing meaningful is done to improve the economic environment. Technology that fosters economic growth will be widely accepted. For instance if the local people, community-based organizations and small businesses are involved in providing technology services and creating content for other businesses, this can generate revenue to help make local technology use sustainable which will in turn have a positive impact on the local economy. However it is important to consider the negative economic effects of the technology. It should be ensured that the technology will have job opportunities for the local people so that they do not go out to look for jobs. Failed community access projects can lead to the rejection of future technology projects as the local people might feel that funds have been drained from the local economy that might have been better used for other things. And in some cases new technologies can replace human labor for example by cutting out middleman – resulting in lost jobs, which also can create negative attitudes towards the technology.

Affordability of Technology and Technology Use

Once it is determined that appropriate technology is available, the question is whether the people can afford or access it and use it in their works and lives. The affordability problem is tied directly to the general conditions of poverty. At a macro level, significant infrastructure investment is needed to bring technology to communities that lack electricity, access to telephone networks, or computer etc. and in very poor communities, which often lack basic necessities such as food, healthcare and sanitation, striking the right balance between technology and other priorities is required. At the micro level, expensive technologies such as ICTs are not easily adopted. Example a computer costs the equivalent of a year's average income for the majority of the people in developing countries. Most parents in developing countries cannot afford the antibiotic used to treat infant pneumonia as the price is above the annual income of most parents. Affordability is an immediate problem, which shifts to a question of sustainability in the long-run. Policy makers and development practitioners need to make realistic choices about introducing expensive technologies in poor communities.

LESSONS LEARNED

According to the New World Encyclopedia (2008) the basic technology of agricultural machines has changed little through the last century. Though modern harvesters and planters may do a better job than their predecessors, the combine of today (costing about US$250,000) cuts, threshes, and separates grains in essentially the same way earlier versions had done. However, technology is changing the way that humans operate the machines, as computer monitoring systems, GPS locators, and self-steer programs allow the most advanced tractors and implements to be more precise and less wasteful in the use of fuel, seed, or fertilizer. In the foreseeable future, some agricultural machines may be made capable of driving themselves, using GPS maps and electronic sensors. Even more esoteric are the new areas of nanotechnology and genetic engineering, where submicroscopic devices and biological processes, respectively, may be used to perform agricultural tasks in unusual new ways.

In recent years scientists have taken advantage of new developments in genetic engineering, or biotechnology. "Biotechnology is in vitro manipulation of whole plant, cellular or molecular materials for the purpose of improving agricultural plants or processes" (Messer and Heywood 1990). Agricultural biotechnology is a range of tools, including traditional breeding techniques that alter living organisms, or parts of organisms, to make or modify products (Ozor and Igbokwe, 2007); improve plants or animals; or develop microorganisms for specific agricultural uses (USDA, 2008). Modern biotechnology today includes the tools of genetic engineering. Recent advances in biotechnology have come out of developments in cellular and molecular biology that help scientists to better understand and manipulate life processes.

Biotechnology provides farmers with tools that can make production cheaper and more manageable. For example, some biotechnology crops can be engineered to tolerate specific herbicides, which make weed control simpler and more efficient. Other crops have been engineered to be resistant to specific plant diseases and insect pests, which can make pest control more reliable and effective, and/or can decrease the use of synthetic pesticides. These crop production options can help countries keep pace with demands for food while reducing production costs. Biotechnology could help solve many problems limiting crops and livestock production in developing countries. Biotechnology-derived solutions for biotic and abiotic stresses, built into the genotype of plants, could reduce use of agrochemicals and water, thus promoting sustainable yields (Spotlight, 1999). The introduction of transgenic crop varieties implies access to a range of genes, techniques and processes. Biotechnology expertise should complement existing technologies and be output-driven. Since biotechnology is often more expensive than conventional research, it should be used only to solve specific problems where it has comparative advantage. Biotechnology can contribute to the conservation, characterization and utilization of biodiversity, thus increasing its usefulness.

Biotechnology is not without hazards apart from moral and ethical objections that people have to tinkering with life forms. Some of the hazards are increased use of herbicides where none are used presently, and the risk of passing on genes to other organisms through natural processes. Use of genes with virus resistance may lead to evolution of viruses capable of attacking a wide range of plants (Conway, 1997). The ability to arrange and rearrange molecular structures in most of the ways consistent with physical law, nanotechnology is creating a growing sense of excitement because we see an opportunity of unprecedented magnitude looming on the horizon (Merkle, 1999). This will have a pervasive impact on how we manufacture almost everything -- what is manufacturing but a way to arrange atoms? If we can arrange atoms with greater precision, at lower cost, and with greater

flexibility then almost all the familiar products in our world will be revolutionized.

It is important to emphasize private-sector experience to highlight some of the successful developments that may be used by the public-sector highway community. The private-sector experience has shown the need for infrastructure to help in identifying innovations, and to create financial and economic capability, as well as human resource capacity for facilitating technology transfer. The private sector has a role to effectively close the gap between those who have an innovation and those who can put the innovation to use. The structure the private sector developed is lacking, in full measure, is public-sector technology transfer efforts

EMERGING CHALLENGES AND RECOMMENDATIONS

To improve agricultural technology development requires strengthening of the enabling environment, including policies, public institutions and regulations. Specifically, there is need for developing economies to draw up transfer of technology policies for the emerging technologies. A few of such countries notably Brazil, India, and South Africa have demonstrated foresight in drawing up national policies on biotechnology and nano-technology and have emplaced mechanisms for implementation. In most other countries, there is a sense of inactivity even where an institution for the implementation is on ground.

Various types of market failure imply that markets, by themselves, will not elicit the optimum amount of technology for farmers. Priorities include more responsive regulations for input supply, support for emerging enterprises, strengthening input marketing, establishing adequate intellectual property protection, and addressing the challenges of biotechnology and nanotechnology.

There is need to build strong partnerships between scientists who develop technological solutions and the communities who use them in order to ensure that the research is relevant. For this to succeed both public and private institutions in developing countries must demonstrate a strong commitment towards supporting such collaborations.

There is need to emplace necessary supporting infrastructures such as electricity, water and accessibility. Equally important is the development of institutional technical support and local capacity. The former can be drawn from researchers and engineering technicians in universities and industry. This calls for close collaboration between universities and industry. Developing local capacity to some extent reduces the job of supervision, promotes a sense of ownership and helps to prevent neglect and in some cases vandalization of the infrastructure

REFERENCES

Al-Thawaad, R. M. (2008). *Technology transfer and sustainability adapting factors: Culture, physical environment and geographic location.* Paper 152, Session IT 305.

Ayoola, G. B. (2001). *Essays on the Agricultural Economy: A Book off Readings on Agricultural Development Policy and Administration in Nigeria.* Ibadan, Nigeria: T. M. A. Publishers and Farm and Infrastructure Foundation.

Bikson, T. K., Law, S. A., Markovich, M., & Harder, B. T. (1996). *NCHRP Report 382: Facilitating the Implementation of Research Findings: A Summary Report.* Washington, DC: Transportation Research Board, National Research Council.

Burns, T. (1969). Models, images and myths . In Gruber, W. H., & Marquis, D. G. (Eds.), *Factors in Transfer of Technology* (pp. 11–23). Cambridge, MA: MIT Press.

Conway, G. (1997). *The Doubly Green Revolution.* London: Penguin Books.

De Walt, B. R. (1978). Appropriate technology in rural Mexico: Antecedents and consequences of an indigenous peasant innovation. *Technology and Culture, 19*(1), 32–52. doi:10.2307/3103307

Donald, A. (1999). Political economy of technology transfer. *British Medical Journal, 319*(7220).

Hall, G. E., & Loucks, S. (1998). Teachers concerns as a basis for facilitating and personalizing staff development. *Teachers College Record, 80*(1), 36–53.

Harder, B. T. (2003). *NCHRP Synthesis of Highway Practice 312: Facilitating Partnerships in Transportation Research.* Washington, DC: Transportation Research Board, National Research Council.

Harder, B. T., & Benke, R. (2005). *Transportation technology transfer: Successes, challenges, and needs: A synthesis of highway practice.* Washington, DC: Transportation Research Board.

Idachaba, F. S. (2006). Good Intensions are not Enough: collected Essays on Government and Nigerian Agriculture: *Vol. 3. Agricultural Research and Uncertainty and Diversification.* Ibadan, Nigeria: University Press.

Igbokwe, E. M. (2004). Rationalizing and streamlining agricultural research institutions in Nigeria. In Legislative and Policy Agenda for Nigerian Agriculture (Vol. 1). Technical Assistance to the House Committee on Agriculture, Briefing Paper No. 4. Enugu, Nigeria: African Institute for Applied Economics.

Igbokwe, E. M. (2005). Strengthening the linkage between agricultural research, extension and the farmer. In Legislative and Policy Agenda for Nigerian Agriculture (Vol. 2). Technical Assistance to the House Committee on Agriculture, Briefing Paper Number 4. Enugu, Nigeria: African Institute for Applied Economics.

International Technology Education Association. (1996). *Technology for all Americans: A Rational and Structure for the Study of Technology.* Reston, VA.

Johnson, S., Gatz, E. T., & Hicks, T. (1999). Expanding the content base of technology and education transfers: A topic of study. *Journal of Technology Education, 8*(2).

Merkle, R. C. (1999). Biotechnology as a route to nanotechnology. *Trends in Biotechnology, 17*(7), 271–274. doi:10.1016/S0167-7799(99)01335-9

Messer, E., & Heywood, P. (1990). Trying technology: Neither sure nor soon. *Food Policy, 15*(4), 336–345. doi:10.1016/0306-9192(90)90076-C

Mock, J. E., Knenkeremath, D. C., & Janis, F. T. (1993). *Moving R&D to the Marketplace: A Guidebook for Technology Transfer Managers.* Washington, DC: J. E. Mock.

New World Encyclopedia. (2008). *Agricultural technology.* Retrieved from http://www.newworldencyclopedia.org/entry/Agricultural_technology

Odigie, H. A., & Li-Hua, R. (2008). *Unlocking the channel of tacit Knowledge transfer.* Retrieved from http:// www.mostc.coms

Ozor, N., & Igbokwe, E. M. (2007). Roles of biotechnology in ensuring adequate food security in developing societies. *African Journal of Biotechnology, 6*(14), 1597–1602.

Pacey, A. (1986). *The Culture of Technology.* Cambridge, MA: MIT Press.

Parayil, G. L. (1992). The green revolution in India: A case study of technological change. *Technology and Culture, 33*(4), 737–756. doi:10.2307/3106588

Pursell, C. (1993). Knowledge innovation system: The common language. *Journal of Technology Studies, 19*(2), 2–8.

Rogers, E. M. (1995). *Diffusion of Innovations* (4th ed.). New York: The Free Press.

Ruttan, V. W., & Hayami, Y. (1975). Technology transfer and agricultural development. *Technology and Culture, 14*(2), 151–199.

Spotlight. (1999). *Biotechnology in agriculture.* Rome: FAO

Swanson, B. E., Sands, C. M., & Peterson, V. (1990). *Analyzing Agricultural Technology Systems: Some Methodological Tools.* The Hague: ISNAR.

Tenkasi, R. V., & Mohram, S. A. (1995). Technological transfer as collaborative learning . In Backer, T. E., David, S. L., & Soucy, O. (Eds.), *Reviewing the Behavioral Science Knowledge Base on Technology Transfer* (pp. 147–167). Rockville, MD: U.S. Department of Health and Human Sciences, Public Health Service, National Institute of Health.

United States Department of Agriculture (USDA). (2008). *Biotechnology.* Washington, DC.

Section 6
Regional Developments

In this section, progress and regional developments on nanotechnology and other emerging technologies are discussed. It covers nanoscience and nanotechnology in Chile, Mexico, Brazil and Argentina. It also discusses innovations, trade policies and management challenges as they affect technology transfer in Africa. An empirical study on technology penetration pattern in a developing nation for a new technology is provided.

Chapter 21
Nanoscience and Nanotechnology in Latin America

Adolfo Nemirovsky
LatIPnet Inc., USA

Fernando Audebert
University of Buenos Aires, Argentina

Osvaldo N. Oliveira Jr.
USP, Brazil

Carlos J. L. Constantino
UNESP, Brazil

Lorena Barrientos
Universidad Metropolitana de Ciencias de la Educación, Chile and Universidad de Chile, Chile

Guillermo González
Universidad de Chile, Chile and CEDENNA, Chile

Elder de la Rosa
Centro de Investigaciones en Óptica, México

ABSTRACT

Latin America (LA) can count some strong research centers with a tradition of research excellence in certain disciplines such as medicine and biology, nuclear technology, metallurgy and materials, among others. Latin American countries have generated networks of researchers across disciplines, centers, etc. within a country, and linking two or more countries in the region (e.g., Argentina-Brazil Bi-National Center for Nanoscience & Nanotechnology, CABN). Additionally, collaborations have extended beyond LA, mainly to the EU and the USA. In general, these programs have been quite successful in the generation of interdisciplinary nanoscience and nanotechnology (N & N) research. The relation between academia and industry has been improving in the last few years, but it is still weak. In particular, funding incentives for N&N efforts have encouraged joint efforts and contributed to new dimensions in collaborations. This chapter reviews the state of nanoscience and nanotechnology in Chile, Brazil, Argentina and Mexico.

DOI: 10.4018/978-1-61692-006-7.ch021

INTRODUCTION

This chapter provides a survey of the status of nanoscience and nanotechnology (N&N) in Latin America (LA), with special focus in the following four countries: Argentina, Brazil, Chile and Mexico. N&N has already impacted the global economy, and it appears to hold a large economic potential that is just emerging. This, in turn, is bringing together strong interest from governments, academia, industry and investors in the most developed countries and some fast growing economies (China, India, etc.). In fact, according to Lux Research (Lux 2008), global funding in N&N during 2008 reached almost $20 billion with over 40% from governments, about 40% of corporate funding, and over 5% from venture capital contribution. Products incorporating nanotech reached over $240 billion with about 65% in manufacturing and materials, 25% electronics and IT and 10% in healthcare and life sciences. Lux predicts that nanotech will touch $3.1 trillion worth of products along the value chain by 2015.

Latin America contribution to total investment and N&N products has been growing but still is quite small. For example, Latin American governments' investment in N&N in 2006 reached just about $50 million (see Table 1).

N&N is a truthfully interdisciplinary endeavor that requires the collaboration of several disciplines such as physics, chemistry, mechanical and electrical engineering, biology, medicine, etc. This, in turn, has fostered partnerships among researchers and practitioner from various fields of expertise either at the same facility (university, lab, etc.) or from different institutions in academia and industry. An important byproduct of the emergence of N&N is the strong enhancement of multidisciplinary collaborations, and LA has not been an exception. In the last few years, the requirement of this field and the steering of funding agencies have fostered the development of networks of researchers from different disciplines and institutions, focused around certain topics of N&N such as optical properties of materials, characterization of materials, biosensors, etc. Some of the efforts are

Table 1. Population, income per capita, R&D and N&N expenditure, Science and Engineering (S&E) and N&N articles and patents to residents, for Argentina, Brazil, Chile and Mexico–compared against selected references

Country	Population in 2006 (millions)	2006 GNI per capita (thousands)	2006 R&D spending (as % GNP)	Government spending (millions) N&N R&D (estimated 2006)	S&E articles 2005 per million people	Nanotech publications per million people (2005)	Patents to residents per million people (2005)
Argentina	39.1	11.7	0.41	2	79.0	4.6	4
Brazil	189.3	8.7	0.98	27-40	53.1	4.7	1
Chile	16.4	11.3	0.61	10	95.6	4.5	1
Mexico	104.2	12.0	0.40	12	37.8	3.9	1
Spain	44.1	28.2	1.11	50	422.5	35.3	53
China	1,311.8	4.7	1.44	220	31.9	7.5	16
India	1,109.8	2.5	0.85	106	13.3	1.6	1
Japan	127.8	32.8	3.15	975	434.0	48.5	857
USA	299.4	44.1	2.68	1,775	692.7	47.6	244

Source: Kay L. & Shapira P. (2009)

GNI is the gross national income per capita at the Purchasing Power Parity (PPP) as defined by the World Bank. PPP exchange rates equalize purchasing power across different countries. The data has been compiled from various sources [Kay L. & Shapira P. (2009)]

involving two or more countries of the region. An example is the Argentina-Brazil Bi-National Center for Nanoscience & Nanotechnology (CABN). The Center organizes international workshops and N&N schools, promoting the cooperation between Brazilians and Argentinean groups.

The study and characterization of nanoscale materials requires a broad range of approaches, materials and tools, and, in particular, of some expensive equipment such as atomic force microscope (AFM), scanning tunneling microscope (STM) and transmission electron microscope (TEM). In some cases a single institution can have some of these facilities, and in many cases national (or international) labs provide shared facilities for expensive equipment. Latin America has suffered from the lack of availability of desired required equipment for nano studies. On the other hand, funding has been provided for some national labs, and large facilities, in particular in Brazil (see Table 1).

Nanotechnology is not a product industry but rather as a set of enabling technologies that supports many existing industries (such as electronics, optics, composite materials, pharmaceuticals). Nanotech impact, in general, is broad but not deep in any one product (e.g., nanotechnology has impacted the aerospace industry, for example, by delivering new materials that are lighter and more resistant, but nanomaterials only represent a tiny fraction, by weight, say of an airplane). Lux Research provides the following classification of nano products–according to the amount of "nano stuff" the final product contains (see below, nanomaterials are 100% nano, final goods have only a tiny fraction of nanomaterials):

- **Nanomaterials**: nanotubes, nanoparticles, nanoporous materials, quantum dots, etc.
- **Intermediate products**: catalysts, coatings, composites, nanostructured alloys, solar cells, sensors, etc.
- **Final goods**: automotive, consumer electronics, cancer treatment, construction, etc.

Also, included as nano-products are the tools for modeling, probing and manipulating structures at nanoscale:

- **Nanotools**: inspection tools (scanning probe microscopes, electron microscopes), fabrication tools (dip-pen nanolithography, nanoimprint lithography), modeling software.

Commercialization requires materials manufactured in a predictable and consistent way at sufficient volumes. Moving from the lab to the market is usually involved, expensive and time consuming. Currently, most of the nano products in the market are for structural applications (ceramics, composites, thin films, powders, metals, coating, etc). Additionally, other markets addressed by nanotech include instruments and tools, cosmetics, and various ICT and biotech applications. In LA, commercialization of nano products has been hindered by several weaknesses such as poor links between academia and industry, lack of patents, lack of consistent and agile frameworks for licensing technologies from universities and labs and limited funding. As shown in Table 1 below, the number of patents in N&N per million inhabitants is much lower than that of developed countries. Nevertheless there are encouraging signs in the last few years, most governments in LA are implementing policies for strengthening IP and supporting its commercialization, encouraging entrepreneurship and academia-industry collaboration and partnerships. Moreover, some startups have been launched by academic researchers, and some of these companies are entering the US and other markets. Additionally, a few mid to large size companies (e.g., Petrobras, Tenaris, etc.) are strategically investing in nano projects and engaged in collaborations with (local and global) academia and labs.

The interdisciplinary nature of N&N is beginning to transform education, not just impacting graduate and undergraduate students in science and engineering but also the general population. Science, math and computer literacy–building critical thinking and learning to model and simulate complex systems-are becoming basic pillars to form our 21st century citizens–allowing people to fully participate of, and contribute to, society (Baum, G., Nemirovsky, A. and Sabelli N., 2008). LA is now responding to this challenge at higher level of education (grad and undergrad levels). For example, Brazil is building its semiconductor industry with strong government support (Ministry of Science and Technology), building key infrastructure (e.g., CEITEC, a foundry and design center in Porto Alegre), and training hundreds of engineers and scientists, engaged with local and global partners from academia and industry. However there still limited efforts in the region in outreach, K-12 education and, in particular, in training and/or retraining the workforce with new skills to build regional advantages to attract (or create new) companies that required skillful and adaptable workers. Nevertheless, there are some recent regional efforts in this direction. For example, the State of Guanajuato (Mexico) is launching an initiative, led by LatIPnet (a not-for-profit organization that leverages the diasporas of Latin professionals in US and EU), for building a N&N-savvy workforce for the 21st century in collaboration with US partners.

NANOSCIENCE AND NANOTECHNOLOGY IN ARGENTINA

Argentina has a long scientific tradition that has led three Argentineans to obtain Nobel Prizes: Bernardo Houssay, in Physiology and Medicine 1947, Luis Federico Leloir, in Chemistry 1970 and César Milstein, in Medicine 1984. Obtaining one Nobel Prize in chemistry and two in medicine is not casual. The disciplines with the longest research tradition in the country are, perhaps, Medicine, Biology and Chemistry. Research has been carried out in public Universities, and later, when

the National Council of Scientific and Technical Research (CONICET) was created (in the 50´), also in CONICET research institutes.

In the 50´s Argentina decided to build up a new research area on nuclear physics, and that effort remained for decades until reaching its own technology to build Fission Nuclear Plants and produce all the technologies associated for the type of the reactors developed. For that achievement it was required to develop the metallurgy technology, which was based on physical metallurgy and materials processing. This development in nuclear physics and metallurgy was led by the National Commission of Atomic Energy (CNEA) created in 1950. The research in metallurgy was later spread in the public universities and CONICET Institutes.

Later on, since the 60´s several areas of research have emerged significantly such as semiconductors, electronics, polymers and ceramics, among others. Of course there are other research areas that have been strongly developed in the scientific spectra in Argentina. This historical brief is just to give an idea on how the basic areas that are the foundations of the N&N in Argentina have been developed. More information on the historical development can be found in Schneider, O. & Sanchez, R. (1980) and in Babini, N & De Asua, M. (2003).

During the last two decades, the government have defined different priority areas for R&D and the funding programs have addressed particularly towards *"information technologies, biotechnology, new materials, renewable energy and nanotechnology"* (Ministry of Education and Science, Argentina (2006)). Thus, N&N in Argentina have grown strongly from chemistry and materials science supported on the strong tradition in science and the strategic programs developed in the last decades.

The current strength of Science in Argentina is based on a population with a high level education, the strong tradition and commitment in the Universities for combining education and research, the prospective studies and the oriented research funding programs carried out on the last decades. The weakness is related with a still low investment in R&D, that is just ~0.6% GDP (public + private), which does not allow the acquisition of the required equipments, such as electron microscopes for a full structural characterization of nanomaterials. However, the current president Cristina Fernandez de Kirchner has set science as one of the main bases for the economic and social development, thus in 2007 she created the Ministry of Science, Technology and Productive Innovation (MinCyT). *Information technology, biotechnology and nanotechnology* have been selected as priority areas, and several grant schemes are being launched for improving the facilities in the stronger R&D Centers and to promote the formation of consortiums among academic or R&D institutions and industry in order to generate innovations. Moreover, "*Innovation*" is promoted at different levels, from the MinCyT (see, www.innovar.gov. ar), regional governments and foundations prizes and start up grants (see www.buenosaires.gob.ar/ emprende and www.empretec.org.ar/contenido/ institucional.php]. This policy has contributed in the last years to introduce projects and innovations in nanotechnology in several companies, not only in international companies such as Tenaris SA, but also in SMEs. It is expected that this policy will be sustained, which would place Argentina as a competitive country in nanotechnology in the next decade.

Table 2. Argentinean research networks on nanoscience & nanotechnology

Network / Main R&D Topics	Partners
Nanostructure Materials and Nanosystems • Nano-magnetism and spintronics • Nanostructured thin films • Surface & coatings • Electronic & opto-electronic properties of nanostructured systems • New instrumentation	CNEA-CAB (Bariloche) CNEA-CAC (Buenos Aires) UBA (Buenos Aires) UNC (Córdoba) UNSL (San Luis)
Nanoscience, Molecular and Supra-molecular Nanotechnology and Interfaces • Modeling and computational simulation • Self-assembly and low dimensional systems • Applications of self-assembly films • Applications of nano/meso-structured systems • Development of new methods of nano/micro-fabrication • Nanobiotechnology	UNRC (Córdoba) UNC (Córdoba) CNEA-CAB (Bariloche) CNEA-CAC (Buenos Aires) UNLP (La Plata) UBA (Buenos Aires) UNSL (San Luis)
Design, Simulation and Fabrication of Micro & Nano Devices and Prototypes • Design of a IMS Prototype • Bench mark for the analysis of fluid flow around MEMs • Modeling the electrokinetic flow in microfluidic devices • Ion mobility spectrometry. Models and Analysis of the detections • Molecular wires • Magnetic measurements in mesoscopic systems using MEMS	CNEA (Buenos Aires) CINSO (Buenos Aires) UBA (Buenos Aires) UNSAM (Buenos Aires) INTEC (Santa Fe) UNER (Entre Ríos) IMBECU (Mendoza) UNNE (Corrientes) CNEA (Bariloche)
Bio-nanostructures for Molecular Information Transmission in Neurobiology and Biological Processes	UNC (Córdoba) CIQUIBIC (Córdoba) UNSL (San Luis) UNT (Tucumán)

CINSO: Center for Research in Solids–CIQUIBIC: Center for Biology Chemistry Research-CNEA: National Commission of Atomic Energy-IMBECU: Institute of Medicine and Biology of Cuyo–INTEC: Institute for Technology Development for the Chemistry Industry–UBA: University of Buenos Aires-UNC: National University pf Córdoba-UNER: National University of Entre Rios-UNNE: National University of North-East-UNRC: National University of Rio Cuarto-UNSAM: National University of San Martín-UNSL: National University of San Luis-UNT: National University of Tucumán.

R&D Structure-Networks

Over the last five years, to accelerate the activity in N&N four research networks have been created through a special grant scheme that involved the main R&D centers and universities along the country (see Table 2).

The networks have been really successful increasing the quantity and quality of scientific publications and developing innovation and also materials and technologies. Moreover, in the last years the R&D networks have included industrial partners achieving patents and industrial products. The increase in the networking in N&N can be observed in Figure 1, where the nodes represent R&D institutions and the size of these is correlated with the number of publications, and the linking lines represent the joint publications, the larger the number of joint publications the thicker the line. It is observed that the University of Buenos Aires (UBA), the National University of La Plata (UNLP) and the CNEA are the institutions that have the highest activity in N&N. Some institutions, such as the Institute for Theoretical and Applied Research in Physics-Chemistry (INIFTA), the National University of Córdoba (UNC) and the National University of Mar del Plata (UNMDP), have notably increased their activity in N&N. Moreover, it is observed that not only have increased the links among the R&D institutions, but also new institutions have started developing R&D in N&N. From 2000 to 2008 the number of institutions that have activities in N&N has increase from 46 to 67 (Ministry of Science, Technology and Productive Innovation, Argentina, 2009).

In order to extend the N&N networks to the international community the "*Argentine-Brazil Binational Center for Nanoscience & Nanotechnology*" (CABN) has been created. The CABN organizes international workshops and N&N schools, promoting the cooperation between Brazilian and Argentinean groups (see http://cabnn.mincyt.gov.ar). Moreover, the International Cooperation Division of MinCyT created the *Argentine Bureau for Enhancing Cooperation* (ABEST) *with the European Community*. The aim of this Bureau is to establish a platform in Argentina in order to improve and expand cooperation activities in the science, technology and innovation area both with the European Union and with its member states. ABEST has been helping research groups and, small and medium sized local enterprises to participate in the FP6 and FP7 European R&D programs, where one of the main areas is N&N (see http://www.abest.mincyt.gov.ar).

Another strong collaboration in N&N is with USA with whom a declaration for R&D cooperation in N&N has been signed. A first workshop in Nanobiotechonology, Inorganic Nanomaterials, Surface & Channels and Dynamic of Materials has already taken place. The program includes student and researcher exchanges and fosters a long cooperation carrying out R&D programs in Nanomaterials (see: cooperación internacional in http://www.mincyt.gov.ar).

Cooperation programs in N&N are also involved in recent agreements for cooperation in R&D in Materials Science and Engineering, signed with the University of Oxford, United Kingdom, and with the Leibniz Institute for Solid State and Materials Research, Dresden, Germany.

R&D Centers and Funding Sources

The government's strategic plan for developing a strong R&D structure enables Argentina to play an important international roll in N&N, and besides the building of N&N networks, the plan includes (i) a grant scheme for providing equipments to the existing R&D groups that have joint research projects; and (ii) the creation of the "*Argentine Foundation for Nanotechnology*" (FAN).

The grants for acquiring equipments allowed upgrading several existing equipments and the acquisition of new ones. However there is still a lack of state of art high resolution microscopes, such as Field Emission Transmission Electron

Figure 1. Networks of Argentinean R&D institutions with activity in nanoscience and nanotechnology; (a) network in 2000, (b) network in 2008 (Ministry of Science, Technology and Productive Innovation, Argentina, 2009)

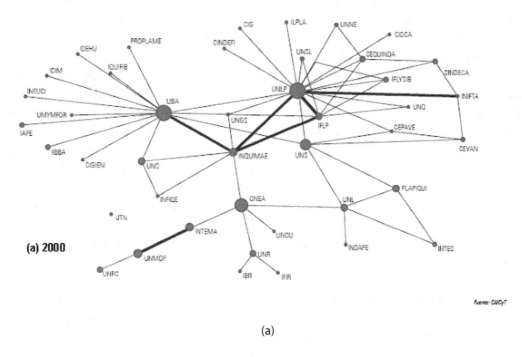

(a)

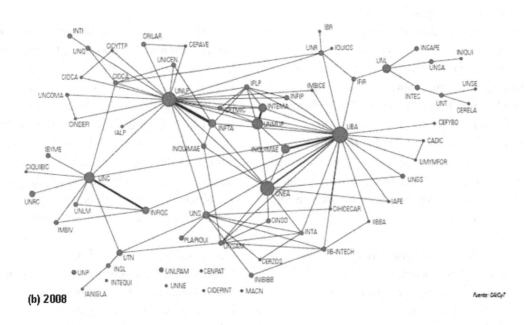

(b)

Microscopes (FEG-TEM) with aberration correctors and FEG Focus Ion Beam equipments. The improvement in the facilities has given an important impulse in the R&D activity in N&N. Therefore, it is expected that new grants for acquiring equipments in the next years will provide sufficient budget for acquiring high cost facilities, which will allow producing higher impact research in N&N.

The *Argentine Foundation for Nanotechnology* (FAN) has been created in 2005 as Private Institution that has as its main objectives (i) to promote the development of technical and human infrastructure in micro- and nanotechnology in Argentina; and (ii) to promote the industrialization with high economic impact of nanomaterials and nanotechnologies. The FAN provides grants to projects for industrializing and commercializing nano- and micro- products or processes that have a high innovation grade and produce an important social and economic impact. It also promotes other activities providing financial support for organizing conferences and meetings. The FAN has already organized two workshops at "MERCOSUR" level where scientists and companies come together with the aim of evaluating the possibilities for increasing the micro- and nanotechnology transfer to the industrial sector (see www.fan.org.ar).

In the last years, perhaps due to the influence of the grants schemes, some institutions have built real and virtual R&D Centers on N&N. Some of these Centers are:

CNEA, INN, Institute for the N&N: this virtual institute is composed of several internal laboratories and R&D departments with facilities and R&D working topics related to N&N. Some of those topics are:

- *synthesis of nanostructured materials,*
- *surface funcionalization,*
- *nanostructured superconductors,*
- *magnetism at nanoscale,*
- *modeling of magnetic and electronic properties,*

- *acoustic photon engineering,*
- *electronic transport in nanoscopic systems,*
- *coatings,*
- *molecules adsorption,*
- *MEMs.*

INTI, Industrial Nanotechnology: the National Institute of Industrial Technology (INTI) is a governmental institute composed by several Departments and Centers with different laboratories. The activities of the INTI are related to the industrial sector, providing technical assistance, developing technologies and accreditation service among other duties. Some of the internal laboratories have merged to form a virtual institute of nanotechnology, with the following topics:

- *nanotechnologies for developing materials,*
- *integration on nano- and micro-systems,*
- *surface modification for metallic biomaterials,*
- *polymer based nanocomposites,*
- *nano- and micro-systems of controlled release,*
- *nanometrology and nanotextile materials.*

CIDIDI, Center for Research, Development, Innovation and Engineering Design: this center has been created recently by agreement between the Faculty of Engineering, University of Buenos Aires (FIUBA) and the Buenos Aires City Government (GCBA). Scientists from FIUBA, industrial partners and government come together in the CIDIDI aiming to develop innovations. Its organization involves an advisory board composed by industrial and financial sectors. Thus the three main actors for developing innovations are working together. The building is still under construction and the Center will be fully opened on mid 2010. The R&D laboratories will contain the state of art equipments for nanomaterials development. The facilities will involve a modeling and simulation laboratory, a solidification laboratory with equipments for rapid solidification, a materials characterization lab with high resolution scanning and transmission microscopes with field emission gun and several detectors, machining and metrology labs. The scheme of the organization includes also the majority of the R&D laboratories and workshop belonging to the FIUBA and the Metropolitan Design Center (CMD). The CIDIDI is an innovative concept where scientists work together with designers.

Currently, there are around 70 R&D institutions that have activities in N&N, among them are some Institutes that were created in association between the CONICET and one university or a research institution [Consejo Nacional de Investigaciones Científicas y Técnicas, Argentina. (2009)], such as:

- CETMIC: *Center of Technology of Minerals and Ceramics* which is associated with the Council for the Scientific and Technical Research of the Buenos Aires Province (CIC);
- CIDEPINT: *Center for R&D in Paint Technology* which is associated with the CIC;
- CIOP: *Center for Optical Research* which is associated with the CIC;
- CINSO: *Center for Research in Solids* which is associated with the IDETEF;
- IFFAMAF: *Institute of Physics* which is associated with the National University of Córdoba (UNC);
- INFITA: *Institute for Theoretical and Applied Research in Physic-Chemistry* which is associated with National University of La Plata (UNLP);
- INQUIMAE: *Institute of Chemistry and Physics of Materials, Environment and Energy* which is associated with the UBA;
- INTECIN: *Institute of Technologies and Engineering Science* which is associated with the FIUBA;
- INTEMA: *Materials Science and Technology Research Institute* which is associated with the National University of Mar del Plata.

The Innovation is strongly promoted through two basic grant schemes, one based on companies associated with R&D academic groups, and another based on academic research with industrial partners. Another promoting activity is the "Innovation Competitions", which are sponsor by an agency of the MinCyT, "*Innovar*". From these financial instruments, innovations on nanotechnology have been generated, such as: (i) Pd Catalyst for molecular synthesis for pharmaceutical purposes, and for hydrogen release from methanol in fuel cells; (ii) intelligent paints containing nanoparticles that just release ions in aggressive environments in order to form a passive film; (iii) microvalve for ocular pressure control; (iv) nanocapsules anti-toxoplasmosis; the nanocapsule transports the drug where is the parasite is using 1000 times lower quantity of the doses; (v) electronic noses based on SnO_2 thin film; it is suitable for using in several Industries, for instance Argentinean beer breweries and mate tea companies are starting to use this equipments [INNOVAR Catalog, (2007); INNOVAR Catalog, (2008); Vago, M. et al., (2008); www.e-nose.com.ar].

R&D Areas in Nanoscience and Nanotechnology

Academic Sector

In Table 2 is summarized the information about the main R&D topics and the partners of the four networks on N&N. There can be observed that no more than 20 groups are involved in the networks. In comparison with the information provided in Figure 1.b, it is worth noting that there are around 50 other institutions/universities working in N&N that are not involved in the networks. The whole contributions of all the groups to the N&N can be classified in: Fundamental Research in Physics & Chemistry, Pharmacology & Medicine, Particles & Materials Development and, Micro- & Nano-Engineering and Technology. The integration of the knowledge development chain from the basic research to the technology suggests a great ability to generate innovations steadily on the next decade (Ministry of Education and Science, Argentina, (2004); Malsch, I., (2008)).

Among the several research topics in N&N developed in Argentina, some of high impact and/ or strong activity can be mentioned. For instance, there are many groups working on carbon nanotubes (CNTs):

1. Production of CNTs and development of polymer-based composites using resins resistant to alkalis and fire, or using biodegradable polymers. Many groups are involved, affiliated to the CNEA Buenos Aires, Faculty of Natural and Exact Science–UBA, INTECIN and INTEMA.
2. Development of metal-matrix composites reinforced by CNTs. Recently, the Advanced Materials Group (GMA), Faculty of Engineering, UBA, have launched a joint project with the University of Oxford and with a Belgium company, NANOCYL.
3. Single wall CNTs funcionalization for biomedical application. There is a group working in this topic in the INTEMA.
4. Application of CNTs to sensors. On this topic two groups have been found, one in

the INQUIMAE and other in the CNEA Bariloche.

Groups at CNEA Bariloche, CNEA Buenos Aires and in the Physics Department, UNT, are working on oxide nanowires (NWs). They study the fabrication processes, magnetic and electric properties for batteries, sensors, field effect transistors, and energy harvesting devices. In particular, photovoltaic devices employing semiconductor nanowires have the potential for lower cost and greater energy conversion efficiency compared to conventional thin film devices due to less material utilization, enhanced photovoltage or photocurrent. Of particular interest are coaxial structures, such as the GaAs nanowire array shown in Figure 2.a, in which a doped nanowire core is surrounded by a shell of opposite doping type, forming a core-shell p-n junction [Caram, J. et al., (2009)]. Another type of nanowire, the one of ZnO, is promising for the development of third generation solar cells, transparent electronics and photonics, spintronics, energy harvesting and sensor applications. The group at UNT is producing and studying these types of NWs. In Figure 2.b is shown a secondary electron image of ZnO NWs grown on a Au nanocluster seeded in a SiO_2 substrate by the vapour transport method [Comedi, D. et al., (2009)].

Figure 2. Array of nanowires (secondary electron images). (a) GaAs [Caram, J. et al., (2009)]; (b) ZnO [Comedi, D. et al., (2009)]

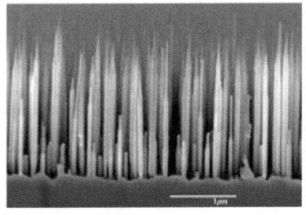 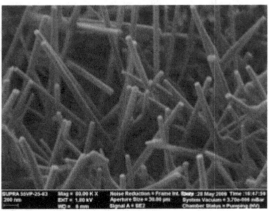

Magnetism and magnetic amorphous & nano-structured materials are other areas with a strong activity, for example,

1. CNEA Bariloche: Some groups study films of magnetic oxides obtained by spray pyrolysis and magnetic properties on different geometries of nanomaterials (nanoparticles, thin films, multilayers), among other specific topics.

2. CNEA Buenos Aires: There is a group that synthesize by means of several techniques (sol gel, microwave or citrate) mixed manganese oxide nanoparticles coated by SiO_2 for cancer therapies.

3. IFFAMAF: in this institute are groups studying magnetic properties of nanoparticles, nanostructured thin films, zeolites, magnetoresistance and piezoelectric properties of composites such as $BaTiO_3$–$BaFe_{12}O_{19}$, memory effect on hematite nanoparticles, bulk magnetic metallic glasses, finemet and Nd-Fe-Al alloys.

4. INQUIMAE: there molecular magnets are studied. They prepare molecules and nanoparticles that behave as permanent magnets and also measure characteristics properties of molecular systems that react to the light and environmental conditions.

5. INTECIN: In this institute associated to the Faculty of Engineering, UBA there are two groups that investigate magnetism. The group at the Chemistry Department prepares and studies ferrites for different applications (electronic, ferrofluids and biomedical). The other group associated with the Physics Department has a long experience in magnetic materials. At present they are studying soft magnetic nanostructured alloys; nanocomposites containing soft or hard magnetic particles prepared by gel casting technique; magnetic nanoparticles and nanowires for several applications (medical therapy, drug delivery, sensors);

Nd-Fe-B alloys for magnets and thin film magneto-resistive nanostructured alloys for biosensors and memory devices. It is worth mention that this group has submitted a project to FAN for industrial applications of magnetic nanomaterials joint to Electropart Córdoba S.A.

6. INTEMA: In this Institute are some groups that study soft ferrites nanoparticles for magnetic storage and magneto-optic sensors, among other specific topics.

7. UNLP: One group in the Physics Department has a long tradition in research on magnetisms & magnetic materials. They are studying metallic and oxide nanoparticles, of various geometries such as ribbons and thin films. A good example of their recent research on magnetostrictive bimagnetic trilayer ribbons for temperature sensing can be found in reference (Mendoza Zélis, P., Sánchez, M., Vázquez J., (2007)).

The networks have addressed the R&D activities targeting certain applications, mainly to sensors, catalysts and different devices. Thus, strong research topics are being developed in nanoparticles, nanopores and thin films. Some of these topics are:

• Study of adhesion of Ag nanoparticles on textile fibers to kill bacteria. This is sought for items of surgery. It seeks to improve adhesion of Ag nanoparticles that do not fall off during washing. This research is led by INQUIMAE and the company Nanotek.

• Encapsulation of nanoparticles in mesopores of oxide materials. This research focuses on the development of materials for controlled diffusion of nutrients and metabolic products, and simultaneously acts as a barrier for other metabolic agents; (INQUIMAE).

• Study of elastic polymer-based nanocomposites containing conductive nanoparticles for developing pressure sensors. As the pressure increases, the electric percolation between particles enhances, which increases the electric current in the nanocomposite film; (INQUIMAE).

- Study of core-shell nanoparticles biosensors, produced on gold nanoparticles, for diabetes diagnosis; (INQUIMAE-CNEA Bariloche).

- Design of nanomaterials for photo-remediation and photo-therapies. When colorant molecules are irradiated active oxygen molecules (O_2*) are formed (Figure 3.a). These active molecules can be used for removing chemical and biological contaminants. Led by INQUIMAE researchers, an automatic self-assembler machine has been developed in order to assemble molecules layer by layer with a good control at the nanoscale (Figure 3.b). This work has been awarded as innovation in 2007.

- Development of nanofilter membranes and mesoporous films for photovoltaic cells, selective sensors and smart windows; (Chemical Department, CNEA Buenos Aires).

There are other important research topics developed by the N&N networks. One is in materials for solid oxide fuel cells, and the second one is on nanostructured Aluminium alloys for mechanical applications. The most active group in nanoceramics for fuel cells is the CINSO. This group is part of one of the networks that also works on materials for sensors.

They synthesized electrolyte solid solutions of the ZrO_2-Y_2O_3, ZrO_2-CaO y ZrO_2-CeO_2 systems with a wide composition range using the gel-combustion method. Also, they synthesized nanostructured dense ceramic based on ZrO_2 (ZrO_2-Y_2O_3 y ZrO_2-CaO) or on CeO_2 (CeO_2-Sm_2O_3 y CeO_2-Y_2O_3) by "fast firing", which avoid grain growth. Complementarily they study (La;Sr)CoO_3 and (Sm;Sr)CoO_3 perovskites for cathode and homogeneous ZrO_2-CeO_2 solid solutions for the anode. In the case of the anode, it is necessary to prepare composed materials with metals such as Ni or Cu to get a better electronic conductivity. Electrodes are prepared from nanocrystalline powders as conductive pastes which are deposited on the electrolytes and later sintered. The Y_2O_3 or CaO doped ZrO_2 or Sm_2O_3 or Y_2O_3 doped CeO_2 nanostructured ceramics are used as solid electrolytes; cobaltites based materials for cathode and compound materials: Ni/ZrO_2-CeO_2 for the anode. Two SOFCs configurations are tested: conventional two-chambers and one-chamber fuel cell. The later cell uses the environmental O_2 to

Figure 3. (a) Representation of the active oxygen molecule release by the irradiation effect on the colorant molecules; (b) Self-assembler machine. [INNOVAR Catalog, (2007)]

produce the partial oxidation of the hydrocarbon and then, the O^{2-} ions coming from the electrolyte react with the H_2 and the CO, generating H_2O and CO_2, respectively. The interest in using a one-chamber cell (operated with air and C_xH_y mixtures as fuel) arises from the fact that the partial oxidation of hydrocarbons can be produced at intermediate temperatures (enabling to reduce the usual high operation temperature of SOFCs), (Figure 4). The CINSO group, since 1981, has obtained twenty patents.

Another group in CNEA Bariloche is also working on ceramic materials for fuel cells. In UBA, there are two groups working on this topic, the one in the Chemical Engineering Department is studying the chemical reactions, and the one in the Mechanical Engineering Department is developing a SOFC in order to test the new nanoceramic materials. Both UBA groups collaborate with CINSO.

With regard to the nanostructured aluminium alloys for mechanical applications, there are three groups, two of which work on the nanoprecipita- tion in conventional Aluminium alloys. One group is at the CNEA Bariloche and the other is at the Institute of Physics of Materials (IFIMAT), Tandil. The third group is the Advanced Materials Group (GMA) at the Faculty of Engineering, UBA that has been working on the development of nano-structured light alloys over the last 15 years. The study has involved molecular dynamic simulation and experimental research to understand the role of several elements in the melt and in the undercooled liquid during the rapid solidification (Audebert F. et al., (2001); Audebert F., (2005); Saporiti F., (2009)). A strong collaboration with the Department of Materials, University of Oxford, allowed the group to have access to electron microscopes and to an unique state-of-the-art research facilities. The collaboration has been extremely successful, with development of *"nanoquasicrystalline Al-based alloys"* with high thermal stability (Figure 5) and high strength at elevated temperatures (275 MPa at 350°C), typically 5 to 6 times higher than the strongest commercial Al alloys (Audebert F. et al., (2002); Galano M. & Audebert F. et al, (2008)

Figure 4. Experimental one-chamber IT-SOFC built at CINSO operated with garbich gases as fuel [De Reca, W., (2007)]

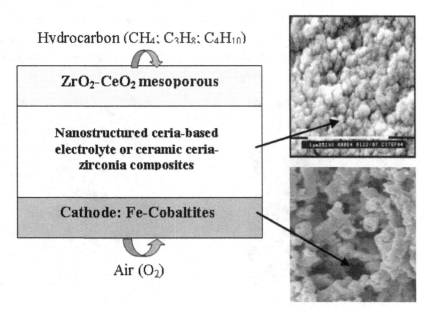

and (2009) part I & II). These alloys can compete with Titanium alloys in the medium temperature range (100 to 400°C), which is relevant considering the Ti price is ~30 times more expensive than Al. They are now working in collaboration with the University of Oxford and an industrial consortium (including Rolls Royce and racing car designers) in a project for industrializing these alloys. Two prototypes, compressors for gas turbines and pistons for racing cars, will be developed. A patent on metal-matrix composites (Smith G.W., (2007)) has been achieved, and the group has begun designing new complex metal-matrix nanocomposites with better set of properties in order to enhance the technical feasibility to substitute titanium products.

Industrial Sector

Some companies are interested in participating in a platform for nanomaterials and nanotechnology development together with the Argentinean research groups: NANOTEK, IAPEL SA, ESSEN SA, Metalúrgica Tandil SA, TENARIS SA [Malsh I., (2008)].

Nanotek

An entrepreneur from the petroleum sector and an academic started Nanotek in 2006 as a spin-off from an academic group which had been developing a catalytic process since 2000. The aim of the enterprise is to identify potential applications: "*Nanosolutions to megaproblems*". 2007 was the first year they had a turnover. The company has developed process: NanoCatox® and ChlorOff®; materials: NanoKupro®, NanoArgen® and NanoFe®, and products: DusTek®, SoilTek® and Asepsis Klima®.

The NanoCatox® process is based on Fenton reactions. It is validated for removing wastes from celluloses pulp, food processing, petrol and petrochemical, photographic wastes, etc. They build a pilot scale process of 3 liters and the client must upscale it. The process takes place inside an industrial reactor. They are extending the use of the process to contaminated sites; it is already approved for PCBs. The reaction products are Carbon Dioxide, Water and Ferric Oxide. The catalyst is nanometallic iron. One of their projects is aimed at treating contaminants including PCBs with iron-oxide nanoparticles in Paraguay. They have a large home market, because under Argentinean

Figure 5. Improvement in the microstructure stability. Both alloys had a similar microstructure as rapid solidified sample /similar to (b). After 66 hs at 400°C, (a) Al-Fe-Cr alloy shows coarsening, precipitation and quasicrystal transformation, however, (b) the microstructure of the improved alloy remains without any changes. [Galano M. & Audebert F., (2008)]

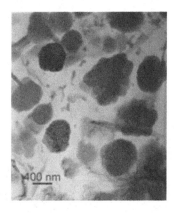

(a)   (b)

law, all PCB contaminated soils must be cleaned before 2010. They also want to develop a method for selective entrapment of heavy metals including lead. This is required due to local regulations in a mining company.

They cooperate with a university for developing an antimicrobial paint using silver nanoparticles. Together with the Paint Company Vilba SA they have developed an antimicrobial latex (Asepsis Klima®). The company also intends to apply silver nanoparticles in polymeric blends as biocide and fungicide for food preservation. Moreover, they also have an initial agreement for a project developing filters, containing silver nanoparticles for air-conditioners in cooperation with a Spanish partner. There is another project at lab scale with INQUIMAE and a manufacturer of specialty textiles for adding silver nanoparticles.

The company is also starting collaboration with the Autonomous University of Barcelona with the aim of enhancing the response of biosensors with metallic nanoparticles or thin films such as copper, silver and iron. Furthermore, they are starting work on zinc nanoparticles and zinc oxide for non-toxic UV barriers in cosmetics (Malsh I., (2008)).

Iapel SA

This Company produces pistons, and is specialized in forged pistons for racing cars. They are also active in the international market. This company collaborates with GMA to develop forged pistons for racing cars using nanostructured Al-based alloys. They are involved, together with GMA, in two international projects.

Metalúrgica Tandil SA

This metallurgy company has cooperated with a research group at IFIMAT for improving the alloy configuration and treatment to enhance the mechanical properties and control the nanoprecipitation.

Essen SA

This company cooperated with GMA for developing wear resistant nanocoatings on Aluminium products.

Tenaris SA

This is an Argentinean international Company that produces seamless tubes for the petroleum industry. The Company has an important R&D sector, with laboratories in Argentina, Mexico, Italy and Japan. The main R&D laboratory is in Argentina; they are generating innovations since many years ago and use N&N mainly on coatings for tribological and corrosion applications. They have few collaborations with the academic sector in topics far from the direct applications but that are required for generating knowledge for their developments.

There are other companies with similar collaborations with the scientific sector, such as IONAR SA. At present the SMEs companies are not much willing to generate innovations using nanomaterials or nanotechnology. Perhaps this is because of the economic recession. After the market recovers, they will probably pursue innovations, and nanotechnology could become the key for entering in the international markets with innovative products.

Production in N&N

Publications

The Argentinean knowledge production in N&N shows an increasing rate when the quantity of the scientific publications is considered. According to the information provided by the Argentinean Center for the Scientific and Technological Information, CAICYT, (2008 & 2009) in 2008, the number of publications in N&N was 4% of the total publications, while in 2003 was only 3%. Half of the publications in N&N included international collaborators. Among these collaborations, as can be observed in Figure 6, the strongest collaboration is with USA followed by Spain, Brazil, France and Germany.

The N&N sector became very active over the last years as is represented by the annual average growth rate of scientific production, which was of 14% in N&N against only the 7% for the total scientific publications in Argentina

When the publications are analyzed by disciplines (considering citations and strength of interdisciplinary collaborations) it is observed that Argentina follows the same world tendency and has a very similar basic structure of interdisciplinary relations. The discipline with major number of citations is Physics, followed by Chemistry, Materials Science, Polymer Science, Engineering, and Biochemistry and Molecular Biology. The three first disciplines are very strongly related in cross-citations. Also, Materials Science is observed to be very strong related with Engineering, and Chemistry with Polymer Science.

Patents

When patents classified by disciplines are analyzed (see Figure 7), it is observed that Argentina shows a strong specialization in Nanomedicine and Nanobiology (BIO) with 82% of its patents, followed by Nanomaterials (MAT) with 18%. The other disciplines, Metrology and Nanoprocess (MET), Nanoelectronics (ELECT) and Optoelectronic (OPT) have small values, around 9%. No patents appeared for Nanoengineering (ENG) and Environment and Energy (ENV). This tendency of active research in Medicine and Biotechnology is perhaps related to the strong scientific tradition in these areas as was previously discussed in relation to the two Nobel prizes in Medicine.

Brazil has a very similar tendency as Argentina with a strong activity in BIO, with 69% of its patents, but presents a higher fraction in MAT, higher than the other Latin American countries, with 44%. Another difference is that Brazil has patents in ENV, while other countries analyzed do not.

Figure 6. Scientific publications (SCI-WOS) in N&N with international collaboration [Argentinean Center for the Scientific and Technological Information, (2009)]

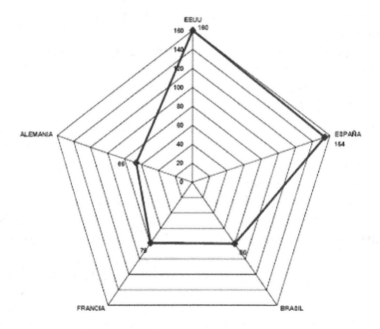

Figure 7. Distribution of patents by disciplines and country ([Argentinean Center for the Scientific and Technological Information, (2009)]

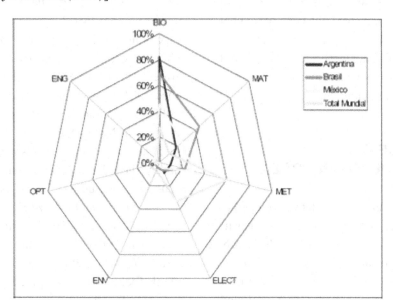

Mexico has a more balanced patent distribution across disciplines, with 45% for BIO, 25% for MAT and 30% for MET. It is the only country among the analyzed which shows patents in ENG, with 5%.

When compared to the world tendency, LA has notable specialization in BIO. The analyzed countries have over 40% patents in this field. This should be contrasted against 27% worldwide, with BIO the third field in order of relevance according to number of patents. MET is the leading discipline worldwide, followed by ELECT with 33%. These two disciplines are vacancy areas for LA.

Final Remarks

Argentina has a strong scientific tradition in Medicine and Biotechnology, which still remains in Nanoscience & Nanotechnology. This tendency is similar in the Latin American region but less pronounced than in Argentina. This asymmetry in the patent distribution by disciplines (contrasted against the worldwide distribution across various fields) highlights the lack of research in material

science and engineering. Probably this fact is also related to the lack of high resolution equipment, particularly electron microscopes. These instruments are necessary for developing nanomaterials, which is strongly related with Physics, Chemistry and Engineering.

The strategy followed by the Ministry of Science, Technology and Productive Innovation based on networks schemes appears to be successful because it has contributed to increase the N&N activities in the country. However, it is still necessary to increase the R&D funding and provide large collaborative grants to build high resolution labs needed to truly leverage the potential of N&N reaching high impact scientific and commercial results.

Centers of excellence that merge academic and industrial sectors, such as the CIDIDI, are indeed required for producing high impact research, promote the knowledge transfer and generate innovations, particularly in Materials and Engineering.

N&N is a now priority area in Argentina, cooperation between academia and industry is stimulated by financial instruments launched by the Ministry, which, in 2010, will initiate a large collaborative grant scheme for fostering to strength scientific-industrial clusters. We expect that, in the coming decade, Argentina will be very active player in N&N, not just bringing significant scientific contributions but introducing innovative N&N products in the global market.

NANOSCIENCE & NANOTECHNOLOGY IN BRAZIL

The academic and scientific system developed in Brazil over the last 5 decades or so was instrumental for establishing extensive research in N&N in the country, especially with the sponsoring of governmental agencies in creating national programs of various kinds. These programs included nationwide networks of excellence in various subareas of nanotechnology, virtual institutes and National laboratories. The advent of nanotechnology, in turn, helped the development and integration of diversified areas. In addition to the expected interdisciplinary research involving physics and chemistry, activities in nanotechnology now encompass many areas of biology and health sciences, taking advantage of methodologies of computer science and mathematics. In this chapter we highlight several of the national and local initiatives for fostering research in N&N, with examples given of important contributions from Brazilian science. Also we comment on the difficulties of establishing an innovation system that could deliver nanotechnology applied to the Brazilian industry.

The realization that N&N may have an impact on the everyday life brought this field to the forefront of research and development initiatives around the world. Brazil was no different in this regard, with many programs being established in this decade. Perhaps the most far-reaching in terms of creating a strong community in the field were the programs of networks of researchers, the sponsoring of virtual institutes dedicated to N&N and the creation of National laboratories for N&N. In 2003, the Brazilian Ministry of Science and Technology (MCT) implemented a Coordinating Committee for Micro and Nanotechnologies, whose mission is to define and foster policies and strategies to develop N&N in the country. At least since 2001 calls for proposals already existed to build or consolidate networks on basic and applied research in N&N. More than 20 other calls for proposals have been opened specifically on N&N through the federal agencies to support science and technology, including CNPq (*Conselho Nacional de Desenvolvimento Científico e Tecnológico*), CAPES (*Coordenação de Aperfeiçoamento de Pessoal de Nível Superior*) and FINEP (*Financiadora de Estudos e Projetos*). These programs are mainly focused on developing an infrastructure of dedicated laboratories, formation of specialized human resources, engaging young scientists in this field and establishment of integrated networks all over the country.

In the strategic planning for industrial policies launched by the Brazilian government in 2004, four areas were considered as key for the technological development of the country, namely semiconductors, pharmaceutical drugs, software and industrial equipment, all of which are directly or indirectly related to N&N. Furthermore, three areas were listed as holding great promise for the near future, viz. biotechnology, biomass and nanotechnology. The specific mention of nanotechnology reflects the importance given to the field. Significantly, the policies of the Brazilian government have been supported by strategic studies conducted by a think tank (*Centro de Gestão e Estudos Estratégicos*, CGEE), with participation of several Brazilian scientists. For instance, a study was produced to discuss the possible technology convergence of key areas, including nanotechnology, biotechnology, information technology, neurosciences and cognitive sciences (www.cgee.org.br). The initiatives may be grouped into 4 categories: implementation of laboratories and networks for N&N, sponsoring of research and infrastructure projects in N&N, sponsoring of partnerships between universities and industry, and creation of national laboratories with facilities to serve the whole country.

In this Section we describe some of the actions and highlight important contributions from Brazilian science. The survey is by no means exhaustive, and the examples given are mere illustrations of the type of work underway in Brazil.

Networks of Researchers

In the first program to establish nationwide networks, 4 networks were created, which were entitled Nanostructured materials, Molecular nanotechnology and interfaces, Nanobiotechnology, and Semiconducting nanodevices and nanostructured materials. All of these networks involved scientists from dozens of universities and research centers across the country, mainly physicists, chemists and engineers. Since the creation of these first 4 networks, similar programs were established at the national level as the Net BrasilNano in 2005, with additional topics being involved such as nanophotonics, nanobiomagnetism, nanostructured coatings, surface probe microscopy, nanocosmetics, carbon nanotubes, computer simulation of nanostructures and nanoglicobiotechnology. The limited participation of biologists, biochemists and professionals from the health sciences has since increased considerably. Just by way of illustration, in 2009 from a single call for proposals by CAPES resulted more than 30 networks specifically for nanobiotechnology (www.capes.gov.br). As it may be expected, the variety of subjects is immense, from fundamental studies with theoretical simulations of biological structures to direct applications in medicine. Also worth noting was the establishment of networks of researchers within universities. For instance, the University of São Paulo, the largest in the country with over 5,000 lecturers and professors, established a network in N&N, which helped the community to get integrated.

A major development in recent years has been the program instituted by Petrobras, the Brazilian company that explores oil in deep waters, with a specific network of researchers to investigate the application of N&N to the oil industry as well as to renewable sources of energy [www.petrobras.com.br]. Many are the possible uses of concepts and materials akin to N&N for this industry. They include novel materials for the coatings and tubes to transport oil and more resistant materials for perforation, materials for catalysis, ultra-sensitive sensors for detecting contaminants and heavy metals that poison the catalysts. Furthermore, research and development in N&N may lead to devices to detect pollutants, and methodologies to separate oil from water, thus reducing the environmental impact of oil spillages [Toma H. E. (2008)].

National Laboratories

The field of N&N in Brazil benefited from the implementation of the Brazilian Laboratory for Synchrotron Light (LNLS), which started functioning in 1997 in Campinas, State of São Paulo. It is the only synchrotron source in Latin America and the first installed in the Southern Hemisphere. Besides the synchrotron light source with X-ray, ultraviolet and infrared beams, which are used in studies of many types of systems akin to N&N, the laboratory also provides infrastructure for characterization of materials, including high-resolution electron microscopy, nuclear magnetic resonance spectrometers, mass spectrometry and micro fabrication. Of special importance for some important contributions of Brazilian science is a high-resolution transmission electron microscope with a resolution of 0.17 nm, coupled to an X-ray diffractometer with micro-analytical capability. In 2008, the Cesar Lattes Nanoscience and Nanotechnology Center (C2Nano) was inaugurated in LNLS, which is devoted to the study of matter at the atomic and molecular levels.

Another National laboratory was created in 2009, namely the National Laboratory of Nanotechnology for the Agribusiness (LNNA), located in São Carlos, State of São Paulo. It is part of EMBRAPA, the Brazilian Agricultural Research Corporation, which has been responsible for much of the research and development in agribusiness in the country, helping placing Brazil as one of the leading countries in agriculture. LNNA is focused on 3 main research fields: i) sensors and biosensors for monitoring processes and products, ii) membranes for separation, and biodegradable, bioactive, and smart packaging and iii) new applications of farming products.

The National Institute of Metrology, Standardization and Industrial Quality (INMETRO), created in 1973, belongs to the Ministry of Development, Industry and Foreign Trade, and is responsible for keeping the standards for industrialized products in the country. Over the last few years, INMETRO has created a sophisticated infrastructure to evaluate new technologies associated with N&N, as is the case of the assessment of organic light-emitting diodes.

Virtual Institutes

Further evidence that N&N has received continued support in Brazil is the creation of virtual institutes to address scientific and technological issues in the area. In the National Program of Millennium Institutes in 2002, which was funded jointly by the World Bank and the Brazilian government, one of the 17 institutes selected was the Institute for Nanoscience and Nanotechnology (INN). It involved researchers from several universities and research centers in the country, dealing with theoretical and experimental aspects of the physics of nanostructures. In addition, some of the other Millennium Institutes also produced research in N&N-related subjects, such as the Institute for polymer materials (IMMP), the Institute for complex materials (IMMC) and the Institute for Micro and Nanoelectronics.

The Millennium program has been replaced by another one in which National Institutes for Science and Technology (so-called INCTs) were established in many areas. Again, N&N received special attention, with an INCT specifically dedicated–continuation of INN–in addition to other ones that have N&N as one of their pillars. Among the latter are included institutes dealing with organic electronics, carbon materials, complex and functional materials, semiconductor devices, pharmaceutical products, biotechnology, catalysis, and computer simulations.

Meetings Involving N&N

The development of N&N field in Brazil has made it a host for Conferences either focused on Materials Sciences with symposia dedicated to N&N or already devoted to N&N. These are the cases of the 9th International Conference on Nano-structured Materials (NANO2008) held in Rio de Janeiro city in 2008 and the 11th International Conference on Advanced Materials (ICAM2009) also held in Rio de Janeiro city in 2009. Other events had their origin within Brazil. These are the cases of the Brazilian MRS Meeting (SBPMat), which is affiliated to the International Materials Research Society (MRS), and the International Symposium on Advanced Materials and Nano-structures (ISAMN), which is focused on N&N. The latter had 4 editions, two of which were held in Chile while the other two were held in Brazil. The next one will be in Mexico, in 2010. ISAMN was conceived to join researchers that already collaborated through integrated networks, but has now acquired the status of a conference bringing together researchers in N&N of several countries.

With regard to meetings involving industrialists and a non-academic audience, in addition to the researchers in the field, one should highlight the Nanotech Fair (Nanotech Expo), which takes place annually in São Paulo city since 2005 (www.nanotecexpo.com.br). There are also forums for discussing the developments and probable impacts from nanotechnology on the society, especially those associated with the environment. The aim is to popularize the topic and offer a balanced view of the opportunities and possible risks in N&N [nanotecnologia.incubadora.fapesp.br/portal].

Some Brazilian Contributions to N&N

The impact of the programs established in the country on the development of N&N can be estimated roughly using scientific databases. For instance, for the query "nano*" (topic) and "Brazil" (address) on 24 August, 2009, in the ISI Web of Science, a total of ca. 4,800 publications in journals appeared. The number of publications per year within the last 20 years (from 1990 to 2009) is shown in Figure 8. It is readily seen that after 2001 the number of publications on "nano*" grew considerably and steadily, which coincides with the policies adopted to induce the development of N&N in Brazil.

As one may expect, with such large number of papers, contributions occurred in many areas of science and technology. An exhaustive, comprehensive analysis of such contributions is outside the scope of the present chapter. Instead, we shall mention only a few works that either had a large impact–in terms of citations in the literature–or

received considerable attention from the media owing to the possible applications of the results reported.

The topmost cited papers in N&N produced in Brazil are related to carbon nanotubes, either with visualization of the nanotubes using high-resolution electron microscopy [Deheer W.A.,et. al (1995), Poncharal P. et. al (1999)] or with the characterization using Raman spectroscopy. By way of illustration, we show in Figure 9 a schematic representation of fullerene molecules (C_{60}) inside single wall carbon nanotubes (SWCN), while Figure 10 depicts resonance Raman data of SWCNs grown by high pressure gas phase decomposition of CO.

Another line of research that also led to a large impact was the use of ionic liquids in the methods of synthesis of nanoparticles [Dupont J. et.al (2002)]. A variety of nanostructures, such as nanowires, were treated theoretically and experimentally, which led to papers with a high number of citations [Caneshi A. et.al (2001), Rodriguez V. et.al (2000)]. A large percentage of the papers could be classified as materials science, physics and chemistry. There is a good balance between works on fundamental aspects of nanoscience and

Figure 8. Number of publications per year found in the Web of Science from 1990 to 2009 (24 August, 2009) using the key words "nano" and the address "Brazil"*

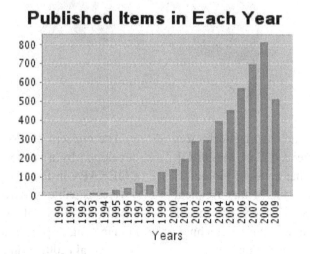

Figure 9. Perspective schematic view of ordered C60 packing (double helix) in a SWCN. Reproduced with permission from: Troche K. S., Coluci V. R, Braga S. F., Chinellato D. D., Sato F., Legoas S. B., Rurali R., Galvão D. S., Nano Letters, 5 (2), 2005, 349-355

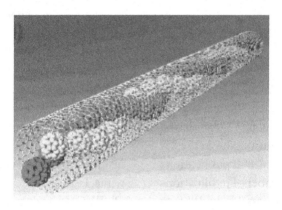

Figure 10. Dependence of the Raman intensity (frequency region of the nanotube radial breathing mode) on different laser excitation energies for SWCNs. Reproduced with permission from: Fantini C., Jorio A., Souza M., Strano M. S., Dresselhaus M. S., Pimenta M. A., Physical Review Letters, 93 (14), 2004, 147406-1-147406-4

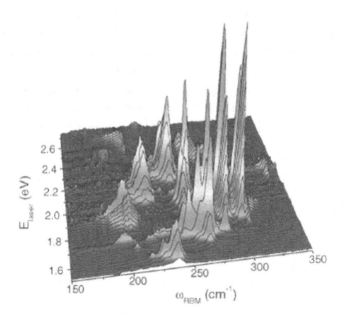

applied science, as in the study of organic devices such as solar cells [Nogueira A.F. et. al (2001)]. It may seem surprising that areas in biological sciences only appeared after the 10 most frequent areas in terms of the number of papers. This means that the effervescent research on nanobiotechnology has not led to many contributions yet.

The concept of electronic tongues has been developed by a consortium of universities and EMBRAPA in Brazil, yielding highly sensitive devices capable of detecting basic tastes with a sensitivity orders of magnitude higher than the human tongue [Ferreira M. et. al. (2003), Riul Jr. A. et. al (2002)], in addition to being able to distinguish between samples of complex liquids, such as wines, orange juice, milk, etc. [Riul Jr. A. et. al (2004)]. This electronic tongue was based on impedance spectroscopy measurements with a sensor array made with nanostructured films. Because a high sensitivity was only attained with nanostructured films, this device is considered as a primary product of nanotechnology. Figure 11 shows a picture of various sensor arrays immersed in different liquids.

Following the seminal work of [Riul et.al (2002)], a number of developments have been made. For instance, the global selectivity paradigm exploited in typical electronic tongues was extended by employing sensing units with materials capable of molecular recognition toward specific analytes, thus imparting selectivity to biosensors. This was performed with the adsorption of antigens on nanostructured films, in an architecture depicted in Figure 12 [Zucolotto et.al. (2007)], with the final sensor array allowing the detection

of antibodies associated with a zoonosis. The latter development holds great promise for clinical diagnosis. Furthermore, the electrical measurements of an electronic tongue can be correlated with the evaluation of professional tasters in an artificial intelligent system, which allows for inferences to be made about the quality of the beverage under study [Ferreira E.J. et.al (2007)].

In terms of technology resulting from the area of N&N, as indicated by the number of patents–either deposited or pending-the Brazilian contribution is modest. This has been a stumbling block in the industrial development of the country, as the relatively intense research activity, among the most intense in developing countries, is not matched by generation of technological innovations. Specifically for N&N, the difficulty in transferring technology appears to apply to other countries as well [Zanetti –Ramos B.G. et.al (2008)]. Nevertheless, successful initiatives in innovation have been reported, including the development of new

Figure 11. Picture of 5 electronic tongues immersed into different solutions. Reproduced with permission from the PhD thesis of Flávio Pandur Albuquerque Cabral, "Desenvolvimento de instrumentação para uso em língua eletrônica", Instituto de Física de São Carlos, Universidade de São Paulo, campus of São Carlos

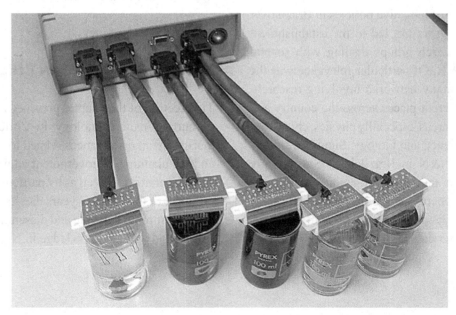

Figure 12. Idealized structure of the PAMAM/proteoliposome layer-by-layer (LbL) films deposited onto interdigitated electrodes. The enlarged area illustrates the exposed portion of bilayers of PAMAM/proteoliposomes, in which interactions with immunoglobulin G (IgG) are expected to occur. Capacitance data were collected as a function of frequency of the ac signal with the electrodes immersed in different solutions containing buffer and the IgG to be detected. PAMAM: polyamidoamine generation 4 dendrimer

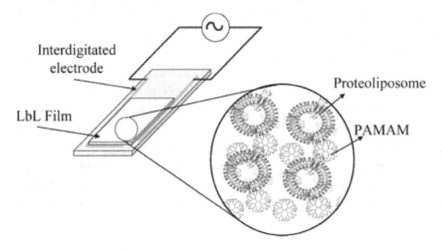

processes for producing graphite used in pencils [Longo E. et.al (2003)], nanoparticles for coatings [Galembeck F. et.al (2006)], chitosan [Da Roz A.L. et.al (2008a)] and silver nanoparticles [Da Roz A.L. et.al (2008b)] for textiles.

Final Remarks

The adoption of inductive policies in Brazil over the last few years has led to the establishment of many research groups dealing with several issues in N&N. Of particular relevance was the creation of many networks involving researchers from different places across the country and from distinct areas, especially physics, chemistry, materials science and biology. Significantly, the research in N&N in Brazil has been strongly connected to that in other countries, through partnerships between groups and institutions, including cooperation programs with countries in South America, North America, Europe, Asia and Oceania. A major challenge for the near future is to transfer knowledge and technology acquired in research projects to the industry, in order to generate products that are useful for the society as a whole. This can only be achieved by an industrial plan with heavy investments in N&N. Furthermore, research into the possible impacts of nanotech products to the environment and human health should be intensified.

NANOSCIENCE & NANOTECHNOLOGY IN CHILE

This section of the chapter presents an overview of current nanotechnology in Chile. We will discuss the main advances related to the funding and the planning of government administrations and universities, as well as by particular research groups. For the past few years these players have been attempting to create and disseminate fundamental and applied knowledge in N&N.

Institutions, National Programs, and Centers

Since 1967, the principal public agency for the development of science and technology in Chile has been the National Commission for Scientific and Technological Research, CONICYT. This agency is an autonomous public institution that encourages the formation of advanced human resources and the growth and strengthening of research and development in the country-the agency also takes regional development and international linkages into account. CONICYT is able to accomplish its goals mainly through the managing of competitive funds in a series of programs: the *National Fund for Scientific and Technological Development* (FONDECYT), the *Fund for the Advancement of Scientific and Technological Development* (FONDEF), the *Centers of Excellence in Research Fund* (FONDAP), the *Non-Formal Science and Technology Education Program* (EXPLORA) and, the most recent, the *Regional Centers of Science and Technology Development*, the *Bicentennial Program of Science and Technology* (PBCT), and the *Basal Financing Program for Scientific and Technological Centers of Excellence* (PFB).

During the last few years, two important new initiatives have substantially strengthened the CONICYT, leading to a much improved National *Science, Technology and Innovation System. They are*, a) The *Millennium Science Initiative* (MSI) --created under the auspices of the Chilean Ministry of Planning (MIDEPLAN) to finance the creation and development of the Millennium Centers of Research-- and b) the *National Council for the Innovation for Competitiveness* (CNIC), which, besides establishing the national policy in science, technology and innovation, has created the *Innovation for Competitiveness Fund* (FIC) managed by the Chilean Ministry of Economy. These two initiatives together with FONDECYT and the Universities (see below) are expected to play a key role in the present and future of nanotechnology in the country.

Indeed, most productivity concerning nanotechnology, as measured by the publication of articles in international journals analyzed in the following sub-section, has been at least partially funded by FONDECYT. Although the budget available for this program is relatively modest, generally enough only for individual researchers or small research groups, the productivity in this area has steadily been growing. The efforts made over the last ten years have succeeded enough in regards to the amount of scientific productivity and formation of qualified human resources to make this discipline competitive in the programs offered by the new initiatives commented in the paragraph above. Considering that such programs provide more resources but simultaneously require the increasing of associative research, the productivity rate in N&N is expected to rise, as is the technological innovation and the social impact of the discipline. During the last years a couple of proposals, presented by groups of senior researchers and/or labs directly or indirectly interested in nanotechnology, have been financed thus permitting the creation of the following scientific entities:

- The MSI *Scientific Nucleus for Basic and Applied Magnetism* constituted by research groups of the Universidad de Santiago (USACH), Pontificia Universidad Católica de Chile (PUC), Universidad Técnica Federico Santa María (UTFSM), and Universidad de Chile (UCH), focused on theoretical and experimental studies of magnetism in different nanostructured systems.
- The *Center for Nanosciences of Valparaíso* gathers scientists from the Universidad Técnica Federico Santa María (UTFSM), the Universidad Católica del Norte (UCN), Pontificia Universidad Católica de Valparaíso (PUCV) and Universidad de Concepción (UdeC).

- The *Center for Development of Nanosciences and Nanotechnology* (CEDENNA) which congregates an important number of physicists, chemists, bioscientists, and medicine experts from the Universidad de Santiago (USACH), Pontificia Universidad Católica de Chile (PUC), the Universidad de la Frontera (UFRO) and the Universidad de Chile (UCH).

As for the progress of the science-industry link in the PBCT program, a series of three National Workshops have recently been conducted.

- *First National Workshop of Nanotechnology*, organized by the UTFSM in Viña del Mar in May 2006, was essentially aimed at diagnosing the current situation of Nanosciences / Nanotechnology in Chile and, in turn, at establishing viable strategies for the development of the discipline with strong links to industry.
- *Second National Workshop of Nanotechnology*, organized by the universities PUC and USACH in October 2006 in Viña del Mar, aimed at identifying the industries and economic sectors that could potentially incorporate tools, products and processes based on the application of nanotechnology and at promoting the use of nanotechnology in industrial processes relevant to sustainable development of the country.
- *Third National Workshop of Nanotechnology: Linking Industry, Science and Technology*, organized by the universities UTFSM, USACH, ECH and PUC in April 2008, aimed at discussing and proposing a strategy for research to develop and apply nanotechnology in Chile in the next ten years.

Moreover, in May 2009, the First National Congress on Nanotechnology took place and over 100 individual technical papers were presented.

Scientific Productivity

The generation of knowledge in nanotechnology in Chile may be illustrated by the scientific publications in journals indexed by ISI written during the past 10 years, which is shown in Figure 13. This productivity almost entirely corresponds to research done in the context of academic programs at Chilean universities; therefore, it mostly concerns basic research. However in the last years the interest in studies related to technological developments and applications has grown. Though there is still no significant activity in the registering of patentable nanotechnology findings there has been some and this is an encouraging response to such efforts. Since the 1980's the number of universities in Chile has been increasing rapidly, but scientific production stems almost exclusively from labs at traditional institutions. Nevertheless, some of the new private universities have begun to show interesting research activity. The institutional distribution of research productivity in nanotechnology can be appreciated in Figure 14. In the figure., the entry "others" is considered the contributions of Universidad Austral, Universidad Católica del Norte, Universidad de Antofagasta, Universidad Nacional Andrés Bello, Universidad de Valparaíso and Universidad Diego Portales. About fifty percent of the work in nanotechnology is carried out in Santiago, the capital of the country, because of the geographical concentration of the population, and the scientific infrastructure and facilities. However, recent initiatives like the program PBCT of CONICYT are trying to change such an injurious tendency. Thanks to such efforts and new research centers in places like Concepción, Valparaiso and Antofagasta the nanotechnology activity is being increasing in the provinces [Abalos J. et.al. (2006)].

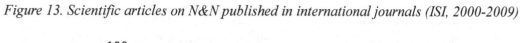

Figure 13. Scientific articles on N&N published in international journals (ISI, 2000-2009)

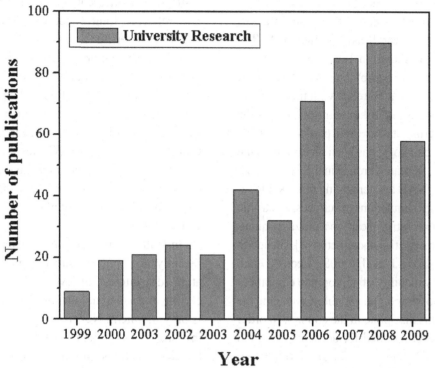

Figure 14. Distribution per institution (%) of international publications on N&N (ISI, 2000-2009)

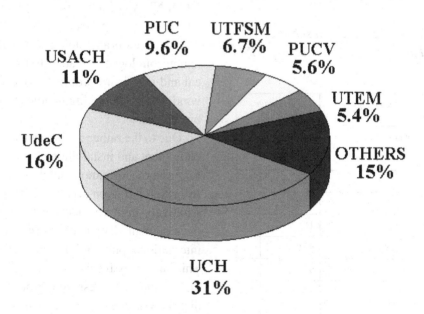

Chile's contribution to research in N&N spreads over a wide spectrum of areas. The distribution of Chilean researchers into sub-areas of nanotechnology research, as classified by *ISI Web of Knowledge*, is shown in Table 3.

In physics a great part of the research concerns theoretical studies that correlate with the historical development of physics in the country. However, the interest in experimental research, particularly in the field of nanotechnology, has been increasing in the past few years. Research groups in the PUC, UCH, UTFSM and UNAB are generating remarkable results in this field (simulations of surfaces, electronic and magnetic properties of carbon nanotubes, quantum dots, and magnetic nanostructures) [Allende S. et.al. (2008); Escrig J. et.al. (2008); Leon A. et.al. (2009)]. The availability of appropriate equipment has been the most important variable in changing the outcomes of this discipline in the country. In order to reach the level required for academic research to become an efficient agent for the development of the country, further policies and resources, both public and private, are required to maintain and significantly improve the park's antiquated equipment [Brunner J. (2001)].

Table 3. Distribution per research sub-area (%) of international articles on nanotechnology (ISI, 2000-2009)

Sub-area/s of nanotechnology research	Percentage (%)	
physics, condensed matter	21.54	
materials science	21.34	
physics, applied	14.03	
chemistry, multidisciplinary	13.04	
chemistry, physical	11.07	
pure nanoscience & nanotechnology	8.50	
polymer science	6.52	
electrochemistry	4.35	
physics, multidisciplinary	4.35	
biochemistry & molecular biology	3.75	

In chemistry there are several research groups distributed in different universities. Some of them have relatively long trajectories, and have attained valuable expertise in areas such as the production of metal nanoparticles in the Universidad de Concepción (UdeC) [Melendrez M. et.al. (2009); Tello A. et.al. (2008)], the preparation of nanostructures or surfaces of low dimensionality by electrochemical methods in the Universidad Católica de Valparaíso, [Cortes A. et.al (2009); Riveros G. et.al. (2008)] and the study of hybrid semiconductor nanocomposites in the Universidad de Chile [O'Dwyer C. et.al (2008); O'Dwyer C. et.al (2006)].

In spite of the fact that bioscience and more recently biotechnology are disciplines which have attained an interesting development, research concerning the use of nanotechnology in these fields is only incipient. Nevertheless, some interesting investigations on the use of nanoparticles as vehicles of peptides important for the treatment of Alzheimer's disease are being undertaken in the Universidad de Chile [Kogan M. et.al (2007); Kogan M. et.al (2006)].

The proposal of the recently formed Center CEDENNA--which has joined physicists and chemists with experience in nanotechnology with medical doctors and biologists to help them apply nanotechnology to advance the treatment of medical and veterinary problems--opens a promising window for integrating nanotechnology into the biosciences.

Due to the rather sophisticated and expensive equipment and installations which are necessary for research in nanotechnology, the productivity which has been attained in this area is comparatively high considering the population and development degree of the country (about 35 international publications per million habitants). This has solely been possible thanks to an active interaction with research groups abroad. The distribution per country of these collaborations is illustrated in Figure 15.

Funding

In Chile, as frequently occurs in developing counties, the capacity for producing scientific and technological knowledge is mostly concentrated in the universities. This knowledge is centered primordially on basic research, and is generally financed directly by public resources and/or by competitive funds, as well as by external resources from the productive sector and from abroad. Indeed, within the national budget for Research and Development (R&D), about 50% is accorded to the universities. Furthermore, an important part of the research costs are provided by the universities themselves in the form of infrastructure and salaries [Abalos J. et.al. (2006)].

The role of the industrial sector in the creation of science and technology is scarce. Nevertheless both private and autonomous public companies contribute significantly to finance expenses in the R&D in the country. In 2004, the contribution amounted to approximately 45%. Such a role--which though important is still considerably less than companies' roles in more developed countries--is expected to increase significantly in the near future because of recent R&D pub-

lic policies. Indeed, the recent creation of new Research Centers and Institutes means that the aforementioned governmental initiatives have the mandatory mission of linking academic research to technological innovation, the initiatives must also begin involving the industrial sector (Leipziger D. et.al. (2004); KAEAX (2009)).

Education and Outreach

Besides the universities' scientific dissemination, this activity has mainly and efficiently been undertaken by CONICYT through the national program EXPLORA. The objective of this program is to contribute to the creation of a scientific-technological culture for children through non-formal education. The idea is that the children develop an understanding for the benefits of these disciplines, specifically in the area of nanotechnology. Since 2007 we have been conducting a series of talks that were included in the National Week of Science and Technology with the aim of introducing the general public to basic concepts and applications of this new area of knowledge.

Moreover, the universities have implemented lectures and experimental courses to introduce stu-

Figure 15. Distribution per country (%) of international collaborations (articles ISI, 2000-2009)

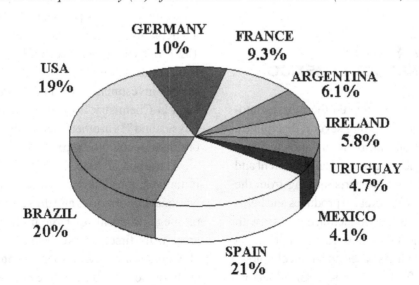

dents at different levels of their educational process to basic concepts of nanoscience, nanotechnology and its applications. Often these institutions also offer special activities to train teachers.

Final Remarks

The synopsis of the state of nanotechnology in Chile described shows that there are about 25 research groups, distributed in the most important universities of the country, which are active in producing knowledge at an international level in this field. Their research has mainly been financed by the universities, by competitive public funds, and is often achieved in collaboration with labs abroad. It is evident that until now this discipline responds more to individual initiatives than to programs specifically designed for the development of this area. However, new governmental programs and funding instruments --which while aiming to foment associative research have led to the creation of research entities like the *Center for Development of Nanosciences and Nanotechnology*, linking basic research, innovative applications as well as technology to the industrial sector-- are promising and could develop nanotechnology more efficiently thereby increasing the social impact of nanoscience and nanotechnology in the near future.

NANOSCIENCE AND NANOTECHNOLOGY IN MEXICO

Nanotechnology is considered the next technological revolution impacting several disciplines, namely information and communication technology, energy, medicine, genetics, environment and various other fields as it starts moving from the laboratories to new markets. It produces and helps to improve products and production processes with better characteristics or new functionalities. In coming years, products based on nanotechnology are expected to affect nearly all-industrial sectors

and will enter the consumer markets. Considering the future prospects of nanotechnology, countries across the world are investing strongly in this sector. The United States, Japan and Germany alone provide ~52% (12,400 M USD) of the total investment around the world, from this 51% is provided by the governments. Underdeveloped countries have also joined this trend as an alternative to economical growth and global competitiveness. However, this emerging field requires expensive infrastructure to be competitive and special strategies to optimize economical resources are required. Countries leading nanotechnology have implemented initiatives or special programs supported by their corresponding governments.

In Mexico, the Programa Especial de Ciencia y Tecnología (PECyT) 2001-2006, a Mexican government program led by the National Council of Science and Technology (CONACyT), considers Nanotechnology as a category within Advanced Materials, which is seen as an emerging, strategic area. The main areas of interest of the program are: catalysis, polymers, nanostructured materials, thin films, semiconductors, metallurgy, biomaterials, optical materials and ceramic advanced materials. It is also mentioned the possibility of implementing a Nanotechnology National Program and a Nanotechnology Network to promote scientific exchange. However, no special programs were implemented in this period. CONACyT reported a federal investment of ~$ 1.4 M USD supporting 152 proposals on Nanotechnology involving 58 institutions within the period of 1998-2004. 53% of such investments were in the area of materials, 14% in Chemistry, 14% in Electronics, 12% in Physics and 7% in others (see http://www.nanored.org.mx/documentos.aspx).

At the end of 2006, CONACyT launched two initiatives, the first one focused on constructing National Laboratories and the second one focused on mega proposals or large projects in strategic areas. In the first case, two nanotechnology proposals were supported with approximately $2 M USD each. In the second case, five pre-proposals were supported with approximately $10 K USD each, this money was to be used to prepare and draft the proposal. However, no specific allocations for nanotechnology were made during this period.

In the PECyTI 2007-2012 Nanotechnology, Mechatronics and Biotechnology are considered strategic areas. At the end of 2008, the Nano-technology Network was finally implemented. Its main purpose is to promote the collaboration between scientists and the formation of multidisciplinary groups working on projects of national interest. No specific nanotechnology allotments have been launched by the Mexican Government. The situation has now worsened given the current Mexican economic problems. In the absence of a National Initiative on Nanotechnology, Universities and Research Centers have defined their own initiatives--in many cases supported by the local government--and submitted proposals for regular grants from different CONACyt programs. It is important to notice that with CONACyT's support many centers have signed agreements with foreign Universities.

Here, we describe some of the actions and highlight Nanoscience and Nanotechnology (N&N) contributions from Mexico.

Universities and Research Centers with N&N Programs

According to the *Diagnóstico y Perspectivas de la Nanotecnología en México (2008)* document, by 2015 more than two million engineers and scientists specialized in nanotechnology will be required worldwide. The document explains that there are 56 Mexican institutions with activities in research and/or education that encompass N&N. There are 12 Undergraduate Programs, 44 Masters in Science Programs, and 43 PhD Programs related to N&N. From those programs, there are 257 PhD and 216 MSc students. There are 450 researchers, 29% are assigned to 14 Centers by the CONA-CyT system, 18% to the National University of Mexico (UNAM), 15% to the Petroleum Mexican

Institute (IMP), 8% to the National Polytechnic Institute (IPN), and 30% are distributed among 20 different institutions.

The UNAM, the largest University in Mexico, concentrates an enormous amount of activity in this emerging field with the leading number of researchers and students working on N&N associated to different Centers and Institutes--in particular, the Center for Nanoscience and Nano-technology (CN&N). This University has some of the best qualifications to perform research on N&N, it is running one the most ambitious projects (PUNTA) related to nanocatalysis for environment with an investment of $1 M USD per year. The IMP depends on the minister of economy and has the most complete capabilities for research on nanocatalysis. Looking at the CONACyT system, perhaps the centers with the stronger N&N programs are: i) the Advanced Materials Research Center (CIMAV), which focuses on nanomaterials with applications in manufacturing, catalysis and energy. ii) The Applied Chemistry Research Center (CIQA), which focuses in molecular, polymer and magnetic nanomaterials. iii) The Potosino Institute of Scientific and Technological Research (IPICyT), which focuses on carbon and magnetic nanostructure. iv)The Optical Research Center (CIO), which has the Nanophotonic initiatives, focusing on optical properties of nanomaterials for applications in lighting, biomedicine, solar cells and optoelectronic devices. v) The Astrophysics, Optics and Electronics Institute (INAOE), which has the nanoelectronic initiative. Many other Universities and Centers are implementing programs in education and research on N&N. However, most of them have limited capabilities to perform this kind of research.

National Laboratories

The purpose of creating National Laboratories was to provide the required infrastructure to promote the research of Nanoscience and Nanotechnology. The laboratories should provide accessibility to

Universities, Institutes and Research Centers in the country, and promote the scientist's collaborative efforts to establish multidisciplinary groups. There are three National Laboratories; all of them are part of the CONACyT system. The Nanotechnology National Laboratory (NANO-TECH) (http://nanotech.cimav.edu.mx/) located at CIMAV in the north of Mexico, was created in 2006 and funded by the Mexican Government ($2 M USD). The main activities are focused on the synthesis and characterization of nanomaterials, supporting training for undergraduate and graduate programs. The laboratory provides the infrastructure for structural characterization including High Resolution Electron Microscopy with a resolution of 0.1 nm in STEM mode and X-ray diffraction.

The Laboratory for Research in Nanoscience and Nanotechnology (LINAN) (http://materials. ipicyt.edu.mx/LINAN/) located at IPICyT in Central Mexico was also created in 2006 and funded by the Mexican Government ($2 M USD). The main activities are focused on the synthesis of nanostructures and nanomaterials, nanobiotechnology, nanocomposites, magnetic nanomaterials, theoretical calculation of electronic properties from nanostructures, and supporting undergraduate and graduate programs. The laboratory provides the infrastructure for structural characterization including High Resolution Electron Microscopy, X-ray diffraction, Raman spectroscopy, and sputtering.

The Nanoelectronic National Laboratory (LNN) (http://www-elec.inaoep.mx/lnn/index. php) located at INAOE in South Mexico was created in 2007 funded by the ministry of economy ($ 1.5 M USD), the State of Puebla ($0.5 M USD), and a donation from MOTOROLA of a complete system for the fabrication of integrated circuits and devices. The main activities are focused on the design and fabrication of devices, integrated circuits and MENS, integrating nanomaterials. The laboratory provides the required infrastructure for those purposes, including clean rooms class 10 and 100.

There are many other National Laboratories where nano-scientists from various disciplines collaborate, just to mention a few: the Genomic National Laboratory where biologists and physicists are using nanomaterials for DNA sequencing, and the National Center of Metrology, which is implementing the protocols to standardize the measurement of nanomaterials' physical characteristics.

Networks and Meetings on Nanotechnology

As a result of PECyTI 2007-2012, CONACyT has promoted the creation of Networks of Scientists across different fields. Four of the pre-proposals supported at the end of 2006 were joined to become the Network of Nanoscience and Nanotechnology (NNN) (http://www.nanored.org.mx/). This network was formally created at the end of 2008. The purpose of this network is foster the development of nanotechnology and integrate scientists from different institutes with a common interest in this emerging field. The formation of multidisciplinary groups or small networks can be expected to resolve problems of national interest in the near future. Already, many scientists from all around the country have joined NNN.

Created in early 2009, the Nanotechnology Network of the State of Guanajuato (RENAG) joined more than 50 scientists from 13 Universities, Institutes and Research Centers. It was funded by CONCYTEG, the Council of Science of the State of Guanajuato, which also supported three proposals that involve collaboration between different groups.

Some Universities have also created internal networks. UNAM has created REGINA (http://www. fisica.unam.mx/nanoifunam/index.htm) since 2003 integrating more than 60 scientists from different institutes and research centers. It launched the Multidisciplinary Research Program (IMPULSA) integrating more than 100 scientists and graduate students from 13 different centers. The UAM has created NanoscienceUAM Network (http://www.nanocienciasuam.com.mx/). The IPN has created the Nanoscience and Micro-Nanotechnology Network. The University of Puebla (BUAP) promotes the International Network of Nanoscience and Nanotechnology. The IPICyT is leading the agreement between CONACyT and the European Union through NANOFORUMEULA an instrument that creates networks for scientists from México and the European Union as part of the seven-framework program. This initiative promotes the formation of consortiums to resolve problems of common interest by using nanotechnology. An interesting initiative has been launched in the State of Nuevo León where the Research and Technology Innovation (PIIT) (http://www.piit.com.mx/?p=acercade) Park has been placed. Located in this place are twelve research centers, seven companies and two business incubators. The areas of interest are Nanotechnology, Health, Mechatronic, Biotechnology and Information Technology.

There are several periodical conferences focused on Nanoscience and Nanotechnology with the aim to provide a forum for researchers and engineers that bring together scientists working on N&N. The oldest event is the symposia *Nanostructured Materials and Nanotechnology* on the International Materials Research Congress organized by the Mexican MRS. The *International Topical Meeting on Nanostructured Materials and Nanotechnology (NANOTECH)* started in 2004 and has been organized by CIO, this year will be the 6th edition. The *Mexican Workshop* organized by the University of Puebla started in 2005. The Workshop on Metastable and Nanostructured Materials started in 2006 and was organized by IPN. The symposium on Nanoscience and Nanotechnology was organized in 2006 by the Nanoscience Division of the Mexican Physics Society. The NANOMEX meetings are organized by UNAM, and this year is the second edition.

Scientific Productivity

As was discussed above, there is strong activity in the Mexican scientific community even in the absence of special federal programs for the development of N&N. The impact can be estimated by quantifying the scientific production. To quantify the effects, the ISI Web of Science was used with the key word "nano*" as the topic and "Mexico" as the address. A total of 4317 published papers in peer-reviewed scientific journals produced about 37980 citations since 1990 to August 25th 2009, with 82% of total publications published in the last ten years. Figure 16 shows the publication and citations per year. It was observed that there was an increment of 27% in publications and 17% in citations in the last ten years, but the growth has been stronger in the last five. A rough analysis suggests an average publication of 9.6 per scientist, citation per item of 8.8, and an h-index of 72 as a country. The top ten areas of research according to the total publications and considering only those with more than 100 inputs are Chemistry and Chemical Engineering (32%), Materials science (27.3%), Physics & Applied Physics (21.6%), Condensed Matter (18.7%), Atomic & Molecular Physics (6.3%), Optics (4.37%), Electrical & Electronic Engineering (3.72%), Coatings & Films (3.7%), Biochemistry & Molecular Biology (3%), Polymer Science (3%).

Table 4 shows the top twenty more prolific institutions. As expected, 42% of total published papers were produced by the biggest University with the largest budget. The University integrates the largest group of scientists working together via internal networks such as REGINA and PUNTA. The latter launched the most ambitious project on nanocatalysis. The other publications are from the 14 Research Centers directly supported by CONACyT. As was mentioned above, there are nine scientists on average per institution.

Figure 16. Published papers and citation per year from 1990 to 29ʰ August 2009. Analysis was performed considering the string, TS=((nano* OR quatum dot* OR quantum wire* OR nanotube* OR carbontub* OR carbon tub* OR buckyball* OR buckyb* OR fullerene* OR self assambled* OR monolayer* OR sers* OR molecular beam epitaxy* OR MBE OR atomic force microscope OR STM OR submicron* OR langmuir blodgett OR plasmonic* OR metamaterials* OR single molecul*) NOT (nanosecond* OR nanogram* OR nanokelvin* OR nanoliter* OR nanovalent*)) AND AD=mexico NOT AD=new mexico, using the Science Citation Index Expanded of the Isi Web of Science

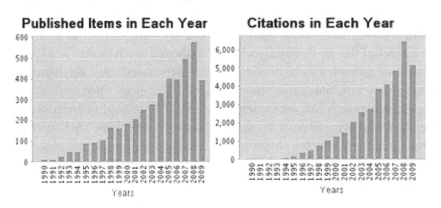

However, the institutions work independently with intermittent collaboration. There is not a network or program to coordinate the efforts of each center and only three of them have received special budgets to support the National Laboratories. Only six of these centers appear in the top twenty, three of them contribute with more than 2% each in the total published papers. The CIO is the only one with an initiative in optical properties of nanomaterials (NANOPHOTONICS) and contributes with 47% of total published papers in the area of optics. These two examples show the importance of scientists with a common interest working together around one specific problem.

Companies

According to the document, *Diagnóstico y Perspectivas de la Nanotecnología en México (2008),* there are no Mexican companies producing nanomaterials or nanostructures or nanosystems for new applications or new devices. Most of the companies are interested in nanotechnology, in particular nanomaterials, to improve properties of

materials used for packaging or to design different components. There are very few Mexican companies that do this but these are just commercializing products from other countries. This is reflected in the absence of patents from Mexican institutions, very few applications (either in absolute term or in terms of patents per million inhabitants) from Mexican institutions have been presented to the Mexican Intellectual Property Institute (IMPI). This should be contrasted against other countries, where the number of patents is quite high in this emerging field. More than 1142 registered patents in México are from foreign countries. Such results highlight the urgency of special programs to promote the interaction between companies and research institutions, and proposals focused to solve specific problems with a well-defined market.

Final Remarks

N&N is believed to be the next industrial revolution and many countries have defined strategies, policies and are investing in this emerging field. Mexico's interest has been highlighted in the

Table 4. Top twenty more prolific Mexican institutions in N&N, from 1990 to 25ᵗʰ August 2009

	Institution	Published papers	% of 4317 papers
1	National University of Mexico (UNAM)	1786	41.37
2	National Polytechnic Institute (IPN)	669	15.48
3	Metropolitan University (UAM)	285	6.58
4	Petroleum Mexican Institute (IMP)	274	6.34
5	University of San Luis Potosi (UASLP)	268	6.20
6	Benemerita University of Puebla (BUAP)	209	4.83
7	Advanced Research Center (CINVESTAV)	207	4.79
8	Potosino Institute of Scientific and Technological Research (IPICyT)*	176	4.06
9	Advanced Materials Research Center (CIMAV)*	105	2.43
10	University of Sonora (USON)	104	2.41
11	Optical Research Center (CIO)*	89	2.06
12	University of Nuevo León (UANL)	87	2.01
13	University of Morelos State (UAEM)	85	1.97
14	Nuclear Engineering National Institute (ININ)	84	1.94
15	University of Guanajuato (U.de Gto.)	72	1.66
16	Applied Chemistry Research Center (CIQA)*	69	1.58
17	University of Guadalajara (U. de G.)	66	1.52
18	University of Michoacan (UMSNH)	53	1.22
19	Scientific Research Center of Ensenada (CICESE)*	36	0.83
20	National Institute of Astrophysics, Optics and Electronics (INAOE)*	33	0.76

* Centers supported directly by CONACyT.

PECyT 2001-2006 and 2007-2012. However, it has not implemented specific N&N programs. Nevertheless, the N&N activity has been ample as can be seen from the high number of publications and citations. The lack of federal policies, of an entity to manage N&N programs and to invest in this emerging field has been substituted by various efforts either regional or by individual institutions. This deficiency hinders capital investments, the generation of human resources and building the required instrumental capacity. It is imperative to define, fund and implement federal N&N policies, perhaps selecting certain market niches (such as medical applications of nanobiotech) to attain global impact according to the country potential and exploit the opportunity working with all, local, regional and global partners.

CONCLUSION

Nanoscience and nanotechnology is already impacting the economy of developed nations and holds an enormous promise to revolutionize existing markets and open new ones. Thus, all developed nations and fast growing countries that leverage the knowledge economy (China, India, Australia, Singapore, Israel, etc.) have launched national efforts to build nano expertise engaging governments, industry, academia, and investing communities. N&N requires collaboration across disciplines, between academia and industry, within countries and regions, and globally. In the knowledge economy, the wealth of nations consists of its people (knowledge workers). In the 21ˢᵗ century a country is not competitive if it does not build a strong science and technology infrastructure that produce first class results and is able to bring some of the generated knowledge to the global market, via technology commercialization, new ventures, etc. Complementarily, a country needs to build a competitive, agile workforce that can work in the industries that are dramatically changing at fast pace. So far, Latin America shows a mixed record.

Latin America counts with some strong research centers and with a tradition of research excellence in certain disciplines such as medicine and biology, agrobio, nuclear technology, metallurgy and materials among others. The multidisciplinary nature of N&N provides a strong opportunity for building in the region, say, nanobio and nano agro-bio global expertise. In fact, we find that all countries in LA have developed network of researchers across disciplines, centers, etc. within a country, and linking 2 or more countries in the region (e.g., Argentina-Brazil Bi-National Center for Nanoscience & Nanotechnology, CABN). Additionally, collaborations have extended beyond LA, mainly to USA and the EU. In general, these collaborative programs have already been quite successful in the generation of interdisciplinary N&N research.

Brazil and Argentina have ministries of science and technology, and in both cases they have played key roles in building a national N&N strategy, and in supporting and funding their N&N efforts. In Mexico and Chile the coordinating role has been mainly played by their respective national research council agencies. All these countries have created interdisciplinary national labs, but still in most countries R&D facilities are not well equipped, and researchers are handicapped to leverage their potential for producing high-impact results in N&N. Brazil, in general, counts with the most and better equipped facilities, and the largest funding.

The relation between academia and industry has been improving in the last years, but it is still quite weak. In particular funding incentives for N&N efforts have encouraged joint efforts and contributed to catalyze these collaborations. Centers of excellence that merge academic and industrial sectors, such as the CIDIDI (Argentina), CEITEC (Brazil), Center for Development of Nanosciences and Nanotechnology (Chile), are indeed required for producing high impact research, promote the knowledge transfer and generate innovations. There are some large companies such as Petrobras (Brazil) with active programs of engagement with academia, or Tenaris (Argentina) which has a solid N&N R&D group that already generated successful nano products for the global market.

A major challenge for the near future is the transformation of knowledge into value, by attracting R&D centers of leading global companies, licensing technologies generated in LA, building new ventures leveraging LA inventions, etc. There are already some success stories of researchers from LA centers that created new ventures to conquest global markets. For example, a team from INAOE, in Puebla, Mexico, led by Prof. Miguel Arias formed Prefixa Inc. Prefixa started 3 years ago and raised initial funds from CONACYT and other sources in Mexico as it was developing its technology and initial products. Additionally it was assisted by the TechBA accelerator–a program from the Mexican government that helps high-potential companies from Mexico to become global. Prefixa has built a solid patent portfolio, validated its technology and products in the US, and later received funding from an US corporate investor which is also a strategic customer & partner. Prefixa, sells subsystems and licenses technology for optical inspection equipment for the semiconductor industry. Prefixa's R&D team, based in Puebla, includes several Ph.D.s, that graduated from Miguel's and other groups at INAOE. From its Silicon Valley office, Prefixa runs its marketing, sales and business development operations working with local (Silicon Valley) experts from the semiconductor industry. To reach US customers and to build the US team, Prefixa leveraged the diaspora of LA professionals in Silicon Valley.

The diasporas of professionals in Silicon Valley, in US, in EU, etc. have played key roles in the development of the software industry of India, the manufacturing industry of China, the semiconductor industry of Taiwan and Ireland, etc. [Saxenian, A. (2006)]. There are large numbers of LA professionals in US and EU in key academic and industrial positions. On the other hand, LA has only taken very limited advantage of this valuable resource. LatIPnet is a not-for-profit organization, with offices in USA (Silicon Valley), Mexico, Argentina and Spain, that leverages the business potential of the Latin diaspora networks of professionals building win-win partnerships. LatIPnet team has a strong experience in establishing industry-academia partnerships, identifying and protecting intellectual property, commercializing technology, and building teams and working with investors for establishing new ventures for the global market. LatIPnet assists researchers in understanding the global market potential of their inventions and defining a commercialization strategy, assists institutions and governments on strategies for leveraging the value of knowledge they generate with global approach for building local wealth, and it has already contributed to the creation and launching of several global companies, such as Prefixa, started by academic researchers in Latin America and Spain.

ACKNOWLEDGMENT

This work was partially funded by FAPESP, CNPq and CAPES (Brazil) and by CONCyTEG (Guanajuato., Mexico). GG likes to thank the Basal Financing Program for Scientific and Technological Centers of Excellence, CONICYT (CEDENNA), and the FONDECYT (Grant 109 0282), and LB thanks the FONDECYT for postdoctoral fellowship (grant 3100088).FA acknowledges Ministry of Science, Technology and Productive Innovation of Argentina, Dr. D. Comedi, Dr. M. Tirado, Dr. E. Calvo and Dr. B. W. de Reca. OO and CC are grateful to Profs. Oswaldo L. Alves, Cylon Gonçalves da Silva and Henrique Toma for providing valuable information about N&N initiatives in Brazil. AN thanks Bo Varga, Nora Sabelli and Simon Goldbard for enlightening discussions. AN also thanks the Secretaria de Desarrollo Económico Sustentable, Government of Guanajuato for partial support, and local researchers, teachers, business leaders and entrepreneurs for useful discussions on the N&N workforce initiative.

REFERENCES

Abalos, J., González, L., & Chervellino, M. (2006). Conceptos, antecedentes históricos y actuales de la institucionalidad nacional y sus instrumentos en ciencia, tecnología e innovación . In *Las regiones de chile ante la ciencia, tecnología e innovación: diagnósticos regionales y lineamientos para sus estrategias* (pp. 24–26). Santiago de Chile, Chile: CONICYT-PBCT.

Abalos, J., González, L., & Chervellino, M. (2006). Diagnóstico y lineamientos estratégicos para el desarrollo de la ciencia, tecnología e innovación en las regiones chilenas . In *Las regiones de chile ante la ciencia, tecnología e innovación: diagnósticos regionales y lineamientos para sus estrategias* (pp. 262–867). Santiago de Chile, Chile: CONICYT-PBCT.

Allende, J., Babul, J., Martínez, S., & Ureta, T. (2005). Publicaciones y patentes. In

Allende, S., Escrig, J., Altbir, D., Salcedo, E., & Bahiana, M. (2008). Angular dependence of the transverse and vortex modes in magnetic nanotubes. *The European Physical Journal B, 66*(1), 37–40. doi:10.1140/epjb/e2008-00385-4

Análisis y proyecciones de la ciencia en Chile (pp. 75-107). Santiago de Chile, Chile: Académia Chilena de Ciencias.

Argentine Center for Scientific Information and Technology. (2008 January). *Nanotechnology: Recent trends and scientific research and technological development*. Buenos Aires, Argentina: CAICYT & MINCYT.

Argentine Center for Scientific Information and Technology. (2009 April). *Boletin Estadistico Tecnologico, No. 3*.

Audebert, F. (2005). *Amorphous and Nanostructured Al-Fe and Al-Ni Based Alloys. NATO Science Series II: Mathematics, Physics and Chemistry, 184. Properties and Applications of Nanocrystalline Alloys from Amorphous Precursors* (Idzikowski, B., Švec, P., & Miglierini, M., Eds.). *Vol. 301*). Dordrecht: Kluwer Acad. Publishers.

Audebert, F., Arcondo, B., Rodríguez, D., & Sirkin, H. (2001). Short range order study in. Al-Fe-X melt spun alloys. *Journal of Metastable and Nanocrystalline Materials, 10*, 155. doi:10.4028/www.scientific.net/JMNM.10.155

Audebert, F., Prima, F., Galano, M., Tomut, M., Warren, P., Stone, I.C., & Cantor, B. (2002). Bulk Amorphous, Nano-Crystalline and Nano-Quasicrystalline *Alloys. Special issue materials Transactions JIM*, 43-8, 2017.

Babini, N., & De Asua, M. (2003, Vol. 26, N°56). La Historia de la Ciencia en Argentina en el último cuarto de siglo. *Revista de la Sociedad Española de Historia de las Ciencias y de las Técnicas*, ISSN 0210-8615, 731-738.

Baum, G., Nemirovsky, A., & Sabelli, N. (2008). La Educación en Ciencia y Tecnología como derecho social en la economía del conocimiento. pp. 133-146 in Propuestas interpretativas para una economía basada en el conocimiento. Argentina, Colombia, México, Estados Unidos, Canadá, Federico Stezano y Gabriel Vélez Cuartas, Editors. 1st Ed. January 2008. ISBN 978-84-96571-58-7

Brunner, J. (2001). Factores comunes o transversales . In *Chile: Informe sobre capacidad tecnológica* (pp. 6–13). Santiago de Chile, Chile: Universidad Adolfo Ibáñez.

Caneschi, A., Gatteschi, D., Lalioti, N., Sangregorio, C., Sessoli, R., Venturi, G., et al. (2001). Cobalt (II)-nitronyl nitroxide chains as molecular magnetic nanowires. Angewandte Chemie-International Ed., 40(9), 1760-1763.

Caram, J., Sandoval, C., Tirado, M., Comedi, Czaban, J., Thompson, D.A., et al. (2009). Electrical characteristics of core-shell p-n GaAs nanowire structures with Te as the n-dopant. Submitted to *Nanotechnology*.

Catalog, I. N. N. O. V. A. R. (2007). 3rd Ed. of the National Innovation Competition. Ministry of Science, Technology and Productive Innovation, Argentina.

Catalog, I. N. N. O. V. A. R. (2008). 4rd Ed. of the National Innovation Competition. Ministry of Science, Technology and Productive Innovation, Argentina.

Comedi, D., Tirado, M., Zapata, C., Heluani, S. P., Villafuerte, M., Mohseni, P., et al. (2009). Randomly Oriented ZnO Nanowires Grown on Amorphous SiO_2 by Metal-Catalized Vapour Deposition. Submitted to *Journal of Alloys and Compounds*.

Consejo Nacional de Ciencia y Tecnología México. (2008, December 16). PECyTI 2007-2012. *Official periodical*. Retrieved from http://74.125.155.132/search?q=cache:9ggPNl8FioMJ:www.siicyt.gob.mx/siicyt/docs/contenido/PECYT.pdf+pecyt+2007-2012&cd=2&hl=es&ct=clnk&gl=mx

Consejo Nacional de Investigaciones Cientiíficas y Técnicas Argentina. (2009). II Nanomercosur Meeting: Opportunities for micro- and Nanotechnology. *CONICET's catalog on Nanoscience & Nanotechnology Institutes*.

CortesA.RiverosG.PalmaJ.DenardinJ.MarottiR. DalchieleE. (2009).

Crespilho, F. N., Pavinatto, F. J., Zucolotto, V., Avansi, W., Barioto, V., Gasparotto, L. H. et al. (2008). *Processo de obtenção de um produto a base nanopartículas metálicas e polímeros para tecidos autolimpantes e auto-esterilizantes e produtos resultantes.* Patented in Brazil, PI 0.802.649-1.

Da Róz, A. L., Pereiro, L. V., Pavinatto, F. J., Crespilho, F. N., Zucolotto, V., Carvalho, A. J. F., et al. (2008). *Produto a base de quitosana e processo de impregnação do mesmo em têxteis.* Patented in Brazil, PI 0.802.290-9.

De Reca, W. (2007). Anales de la Academia Nacional de Ciencias Exactas, Físicas y Naturales. *Nanostructured Materials: Synthesis, Characterization . Properties and Applications, 59*, 59–94.

Deheer, W. A., Chatelain, A., & Ugarte, D. (1995). A carbon nanotube field-emission electron. *Science, 270*(5239), 1179–1180. doi:10.1126/science.270.5239.1179

Dresselhaus, M. S., Dresselhaus, G., Jorio, A., Souza, A. G., & Saito, R. (2002). Raman spectroscopy on isolated single wall carbon nanotubes. *Carbon, 40*(12), 2043–2061. doi:10.1016/S0008-6223(02)00066-0

Dupont, J., Fonseca, G. S., Umpierre, A. P., Fichtner, P. F. P., & Teixeira, S. R. (2002). Transition-metal nanoparticles in imidazolium ionic liquids: recycable catalysts for biphasic hydrogenation reactions. *Journal of the American Chemical Society, 124*(16), 4228–4229. doi:10.1021/ja025818u

Escrig, J., Lavin, R., Palma, J., Denardin, J., Altbir, D., Cortes, A., et al. (2008). Geometry dependence of coercivity in Ni nanowire arrays. *Nanotechnology, 19*(7). Retrieved from DOI: 10.1088/0957-4484/19/7/075713.

Ferreira, E. J., Pereira, R. C. T., Delbem, A. C. B., Oliveira, O. N. Jr, & Mattoso, L. H. C. (2007). Random subspace method for analyzing coffee with electronic tongue. *Electronics Letters, 43*(21), 1138–1140. doi:10.1049/el:20071182

Ferreira, M., Riul, A. Jr, Wohnrath, K., Fonseca, F. J., Oliveira, O. N. Jr, & Mattoso, L. H. C. (2003). High-performance taste sensor made from Langmuir-Blodgett films of conducting polymers and a ruthenium complex. *Analytical Chemistry, 75*(4), 953–955. doi:10.1021/ac026031p

Galano, M., Audebert, F., García Escorial, A., Stone, I. C., & Cantor, B. (2009). Nanoquasicrystalline Al-Fe-Cr-based alloys: Part II mechanical properties. *Acta Materialia, 57*(17), 5120–5130. doi:10.1016/j.actamat.2009.07.009

Galano, M., Audebert, F., Stone, I. C., & Cantor, B. (2008). Effect of Nb on nanoquasicrystalline Al-based alloys. *Philosophical Magazine Letters, 88*(4), 269. doi:10.1080/09500830801935277

Galano, M., Audebert, F., Stone, I. C., & Cantor, B. (2009). Nanoquasicrystalline Al-Fe-Cr-based alloys: Part I phase transformation process. *Acta Materialia, 57*(17), 5107–5119. doi:10.1016/j.actamat.2009.07.011

Galembeck, F., & De Brito, J. (2006). *Aluminum phosphate or polyphosphate particles for use as pigments in paints and method of making same.* U.S. Patent 2006045831.

KAWAX. (2009). *Observatorio Chileno de CTI.* CONICYT.

Kay, L., & Shapira, P. (2009). Developing nanotechnology in Latin America. *Journal of Nanoparticle Research, 11*, 259–278. doi:10.1007/s11051-008-9503-z

Kogan, M., Bastus, N., & Amigo, R. (2006). Nanoparticle-mediated local and remote manipulation of protein aggregation. *Nano Letters, 6*(1), 110–115. doi:10.1021/nl0516862

Kogan, M., Olmedo, I., Hosta, L., Guerrero, A., Cruz, L., & Albericio, F. (2007). Peptides and metallic nanoparticles for biomedical applications. *Nanomedicine; Nanotechnology, Biology, and Medicine, 2*(3), 287–306.

Leipziger, D., Motta, M., & Dahlman, C. (2004). Chile new economy study (Report No. 25666-CL). In *Documents of the World Bank (Vol. 1): Executive Summary and Policy Recommendations* (pp. 1-31).

Leon, A., Barticevic, Z., & Pacheco, M. (2009). Coupling and chemical shifts in carbon nano-structures for quantum computing . *Chemical Physics Letters, 470*(4-6), 249–254. doi:10.1016/j.cplett.2009.01.052

Longo, E., Paskocimas, C. A., Leite, E. R., Magnani, R. A., Pontes, F. M. L., & Barroso, V. (2003). *Método para transformar carbono amorfo em grafite cristalino em minas de grafite utilizando nanopartículas de metais de transição*. Pending patent in Brazil.

Lux. (2008). *Nanomaterials State of the Market Q3 2008*. Lux Research.

Malsch, I. (2008). *Nanotechnology in Argentina*. Utrecht, Netherlands: Malsch TechnoValuation.

Meléndrez, D., Motta, M., & Dahlman, C. (2004). Synthesis and aggregation study of tin nanoparticles and colloids obtained by chemical liquid deposition. *Colloid & Polymer Science, 287*(1), 13–22. doi:10.1007/s00396-008-1950-7

Mendoza Zélis, P., Sánchez, M., & Vázquez, J. (2007). Magnetostrictive bimagnetic trilayer ribbons for temperature sensing. *Journal of Applied Physics, 101*, 034507. doi:10.1063/1.2422905

Ministry of Education and Science. (2004). *PAV Projects Report. SECYT*. Argentina: ANPCYT.

Ministry of Education and Science. (2006). *Science and Technology Indicators*. Argentina: SECYT.

Ministry of Science, Technology and Productive Innovation, Argentina. (2009 April/June). *Statistic Bulletin on Technology No. 3*.

Nogueira, A. F., Durrant, J. R., & De Paoli, M. A. (2001). Dye-sensitized nanocrystalline solar cells employing a polymer electrolyte. *Advanced Materials (Deerfield Beach, Fla.), 13*(11), 826. doi:10.1002/1521-4095(200106)13:11<826::AID-ADMA826>3.0.CO;2-L

O'Dwyer, C., Lavayen, V., Fuenzalida, D., Lozano, H., Santa Ana, M., & Benavente, E. (2008). Low-dimensional, hinged bar-code metal oxide layers and free-standing, ordered organic nanostructures from turbostratic vanadium oxide. *Small, 4*(7), 990–1000. doi:10.1002/smll.200701014

O'Dwyer, C., Navas, D., Lavayen, V., Benavente, E., Santa Ana, M., & González, G. (2006). Nano-urchin: The formation and structure of high-density spherical clusters of vanadium oxide nanotubes. *Chemistry of Materials, 18*(13), 3016–3022. doi:10.1021/cm0603809

Poncharal, P., Wang, Z. L., Ugarte, D., & de Heer, W. A. (1999). Electrostatic deflections and electromechanical resonances of carbon nanotubes. *Science, 283*(5407), 1513–1516. doi:10.1126/science.283.5407.1513

Red de Nanociencias y Nanotecnologioa. (2008). *Diagnóstico y Perspectivas de la Nanotecnología en México*. Retrieved from http://www.nanored.org.mx/documentos.aspx

Riul, A. Jr, De Sousa, H. C., Malmegrim, R. R., Dos Santos, D. S. Jr, Carvalho, A. C. P. L. F., & Fonseca, F. J. (2004). Wine classification by taste sensors made from ultra-thin films and using neural networks. *Sensors and Actuators. B, Chemical, 98*(1), 77–82. doi:10.1016/j.snb.2003.09.025

Riul, A. Jr, Dos Santos, D. S. Jr, Wohnrath, K., Di Tommazo, R., Carvalho, A. C. P. L. F., & Fonseca, F. J. (2002). Artificial taste sensor: efficient combination of sensors made from Langmuir-Blodgett films of conducting polymers and a ruthenium complex and self-assembled films of an azobenzene-containing polymer. *Langmuir, 18*(1), 239–245. doi:10.1021/la011017d

Riveros, G., Vasquez, J., Gomez, H., Makarova, T., Silva, D., Marotti, R., & Dalchiele, E. (2008). Single-step electrodeposition of polycrystalline CdSe microwire arrays: structural and optical properties. *Applied Physics. A, Materials Science & Processing, 90*(3), 423–430. doi:10.1007/s00339-007-4318-9

Rodrigues, V., Fuhrer, T., & Ugarte, D. (2000). Signature of atomic structure in the quantum conductance of gold nanowires. *Physical Review Letters, 85*(19), 4124–4127. doi:10.1103/PhysRevLett.85.4124

Saporiti, F., Boudard, M., & Audebert, F. (2009). Short range order in Al-Fe-Nb, Al-Fe-Ce and Al-Ni-Ce metallic glasses. Submitted to *J. Alloys & Compounds*.

Saxenian, A. (2006). *The New Argonauts: Regional Advantage in a Global Economy*. Cambridge, MA: Harvard University Press.

Schneider, O., & Sanchez, R. (1980). Evolución de las ciencias en la Republica Argentina 1923-1973: *Vol. 8. Geofisica y Geodesia*. Buenos Aires, Argentina: Sociedad Científica Argentina.

Single-Crystal Growth of Nickel Nanowires.. Influence of deposition conditions on structural and magnetic properties. *Journal of Nanoscience and Nanotechnology, 9*(3), 1992–2000. doi:10.1166/jnn.2009.374

Smith, G. W., Audebert, F., Galano, M., & Grant, P. (n.d.). *Metal Matrix Composite Material*. Patent Application N^{er}: PCT/GB2007/004004.

Tello, A., Cardenas, G., Haberle, P., & Segura, R. (2008). The synthesis of hybrid nanostructures of gold nanoparticles and carbon nanotubes and their transformation to solid carbon nanorods. *Carbon, 46*(6), 884–889. doi:10.1016/j.carbon.2008.02.024

The National Council of Science and Technology in Mexico. (2006). *Programa Especial de Ciencia y Tecnología 2001-2006*. Retrieved from http://www.siicyt.gob.mx/siicyt/docs/Programa_Nacional_de_C_y_T_1970-2006/documentos/PECYT.pdf

Toma, H. E. (2008). '08 Workshop of the Thematic Net of Petrobras: Research on molecular interfaces applied to oil and gas nanotechnology.

Vago, M., Tagliazucch, M., Williams, F. J., & Calvo, E. J. (2008). Electrodeposition of a palladium nanocatalyst by ion confinement in polyelectrolyte multilayers. *Chemical Communications, 44*, 5746. doi:10.1039/b812181h

Zanetti-Ramos, B. G., & Creczynski-Pasa, T. B. (2008). Nanotechnology development: worldwide and national investments. *Rev. Bras. Farm., 89*(2), 95–101.

Zheng, M., Jagota, A., Strano, M. S., Santos, A. P., Barone, P., & Chou, S. G. (2003). Structure- based carbon nanotube sorting by sequence-dependent DNA assembly. *Science, 302*(5650), 1545–1548. doi:10.1126/science.1091911

Zucolotto, V., Daghastanli, K. R. P., Hayasaka, C. O., Riul, A. Jr, Ciancaglini, P., & Oliveira, O. N. Jr. (2007). Using capacitance measurements as the detection method in antigen-containing layer-by-layer films for biosensing. *Analytical Chemistry, 79*(5), 2163–2167. doi:10.1021/ac0616153

Chapter 22
Technological Innovations and Africa's Quest for Development in the 21st Century

Evans S. C. Osabuohien
Covenant University, Nigeria Lund University, Sweden

ABSTRACT

Technology is generally seen as a significant tool for development while technological innovations connote better ways of achieving results. This chapter assesses different areas countries can experience technological innovations and notes that most African countries are lagging below expectations in this regards using secondary data sourced from International Telecommunication Union (ITU), United Nations Statistical Divisions (UNSTAT), among others. From the analytical perspective, the chapter established that the low levels of technological innovations in Africa is one of the major reasons why the continent remains in the low developmental echelon compared to other regions of the world. Thus, this chapter submits that adequate efforts should be placed on functional education, health system and technology related innovation programs. Besides, Africa and indeed all developing world must revamp their infrastructures, especially transportation, power and communication towards development in the 21ˢᵗ century.

INTRODUCTION

Technologies have been seen as key tools for economic growth and development across the countries of the world especially the developed ones (UNDP, 2008). Technology is simply viewed as the techniques of doing things, while technological innovations imply better ways of getting

DOI: 10.4018/978-1-61692-006-7.ch022

any task done. Technological innovations are often conceptualized in a general parlance as getting higher level of output with same inputs or same level of output with lower level of inputs. This denotes a better management/utilization of scarce resources in the most efficient manner (Córdova, 2009). There are several areas that a country can experience technological innovations such as education, health, security and governance, banking, transport and communication, among others.

There are various forms of technological innovations (development) such as: Nanotechnology, Biotechnology, Bioinformatics, Information and Communication Technology (ICT), and so on. One of the standpoints of development has been the need for developing countries to improve their technological innovations (Ekekwe, 2009). This viewpoint is usually supported by the fact that most countries that have attained some measures of development followed the path of technological innovations, *inter alia,* as one of the vital tools. However, the situation in Africa has not been as desired. For example, the average growth rate of total phone per 1000 population in Africa was 6.6 between 2001 and 2006 (African Development Bank- AfDB, 2008). While the number of internet users per 100 inhabitants in Sub-Saharan Africa (SSA) increased from 1 in 2000 to 3 in 2006 compared to world average that increased from 7 to 18 (UNDP, 2008; World Bank, 2008a).

In many African countries, most of the technological development indicators have not been impressive. As at 2005, the total number of telephone (mainline and mobile) subscribers per 1000 persons was 2.8 in Liberia, 7.9 in Guinea-Bissau, and 14.4 in Chad. Access to mobile lines per 1000 persons was as 5.8 for Ethiopia and 9.2 for Eritrea. While that of access to the internet per 1000 persons was 2.1 for Niger, 2.4 in Democratic Republic of Congo (DRC) and 2.7 for Central African Republic-CAR (World Bank, 2008b). Access to electricity, which is paramount for any meaningful technological innovation, is not given serious attention in most African countries. The highest recorded figure on access to electricity as percentage of total population in 2006 was about 75.2% in Gabon. It was as low as 4.3% in Chad and 5.7% in Rwanda. For the rural area there is very little access. It was 0.3% in Chad, 0.8% in Burkina Faso and Lesotho (World Bank, 2008b). This is a great challenge to technological development given the role of power supply in technological revolutions.

Another issue of concern for Africa in this 21st century, especially given the global meltdown where most of the countries in Africa that are primary export dependent are being further 'wounded', is the low level of human capital formation. This is usually witnessed from the low priorities education and health sectors are given in Africa. For example, the total expenditure as percentage of gross national income (GNI) on education was 4.4% in 1980 and did not significantly change as it remains at 4.9% in 2005. While the total health expenditure as percentage of gross domestic product (GDP) was 5.6% from 2003 to 2005, and the average proportion of population with access to sanitation was 45% between 2004 and 2006 (AfDB, 2008; World Bank, 2008b).

Based on the above backdrops, this chapter explored the level of technological innovations and development in Africa. The specific objectives include: to document the extent of technological innovation in Africa; to relate some development indicators to technological innovations. The above objectives were achieved using descriptive analysis with secondary data sourced from, among others, International Telecommunication Union (ITU), United Nations Statistical Divisions (UNSTAT), World Bank Trade Indicators, and International Financial Statistics (IFS). Ten (10) countries were selected across Africa- two (2) from five of the regions. They include: Cameroon and DRC (Central Africa); Ethiopia and Tanzania (East Africa); Egypt and Morocco (North Africa); Lesotho and South Africa (Southern Africa); and Ghana and Nigeria (West Africa). The countries were selected based on the two most populated countries in each of the regions in order to have a fair representation. The selected countries represent over 56% of the continent's population (UNCTAD, 2008; UNSTAT, 2008).

The rest of the chapter is organized as follows: next to this introductory part is technology and technological innovations. Development and its indicators are in part III, relevance of technological innovations is in IV. Part V discusses some stylized facts on technological innovations and development in Africa. The last part is the conclusion.

TECHNOLOGY AND TECHNOLOGICAL INNOVATIONS

Technology and Technology Innovations

Technology is a broad concept that deals with the usage and knowledge of tools, crafts and how it affects man's ability to control and adapt to his environment. It is quite difficult to have a generally acceptable definition of technology. By and large, it simply refers to material objects of use to humanity, such as machines, hardware or utensils and many others (Franklin, 1989). It can equally embrace broader terms such as systems, methods of organizing things, and techniques. Technological innovations, on the other hand, connote improvements of previous technologies to a better one. Technological innovations encapsulate the following: printing press, telephony, and internet, information and communication technology (ICT), and so on. Technological innovations have reduced barriers to communication, which have made people interact more freely within and across countries on a global space (National Science Foundation, 2002).

Technology and Technological innovations have greatly influenced human existence and its interaction within the environment in diverse ways. In many societies, it has helped to improve the general welfare of mankind with respect to providing better alternatives to undertaking task. Technology can also connote the relationship that society has with its tools and crafts, and to what extent the society can control its environment (immediate and remote). Thus, technology is seen as the realistic application of knowledge especially in a given area (Franklin, 1989). It can also be conceptualized as a specific field of

technology- for instance information technology, space technology, internet technology, water technology, and so on. In this wise, technology denote tools and machines that may be used to solve practical problems.

Varieties of Technological Innovations

There are varieties of technology and Technological Innovations that have been developed across the world with different shades of application. The varieties are quite many but some of interests are discussed herein.

Nanotechnology

The term 'nanotechnology' is derived from the prefix 'nano'- the Greek word for 'dwarf'. Thus, nanotechnology involves the world of the very small and a nanometer is one-billionth of a meter, while a sheet of paper contains about 100,000 nanometers in terms of thickness (Buzea, 2007). It is often christened as *nanotech*. Nanotechnologies are concerned with the development of new technologies at the size scale of about 1 to 100 nanometers (Ekekwe, 2008). Nanotechnologies involve creating structures and devices that have original properties, applications, and behaviour because of their size. Nanotechnologies are based on the ability to monitor the devices on the atomic/near-atomic size (Buzea, 2007). It generally deals with structures of the size 100 nanometers or smaller, and involves developing devices within that size. Nanotechnologies ranging from the extensions of traditional device in physics, to new approaches based upon molecular self-assembly, to developing new materials with dimensions on the nano-scale, to speculation on whether matter on the atomic scale can be monitored.

Nanotechnology has the potential to create many new materials and devices with varieties of usage including medicine, electronics, and energy production and so on. On the other hand, nanotechnology raises many of the same issues as with any introduction of new technology, including concerns about the toxicity and environmental impact of nano-materials (Ekekwe, 2008; 2009). For instance, in 2000, the United States National Nanotechnology Initiative was established to coordinate Federal Nanotechnology Research and Development (Centre for Nanotechnology and Society, 2008). Other applications of Nanotechnology range from new electronic devices and the process of constructing them to materials for health and environmental uses. Unfortunately, this variety of technological innovation has not been given attention in Africa. For instance, only South Africa has an operational nanotechnology strategy in the continent unlike countries in other continents (especially China, India, USA) that are greatly harnessing its potentials (Parker, 2008; Ekekwe, 2009).

Bioinformatics

Bioinformatics entails the application of information technology to the field of molecular biology. It was derived from Hogeweg (1978) for the study of informatic processes in biotic systems. In recent times, bioinformatics involves the creation and advancement of databases, algorithms, computational and statistical techniques, as well as theoretical foundations for solving practical issues emanating from the management and analysis of biological data. Bioinformatics comprises mathematical and computing approaches used for understanding biological processes. General operations in bioinformatics include mapping and analyzing Deoxyribonucleic Acid (DNA), protein sequences, the alignment of different DNA and protein sequences for comparability as well as the creation and viewing 3-D models of protein configuration (Baldi and Brunak, 2001).

Some sub-disciplines within bioinformatics are: development and implementation of tools that helps in efficient use and management of information; and development of new algorithms and

statistics that are useful in assessing relationships between data sets such as methods for locating a gene within a sequence. Bioinformatics is quite different from other approaches being the fact it is focused on developing and rigorously applying computational techniques in data mining, and machine learning algorithms. Some milestone research efforts in Bioinformatics include: gene finding, sequence alignment, alignment of protein configuration and genome assembly. Others are: forecasting protein structure, predicting gene expression, and protein-to-protein interconnection, genome-wide connection studies as well as evolution modeling.

Biotechnology

Biotechnology is the kind of technological innovation that is based on biology, especially with regards to agriculture, food science, and medicine. It is regarded as any technological application that uses biological systems, dead organisms, or their derivatives in making or modifying products or processes for given usage (United Nations Convention on Biological Diversity, 2002). Biotechnology is at times referred to as genetic engineering technology of the 21st century, though it embraces a broader range of procedures for modifying biological organisms depending on desired needs. Given the development of new approaches and modern techniques in conventional biotechnology, industries have started acquiring new horizons to help them in improving the quality of their products and overall productivity.

Prior to 1971, biotechnology was mainly applied in agriculture and agricultural firms. However, it is now used in some countries to imply laboratory-oriented techniques developed in biological research such as recombinant DNA, horizontal gene transfer in living plants, and tissue culture-based techniques. It can also be used in describing the range of methods that are helpful in manipulating organic materials to boost food production. Biotechnology combines some disciplines such as genetics, molecular biology, biochemistry, embryology, and cell biology, and how they are linked with the exploration of pathogens or pathogen derived compounds for the good of humanity. Biotechnology is also useful in the production of fertilizer, restoration of nitrogen, crop breeding, and control pests. A key factor influencing the biotechnology application and wide use is the improvement of intellectual property rights (IPRs) legislation and enforcement across the globe especially in developed countries. Increased demand for medical and pharmaceutical products in today's world has equally contributed its popularity.

Some classifications of biotechnology based on specific areas are: 1) Blue biotechnology -used to describe the marine and aquatic applications of biotechnology, but its usage is rare. 2) Green biotechnology - applied to agricultural processes e.g. the selection and domestication of plants via micro-propagation and designing of transgenic plants to grow under specific environmental conditions. 3) Red biotechnology - applied to medical processes such as the designing of organisms to produce antibiotics and the engineering of genetic cures through genomic manipulation. 4) White biotechnology or industrial biotechnology- applied to industrial processes like designing of an organism to produce a useful chemical and the use of enzymes as industrial catalyst (Bunders, Haverkort, and Hiemstra, 1996). The investments and economic output of all of these types of applied biotechnologies is described as *bioeconomy.*

Information and Communication Technology-ICT

Information and Communication Technology (ICT) basically involves the use of electronic gadgets especially computers for storing, analyzing and distributing data. ICT involves computer systems, telecommunication, networks, and multimedia applications (Frenzel, 1996). ICT goes beyond the role of support services or only electronic data processing to the applications of global and unlimited services such as the Internet through the World Wide Web (www) and modern computer email facilities, which has strengthened early innovations like the telephone and fax (Osabuohien, 2008). With ICT there is the convergence of computer and telecommunication because it involves applying computers to regular data processing, information storage and retrieval which has brought a somewhat revolution in societies resulting in transformation of activities such as banking, education, health, among others. This form of technological innovation is being appreciated in most African countries compared to other varieties that have been discussed. However, the low level of infrastructural development have not made most of the African countries to benefit much coupled with securities issues that are associated with it.

DEVELOPMENT AND ITS INDICATORS

The Concept of Development

Development[1] can be defined as a sustainable increase in the standard of living that integrates material consumption, education, health and environmental protection (Ghatak, 2001). This buttresses the fact that development encapsulates more equality of opportunities, political and civil freedom. Thus, when growth in capacity is foisted on advancing technology, varying of institution, adjustment in ideologies that are imperative for achieving expansion and brighter opportunities, such a country is experiencing economic development (Cypher and Dietz, 2004).

Development incorporates for ideal structural and institutional changes to economic growth i.e. enhancement of countries capacity in other aspects of human, socio-cultural and infrastructural development. Using the United Nations Development Programme (UNDP) aphorism, there is another brand of development that is generally referred to as sustainable development, which was first used in 1980. Sustainable development is generally defined as a kind of development that involves 'meeting the desires of the present generation without compromising the ability of the future generations to meet their own aspirations' (UNDP, 2001). In this wise, a country should not just strive to attain economic growth but pursue to develop sustainably because every responsible generation should make its shoulder available for the next generations.

The main objective of sustainable development is to enhance economic development in order to increase and conserve the stock of environment, human and physical capital without making the future generations worse-off (The World Conservation Union, 2007). Sustainable development involves the following components, viz: improving the citizen's health and access to education; giving equal opportunities for participation in public domain; provision of clean environment; and promotion of equity that are trans-generational in nature (Human Development Report, 2006; UNDP, 2007). Thus, economic development incorporates for ideal structural and institutional changes to economic growth, which enhances a country's capacity in other aspect of human, socio-cultural and infrastructural development.

Indicators of Development

There are some indicators of levels of development in a country, which are useful in comparing the level of development across countries/regions of the world. Some of them are presented herein:

The Size and Growth of the Economy

Generally, this indicator is captured with a country's size of gross domestic product (GDP) and/or real GDP per capita (GDP divided by population size). On this indicator, countries of the world are classified into high, medium and low-income

economies (UNCTAD 2008), developed countries and less developed countries (Ghatak, 2001). For instance, the GDP per capita of Nigeria and Singapore in 2004 were $500.08 and $25,865 respectively (International Financial Statistics-IFS, 2008). From the above, Singapore with high GDP per capita can be classified as a developed (or high income) country, while Nigeria as less developed (or low income) country. One of the shortcoming of this indicator is that it does not reflect other salient issues like income distribution and welfare about the country besides the monetary values from the GDP and population.

Distribution of Income and/ or Consumption

The pattern of income distribution amongst households/citizens of a given economy is equally a reflection of its level of development. Income disparity would be less pronounced in developed countries than the less developed ones. This is predicated on the fact that unequal opportunities would be higher in the less developed countries. The gini-coefficient of distribution, which varies from 0 (complete equality) and 1 (complete inequality) is normally used to capture it. The figure is always smaller in developed countries indicating a better income distribution than the less developed ones. Sala-i-Martin and Subramanian (2003) showed that the income distribution of the top 2% and bottom 55% of the Nigerian population were earning the same total income in year 2000. This is an indication of a severe disparity in income distribution, which is a characteristic of most less developed countries especially in Africa.

Level of Educational Development

The commitment to education by a given country also gives indication about her level of development. The components that are usually built in education when using it to compare the level of development in various economies include: public expenditure on education as percentage of Gross national product-GNP; number of enrollment figure of relevant age group; and expected years of schooling. For example, in 2007, the secondary enrolment rate in Organization for Economic Co-operation and Development (OECD) countries was 95.65%, while the values for Sub-Saharan African (SSA) countries was about 33.6% compared to the world average of about 70% (World Bank, 2008a). The average value for total expenditure as percentage of GNI in Africa was 4.9% in 2005 (AfDB, 2008).

Quality of Health Service

The level of health provision in a country is also an indication of the level of development in such a country. Little wonder it is argued that a healthy worker is a wealthy worker and by extension a country with health citizens would be wealthy, *ceteris paribus*, (Adebayo and Oladeji, 2005). Health provision as an indicator of development can be viewed via some parameters such as: public expenditure on health; access to safe water; level of sanitation; infant mortality rate per 1000 births; among others. In less developed countries the values for public expenditure on health; access to safe water; and sanitation is usually lower than the developed countries. For example, population with access to safe water in United Kingdom and United States were 100% and 73%, respectively, the average for Africa 62% from 2004 to 2006. In addition, under five mortality per 1000 was 140 in Africa in 2007, while total health expenditure as percentage of GDP was 5.6% (2003-2005). The level of maternal mortality rate per 100,000 live births was as high as 724 in Africa between 2005 and 2007 (AfDB, 2008).

Environmental Issues

As mentioned earlier, one of the main objectives of sustainable development is to have development that embraces environmental protection

(The World Conservation Union, 2007). In this regards, the nature of the environmental protection in a country entails: pattern of land use, water usage; protection of the forest/vegetation, among others, can be seen as indicators of development. Given the relevance of environmental protection in meaningful development, the Millennium Development Goals (MDGs) has environment as one of the targets (Human Development Report, 2006). The nationally protected area as percentage of total area in developed countries such as Austria was 28.3%, United Kingdom was 20.9%, while those of the developing economies like Nigeria was 3.3%, and the average for SSA countries was 6.2% (UNDP, 2007)

Power and Transportation

Another indicator of development in a given country is the level of power generation, distribution and consumption, as well as the extent to which its transport system is developed. The power generation in a country speaks volume of the productive activities that can be carried out in such an economy due to its relevance in energy supply for both the industrial and home use, while the transport system (road, water and air) will determine the link between the urban centres and the hinterland. If a country has good transport system, people and goods can move easily at a minimized cost from one part of the country to the other, which will have multiplier effects on the national economy. However, in the developing countries, power and transportation are either in low supply or under-developed. For instance, the electric power consumption per capita in kilowatt-hour for SSA was 439 in 1998. It was 433 for low income countries, while the average for lower middle income and upper middle income countries were 1,771 and 2,106, respectively. On the other hand, paved roads as percentage of total roads for SSA and low income countries were as low as 16% and 19%, respectively compared to 51% and 92% for middle and high income countries (Human Development Report, 2006).

Communications, Information and Technology

Information and Communication Technology (ICT), which encompasses the use of computer systems, telecommunication, networks, and multimedia applications for the execution of a given task, can also be an indicator in the assessment of the level of development of a country (Adeoti, 2005; Osabuohien, 2008). The level of usage of ICT in a given country can determine, to a reasonable extent, such country's level of development because the use of ICT necessitates the integration of different economic units/agents as well as sectors such as banking, commerce, education, insurance, and so on. In most developing countries, ICT usage (though is improving in recent times) has not been remarkable compared to those observed for advanced countries (Adeoti, 2005).

In using this indicator, the following parameters are usually considered: telephone main lines, mobile phones, personal computers, internet host per 10,000 people, among others. In 2007 for example, the number of telephone (fixed and mobile) lines per 100 inhabitants SSA countries was 17 compared to world average 67. Also personal computer per 100 inhabitants in SSA was 2.4 in 2007, while the world average was 14.7. With regards to internet users, the number was 4.46 per 100 inhabitants in 2007 compared to the average for the world of 22.75 in the same period (ITU Dataset, 2008; World Bank, 2008a). In addition, Africa had cell density of 0.17 in 2002 comapred to other regions like Europe with 4.57, the Americas with 6.92 and Asia of with 1.35 (Olayiwola, Aderinkomi and Atunwa, 2004; ITU Dataset, 2008).

Structure of Output and Demand

The structure of output in this context connotes how output of an economy is distributed among the various sectors such as: agriculture, industry and manufacturing, and services. From the Rostow's stages of growth maxim countries' at the developing stage usually have bulk of their products from the agricultural (traditional) sector, while advanced countries have services contributing more to their output than any other sector (Cypher and Dietz, 2004). For instance, the share of services to total output in low income countries was 38% compared to 57% from the developed economies, in 1998. The same can be said about the structure of demand where most of the goods and services produced in developing countries are private consumption and general government consumption with little level of exportation, except those that are endowed with natural resources (Fosu, 1996, Osabuohien, 2007).

Private Sector Finance

Another indicator of development is the extent to which the private sector of a country plays key roles in the economy when compared to the public sector. The level of private sector participation in less developed countries is lower than the developed ones. For example, the private sector finance as percentage of gross fixed investment in Canada was 86.3%, in UK and United States, it was 87% and 85.9% respectively. In Nigeria it was 44.0% and the average for low income countries was 55.5% (World Bank, 2008b). This is also worsened by the low market capitalization of their stock markets as well as the spread of their interest rate. In UK it was just 2.7%, while in Nigeria it was 13.1%; it was even as high as 18.7% in Zambia. The above points out that the private sector is not playing active roles in most African economies, while the public sector takes more than required in the provision of investment and production

TECHNOLOGICAL INNOVATIONS AND DEVELOPMENT

The relevance of technological innovations filters into almost all spheres of human endeavours and it may not be possible to exhaust areas of application. Few areas are discussed herein. In education, it is applied in teaching using electronic board. The use of internet has resulted in e-learning, e-library as well as when surfing for materials for research purposes. On the other hand, technological innovations have enhanced the review process of articles and thus reducing the time and cost spent in publication of journal and books.

The technological innovations in the banking sector have impacted banks' methods of transactions in terms of payment systems such as: Automated Teller Machine, Electronic Fund Transfer, Clearing House Automated Payments, Electronic Purse, Automated Cheque Sorter, and Electronic and Transfer at Point of Sale. The payment systems have made transactions easy and convenient (Adeoti, 2005; Osabuohien, 2008). The use of ICT has made the banking sector turn away from the traditional method of banking to presumably better methods given the technological innovations that improve efficiency. Thus, banking activities has become more technologically driven given the level of innovations, as can be seen by its application to almost all areas of banking activities (Osabuohien, 2008). Thus, banks have improved customer services as a result of faster mode of operations.

Also in the field of agriculture, technological innovations have helped in the production of disease resistant species of crops. The production of crops that attain early maturity as well improved market and nutrient values has been made possible by technologies like green biotechnology. Similar improvement has been witnessed in animal husbandry where animals with higher nutrient and market values are being raised. The field of medicine has equally enjoyed the improvement of technological innovations given the fact that many discoveries have been made. In this regards, better ways of treating ailments such as cancer and kidney problems have been made possible. The use of scans has also enhanced the treatment of patients in better and less painful ways. All of these have helped to improve the quality of living condition and life altogether.

In mining and manufacturing sector, technological innovations have helped in easier and better ways of prospecting for natural resources. This has, in no small measure, reduced the time and monetary involvement in the process of mineral exploration, production, which will in the long-run improve the economic viability of such exploration and production. Technological innovations have also improved the performance of transport and communication systems in modern day societies. For example, without physical appearance, passengers can book for their flight tickets and make adjustment when exigencies emanate, which is made realistic by modern technology like internet and mobile phones. These have improved the efficiency of the transport and communication sectors.

In addition, technology is relevant in modern day governance (E-governance) as well as security and crime prevention. For example, the use of bioinformatics in obtaining biometrics such as finger print, eye reading and so on have helped in controlling crime because the security agencies can easily trace the identity of individuals. Also in most countries' embassies technological innovations have helped to control falsification of documents and since people can book for interview online ahead of time, it has also reduced the crowd usually witnessed in the embassies.

The enormous relevance of technological innovations notwithstanding some negative effects have resulted just like any human activities that have both advantages and disadvantages. One of it is the issue of environmental hazard that can occur from the emission of fumes and toxic substance into the environment, which may be injurious to human life. Another area is the emergence of financial fraud arising from ATM fraud, internet scam and the likes. For instance, in countries where there is a wide spread of uncensored internet access there is the possibility for people to surf the internet and interact with people across the globe with some ill-intentions. The aforementioned *side effects* notwithstanding, the level of relevance of technological innovations are far greater. More so, with proper monitoring such the side effects can be minimized.

SOME STYLIZED FACTS ON TECHNOLOGICAL INNOVATION AND DEVELOPMENT IN AFRICA

Having discussed some areas of technological innovations and the indicators of development, this part presents some stylized facts from 10 selected African countries. The countries selected (two from each of the five regions in Africa) include: Cameroon, Democratic Republic Congo (DRC), Ghana, Egypt, Ethiopia, Lesotho, Morocco, Nigeria, South Africa (SA), and Tanzania[2]. This is in order to buttress some of the arguments discussed

previously and further achieve the objective of the chapter. Focus was made on the period 2000-2007, given the fact that the issues were examined with regards to the 21sth century.

In Table 1.0, some macroeconomic indicators like real per capita gross national product (PGDP) measured in US dollars at 1990 constant price were presented. The values show that South Africa had the highest real per capita income in 2007. It was distantly followed by Morocco, Egypt, Nigeria and Cameroon. DRC had the lowest value of US$15. Ethiopia had US$201 while Tanzania had US$417. Thus, using the level of real per capita as an indicator of economic development, it can be said that the average South African live better than citizens from other countries, while the average Congolese had the lowest. In addition, with a view to assessing the countries' global strength, some global indicators like degree of trade openness and global competitiveness ranking were equally presented in Table 1.

From the degree of openness, which is measured as the ratio of total trade to GDP, Lesotho had the highest value between 2000 and 2007. This is followed by Morocco, Nigeria, Egypt, South

Table 1. Some macroeconomic and global competitiveness indicators in selected African countries

Countries	Population (million)		Real PGDP($US)		Degree of Trade Openness		Global competitiveness Ranking	
	2000	2007	2000	2007	2000	2007	2006/07 (out of 122)	2007/08 (out of 131)
Cameroon	15.86	18.17	586	1111	0.58	0.66	99	116
DRC	50.69	60.64	104	151	0.40	0.43	n.a	n.a
Egypt	66.53	74.17	1497	1770	0.48	0.77	77	77
Ethiopia	69.39	81.02	113	201	0.22	0.29	116	123
Ghana	20.15	23.01	247	647	0.60	0.60	n.a	n.a
Lesotho	1.89	1.99	452	797	1.26	1.42	102	124
Morocco	28.83	30.85	1272	2316	0.74	1.03	64	65
Nigeria	124.77	144.72	372	1169	0.54	0.87	95	95
SA	45.40	48.28	2927	5826	0.61	0.72	36	44
Tanzania	33.85	39.46	260	417	0.38	0.63	97	104

Note: n.a. means not available.

Sources: UNSTAT (2008), IMF-IFS (2008), World Economic Forum (2008).

Africa, Cameroon, Tanzania and DRC. The global competitiveness ranking[3] indicates that South Africa had the highest rank in the two periods presented followed by Morocco, Egypt, Nigeria, Cameroon. The low rank on the global competitiveness may be as a result of low technological innovations where the participation of a country especially in terms of trade will be influenced by the level of technological advancement.

Examining the level of technological innovations in the selected countries, the chapter presented some measures of technological development in Table 2.

From the Table, one could observe that the number of personal computers per 100 inhabitants in 2000 and 2007 was highest in South Africa with the value of 8.46 followed by Egypt with the value of 8.09. It was as low as 0.02 in DRC, 0.07 in Lesotho and 0.40 in Ethiopia. The values appear high for South Africa and Egypt; they were lower than the world average even though they were far higher than the average for SSA.

In similar vein, the number of internet users per 100 inhabitants in the countries indicates that the highest was in Morocco in 2007 with the value of 20.00, which was distantly followed by South Africa with the value of 10.88 and Nigeria with 5.53. The lowest was in Ethiopia with just 0.22 and DRC with 0.30. Only the values for Morocco, South Africa and Nigeria were above the SSA average and yet they were all far below the global average. Using the number of telephones lines (fixed and mobile) per 100 inhabitants, only South Africa managed to scale above the world average in 2007. The deduction here is that the level of technological innovations based on technological infrastructures and indicators mentioned above is still very low and are far below the global average. This implies that African countries are lagging behind in technological innovations. This may be one of the reasons why the level of value added by transport, communication and storage sector to the GDP in most of the countries are low (see column B in Table 2). This indicator

Table 2. Some indicators of technological innovations and development selected African countries

Countries	Personal computers per 100 inhabitants (A)		Value Added by transport, communication and storage sector in GDP ($US) (B)		Internet users per 100 inhabitants (C)		Telephones (fixed + mobile) per 100 inhabitants (D)	
	2000	2007	2000	2007	2000	2007	2000	2007
Cameroon	0.32	1.12	1158	1844	0.25	2.04	1.00	17.97
DRC	0.35	0.02	326	779	0.01	0.30	0.00	7.30
Egypt	0.68	8.09	5582	8043	1.20	4.26	10.29	38.84
Ethiopia	0.09	0.40	520	925	0.02	0.22	0.00	2.06
Ghana	0.30	0.57	608	914	0.15	2.65	1.70	24.18
Lesotho	0.06	0.07	64	102	0.21	2.60	2.32	20.60
Morocco	1.23	3.02	3440	4456	0.70	20.00	13.23	56.63
Nigeria	0.60	0.85	980	3224	0.06	5.53	0.00	23.50
SA	6.59	8.46	13505	20113	5.45	10.88	30.23	82.51
Tanzania	0.30	0.93	13	28	0.12	1.00	0.84	15.01
SSA Average	1.53	2.40			0.88	4.56	5.25	17.00
World Average	10.22	14.70			7.67	22.75	36.00	67.00

Sources: World Bank (2008a) World Trade Indicators and ITU (2008) Dataset, UNSTAT (2008)

gives the economic relevance of technology because technology usually impact on transport, communication and storage sectors. As countries with higher technological innovations will have higher contribution to the economy (Ayogu, 2007). Consequently, if African countries desire to be relevant and reckoned with in this 21st century efforts are to be made to improve their level of technological infrasturcures and innovations. This will enable them utilize the benefits that are inherent in technological innovations as pointed out in the previous section. This is *sine qua non* for the current global village whose vehicle is fueled by information and knowledge adaptation.

From the foregoing, the level of infrastructural development in Africa has not been impressive. Thus, the chapter assessed the proportion of paved roads as percentage of total roads between 2000 and 2004 as presented in Table 3. The Table indicates that only 2% of total roads in DRC are paved. In Tanzania and Cameroon, it was about 9% and 10%, respectively. The highest was in Egypt with about 81%. In addition, the total electricity production, which comprises thermal, hydro, nuclear and geothermal sources, was quite low in Africa.

It was only 2873 million Kilowatt in Ethiopia in 2005. This is a far cry from what obtains in other regions of the world (World Bank (2008a).

Given the fact that neither technological innovation nor development can occur without a commensurate human factor, this chapter went forward to present some human capital indicators-health and educational development in Table 4. This is because they are quality of life especially within the context of human capital formation and utilization. In Table 4, some of the indicators of health in the selected countries were presented. The value for Human Development Index (HDI) was an average of 0.514 for Africa in 2005 and 6 out of the 10 selected countries had little above the average. It was below 0.50 in DRC, Ethiopia, Nigeria and Tanzania. This implies that the human development components are still low in Africa. Close to this indicator is life expectancy at birth. The average for the continent was 54 years. The highest was in Egypt and Morocco at 71 years. The North African countries appear to have better indicators than other regions. The reason for this is can be taken up in another research. The lowest in the group was Lesotho with just 43 years. The

Table 3. Paved roads and electricity production in selected African countries

Countries	Paved Roads as percentage of Total (%)	Total Electricity Production (Million Kilowatt	
	2000-2004	2001	2005
Cameroon	10	3541	4145
DRC	2	5898	7419
Egypt	81	76629	108138
Ethiopia	19	1833	2873
Ghana	18	7864	6793
Lesotho	n.a	n.a	n.a.
Morocco	57	15409	19037
Nigeria	15	15453	20468
SA	17	242257	278746
Tanzania	9	2753	3036
Total for Africa		480676	589843

Note: n.a. means not available
Source: World Bank (2008a)

implication of this is that the utilization of human capital in the countries is low and the years to recoup investment in human capital were rather small. This may be due to the ravaging effects of malaria, HIV/AIDS and other diseases as well as socio-political and ethno-religious violence that usually occur in many Africa countries.

This further buttressed by under 5 mortality and maternal mortality rates, which were quite high. The average in the continent between 2000 and 2007 was 140 per 1000 and 724 per 100000, respectively. DRC, Lesotho, Nigeria had figures above the average. This scenario is worrisome when one takes a look at the total health expenditure as percentage of GDP[4]. The values between 2003 and 2005 in DRC were 4.0%, 4.2% in Cameroon, 4.3% in Tanzania and Lesotho, 3.7% in Morocco. The highest was 8.4% in South Africa. This speaks volume of the low attention placed on health care provisions and services in most African countries.

To complement the health indicator, measures of educational quality and development were presented in Table 5. This is because education and health are key ways of human capital formation (Osabuohien, 2009). Between 2000 and 2005, the share of public expenditure on education as percentage of GDP was 2.0% in Cameron and 5.0% in Ethiopia and Ghana. Akin to this is secondary school enrolment rate[5] which was quite low. The next indicator used was adult illiteracy rate. This average for the continent in 2007 was 33.3%. It was 52.5% in Ethiopia and 44.9% in Morocco. What this connotes, *inter alia,* is that the level of technological appreciation and adoption would be very low given high adult illiteracy rate in the selected countries.

To sum the discourse, some indicators of quality of life were presented in Table 6. The Table shows that majority of African population dwells in rural areas (over 61.4%). It was even over 80% in Ethiopia and 75% in Lesotho and Tanzania. The major challenge is that rural areas in most African countries have limited social amenities like electricity supply, good roads, and safe water. Population with access to sanitation and safe water between 2004 and 2006 were quite low except for Egypt and Morocco. Another astonishing factor is

Table 4. Some health indicators in selected African countries

Countries	HDI Value (2005)	Life Expectancy at Birth (2007)	Maternal Mortality Rate Per 100,000 Live Births (2005-2007)	Under 5 Mortality Rate Per 100 (2007)	Health Expenditure as % of GDP (2003-2005)
Cameroon	0.532	50	669	144	4.2
DRC	0.411	46	1100	196	4.0
Egypt	0.708	71	130	34	5.1
Ethiopia	0.406	53	673	145	4.4
Ghana	0.553	60	560	90	8.1
Lesotho	0.549	43	762	98	4.3
Morocco	0.646	71	227	36	3.7
Nigeria	0.470	47	1100	187	4.7
SA	0.674	49	110	66	8.4
Tanzania	0.467	53	576	118	4.3
Average for Africa	0.514	54	724	140	5.4

Source: World Bank (2008a)

Table 5. Some indicators of educational development in selected African countries

Countries	Primary School Enrolment (2005-2007)	Secondary School Enrolment (2004-2007)	Adult Illiteracy Rate (%) (2007)	Primary Student-Teacher Ratio (2005)	Education Expenditure as % of GDP (2000-2005)
Cameroon	106	41	21.2	48	2.0
DRC	n.a	22	29.5	n.a	n.a
Egypt	102	86	39.4	n.a	n.a
Ethiopia	98	34	52.5	72	4
Ghana	92	46	21.1	33	5
Lesotho	114	37	13.5	42	13
Morocco	107	49	44.9	27	7
Nigeria	93	32	26.9	37	n.a
SA	103	95	12.2	n.a	5.0
Tanzania	110	n.a	18.2	56	n.a
Average for Africa	96	44	33.3		

Note: n.a. means not available
Source: World Bank (2008a)

Table 6. Some human development and quality of life indicators in selected African countries

Countries	Population with Access to Sanitation % (2004-2006)	Population with Access to Safe Water % (2004-2006)	Rural Population (%) (2007)	Economically Active Population % (2005)
Cameroon	51	51	44.0	42
DRC	30	46	72.0	41
Egypt	70	98	57.3	39
Ethiopia	13	22	83.4	44
Ghana	18	75	50.7	51
Lesotho	37	79	75.2	40
Morocco	73	81	44.3	41
Nigeria	44	51	52.3	40
SA	65	85	39.6	42
Tanzania	47	62	75.0	53
Average for Africa	45	62	61.4	44

Source: World Bank (2008a)

the percentage of economically active population which is 44% for the continent. The other countries presented were all below 50% except for Ghana and Tanzania that marginally exceeded it.

CONCLUSION

Technology is seen as a significant tool for development while technological innovations is taken to connote better ways of achieving results in a given task. Countries can experience technological innovations in different spheres like education, health, governance, banking, transport and communication, among others. However, in most African countries the issues of technological innovations and development have been below expectations. It was based on the above that this chapter explored the level of technological innovation and development in Africa using 10 selected countries. The levels of technological innovations in the selected countries were analytical assessed with regards to some development indicators. The above was achieved employing data sourced from International Telecommunication Union (ITU), United Nations Statistical Divisions (UNSTAT), International Financial Statistics (IFS), World Bank Statistics on African countries, African Development Bank Database and World Economic Forum.

From the analytical perspective discussed in the chapter, it was observed that the levels of technological innovations and development are low in most African countries, which might have been one of the reasons they are in low technological and developmental echelon compared to other regions of the world. The chapter submits that serious emphases should be placed on functional and result-oriented educational system. This is akin to urgent attention that should be accorded the health sector given its unique role in a country's developmental processes. Another suggestion is that the level of infrastructures especially those that relate to transport and communication should be improved in Africa. The provision of reliable electricity supply in African countries would equally enhance their technological development. Thus, this chapter concludes that African countries should take the issue of technological innovations more seriously by through frantic steps because development does not emanate from the *blues* but it is usually made to happen by conscious efforts of man.

ACKNOWLEDGMENT

The author acknowledges the useful comments of the anonymous reviewer as well as the assistance of Uche Efobi of Department of Accounting, Covenant University, Ota, Nigeria. The philosophical direction of Prof Gote Hansson of Department of Economics, Lund University during my fellowship in Sweden is appreciated. The fellowship award by Swedish Institute as guest researcher 2009/2010 is acknowledged.

REFERENCES

Adebayo, A. A., & Oladeji, S. I. (2005). *Health Human Capital Condition: An Analysis of the Determinant in Nigeria.* Paper Presented at African Conference 2005 on African Health and Illness, Austin, The University of Texas, March 25-27.

Adeoti, J. O. (2005). Information Technology Investment in Nigerian Manufacturing Industry: The Progress So Far. In *Selected Papers for the 2004 Annual Conference* (pp. 213-244). Ibadan, Nigeria: Nigerian Economic Society.

African Development Bank- AfDB. (2008). *Selected Statistics on African Countries* (*Vol. 27*). Tunis: African Development Bank.

Ayogu, M. (2007). Infrastructure and economic development in Africa: A review. *Journal of African Economies, 16*(1), 75–126. doi:10.1093/jae/ejm024

Baldi, P., & Brunak, S. (2001). *Bioinformatics: The Machine Learning Approach* (2nd ed.). Cambridge, MA: MIT Press.

Bunders, J., Haverkort, W., & Hiemstra, W. (1996). *Biotechnology: Building on Farmer's Knowledge.* London: Macmillan Education Ltd.

Buzea, C. (2007). Nanomaterials and nanoparticles: Sources and toxicity. *Biointerphases, 2*(4), MR17–MR71. doi:10.1116/1.2815690

Centre for Nanotechnology and Society. (2008). *Nanotechnology and Society.* Santa Barbara, CA: University of California. Retrieved from http://www.cns.ucsb.edu/nanotechnology-society

Córdova, V. S. (2009). Multicultural and Creative On-Line Learning. *International Journal of Education and Development using ICT, 5*(2), 1-10.

Cypher, J. M., & Dietz, J. L. (2004). *The Process of Economic Development* (2nd ed.). London: Routledge.

Ekekwe, N. (2008). Neuromorphs: Replaceable human organs of the future? *IEEE Potentials, 27*(1), 8–25. doi:10.1109/MPOT.2007.913683

Ekekwe, N. (2009). *Towards Competitiveness and Global Outsourcing: Practical Model for Microelectronics Diffusion in Africa.* Paper Presented at International Conference on Industry Growth, Investment, Competitiveness in Africa (IGICA), Abuja, Nigeria, 8-10th June.

Fosu, A. K. (1996). Primary exports and economic growth in developing countries. *World Economy, 19*(4), 465–475. doi:10.1111/j.1467-9701.1996.tb00690.x

Franklin, U. (1989). *Real World of Technology.* Toronto, Canada: House of Anansi Press.

Frenzel, C. W. (1996). *Information Technology Management.* Cambridge, UK: Thomson Publishing Company.

Ghatak, S. (2001). *An Introduction to Development Economics* (4th ed.). London: Routledge.

Hogeweg, P. (1978). Simulation of cellular forms. *Frontiers in Systems Modeling Simulation, 31,* 90–95. doi:10.1177/003754977803100305

Human Development Report. (2006). *Beyond Scarcity: Power, Poverty and the Global Water Crisis.* New York: United Nations Development Programme.

International Monetary Fund (IMF). (2008). *International Financial Statistics.* Washington, DC: IMF.

International Telecommunication Union (ITU). (2008). *Information and Communication Technology Statistics.* Geneva.

National Science Foundation. (2002). *Science and Engineering Indicators* (*Vol. 2*). Washington, DC.

Olayiwola, K., Aderinkomi, A., & Atunwa, O. (2004). Information Technology and Business Development in the Nigerian Banking Sector. In R. J. Bauerly, P. C. Thistlethwaite, & D. W. Schofield (Eds.), Emerging Issues in Business and Technology (pp. 214-223). Macomb, IL: Western Illinois University.

Osabuohien, E. S. (2007). Trade openness and economic performance of ECOWAS members: Reflections from Ghana and Nigeria. *African Journal of Business and Economic Research, 2*(2-3), 57–73.

Osabuohien, E. S. (2008). ICT and Nigerian banks' reforms: Analysis of anticipated impacts in selected banks. *Global Journal of Business Research, 2*(2), 67–76.

Osabuohien, E. S. (2009). *Industrial Conflicts and Health Care Provision in Nigeria.* Paper Presented at the Faculty of Social Sciences Discourse Series in Conjunction with Sociology Dept, University of Lagos on Human Resources, the Moral Domain and the Human Condition, 14th - 16th April.

Parker, R. (2008). *Leapfrogging Development through Nanotechnology Investment: Chinese and Indian Science and Technology Policy Strategies.* Presented at China-India-US Workshop on Science, Technology and Innovation Policy.

Sala-i-Martin, X., & Subramanian, A. (2003). *Addressing the Natural Resource Curse: An Illustration from Nigeria.* Mimeo, Columbia University.

The World Conservation Union. (2007). *The World Conservation Union - About IUCN*. Retrieved from http://www.iucn.org/eng/about

UNCTAD. (2008). *Handbook of Statistics*. Washington, DC: United Nations.

UNDP. (2001). *Human Development Report*. New York: United Nations Development Programme.

UNDP. (2007). *Making Globalization Work for All*. United Nations Development Programme Annual Report. Retrieved from http://www.undp.org/publications/annualreport2007/IAR07-ENG.pdf

UNDP. (2008). *Millennium Development Goals (MDGs)*. Retrieved March 15, 2009, from http://www.undp.org/publications/MDG_Report_2008_En.pdf

United Nations Convention on Biological Diversity. (2002). *Convention on Biological Diversity*. Montreal, Canada: Secretariat of the Convention on Biological Diversity.

United Nations Statistics Division (UNTSAT). (2008). *Country Table*. Retrieved from http://unstats.un.org/unsd/snaama/SelectionCountry.asp

World Bank. (2008a). *World Trade Indicators*. Washington, DC: World Bank.

World Bank. (2008b). *Africa Development Indicators 2007*. Washington, DC: World Bank World Economic Forum. (2008). *The Global Competitiveness Report*. Retrieved February 19, 2008, from http://www.weforum.org/en/initiatives/gcp/Global%20Competitiveness%20Report/index.htm

ENDNOTES

[1] Development, in most cases, embraces many areas of human endeavours like economic, political, and social and so on. However, in this chapter 'development' and 'economic development' are used almost synonymously because economic development encompasses more issues than economic growth. The economic growth only defines increase in GDP growth rate, while development involves quality of life and stand of living of a country's citizenry.

[2] The countries were selected based on the two most populated in the region and they are arranged alphabetically for analytical convenience.

[3] This is computed by World Economic Forum (2008). It has about 12 components, which shows how competitive an economy is in the global economy with respect to maximizing the opportunities there in. The rank ranges from one, which is most competitive.

[4] The values are considered very minimal given the fact that the private sector that would have augmented the public expenditures are not developed in many African countries.

[5] This was used because decisions about future plan with regards to occupation and vocation are assumed to be made at this level.

Chapter 23
Emerging Technology Transfer, Economic Development and Policy in Africa

Alfred Kisubi
University of Wisconsin, USA

Chi Anyansi-Archibong
North Carolina A&T State University, USA

Ngozi C. Kamalu
Fayetteville State University, USA

Johnson A. Kamalu
Alabama A&M University, USA

Michael U. Adikwu
World Bank-Step-B Project and University of Nigeria, Nigeria

ABSTRACT

No nation can succeed economically without a strong and solid scientific educational base particularly in this era of knowledge economy. In many developing nations, the resources to develop both the human capital and infrastructure for education are inadequate. Specifically, in Africa, the intellectual capabilities on nanotechnology and microelectronics research and education are still evolving and some foundation technologies like electricity and ICT needed to drive and support them are not available. Lack of management efficiency and good governance continue to stall progress in the continent. In these matrixed four sub-chapters, these issues are discussed including a new model, Generic and Incremental Value (GIV), proposed for African development.

DOI: 10.4018/978-1-61692-006-7.ch023

THOUGHTS ON NANOTECHNOLOGY TRANSFER IN AFRICA- AN INTRODUCTION

Technological and social planning often have unanticipated outcomes. It is already very evident that science and technology would be very important in the search for solutions to the multitude of problems facing Africa. Nanotechnology has the potential to ameliorate the problems of hunger, disease and communication by making possible for African villages to have new suitable facilities in agriculture, healthcare and education. For success in the adoption and diffusion of this technology, it is very imperative that Africa develops indigenous capability. This will require huge commitment of financial and human resources to research and development by African governments, the private sector donors and nongovernmental organizations.

There are two outcomes in the developments and applications of nanotechnology in various areas that include information and storage, agricultural advancement, human enhancement, medical facilitation and drug administration, clean energy, sports, smarter cars, durable clothes, among others. They are intended outcomes (manifest functions) and the unanticipated outcomes (latent functions). A manifest function is the intended outcome of a technological change or visible act or outcome of a social behavior. But there are also unintended or unanticipated outcomes. These unanticipated outcomes, if they begin to significantly influence or change parts of a social system or society, are called latent functions. Latent functions are not often recognized when they occur and often not recognized. It is the duties of the African Union to help member nations to ensure nanotechnology is adopted and diffused safely and responsibly in the continent.

The critical shortage of the scientific and technical manpower on the continent calls for an arrangement that will enable the small groups of scientists and technologists in individual countries to collaborate effectively through the concept of network. With the close collaboration of the governmental and non-governmental institutions of higher learning Africa could establish effective working relations without having to spend a lot of money on new scientific and administrative facilities. Database technology must be used to increase access to sources of knowledge and information, provide access to a wide diversity of sources, and assist in the generation of new knowledge through increased access. The advent of information technology offers a competitive edge in analysis, full scale planning, and proposal competition, project development, marketing and sales. Therefore, Africans must work to make information technology affordable to individuals, grassroots organizations, schools and their economic institutions. Technology is power and Africa must exercise it.

There is a continuing acceleration in scientific and technological knowledge. It has been estimated the last 170 years have seen more advances in scientific/technological knowledge than took place in all of recorded history up to 1850. There have also been incredible advances in the application of this knowledge in industry, business, medicine, and social support systems such as utilities, fire and police. Hopefully, as emerging technologies become more available in Africa; Africans will harness the abundance of solar power for household consumption, industrial power and medical adaptive systems. The transcendental days, when Africans worshipped the Sun God, will be replaced by worship of nature.

A large part of the modern society is dependent upon technology. Our social structure rests on a foundation of complex scientific/technological support systems driven by fossil fuels. Dependency of this magnitude requires ever broadening technical applications of increasing efficiency, such as nanotechnology. In order to gain efficiency in application scientific/technological achievements has led to specialization. Efficiency and specialization lead to a consolidation in technology in which fewer component parts handle greater loads. More parts of the social support system which once were 'independent', if you will, become part of one technology. The social structure then becomes dependent upon fewer parts of a specialized technology. The world is evolving into a technological ecology as delicate to the maintenance of human species as is the ecology of nature to the maintenance of the food chain. However, this man-made ecology has been found to be harmful to the environment, so nanotechnology comes with both problems and solutions. Africa must plan and ensure failed attempts on technology adoptions of previous technologies are not repeated with respect to nanotechnology.

Nanotechnology promises to enhance African appreciation of its own beauties; nonetheless, it is also noteworthy to point that earlier Western technologies and capitalism have also strengthened Africa's capacities as well as propensity to destroy them. The continent must learn from past mistakes so that nanotechnology succeeds in its manifest functions. Nanotechnology is an opportunity to rise high above historical technological subordination and dependence. African scientists should popularize and disseminate knowledge in common terms for common service, and in the language of the people. Nanotechnology will carry out sustainable services to the poor and hungry, empower the people to meet the needs that confront them, respond to the crises of drought and floods and also to the multiple appeals from individuals and communities who normally lack basics that the rest of the world including the African elite take for granted - sufficient food, adequate shelter and the opportunity for people. Africa, the cradle of civilizations, can no longer afford to continue serving the grim role of graveyard of cultures. Nanotechnology and indeed technology must be the new culture.

SUSTAINABLE DEVELOPMENT IN AFRICA: TECHNOLOGY TRANSFER AND MANAGEMENT CHALLENGES

Economic development is the promotion of more intensive and more advanced economic activity through such means as education, improved tools and techniques, more available financing, energy, better transportation facilities, and creation of new businesses. It is a process of wealth creation for the benefit of the nation, region, or continent. It is the process that challenges leaders/governments to improve both standard of living and the quality of life for its people.

The role of public and private institutions, resource development, and technology transfer in the process of achieving sustainable economic development is the focus of this section. This process thus, calls for efficient and effective functioning of institutions designed to promote economic activities. Emphasis is on Africa where studies have shown that poor governance, policies, management of resources, and absence of appropriate functioning institutions inhibits development. Often times, organizations and nations, in the case of this article, fail to succeed simply because of their inability to innovate and re-imagine existing processes. Most African nations continue to dwell or maintain development models established by the colonial rulers. African leaders must engage in the innovation of economic development models that employ its comparative advantages to serve its societies, that is, engage in collective and collaborative actions and processes that enhance standard of living and quality of life.

Technology Transfer and Economic Development

Technology can be broadly defined as the "entities, both material and immaterial, created by the application of mental and physical effort in order to achieve some value". In this usage, it refers to tools and machines that maybe used to solve real world problem. Could these tools and machines be applied to solve Africa's economic development problems? How will the governments develop the appropriate entities? How does appropriate technology transfer impact economic development?

To understand the relationship between technology transfer and development, it is necessary to identify the meaning of technology transfer as it applies in this article. Technology transfer is defined as "the process of sharing skills, knowledge, technologies, and methods of manufacturing, samples of manufacturing, ideas, facilities, etc, among governments and other institutions to ensure that scientific and technological developments are accessible to a large number of users. The users are those who can then further develop and enhance the technology into new products, processes, applications, other materials and services.

Emphasis is on determination of needs of the society and the choice of technology transfer. Many institutions and government organizations today have an office of technology transfer whose responsibility is to identify best technology and how to exploit it to achieve the objectives of the organizations. This process designed to exploit scientific research or natural resources varies and includes licensing agreements, joint ventures and partnerships that may share risks and rewards. Other possibilities include spin-outs which are often used when the host organization lacks the resources or skills to develop new technology.

Societies must distinguish among the various types and the mechanisms of technology transfer especially transfer from developed to developing nations. These distinctions include but not limited to general, system-specific, product and process technology as well as embodied and disembodied vertical and horizontal, material, design, and capacity transfers.

A major limitation in the efforts to economically develop African nations rests with the institutions design to effect development. Instability in governments of this continent further complicates

the potential complexity of technology transfer process. Many technology transfer institutions are multidisciplinary and include such experts as economists, engineers, attorneys, marketers, scientists, and managers. In Africa, political motives, personal agenda and poor governance makes it difficult to put the necessary forces together.

The inabilities of the society to create a common goal and to maintain continuity are some of the major issues in developing the African continent. China, although a country and cannot be compared to the African continent, has successfully worked with the process of licensing, joint-ventures, and partnerships. The country has succeeded and is making progress with policies that attract and sustain needed technologies for its fast developing economy. African countries and leaders need to reexamine its communities, identify the resources available, and define the best technological processes (type of education, training, skills development, products, services, partners and alliances, etc) that will be appropriate to for sustainable economic development. Ability to identify and select appropriate technology is a critical form of resource development.

There is a need to develop effective relationships among the institutions for effective management of resources and technology development. This is vital as effective resource development is a multifaceted concept. It includes both the acquisition of funds and the optimization of the innate potentials of the nation or the continent, its personnel (human resources), and its clientele or potential partners. It also relates to the identification, solicitation, and acquisition of all available resources needed for effective economic development. This concept is inseparable from the functions of planning or institutional development. Resource development is a planned, deliberate, and conscious process of assessing the needs of the continent and its potential partners in economic development, determining the extent to which such needs fit within the established or legislated mission of the continents development. It can also be viewed to be assessing the extent to which the continent is meeting or can meet the these needs with its current resources as well as determining what additional resources will be required for the identified need to be satisfied; and then identifying, soliciting, and acquiring the extra resources to meet these needs. Based on the foregoing, we propose a model for development and management in Africa.

Proposed Model

There is a strong relationship among institutions, resources, technology and sustainable economic development. Also recognition of the fact that one model of economic development does not fit all is a viable premise for Africa's development. The figure below depicts and explains the relationships among institutions, resource development, appropriate technology transfer, and management of growth. The model, Generic and Incremental Value (GIV), which proposes a generic and incremental value provides a holistic view of the society and its needs, appropriate institutional relationships to guide and manage resources (all factors of production), allows African nations and the continent as a whole to focus on its areas of comparative advantages. The model in a different form has been applied successfully in emerging nations such as China and India and Africa can learn vital lessons from it.

India, for example, invested early in technical education. Institutes of Information Technology initiated several decades ago have spawn major technological operations such as out -sourcing in Bangalore. India used its personnel and skills which offers competitive advantages to compete successfully in the global society. India further refused to allow its diverse political and religious ideologies to interfere in its economic development goals.

China on the other hand started with adoption of the then Soviet's economic development strategies with little variations. Objectives were to "learn from the master". In the set of agreements starting in 1950, the then Soviet Union agreed to transfer to the People's Republic of China, in course of three five- year plans (1953-1967), a total of three hundred (300) industrial plants, a spectrum of Soviet technology, and administrative techniques and skills needed for managing the modern technology.

China followed this effort for resource building and technology transfer with the 7th five-year development plan (1956-1990) which involved wide variety of officials, scholars, technical experts within and outside the normal planning mechanism. Goals were to initiate basic changes in structure of the economy through support for growth of heavy industries transportation, communication, raw and semi-finished materials. This plan was based squarely on the assumption that China will remain open through the 1990s and beyond.

By the mid 1989, the Special Economic Zones (SEZ) was established to attract foreign investors, encourage development of entrepreneurship, and private enterprises from the scratch and promote export. China currently graduates over 350,000 engineers per year, fueling massive economic growth and global competitiveness. The same can be said of Korea and other emerging giants in Asia.

There is an urgent need for a systematic and master plan in Africa with capacity building in engineering, technology, human resource training, and infrastructural development (resource building and allocation) to encourage entrepreneurial start ups and knowledge based enterprises. The continent and the national leaders should identify areas of comparative and competitive advantages (mostly in natural resources and cheap labor) and use those to become major players in the global market.

A Framework for Effective GIV Master Strategy

This model proposes a GIV (Generic and Incremental Value) framework with four distinct and related phases (Figure 1). Phase one is the establishment of economic development visions mission, goals, and objectives by Government and other related public institutions. This phase also establishes the premise for development and focuses on the development of target societal outcome and institutions for the jobs that needs to be completed. Phase two is the Logic and Architecture which focuses on resources development and technology selection. This phase creates the blueprint and infrastructure for effective economic growth and potential sustainability. This phase is

built on the identifiable opportunities and possible threats in the environment. Focus on relationship building among institutions needed to not only develop and use resources but to provide the appropriate technologies (process/product, knowledge, skills, material/design, techniques, etc) is the responsibility of the third phase. Investments in Research and Development (R&D), building University-Industry research collaborations, and acquiring licenses or pioneering home grown technologies are a few of the activities of this phase. Phase four deals with detailed resource allocation building on the activities of phase two and with the goal of maintaining continuity.

Emphasis is on the development and sustenance of both congruity and consistency among the various elements of the phases. For example

Figure 1. Generic and incremental value model for development in Africa

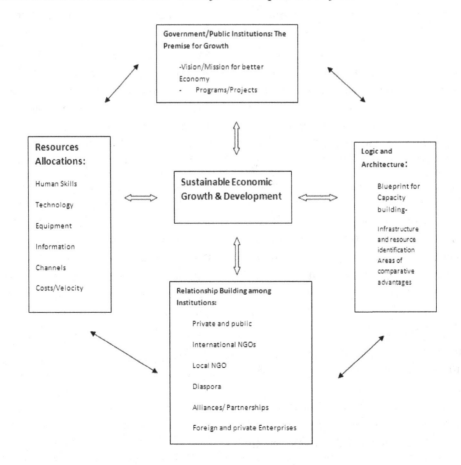

congruity exists when goals are consistent with objectives, and objectives are consistent with available resources and planning technology. These in turn should be consistent with the strategies and implementation processes - the blueprint, the relationships and partnerships, and type of technologies developed or acquired.

The model challenges the respective nations, regions, and the continent to set constructive and achievable goals designed to inspire and sustain economic development. This process should be strategic in nature and includes the society (scholars, technical experts in the Diaspora and within), systems (government and private institutions, foreign investors), and outcome expectations (time horizon and deadlines, accountability, continuity). These activities should include specific considerations for staff and technical support personnel needs, programs and portfolios development, as well as the status of existing programs. Identification and selection of programs at this stage serve the purpose of raising the question of how values are to be added as the planning progresses.

Given the proposed model, strategic planning for sustainable development is defined in this paper as "The continuing process of creating the vision and mission statements, scanning the environment (internal and external), formulating and adapting strategies, rearranging and maintaining desired goals (value chain) in the face of a fast changing and technology driven global environment."

Table 1 attempts to present a more detailed plan of action and potential responsibilities and accountability among the nations and institutions. It is an extension to the GIV model. Emphasis is on identification of most appropriate Agency or institution which should initiate and account for the various actions under the GIV model.

It should be noted that this is only a grand (master) plan and should be further broken into specific "action plans" or short term strategies for effective implementation. Further, all activities should be consistent and congruent within and among the institutions as proposed in the model.

Value cannot be added to projects and processes if the responsible institutions continue to act in isolation or discontinue activities because of change in leadership (through elections and changes in organizational leadership).

A closing thought and question is- How can African nations develop economically and otherwise when they chose to settle as consumer economies in the midst of a global competitive society? Natural resources such as petroleum and other rich minerals are being explored and "exploited". The raw materials are shipped to outside the country for processing and then sold to the governments and the citizens (Values are added outside of the continent) at a higher price. The continent should take a page from the Chinese government - Investors (foreign, Diaspora, entrepreneurs) should be encouraged to (create incentives and viable private enterprise environment) to go into Joint ventures with local firms or create a direct investments in manufacturing, production of goods and services for exports. Governments and broader institutions should equally (another page from Chinese), seek opportunities to purchase expired or new patents from developed nations- this is the fastest and viable approach to appropriate technology acquisition (transfer) and possible economic development.

Final Remarks

The need to understand the impact of appropriate institution, adequate technology transfer, and resource development on sustainable economic development in Africa is very critical for future planning in the continent. Researchers, Foundations, and Non-Governmental Agencies define capacity building as a long term process aimed at systematic development of sustainable skills, infrastructures, resources, and commitment to education and training of personnel. "It is a nation's or continent's ability to carry out its mission using best practices and having the resources to do so". It is a systematic investment in develop-

Table 1. A ten-year strategic master plan for economic development in Africa

Supervising Institutions	Major Goals	Objectives	Tactics	Estimated Time line
African Union (AU)	-Vision &mission statements: 1. unite African Nations 2. Stabilize Africa: political, economic, social	* Minimize and eliminate: - Political risks - Economic risks - Property-rights risk - corruptions -illegal activities * Increase visibility of Africa as a Business destination * Increase visibility of Africa as a viable world economy	* Develop Strategic Alliances to build and promote image * Identify and Deconstruct inhibiting economic and political architecture inherited as colonies * Develop strong legal and regulatory institutions to enforce agreements among nations and international allies * Develop strategic task force to assess the effectiveness of current political and economic institutions and make recommendations	2010- 2014
African Economic Council (AEC)/ Abuja Treaty	Create a Viable environment for economic Development and increase capacity for growth	*Facilitate trade among African nations *Reduce barriers to trade *Create Regional trade center *Develop enforceable trade agreements	*Create interregional infrastructures including physical, commercial, natural, and legal infrastructure *Educate individuals and leadership in implementing the five stages of integration *Identify all non value added stages and replace with a more meaningful approach, example, how feasible is it for Africa to reach Economic union stage? *And how feasible is custom union (Mobility of labor and capital	2010-2030
Regional Economic Community (REC) National Governments/ AEC	Create and Improve Mutual Economic relationships	*Increase trade among members *Attract international Investors *Increase relations with Social entrepreneurs	*Create institutions to enforce trade agreement *Assess the details involved in the 5 integration stages proposed *Eliminate trade barriers *Advance to custom union and involve members at all levels for implementation *Identify comparative advantage	2010-2030
National Governments and local institutions and LAP	Create viable environment and opportunities for economic growth and development	*Improve all national infrastructure (Physical, natural, Commercial, etc) *Establish systems and processes to improve communications, productivity *Increase bases for manufacturing and exports *Improve relations with Diaspora organizations *Reduce social division and conflict *Improve leadership effectiveness and transparency *Improve Standard of living and quality of life	*Restructure educational systems and programs to include technical, entrepreneurship, engineering, technology, telecommunications, etc. *Create incentives to attract both local and Diaspora investors * Seek and acquire expired patents for manufacutring *Create institutions for research and innovation *Integrate research with local resources *Collaborate with foreign investors to establish operations that add value to the resources, *Develop a culture of national identity, unity, and pride *Establish law and order in the society *Improve health and other social systems	2010-2015

ing the "Continents internal and external system so that it can realize its mission. This involves taking the steps in validating goals, evaluating the vision, reaffirming values, building resources, setting strategies and policies. It is the "Creation of enabling environment with appropriate policies and legal framework, institutional development, including community participation, human resource development and training (UNDP). Effective capacity building therefore embodies the creation and management of institutions, resource and technology

In broad terms, capacity building activities for sustainable economic development should focus on developing knowledge, skills, attitudes of the society, and creating incentives with the objective of strengthening economic, government, institutions, and stakeholders through education, training, mentoring, infusion of resources to develop secure, stable, and sustainable structures and systems for inspiring survival and improved quality of life in a fast changing world.

Although the continent leaders have made major strides in the various institutions and programs designed for collective development, the initiatives lack many of the key ingredients needed for economic development. On the positive side, The African Economic Community, (AEC) among the many institutions, is the most promising.

Critical limitations and causes for failed development strategies among African nations are the absence of or inadequate and systemic action plans and implementations processes for planned programs. Strategies must be formulated, assessed and ranked. Policies and procedures for operations, evaluation, and control must be set and enforced. The significance of resource availability and its allocation cannot be ignored nor should the selection of the most efficient and effective method (technology) and integration of resources (including training, skills development, cost analysis and velocity) be neglected.

Another area where African nations and African collective strategies failed is in its inadequate assessment and management of major factors of production and development. An examination of major factors of production (land, labor, capital, entrepreneurship, and knowledge) show that Africa and its States are rich in most of these factors. Major problems are management and control of resources. The government has not invested in developing incentives and the necessary infrastructure (energy, roads, water, fiscal, etc.) that will attract knowledgeable and skilled African entrepreneurs in the Diaspora to invest in the continent. There is the need for entrepreneurial spirited individuals and knowledge workers to help build capacity for (lessons from China, India, Korea, etc) economic development.

Emphasis is on recognition of change, differences in environment, culture, economy, etc. and the development of technology infrastructure, good governance, enforcement of rules and regulations, etc. to compete in the global society. The is also a need to design techniques to develop cost effective institutions and invest in programs that will eventually enhance the standard of living and the quality of life for Africans. For African leaders, the devil is in the details.

FACTORS FOR NANOTECHNOLOGY AND MICROELECTRONICS TRANSFER TO AFRICA

Several factors are necessary for effective acquisition, adaptability, utilization and diffusion and transfer of microelectronics and nanotechnology to Africa. These factors are as follows:

Government Plans and Policies

Government role is very critical for technology transfer to take place. A few have made the necessary investment in technology creation infrastructure while others have failed. However, the decision as to whether to invest in any type of technology development depends of such factors as content knowledge of the type of technology (government awareness) and the financial position of the country in question. Nevertheless many African countries have not yet realized the extent and potential of the impact technology can have on the politics, culture and economies of their societies. .

In view of the burden placed on the shoulder of government by technology development challenges, UN Commission of Science and Technology Development (2006) proposed the following sets of initiatives that governments, including those in Africa may wish to follow:

a. Undertake needs assessment exercises to determine whether or not existing science, technology and innovation policies effectively serve the needs of national development goals.

b. Involve representatives from industry, academia and public sector in carrying out comprehensive technology foresight exercises with a view to identifying technologies that are likely to help address pressing socioeconomic needs and establish priorities in science and technology policy and government programs on research and education.

c. Strengthening linkages between public research and private industry and tap into regional and international research and development networks.

d. Improving national mechanisms for the promotion of knowledge-based and innovative enterprises through various interventions and incentives.

e. Setting up centers of excellence, technology incubators and science parks to apply knowledge and to facilitate commercialization and diffusion of technology

f. Adopting special measures to retain and attract young and talented scientists and engineers and

g. Encouraging venture capital from both public and private sources to assist product development and commercialization of new and emerging technologies.

Economic Situation

A country with adequate economic resource not only is capable of acquiring a particular type of technology, but it can build, maintain and sustain it. Many African countries have weak economies. Even if the technology were in place, the average African in his or her remote village would not necessarily have the money to pay for it. Hence, the problem of lack of financial resources is now compounded with the high cost of some emerging technologies. In situations such as these, Orszag (2000) believes that the government funds should be deployed to provide: (a) new capacity building programs and projects that support growth of businesses in local communities (b) strengthen the economic infrastructure of rural communities (c) enhance technology infrastructure (d) develop opportunities in such communities (e) support sustainable economic development opportunities in low income neighborhoods (f) develop communities' natural resource-based economies (g) support workforce development programs and (g) facilitate skill-training and distance learning facilities (h) offer federal subsidies to reduce the cost of basic telephone services (i) attract outside capital and locate basic commercial business operations in rural communities. The economic situation is a major factor that has affected the development of infrastructure which could facilitate technology transfer, penetration and clusters.

Education and Training

Another means of acquiring technology knowhow is through education and training. Many African universities, colleges and polytechnics offer low quality technical programs which generally undermine the capacity of their students to graduate with world class skills. Lack of these 21st century skills will continue to undermine the level of competitiveness and abilities to transform their nations and regions.

A key part of improving technological infrastructure involves improving investment in educational system which remains the best organic means to build capacity in a nation. Sound education will also facilitate the transfer process from developed nations to developing ones.

Donor Agencies

The sale and purchase of technology equipment by suppliers and other vendors is one way of transferring technology from one country to another. These attempts to purchase technical assistance are primarily used by African and other developing countries to fill the gaps in vacant technical positions. But, one of the main setbacks is the tendency of host nations to rely on foreign consultants and expatriates to build up domestic expertise and capabilities. Limited success has been registered with this practice, as many expatriate technical advisers and assistants often do not stay throughout the life of their projects, but, rather depart without fully training their replacements/ users or in some cases leave before finishing their expert contractual obligations. The use of African emigrants (Diaspora) for technical assistance is desirable. Unfortunately, several African governments employ foreign expatriates at higher salaries than their indigenous counterparts, regardless of their qualifications and experience.

Even, the few indigenous graduates hired are often under-employed. That is, they are employed in jobs that do not necessarily utilize the skills and expertise directly related to their training. Appropriate technical assistance programs will be the only way out. This is because, the indigenous African country will be able to provide adequate employment opportunity to its indigenes, while at the same time, able to accomplish the required task at a smaller premium or cost compared to a foreign expert. This thinking is consistent with the finding of Mckinley (2000), who concluded that

African migrant experts are more likely to trade off their gains, in terms of salaries and other benefits, just to return home, primarily because of their social and cultural ties or links when compared with their foreign counterparts.

The result has been the failure of the consultants to pass on the skills to the users. Perhaps, this type of problem has caught the browsing eye of the United Nations Development Program (UNDP) which has in recent years with some targeted developmental aid, instituted a comprehensive program that assists developing countries to attract home, much of their skilled manpower trapped overseas after completion of their studies. However, the program met with limited success with a significant number of the people attracted back to their nations of origin only stayed for a short period of time and then returned to their country of study. These suppliers (donor or aid agencies) have also played a critical role in the technological transfer process. They do so through the transfer of knowhow which is acquired through the training they provide to the users in preparation for the set up, operation, repair and maintenance of the said equipment.

Organization and Management

Management plays a critical role in the process of technological transfer. In many African countries, management has been reluctant to embrace change and innovation, and in some cases has even resisted technology and change. Sometimes, different administrations have acquired and adopted certain kinds of technology based on narrow and selfish interests, such as adaptability to spare parts, and reliance on external experts and consultants. In others, management has neglected to invest in manpower training, policy output, implementation and feedback that have greatly contributed to the depressing level of computer literacy and information awareness among its management cadre. In fact, many African bureaucracies have experienced little progress in the area of emerging technology transfer and capacity to use technology to create wealth and lift millions out of poverty.

Final Remarks

The lag in Africa's socio-economic development is directly related to the low technology development in the continent in core technology areas like microelectronics. Though progress is being made in using technology, Africa remains a consumer economics. Consequently, Africa is unable to produce knowledge and also lacks the means to process already developed ones. This technology- divide, in essence has stunted economic development in several African countries, and thus makes it difficult for the continent to enter into the global innovation age.

One of the factors that constrain the ability of African nations to acquire, utilize and process technology is the poor nature and quality of technology that they attract. The continent is still positioned at the lowest cadres of knowledge creation and lacks the inventive capability to move to the upstream sector of technology development. In essence, the structural impediments facing African governments in pursuit of their development goals are: poor government plans and policies; inadequate resources because of bad economy; poor education and training; overreliance on the services of aid/donor agencies and expensive foreign consultants and expatriates to build up domestic expertise and capabilities; socio-cultural attitudes which feed negative perception of technologies as luxury goods, and ignorance regarding the need and possibilities in human development of technology. Others are inadequate technology infrastructure, such as outdated technology, inadequate power sources, etc. All these limit Africa's path to effective utilization, adaptation and transfer of technology. Because some foundation technologies such as electricity and water supplies have not been properly adopted, it is going to be difficult to adopt many emerging ones like microelectronics and nanotechnology that require some of them to pre-exist.

The solution to the problems of technology acquisition, utilization and diffusion facing African nations can only be solved through appropriate and targeted technological and human capital investment policies and strategies, such as - developing human capital and skills through training and education for local capacity building; acquiring, utilizing and locally adapting existing and emerging technologies to domestic needs and resources; building, promoting and sustaining a solid indigenous scientific base. The ability of African governments to implement these development initiatives will largely depend on their ability to pay for them and their costs. Hence, there is need for African leaders to improve their national economies as well as run effective and good governments capable of maintaining and sustaining their individual nation's development goals.

RECENT POLICIES IN SCIENCE AND TECHNOLOGY DEVELOPMENT IN AFRICA: THE CASE OF STEP-B, NIGERIA

Nigeria is Africa's most populous country, and with some 70m living in poverty, only China and India have more poor people The World Bank project appraisal study on basic education showed that at the basic level in Nigeria, there is currently an overwhelming demand for more and better-quality education and training both for pupils and teachers. A number of studies suggest that Nigeria's low-level of institutional capacity reduces its overall global competitiveness and ranked Nigeria 101 out of the 125 countries of the countries evaluated on the strength of their post-basic education.

To increase this competitiveness and better leverage Nigeria's inherent resources, education must therefore be improved. Change is required from several sectors: increased teacher knowledge of their content as well as enhanced pedagogy skills to translate this content knowledge in their classrooms. To be able to bring in some changes, money is required (Adikwu, 2007a; Adikwu, 2007b).

The STEP-B Project

STEP-B is simply the acronym for Science and Technology Post-Basic. The STEP-B Project is a World Bank-assisted Project but a Federal Government of Nigeria intervention in the Science and Technology Education and Research Sub-sectors targeted at the Post-Basic level. The STEP-B Project is a credit facility worth one hundred and million dollars (USD180m). In the Nigeria education system, two levels can be clearly defined: the basic level and the post-basic level (Obioma, 2007).

The basic level talks about the first nine years of a child's education career which comprises the Primary School and the Junior Secondary School. A redefinition of the basic level of education became necessary when it became clear that children in Nigeria could no longer be said to be literate after six years of primary education since most could not read and write. As such they cannot be certified as literate until after nine years.

The post-basic level of education includes the three years of senior secondary school as well as the tertiary education system which comprises the colleges of education, monotechnics, polytechnics, colleges of education and the universities. The Project is subdivided into three main components (PAD, 2007).

Component One

Competitive Fund for Quality Enhancement and Innovation (US$81 million; 45% of total Credit). *Component 1 supports a facility to provide peer-reviewed and competitively-awarded quality improvement grants to federal Post-Basic Education S&T Institutions (PBEIs) and other federal S&T agencies.* Fundamental to the orientation of

Component 1 is the peer-review process through which funds are allocated. Competitive allocation of resources based on transparent criteria for quality, relevance, and impact represents an important culture shift and innovation in federal post-basic education in Nigeria. Component 1 comprises two sub-components: 1A supporting research and development; and 1B supporting teaching and learning.

Component Two

Support for the emergence of Centers of Excellence in post-basic S&T (US$54 million; 30% of total Credit). The objective of Component Two is to inject a small number (6-8 over the life of the project) of promising STEP-B institutions with the resources necessary for them to emerge into Centers of Excellence, raising the national competence of Nigeria in the specific S&T area addressed by the Center of Excellence and increasing the economic benefits derived from it. This component supports the development of Centers of Excellence in S&T in selected tertiary educational institutions.

Component Three

Support for strengthening strategic planning, management, and M&E in post-basic S&T education (US$27 million; 15% of total Credit). Component 3 supports complementary initiatives that aim to have a sub-sector-wide impact on quality, relevance, and access to federal S&T education at the post-basic level and it thus supports project implementation and management. Component 3 consists of two sub-components, 3A and 3B:

The Nigerian education sector is facing manifold challenges which must be adequately addressed to enable the sector to serve the expected role as 'the key bridge' for the national prosperity. The STEP-B is driving means to improve the quality of teaching and learning in federal post-basic S&T institutions, strengthen

S&T research capacity in tertiary institutions, including skills to transition from research output to commercializable products and services, support a few select institutions that have the potential to eventually become centers of excellence in key areas deemed critical to national development and address cross-cutting sector-wide issues in S&T policy, management and development that have a national character and whose implementation will have an impact beyond the lifetime of the project.

REFERENCES

Adikwu, M. U. (2007a). *Highlights on Science and Technology Education Post-Basic (STEP-B)*. Presented at the Digital Bridge Institute at a Conference organized by the Nigerian ICT Forum, Abuja, July 24.

Adikwu, M. U. (2007b). *Options for Financing the Change-Management System*. Presented at the Stakeholders' Conference on Engineering and Technology Academic Programmes, Transcorp Hilton, Abuja, November 10.

Anyansi-Archibong, C. (1995). Planning . In *Developing Countries*. Chicago, IL: The Planning Forum.

Anyansi-Archibong, C. (2005). *Role of Women in the Socio-Economic development of the Continent African Forum. Global Leadership Institute, UVI-St*. USA: Thomas.

Anyansi-Archibong, C. (2006). *Entrepreneurship as the Missing Factor in the Economic Development of Developing Nations, Global Entrepreneurship Monitor*. New Zealand: GEM.

Brinkerhoff, J. M., Smith, S. C., & Teegen, H. (2007). Beyond the Non: The Strategic Space for NGOs in Development . In Brinkerhoff, J. M., Smith, S. C., & Teegen, H. (Eds.), *NGOs and the Millennium Development Goals, Citizen Action to Reduce Poverty* (pp. 53–79). New York: Palgrave - Macmillan. doi:10.1057/9780230604933

Giddens, A. (2003). *The Runaway World- How Globalization is Reshaping Our Lives*. New York: Routledge.

McKinley, B. (2000). Statement to the Migration in Africa conference, Addis Ababa, October 18, 2000. International Organization for Migration (IOM).

North, D. C. (2000). Building Institutions: Complement, Innovate, Connect, and Compete . In *World Development Report 2002: Building Institutions for Markets*. Washington, DC: Oxford University Press.

Obioma, G. (2007). *The 9 year Basic Education Curriculum (Structure, Content and Strategy for Implementation): Nigeria's Experience in Educational reform*. NERDC.

Orszag, J. (2000). *Statement before the U.S. Senate on Indian Affairs*. Retrieved from http///www.education.elibrary.com.au

Sarr, M. D. (2008 September). Africa's Search for Collective Development Strategies. *The African Executive, 179*.

UN ICT Task Force. (2002). *Information and Communication Technologies in Africa – A Status report*. Third Task Force Meeting, United Nations headquarters, 30 September – 1 October 2002.

United Nations. (2006 May). *Bridging the Technology Gap between and within nations*. Commission on science and Technology for development Report. Geneva.

Van Dijk, J., & Hacker, K. (2003). The digital divide as a complex and dynamic phenomenon. *The Information Society, 19*, 315–327. doi:10.1080/01972240309487

World Bank. (2007). Project Appraisal Document (PAD) for Science and Technology Education Post-Basic Program.

Chapter 24
Trade Policies and Development of Technology in Africa

Louis O. Osuji
Chicago State University, USA

ABSTRACT

Trade between nations is very crucial in the process of economic and technological growth. Directly or indirectly, trade facilitates the process of technology innovation, transfer and diffusion. It offers the trajectory to evaluate and understand how technology penetrates economies and remains a good indicator to measure national progress on technology creation and assimilation. The growth link between international trade and economic development could be traced to the classical trade theory of Adam Smith, and David Ricardo and the modern neoclassical trade model of Heckscher-Ohlin (H-O). While there is no single model that captures the route to economic development, this chapter explores how African countries working closely can harness and utilize technological advancements to improve their share of global trade so as to accelerate their overall economic growth and development.

INTRODUCTION

Trade between nations is very crucial in the process of economic and technological growth. The growth link between international trade and economic development could be traced to the classical trade theory of Adam Smith, and David Ricardo and the modern neoclassical trade model of Heckscher-Ohlin (H-O) model. The basic tenet of the trade theory is that welfare is generally en-

hanced through improvement in productivity and sectoral allocation of factors of production. Some of the benefits associated with trade include foreign exchange earnings needed in the importation of intermediate goods required in the industrial sector, increased competition, product specialization, and a broader avenue for technology transfer (World Bank, 1991). Studies by Bhagwati, (1978, Bhagwati and Srinivasan, 2002); Balassa (1982, 1985); Krueger (1978, 1990, 1997); and Edwards, (1993) confirm the positive relationship between exports and economic growth. In the African continent,

DOI: 10.4018/978-1-61692-006-7.ch024

however, the trade policy of many countries was based on the doctrine of import-substitution industrialization. That policy gained wide acceptance as a viable policy package that would help many of these countries achieve structural transformation and subsequently lessen their dependence on primary products (UNCTAD, 2008). According to UNCTAD 2006 Report, African countries have not diversified their exports towards more dynamic primary commodities and manufacturing goods which are less prone to the vagaries of international markets and that Asia dominates trade between developing countries, accounting for more than half of all South-South commodities exports (UNCTAD, 2006). The impact of this weak export situation is more noticeable in the class the majority of African countries fall into according to World Bank categories of nations.

The World Bank has classified economies based on gross national income (GNI) per capita as low income, middle income, or high income World Bank (1993). The middle income group is further subdivided into lower middle and upper middle income. The low income economies are characterized by GNI per capita of $935 or less; lower middle income from $936 to $3,705; upper middle income, $3,706 to $11,455; and high income, $11,456 and above (World Bank, 2009). According to 2007 GNI per capita calculations, out of 47 Sub-Saharan African countries, only eight were classified as lower-middle-income, and three as upper-middle-income economies. In effect, only three countries South of Sahara, namely Gabon, Seychelles and South Africa meet the criterion of GNI per capita of $3,706 to $11,455 range. The situation should not surprise many people if technological innovation and application in most African countries is low.

While there is no single model that captures the route to economic development, this chapter explores how African countries working closely can harness and utilize technological advance-

ments to improve their share of global trade so as to accelerate their overall economic growth and development. In that regard, one cannot but agree with Ekanem (1997) that African countries need to address their problem using a three-pronged approach, namely, establishing a stable sociopolitical environment, strengthening their economies, and the need to establish a strong scientific and technological base. Echoing the importance of technology Edoho (1997) argues that the failure of Africa to avail itself of access to technology would lead to increased marginalization by advanced industrial countries. Since that observation was made more than ten years ago, it becomes necessary at this point in time to assess the state of affairs as a guide to future action. In this chapter, it would be extremely difficult to cover all the countries in Africa hence most materials would come from the Sub-Saharan African (SSA) and North African countries. Where possible, however, trade and economic policies of member and non member countries in the two regions or Africa as a whole will be analyzed based on availability and relevance of data. The emphasis on SSA countries derives from the fact that the majority of member countries as opposed to the North African region fall into low income economies. This chapter is arranged in the following sequence: a brief review of Africa's economies in the late 1960s and after independence is followed by trade and economic policies of the 1970s and 1980s. That is closely followed by assessment of export trade performance of selected African countries after trade liberalization policies. External and internal problems militating against some of these nations were also identified. Policy options aimed at improving trade through application of sustainable economic and technological innovations are discussed, concluding with summary and recommendations.

CHARACTERISTICS OF AFRICA'S ECONOMIES FROM THE LATE 1960S TO 1990S

Agriculture was the mainstay of Sub-Saharan African countries' economies for more than sixty years contributing a sizeable percentage of gross domestic product. Sometimes this contribution has been as high as 55 percent for Nigeria between 1965 and 1970; and as low as 20 percent in Zaire, (Ekanem, 1997). This was not surprising as prior to the early 1960s, African trade policy was dictated by the colonial powers whereby primary commodities were exported and manufactured products imported.

However, the economies of most of African countries in the 1970s were constrained by a lot of internal and external environmental factors. The trade policies were characterized by extensive involvement of the state both in production and in marketing. Some of the internal issues included high tariffs, political instability which gave rise to vicious cycle of low human labor output, and abysmal productivity in the agricultural sector. Not only was the agricultural sector weak in the production of domestically consumed food items, its primary commodities or the export sector faced deteriorating terms of trade and purchasing power parity disadvantage in the world market. According to Svedberg (1993: 21) export earnings fell from U.S $50 billion in 1980 to about $36 billion at the beginning of the 1990s. Between 1970 and 1988, Sub-Saharan African (SSA) countries' share of primary commodity in the world market fell from 7.0 percent to 3.7 percent. Osuji (1997) records that in the market for coffee, cocoa beans, refined copper, cotton, timber, and sugar, the decline ranged from 38 percent to 22 percent; for iron ore it was 50 percent and 84 Percent for vegetable oils. Other external factors included shocks arising from their current account imbalance and the debt crises that are yet to be completely resolved. During this period, and especially between 1980 and 1989, it is also fair to say that the contribution

of manufacturing to GDP was very insignificant. For a country like Mali, manufacturing export as percentage of GDP was 0.1 percent, for Benin and Burkina Faso, 0.5 percent each and for Ghana, 0.3 percent (UNCTAD Report, 2008). The structural adjustment programs with their attendant problems were adequately articulated in the views of Green (1990) that those programs aimed more at production recovery rather than human condition recovery.

What economic policies did most of these African countries pursue that led to the continental catastrophe? It is relevant to discuss the policies that contributed to this economic disaster even though some structural changes have taken place since then.

ECONOMIC POLICIES OF THE 1970S AND 1980S

Most of the SSA countries after independence, as said earlier, embarked upon economic policies that were supposed to transform their highly dependent agricultural economy into modern diversified and self-sufficient ones. One of the flaws of this approach was the failure on the part of policy makers to recognize the weakness of the private sector to provide the capital needed to fund such transformation. The consequence was that the governments of those countries were forced to assume the responsibility of funding, which essentially was described as demand-driven especially between 1970s and 1980s, but exacerbated by the energy crisis of 1973. The high energy cost resulted in a surge of imports in fuel for such countries as Kenya and Morocco. According to (IMF, 1992), energy cost rose from 10 percent in 1965 to 32 percent in 1990 in Kenya and Morocco from 5 percent to 15 percent during the same period. The jolt caused by high energy cost resulted in many countries including Zaire, Egypt, Kenya and Morocco running huge budget deficits which ranged from 2 percent of GDP for

Morocco, 5 percent for Kenya to about 9 percent for Zaire (IMF, 1992). To finance this growing budget deficit, these governments resorted to large scale borrowing from the banking system and money creation which ballooned from 14 percent in Nigeria to over 69 percent in Zaire (Ekanem, 1997 p 56). Other contributory factors included overvalued exchange rate, and government subsidies which encouraged import demand of both industrial and luxury goods. To finance these external deficits, governments were forced to tap into their reserves which eventually became overdrawn leading to the adoption of the structural adjustment programs (SAP) that were supposed to realign the micro and macroeconomic policies of the recipient countries toward achieving fiscal balance. This was to be achieved by controlling spending, reducing inflation, and through export promotion. Other conditionalities of the lending body worthy of mention are trade liberalization and privatization of state-owned enterprises. Conditionality in this context according to (Callaghy, 1990); (Callaghy and Ravenhill, 1993, 1994) refers to the agreements reached between donors and recipients that exchange financial transfers (in the form of grants or loans) by the donors for policy changes which should be embarked upon by recipients.

To what extent have the numerous steps and programs taken by the various multinational organizations such as the International Monetary Fund (IMF), the World Bank, and the United Nations Conference on Trade and Development (UNCTAD) improved the economic landscape of Sub-Saharan and other African countries? It is refreshing to observe that the World Bank's Assessment Report (World Bank, 1994) mentioned countries like Gambia, Ghana, Tanzania, and Burkina Faso as doing well in their efforts to develop their economies. Cameroon, Zambia, and Sierra Leone were among those countries reported as not doing well; in other words they were doing poorly while countries like Benin, and Malawi were in between. What is the impact of trade liberalization policy on the countries that adopted that prescription? Specifically, the export sector of some of the adopting countries is examined in subsequent section.

EXPORT PERFORMANCE OF SSA COUNTRIES AFTER TRADE LIBERALIZATION POLICIES

The rationale for trade liberalization could be attributed to the low performance of commodity-based exports in the continent and the essence was to reallocate resources towards the production of exports including manufactured products. In other words and more specifically, the trade liberalization policies implemented by some of the SSA countries in the 1980s were designed to revamp the manufacturing sector and ward off the existence of various forms of trade protections prevalent in the 1970s and 1980s. According to UNCTAD Report (UNCTAD 2008: 53) forms of trade protection included high import tariffs, quantitative restrictions on competing imports, and high levels of tariffs on inputs and capital goods. Export liberalization measures include eliminating foreign exchange rationing, export licensing and export taxes, and dismantling marketing boards. Attempts to increase export in manufacturing have been justified by four factors. First, the positive role played by the manufacturing sector in the development of East Asian countries as well as other regions could be emulated by African countries. Second, the dependence of most African countries on primary commodity exports with its vicissitudes calls for a new export strategy that would encourage the much needed diversification in this sector. Third, the potential for growth is more attainable via manufactured products rather than through the traditional primary commodities export. Finally, the small size of the domestic markets in most of the African countries makes a good case for export markets needed to absorb the additional products that would result from structural transformation and its attendant economic growth. The magnitude of the external markets as contended by the crafters of the policy could help firms in Africa to realize economies of scale desperately needed to achieve competitive advantage in the global market place. The extent to which this strategy achieved the intended objectives is subsequently discussed.

Objective assessment of SSA countries following trade liberalization can be difficult, arbitrary, and complex as the definition of trade liberalization itself. This is because according to UNCTAD 2008 Report, among other factors, the definition of liberalization involves different types of data from different sources, some of which may require a certain level of subjective interpretation. This problem notwithstanding, the Report does not give any impression of African countries performing well in terms of level and composition of the continent's export performance. According to the Report, Africa's export as a whole lost its export market share from 6 percent of world exports in 1980 to about 3 percent in 2007. Average tariffs in SSA countries were nearly halved between 1995 and 2006 (UNCTAD, 2008c). However, trade liberalization policy impacted export trade and manufacturing activities of many countries in a number of different ways.

Manufacturing Sector after Trade Liberalization

As said much earlier, the performance of export manufacturing sector in many African countries has been anything but satisfactory. UNCTAD Report (2008: 54) shows how manufacturing exports to GDP over the last 25 years have remained very small for most countries. Table 1 contains African countries' average manufacturing exports as percentage of GDP.

A breakdown of Table 1 into three time periods, namely, 1980-1989, 1990-1999, and 2000-2006, reveals that only three countries, Mauritius (25.2%), Morocco (6.0%) and Tunisia (11.7%)

Table 1. African countries' average manufacturing exports (as GDP percentage)

Country	1980-1989	1990-1999	2000-2006
Benin	0.5	1.5	1.3
Botswana	..	..	35.7
Burkina Faso	0.5	..	1.4
Burundi	..	0.2	0.4
Cameroon	1.3	1.5	0.9
Cape Verde	..	1.3	1.4
Cote d'Ivoire	3.4	6.3	7.8
Egypt	2.4	2.3	2.1
Ethiopia	..	0.4	0.8
Gabon	2.8	1.4	4.0
Gambia	0.4	1.4	0.6
Ghana	0.3	3.1	4.5
Guinea	..	4.5	6.3
Kenya	2.0	4.6	3.5
Madagascar	0.8	3.1	6.3
Malawi	1.6	2.5	2.6
Mali	0.1	1.4	8.8
Mauritius	25.5	28.5	26.1
Morocco	6.0	10.0	14.0
Mozambique	..	0.9	1.1
Namibia	..	..	17.2
Niger	0.4	3.5	1.8
Nigeria	0.0	0.6	0.7
Rwanda	..	0.2	0.2
Senegal	3.5	7.2	7.5
Seychelles	1.2	2.3	2.3
Sierra Leone	3.8	..	0.4
South Africa	4.4	9.1	13.2
Sudan	0.2	0.3	0.3
Swaziland	..	..	46.9
Togo	2.5	4.0	13.7
Tunisia	11.7	21.6	25.9
Uganda	..	0.7	1.0
United Republic of Tanzania	..	1.2	1.9
Zambia	..	4.4	4.4

Source UNCTAD Report, 2008.

had manufacturing as GDP percentage above 5 percent during the 1980-1989 period. The number of countries that met this criterion (i.e. 5% and above) increased to six countries between 1990 and 1999 with the number doubling between the 2000-2006 period. The countries with remarkable export percentage include Swaziland (46.9%), Botswana (35.7%), Mauritius (26.1%), and Tunisia (25.9%). Others include Namibia (17.2%), Morocco (14.0%), Togo (13.7), and South Africa (13.2%). In all, only eight countries' export as percent of GDP were above 10 percent, while five other countries' percent ranged from 10 to 5 percent. African countries' manufacturing export share averaged 26 percent of total world merchandise exports. This figure pales in comparison with the figures from other regions like East Asia (91.9%), South Asia (56.3%), and South America (54.5%). What this means sadly is that even as recent as 2006, Africa continued to be an inconsequential player in the export of manufactured products both before and after trade liberalization. According to the UNCTAD 2008 Report, export performance in African countries following trade liberalization has been very disappointing and that the overall trade balance in African countries has deteriorated since liberalization (UNCTAD, 2008).

The low level of Africa's manufacturing exports has been blamed on low manufacturing production. According to experts, Sub-Saharan African countries manufacturing value added from 1965 to 2005 did not improve from its original value of 15 percent of GDP in the 1960s. What are the explanations for this low performance in the global export manufacturing? A number of arguments have been advanced including low level of exportable manufacturing production which according to UNCTAD Report (2008) "leads to the failure to take advantage of available manufacturing export opportunities in the world economy". Three explanations have been advanced to account for this low level of manufacturing production in Africa. The first is the heavy and massive infusion of capital required in the manufacturing sector

which is very difficult to meet in Africa in light of the risk prevalent in many countries' business environment. This issue becomes relevant when considering the content of Table 2 which shows foreign direct investment inflow to Africa. Without any doubt, investment inflow between 2000 and 2006 in most of the countries has not been substantial by any measure. Not only was FDI inflow a pittance in some countries, nations like Angola and South Africa experienced capital outflow in 2006. The situation becomes all the more tenuous in the face of the current world economic meltdown and the impact this has on global investment by trans-national corporations (TNCs). According to UNCTAD Press Release (2009), global foreign direct investment (FDI) inflows and cross-border mergers and acquisitions (M&As) drastically declined in the last quarter of 2008, and the fall has continued into 2009. It should be pointed out that cross-border mergers and acquisitions are the main mode of FDI. According to the Report, FDI outflows fell by 57% while M&As decreased by 77% in value for all countries in the first quarter of 2009 when compared with the figures of 2008. The second explanation is technological. Africa is said to lack the technological capabilities needed to launch a successful industrialization process. As a result, firms in Africa do not export high-value manufacturing products because of their lack of the technical efficiency essential in innovating and creating new products that could be competitive in global marketplace. The final explanation centers on the comparative advantage argument that Africa is supposed to have in the production and export of primary commodities. This incidentally is said to have seriously inhibited the development and diversification of export-oriented manufacturing sector.

The veracity or otherwise of the lack of technology argument could be viewed indirectly through the demographic dimensions of some African countries shown in Table 3. To a large extent technological advancement is closely related to a country's educational level and the percentage

Table 2. Selected African countries' foreign direct investment inflows ($m), 1990-2006)

Country	1990-2000(annual average)	2003	2004	2005	2006
Angola	602	3,305	1449	-1303	-1140
Benin	40	45	64	53	63
Botswana	20	419	392	281	274
Burkina Faso	10	29	14	34	26
Burundi	3	-	-	1	290
Cameroon	40	383	319	225	309
Chad	31	713	495	613	700
Congo	121	321	-13	724	344
Cote d'Ivoire	206	165	283	312	253
Democratic Republic of Congo	5	158	10	-79	180
Gabon	-124	263	219	321	268
Gambia	21	15	49	45	70
Ghana	108	137	139	145	435
Guinea	21	83	98	102	108
Guinea-Bissau	4	4	2	9	42
Kenya	29	82	46	21	61
Lesotho	24	42	53	57	57
Liberia	15	16	10	11.5	12.5
Malawi	15	7	22	27	30
Mali	30	132	101	224	185
Mozambique	97	337	245	108	154
Namibia	97	149	226	348	327
Nigeria	1477	2171	2127	3403	5445
South Africa	854	734	799	6251	-323
Togo	23	42	69.7	57.2	66.9
Zambia	140	172	364	380	350
Zimbabwe	88	4	9	103	40

Source: UNCTAD World Investment Report 2007

of that nation's literacy level. Literacy rate affects population growth, and productivity which in turn influences per capita gross national income (GNI), availability and distribution of economic infrastructure. For instance those with high income can afford to buy and maintain cellular phones for business or personal use. Technological innovation and application can effectively be utilized by people with education. The implication is that a country with high literacy level is more likely to

adopt new technology in for personal or business purposes.

Table 3 displays three striking characteristics found among most of the selected African countries. The first is high projected population growth rate of between 16.4 percent and 69.9 percent for the sixteen countries in the table with South Africa as the only exception with 7.9 percent decrease. Such a high increase in population growth rate could negatively impact the economic and so-

Table 3. Demographic dimensions of selected African countries

Country	2010 Projected Population (000s)	2025 Projected Population (000s)	2010-2025 Projected Population Change (%)	2006 Population Density (per square mile)	2006 Percent of Population in Urban Areas	2006 GNI Per Capita ($ U.S)	Estimated Literacy Percent
Algeria	34,555	40,255	16.5	36	58	3,030	69.9
Cameroon	19,294	25,522	32.3	97	53	1,080	67.9
Egypt	84,440	103,573	22.7	205	43	1,350	71.4
Ethiopia	81,754	107,804	31.9	173	16	180	42.7
Ghana	24,279	30,536	25.8	252	44	520	57.9
Kenya	40,047	51,261	28.0	163	19	580	85.1
Libya	6,447	8,323	29.1	9	85	7,380	82.6
Madagascar	21,282	32,431	52.4	84	26	280	68.9
Morocco	35,301	42,553	20.5	193	55	1,900	52.3
Mozambique	22,061	28,893	31.0	68	35	340	47.8
Nigeria	145,032	206,166	42.2	375	44	640	68.0
Somalia	9,922	14,862	49.8	37	34	-	37.8
South Africa	43,333	39,906	-7.9	94	53	5,390	86.4
Sudan	41,980	57,462	36.9	42	41	810	61.1
Tanzania	41,893	53,428	27.5	113	23	350	69.4
Uganda	33,399	56,745	69.9	379	12	300	66.8

Source: Perreault et al 2009

cial policies in those countries. The second is a preponderance of low per capita gross national income of below $5000.00; with the exception of Libya and South Africa with $7,380 and $5,390 respectively. Countries like Ethiopia, Madagascar, Tanzania, and Uganda each has annual GNI per capita of below $400. Such low rates undoubtedly will affect allocation of resources needed to improve or fund education, invest in technology or any other economic sector.

The third striking observation is the estimated level of literacy among these countries. Out of the 16 countries, only three countries, namely, Kenya (85.1%), Libya (82.6%), and South Africa (86.4%) met the criterion of above 80 percent literacy rate. Urbanization is another factor that can affect technological development. It is much easier and more cost effective to provide urbanized areas with social and economic amenities which in turn could provide basis for economic growth

and technological progress. The low percentage urban population as contained in Table 3 should be seen as a serious hindrance and handicap to Africa's technological aspiration and not a death sentence as policies designed to address these issues are treated in the later part of the chapter. Meanwhile a glimmer of hope and optimism still exist in the sense that SSA's and in fact Africa's average annual growth of exports by product has been growing as shown in Table 4. The table shows that average annual growth of exports in Africa was 10.0 percent overall, but was 8.9 percent in manufactured goods. The percentage for SSA including Sudan was 9.2 percent, North Africa 11.8 percent. The average annual export growth rate for non-fuel primary commodities was 5.5 percent for Africa and Sub-Saharan African countries. The table however, is not a substitute for, nor does it explain the relationship between liberalization and the rate of investment in Africa.

Table 4. Regional average annual growth of exports by product category, 1995-2005

ECONOMY	Total all products	Primary commodities including fuels	Non-fuel primary commodities	Manufactured goods
WORLD	6.6	7.2	3.7	6.3
Developed economies	5.2	5.3	3.1	5.1
Developing economies	9.1	8.9	4.6	9.3
Developing economies: Africa	10.0	11.2	5.5	8.9
North Africa excluding Sudan	11.8	13.0	5.5	8.2
Sub-Saharan Africa including Sudan	9.2	10.2	5.5	9.4
Developing economies: America	7.9	7.4	5.1	8.5
Developing economies: Asia	9.3	9.0	4.3	9.4

Source: UNCTAD secretariat computations based on UNDESA, Statistics Division

It has been estimated that African countries need to increase their investment-to-GDP ratio to about 34 percent for it to attain the Millennium Development Goals (UNCTAD, 2007). Unfortunately, as contained in Table 5, Africa's response to investment after trade liberalization has been very weak. The table reveals that before trade liberalization, investment in Africa was 17.30 percent of GDP; for Sub-Sahara Africa, 16.44 percent. After liberalization, the figures were for Africa, 19.47 percent, and SSA, 18.87 percent; a modest increase of 12.54 and 14.78 percent respectively. This may go a long way to show that other factors may be responsible for the low level of investment in the continent. According to Ndikumana (2000), the amount of credit to the private sector and total liquid liabilities can predict rate of investment and this is consistent with the lack of financial resources considered the major investment constraint in Africa. Other contributing factors according to the UNCTAD 2007 Report include, "poor infrastructure, high entry costs, labor market constraints, low investor protection, and difficulty of accessing credit". The reliance of African firms on internal sources to finance their investment, and the costly nature of external sources of finance has immense policy implications.

Commodities Export after Trade Liberalization

It is interesting to assess the impact of trade liberalization on Africa's primary commodities export.

Africa's terms of trade in the primary commodities declined from 1965 to 2004. Svedberg (1993) reported that between 1970 and 1998, the SSA share of primary commodity fell from 7.0 percent to 3.7 percent. Deteriorating terms of trade facing these countries were to blame. Calculations in Ocampo and Parra (2003) show that over a longer period of time, (1900-2000), the prices of Africa's major export commodities like cocoa, coffee, cotton, sugar and tea declined by an annual average of 1 percent. Ackah and Morrissey, (2005) report that most African export prices in 2002 were a fraction of what they

Table 5. Trade liberalization and investment in Africa (Investment to GDP ratio)

Region	Before	After	Change (%)
Overall	19.31	20.41	5.70
Africa	17.30	19.47	12.54
Sub-Saharan Africa	16.44	18.87	14.78
Non-Africa	20.42	20.83	2.01

Source: World Bank, 2008a.

were in 1995 with coffee losing two thirds of its value, and sugar with cotton half of their value. The disadvantage of market concentration in importing countries where a small number of big companies act as processors, wholesalers and retailers share in the blame for Africa's commodities low price in the world market. In fact the UNCTAD 2008 Report identifies Africa's weak supply response as the most important obstacle to the continent's export performance. According to the highlights of the report, Africa lost its export market share from 6 percent of world exports in 1980 to about 3 percent in 2007. Moreover, the trade structure of most of the countries did not change much following trade liberalization as the majority of them are still primary product exporters. According to the Report, only a handful of countries like Lesotho, Mauritius and Tunisia had a significant part of their export revenue from manufactured products. The situation was exacerbated by the export concentration index following trade liberalization that increased by 80 per cent, from a value of 0.21 in 1995 to 0.38 in 2006. The implication of this high export concentration index is that African countries have become more dependent on a limited number of commodities than other developing regions of the world.

In light of the foregoing, it is clear that neither the traditional comparative advantage argument that tended to limit Africa to the export of primary commodities, nor the export of low-productivity goods currently in practice by some countries will provide the much needed path to sustained economic growth. There is need for policy makers, therefore, to harness and reinforce existing strategies and where practicable, chart a new economic path. In this regard, we are reminded of the positive externalities latent in the accumulation of human capital and application of technology which fortunately is the focus of this book. Put in another way, how can strategies for exporting high-valued manufacturing products converge with appropriate technology and policy to enhance Africa's development as the twenty-first century global competition relentlessly matches on.

POLICIES TO ACHIEVE ECONOMIC AND TECHNOLOGICAL ADVANCEMENT

Regional economic cooperations or unions are seen by scholars and policymakers as very important institutions in regional and global development. This helps to explain why such world organizations as United Nations Conference on Trade and Development (UNCTAD), United Nations Development Program (UNDP), Economic Commission for Africa (ECA) within their mandates are involved in providing training, technical and financial assistance to developing African economies for them to be less marginalized in global business and technological environment.

The views expressed by the Secretary-General of UNCTAD Supachai Panitchpakdi that regional economic integration should be part of African governments' policies for spurring and diversifying economic growth and competitiveness should not be taken lightly by policy makers and implementers in the continent of Africa. On June 24, 2009 Economic Development in Africa Report was presented in New York by Mr. Panitchpakdi in which he said among others, that "regional integration can enhance economies of scale, increase market size and attract more foreign direct investment, thus enhancing the capacity of African economies to diversify production and become more competitive." The need and rationale for formation of regional economic cooperation have been expressed by many writers. Those who favor regional economic cooperation in order to expand domestic market include Luma (1990); Lyakurma (1997); Ndagijimana & Musonera (2006), and Osuji and Chukwuanu, (2009). Other potential benefits of regional economic cooperation according to these writers include possible increased capital formation through technology transfer and externalities from export growth. Some of the enumerated benefits of regional cooperation are within reach if trade between member states could progressively grow beyond what the current situation is as observed in Table 6.

Table 6. Intra-trade of trade groups as percentage of total exports of each group (1980-2006)

Trade Group	1980	1990	2000	2005	2006
AFRICA					
CEPGL (3)	0.1	0.5	0.8	1.2	1.3
CEMAC (formerly UDEAC)(6)	1.6	2.3	1.0	0.9	0.9
COMESA (19)	1.8	4.7	4.6	4.5	4.2
ECCAS (11)	1.4	1.4	1.1	0.6	0.6
ECOWAS (16)	9.6	8.0	7.6	9.3	8.3
MRU (3)	0.8	0.0	0.4	0.3	0.3
SADC (14)	0.4	3.1	9.4	9.2	9.1
UEMOA (8)	9.6	13.0	13.1	13.4	13.1
AMERICA					
ANCOM	4.1	4.0	7.7	9.0	8.4
CACM	24.4	15.3	19.1	18.9	16.8
CARICOM	5.4	8.0	14.6	11.6	11.3
FTAA	43.4	46.6	60.7	60.2	58.4
LAIA	13.9	11.6	13.2	13.6	14.3
MERCOSUR	11.6	8.9	20.0	12.9	13.5
NAFTA	33.6	41.4	55.7	55.8	53.8
OECS	9.0	8.5	13.7	15.1	11.2
ACP	4.0	6.3	10.4	11.0	10.9

Sources: UNCTAD secretariat calculations based on International Monetary Fund (IMF) Division of Trade Statistics, 2008

Intra-Trade of Sub-Saharan African Trade Groups

Table 6 shows that trade between member states of the various economic groupings (intra-trade blocs) in SSA was very low between 1980 and 1990. This was followed by a near steady increase among some groups in subsequent years and especially after 2000. The performance of Economic Community of West African States (ECOWAS), South African Development Community (SADC), and Common Market for Eastern and Southern Africa (COMESA) which showed annual growth in trade between member groups is very encouraging and requires emulation by other regional groups. The increase in percentage displayed by these groups however, pales in comparison with figures from South American trade blocs which during the same period showed higher intra-trade figures. Between 2000 and 2006, intra-trade of the various regional economic cooperations ranged from 7.7 percent for ANCOM (Andean Common Market) in 2000 to 58.4 percent in 2006 for FTAA (Free Trade Area of the Americas) members. It is pertinent to mention that the annual growth rates of many of SSA's regional groupings' real GDP according to Osuji (2009) have been growing since 2000. For this positive development to continue and be sustained requires removal of some of those problems that confront these organizations in the form of structural disequilibria, poor transportation and communications infrastructure (Osuji 1997), Aly 1994). Improving regional cooperation has prompted Ravenhill (1990), Arhin (1990), and Wright (1993) to advise increased political will to support regional trade blocs, as well as creating enabling environment. Improvement in regional cooperation involves upgrading some facilities in communications and transportation essential in checking unrecorded transborder trade (UTT) which poses a big problem in many SSA countries. Hardy (1992) reports that two major groups carry out UTT; the first group comprises numerous small-scale operators of every gender, age and class who smuggle goods across borders and the second group consists of organized traders who in collusion with customs and other government officials smuggle large quantities of goods across borders. Application of modern technology can cut down this if sensors are deployed at checkpoints and border posts.

With regard to regional cooperation, the Consultative Meeting of the Regional Economic Communities under the aegis of the Economic Commission for Africa (ECA) and the African Union Commission resolved the multiplicity of regional economic communities with overlapping institutions which were prevalent in the continent. This was achieved through rationalization of these communities from fourteen to eight economic communities (RECs) (AUC, 2006). This move is seen by many as a sound policy that recognizes the role regional economic integration can play in the development of a region and this is consistent with the views of UNCTAD Secretary- General Panitchpakdi that Africa must do more for itself. Cross-border trade infrastructure is encouraged and recommended by UNCTAD (2008) as the smallness of individual African markets, coupled with the difficulty most firms face to access the market of industrialized countries favors the argument of expanding the intra-African trade. This is buttressed by opportunities that exist for intra-regional trade in food items such as maize, cassava and cassava products, fish and live animals observed in informal cross-border trade in the West African region. In Southern African region, the potential for increasing the ongoing intra-regional trade in water, electricity and other services such as game parks and tourism should be tapped.

At the global level, however, several initiatives have been undertaken to assist developing countries including Africa, in the fields of training and capacity-building. Example of such training is the "train-the-trainer" workshop held in Dublin for port managers from Ghana, and the United Republic of Tanzania. A similar "TrainForTrade" port-training-program was conducted for French-speaking and Portuguese-speaking port communities of Bissau, Conakry, Cotonou, Luanda and Marseille all in the African continent. These measures were due to earlier concerns raised by UNCTAD in 2007 about technological gap that exists between developed and developing countries.

The 2007 UNCTAD Report raised issues relating to science, technology and innovation (STI)policy and how STI policies geared towards technology can be integrated in the development and poverty reduction strategies, and how aid for STI (as part of official development assistance can support learning and innovation in recipient countries (UNCTAD, 2007). Other policy initiatives listed in the report to benefit African countries include commodity development and training programs. The training activities were designed

to strengthen the links between policymakers and academics; and to encourage the integration of course materials into participating countries' academic programs. Development of services in the areas of transport and transit, information and communication technology, and e-business were also covered. While many regional workshops on productive capacities, economic growth and poverty reduction in Africa was held in Addis Ababa, Ethiopia in February 22-23, 2007, there is need for more of such workshops. Another initiative worth mentioning is the Virtual Institute (VI) on Trade and Development designed to assist universities in member countries to improve their ability to deliver trade-related academic programs. Unfortunately low infrastructural facilities have prevented full participation by most of the countries in Africa. However, African institutes can benefit from research and various professional developments extended to other member countries. It is also necessary to use workshops to promote not only more intra-bloc trade and cooperation between member countries, but also cooperation in research and technology in their higher institutions. Through such cooperation, for example, international workshops organized in one member country's university, could be attended by scholars and scientists from other countries.

One such example is the United States/Africa International Workshop on Nanotechnology and Nanosciences Education which was held at the University of Nigeria, Nsukka Enugu State Nigeria in April 20-23rd, 2008. That workshop was in collaboration with the Nigerian Nanotechnology Initiative (NNI) with one of the objectives being, "to bring together researchers, educators, development partners and government representatives, to work on a future agenda for the strengthening of emerging US/Africa collaborations". Such a workshop in future could expand in coverage to include scientists and educators from other ECOWAS and SADC member nations. Moreover, other aspects of workshop on technology, policy analysis and research could be organized on regular basis with the assistance of not only international organizations like the United Nations (UN), but also New Partnership for Africa's Development (NEPAD) and other advanced industrialized countries like members of G-8.

This is where UNCTAD's Cluster XV. (Science, Technology and Innovation Policy becomes very relevant for the African continent in the context of the role science, technology and innovation (STI) can play in supporting and strengthening trade and other economic issues in the development process. Under this cluster, programs like capacity-building in developing countries, technology and innovation, and support for technology transfer would be beneficial to African countries. The cluster becomes all the more crucial to the cause of Africa as South-South cooperation in science and technology is encouraged.

CONCLUSION

This chapter has examined the role trade can play in a nation's economic and technological growth process. Part of this occurs when a nation exports and acquires foreign exchange which is used to import both consumer and industrial goods needed for improved productivity and overall benefit of citizens. Unfortunately that has not been the case with most African countries where after their independence pursued import-substitution industrialization policy that did not produce the intended result. Consequently the World Bank and International Monetary Fund (IMF) intervened with their Structural Adjustment Programs that aimed at stabilizing the economies of most of the Sub-Saharan African countries. Trade liberalization was one of the policy instruments which unfortunately did not achieve much in Africa. Rather most of the countries became more dependent on limited number of commodities than their counterparts in other developing regions. However, countries like Lesotho, Mauritius and Tunisia increased their export revenue from

manufactured products. Policy makers in other African countries should emulate the success of these nations.

Some other policy options essential for SSA countries to compete in the global arena include investment in education and infrastructure like transportation, and working cooperatively with one another under the aegis of the regional trade blocs in existence. Such regional cooperation in the areas of trade and manufacture becomes all the more important when the smallness of individual African markets is considered a major obstacle toward the growth of the manufacturing sector. The expansion of intra-bloc trade among African countries has the unique advantage of widening markets outside national boundaries. Other areas of cooperation are at the university and tertiary levels where research and training should be encouraged among different African universities and research institutions. Regional workshops sponsored by international organizations in the areas of science, technology and innovation (STI) as well as in trade and investment management for government officials, universities and research institutions are highly recommended for capacity building and retention. The African Development Bank and NEPAD should play very prominent role in mobilizing resources and financially supporting those institutions essential for Africa's economic and technological development.

In concluding this chapter, one cannot but reiterate an aspect of UNCTAD 2007 Report that, "although Africa's less developed countries (LDCs) are highly integrated into the global economy in terms of trade and investment flows, they are marginalized in terms of the international diffusion of knowledge and technology. Brain drain from African LDCs is also accelerating". To say that the brain drain has weakened Africa's institutional and human capacities, which are prerequisites for economic and technological development is saying the obvious. While that is no excuse, there is an urgent need for African countries to harmonize their trade policies and work cooperatively toward technological advancement in order to achieve some of the African Union's agenda.

REFERENCES

Ackah, C., & Morrissey, O. (2005). *Trade Policy and performance in Sub-Saharan Africa since the 1980s*. Economic research working paper 78. Tunis. African Development Bank.

Aly, A. H. M. (1994). Economic Cooperation in Africa . In *Search of Direction*. Boulder, CO: Lynne Rienner Publishers.

Balassa, B. (1982). *Development Strategies in Semi-industrial Economies*. Baltimore, MD: Johns Hopkins University Press.

Balassa, B. (1985). Export policy choices and economic growth in developing countries after the oil shock. *Journal of Development Economics, 18*, 23–35. doi:10.1016/0304-3878(85)90004-5

Bhagwati, J., & Srinivasan, T. N. (2002, May). Trade and poverty in the poor countries. *The American Economic Review, 92*(2), 180–183. doi:10.1257/000282802320189212

Bhagwati, J. N. (1978). *Foreign Trade Regimes and Economic Development: Anatomy of Consequences of Exchange Control Regimes*. Cambridge, MA: Ballinger.

Brinkman, H.-J. (1992). *Economic Performance, Exogenous Factors, and Adjustment Policies in Africa*. Paper presented at the African Finance and Economics Association at Howard University, Washington, D.C.

Callaghy, T. M. (1990). Lost Between State and Market: The Politics of Economic Adjustment in Ghana, Zambia and Nigeria . In Nelson, J. M. (Ed.), *Economic Crisis and Policy Choices* (pp. 257–319). Princeton, NJ: Princeton University Press.

Callaghy, T. M., & Ravenhill, J. (1993). Towards State Capability and Unbedded Liberalism in the Third World: Lessons for Adjustment . In Nelson, J. M. (Ed.), *Fragile Coalitions: The Policies of Economic Adjustment*. Oxford, UK: Transaction Books.

Chazan, N. (1992). *Politics and Society in Contemporary Africa*. Boulder, CO: Lynne Rienner Publishers.

Economic Commission for Africa and African Union. (2007). *Economic Report on Africa 2007: Accelerating Africa's Development through Diversification*. Addis Ababa, Ethiopia: United Nations Economic Commission for Africa.

Edoho, F. M. (1997). International Technology Transfer in the Emerging Global Order . In Edoho, F. M. (Ed.), *Globalization And the New World Order, Promises, Problems, and Prospects for Africa in the Twenty-First Century* (pp. 99–126). Westport, CT: Praeger Publishers.

Edwards, S. (1993, September). Openness, trade liberalization, and growth in developing countries. *Journal of Economic Literature, 31*, 1358–1393.

Ekanem, N. F. (1997). Economic Instability and Africa's Marginal Role in the New World Order: A Study of Selected African Countries . In Edoho, F. M. (Ed.), *Globalization And the New World Order* (pp. 47–60). Westport, CT: Praeger.

Green, C. J., Kimuyu, P., Manos, R., & Murinde, V. (2007). How do small firms in developing countries raise capital? Evidence from a large-scale survey of Kenyan micro and small-scale enterprises. *Advances in Financial Economics, 12*, 379–404. doi:10.1016/S1569-3732(07)12015-6

Hardy, C. (1992). The Prospects for Intra-Regional Trade Growth in Africa . In Stewart, F. (Eds.), *Alternative Development Strategies in Sub-Saharan Africa* (pp. 426–442). London: Palgrave Macmillan.

(1994). Hemmed . InCallaghy, T. M., & Ravenhill, J. (Eds.), *Responses to Africa's Economic Decline*. New York: Columbia University Press.

IMF. (1992). *International Financial Statistics*. Washington, DC: International Monetary Fund.

IMF. (2005). *Review of the IMF's trade restrictiveness index. Background paper to the Review of Fund Work on Trade*. Washington, DC: International Monetary Fund.

Jovanovic, M. N. (1992). *International Economic Integration*. London: Routlege Publishers.

Krueger, A. (1978). *Foreign Trade Regimes and Economic Development: Liberalization Attempts and Consequences*. Cambridge, MA: Ballinger Publishing Company.

Krueger, A. (1990). Import Substitution Versus Export Promotion . In King, P. (Ed.), *International Economics and International Economic Policy: A Reader* (pp. 155–165). New York: McGraw Hill.

Krueger, A. (1997, March). Trade policy and economic development how we learn. *The American Economic Review*, 87.

Lyakurwa, W. (1997). Regional Integration in Sub-Saharan Africa: A Review of Experiences and Issues . In Oyejide, A., Elbadawi, I., & Collier, P. (Eds.), *Regional Integration and Trade Liberalization in Sub-Saharan Africa*. New York: St. Martin's Press Inc.

McCalla, A. F., & Nash, J. (2007). *Reforming Agricultural Trade for Developing Countries, Key Issues for a Pro-development Outcome of the Doha Round* (*Vol. 1*). Washington, DC: The World Bank.

Mengistae, T., & Pattillo, C. (2004). Export orientation and productivity in Sub-Saharan Africa. *IMF Staff Papers*, *51*(2), 327–353.

Ndikumana, L. (2000). Financial determinants of domestic investment in Sub-Saharan Africa: evidence from panel data. *World Development*, *28*(2), 381–400. doi:10.1016/S0305-750X(99)00129-1

Ocampo, J. A., & Parra, M. A. (2003). The terms of trade for commodities in the twentieth century. [Santiago, Chile: United Nations Economic Commission for Latin America.]. *CEPAL Review*, 79.

Ostergaard, T. (1993). *Classical Models of Regional Integration- What relevance for Southern Africa?*Uppsala, Sweden: ODEN.

Osuji, L. O. (1997). Regional Trade Blocs and Economic Development: A Study of the West African Subregion . In Edoho, F. M. (Ed.), *Globalization And The New World Order* (pp. 171–182). Westport, CT: Praeger Publishing Company.

Osuji, L. O., & Chukwuanu, M. (2009). *Regional Trade Agreements and South-South Trade: Analysis of the experience of the major Sub-Saharan African Groupings*. Paper delivered at the MBAA Conference, Drake Hotel, Chicago Illinois.

Panitchpakdi, S. (2009, June 24). *Economic Development in Africa*. Report presented in New York to UNCTAD.

Perreault, W., Cannon, J., & McCarthy, E. J. (2009). *Basic Marketing: A Marketing Strategy Planning Approach*. New York: McGraw-Hill Publishing.

Ravenhill, J. (1990). Overcoming Constraints to Regional Cooperation in Africa: Coordination rather than Integration. In *Long-Term Perspective Study of Sub-Saharan Africa, Vol. 4.: Proceedings of a Workshop on Regional Integration and Cooperation* (pp. 81-85). Washington DC: World Bank.

Santos-Paulino, A., & Thirlwall, A. (2004). The impact of trade liberalization on exports, imports and the balance of payments of developing countries. *The Economic Journal*, *114*, 50–72. doi:10.1111/j.0013-0133.2004.00187.x

Smith, A. (1998). *An Inquiry into the Nature and causes of the Wealth of Nations*. Washington, DC: Regnery Publishing. (Original work published 1776)

UNCTAD. (2003a). *Economic Development in Africa: Trade Performance and Commodity Dependence. United Nations publication No. E.03.11.D.34*. New York: United Nations.

UNCTAD. (2007). *Economic Development in Africa: Reclaiming Policy Space-Domestic Resource Mobilization and Developmental States*. New York: United Nations.

UNCTAD. (2008b). *The Changing face of commodities in the twenty-first century*. TD/428. Note prepared by the UNCTAD secretariat, UNCTAD X11, Accra, Ghana, 20-25 April.

World Bank. (1981). *Accelerated Development in Sub-Saharan Africa: An Agenda for Action*. Washington, DC: World Bank.

World Bank. (1987). *World Development Report 1987*. New York: Oxford University Press.

World Bank. (1993). *World Development Report*. New York: Oxford University Press.

World Bank. (1994). *Adjustment in Africa: Reforms, Results, and the Road Ahead*. Washington, DC: World Bank.

World Bank. (2008a). *World Development Indicators*. Washington, DC: The World Bank.

World Bank. (2009). *World Development Report*. Washington, DC: The World Bank.

Wu, Y., & Zeng, L. (2008). *The impact of trade liberalization on the trade balance in developing countries*. IMF working paper WP/08/14. Washington, DC: IMF.

Chapter 25
Emerging Technology Penetration:
The Case of Solar Electricity in Nigeria

Olalekan A. Jesuleye
National Centre for Technology Management, Nigeria and Obafemi Awolowo University, Nigeria

Williams O. Siyanbola
National Centre for Technology Management, Nigeria and Obafemi Awolowo University, Nigeria

Mathew O. Ilori
Obafemi Awolowo University, Nigeria

ABSTRACT

Considering the huge wastage associated with the present energy production and consumption pattern in Nigeria, solar electricity (SE) is acclaimed to be of great potentials as a viable alternative to fossil fuels and is being considered by policy makers to contribute to improving energy efficiency, security and environmental protection. The veracity of such claim is being ascertained in this study through analysis of solar electricity utilization for lighting, refrigeration, ventilation, water pumping and others by just 5% of about 100 million Nigerian rural dwellers who lack access to national grid. The study deduced that increase in rural access to SE will yield tremendous carbon credits for Nigeria under the clean development mechanism and that generating more SE at cheaper cost will enhance policy support for green energy. This connotes a great future for microelectronics and nanotechnology in processing high efficiency multi-junction solar cells and nanosolar utility panel being optimized for utility-scale solar electricity systems.

CURRENT STATUS IN ELECTRICITY ACCESS IN NIGERIA AND THE INCESSANT ENERGY CRISIS

The per-capita electricity consumption in Nigeria ranged from 68 to 95 kWh between 1980 and

1997, which is about 17% the African average and 2% that of South Africa (NCP, 2001) and (PHCN, 2005). Indeed, with the exception of relatively poor countries such as Benin, Sudan, Mozambique and Angola, per capita electricity consumption in Nigeria was very far below that of much of Africa during the period (Hart, 2000). Although further increase in the per capita con-

DOI: 10.4018/978-1-61692-006-7.ch025

sumption was achieved in Nigeria between 1998 and 2003, nevertheless only 34% and 45% of the population had access to electricity respectively during this period (Nigeria's Electricity Sector, Executive Report, 2006). So also, as observed by Energy Commission of Nigeria in a study (ECN 2004), electricity access is more pronounced in the urban areas (81%), than in the rural areas (18%).

This situation is rather unfortunate because it contrasts sharply with the impressive primary energy profile the country parades. Chiefly among these are over 33 billion barrels of proved recoverable crude oil reserves (the world's sixth largest) (Kupolokun F., 2007), natural gas proved recoverable reserves of 187 trillion cubic feet (the world's ninth largest), about 2.7 billion coal and lignite reserves, 10,000 MW of large scale and 734 MW of small scale hydro-electricity power exploitable capability. The fuelwood reserve is estimated to be 13,071,464 ha. The country also has 31 billion Tar Sands reserves, 61 million tonnes/year and 83 million tonnes of animal waste and crop residue reserves respectively, 3.5-7.0kWh/m²·day of solar radiation potentials and 2-4 m/s annual average of wind reserves (ECN, 2005)).

In spite of the country's very rich energy resource endowment, Nigeria is still greatly plagued with acute energy crisis which permeates all sectors of the economy. Close to 100 million Nigerians remain "in the dark" without access to electricity (NBS 2006) and estimates of the new connections to the national utility grid system is well under 50,000 per year (ESMAP 2005). The country requires over 10000 MW of electricity to meet present demand, has deteriorating installed capacity of 6000 MW, but her current output is less than 3000MW, much of which is not put to use due to poor power transmission and distribution infrastructure. Only 18% of the 70% Nigerians living in rural areas have access to electricity, which shows that the country's rate of electricity access is far less than population growth rate. The country's 2003 per capita electricity consumption is put at 106 Kwh per year, far lower than that

of South African which is 4546.4 Kwh per year (IEA, 2003).

Most remote rural settlements are placed at a very serious disadvantaged due to relatively high grid (generation, transmission and distribution) expansion cost. Obvious potentials of solar technology as readily deployable stand-alone facilities for provision of electricity, with long service life (about 30yrs) and low maintenance cost suitable for rural areas are yet to be fully harnessed. Applications of solar PV technologies such as mini-grid systems; solar home systems; solar water pumping; street and traffic lightings; solar vaccine refrigeration and solar PV power back-up for communication gargets only exist in skeletal urban and rural demonstrated pilot projects. Thus, the real impact of solar electricity (for instance) for lighting is yet to be felt as good substitute for kerosene (Siyanbola et al 2004) and palm oil in the rural areas of Nigeria (NBS 2006).

From the foregoing, it becomes obvious that living within the barriers of the old conventional fossil fuel energy systems may not be sufficient in providing solution to the rural energy problem. Therefore, assessing constraints to rural solar electricity utilization as basis for determining its optimal contribution to the rural energy mix become imperative due to dart of information in this area in Nigeria. This also offers a means to ascertain the trajectory of an emerging technology penetration in a developing economy.

SOLAR ELECTRICITY OPTION FOR RURAL ACCESS TO ENERGY SERVICES AND CRITICAL ISSUES FOR POLICY

Consequence upon the challenges posed by the various energy problems in the country, certain pertinent research questions constantly agitate the minds of policy makers, energy planners, researchers and energy analysts in their search for optimal solutions to the Nigerian Energy problems. Some of these are:

a. To what extent can solar electricity contribute to improved access, improving energy efficiency, energy security and environmental protection of the country?

b. What is the current percentage share of solar electricity in the total renewable energy mix of the country at the useful energy level?

c. What is the present and potential future share of solar electricity in the grid expansion programme for rural electrification in the country?

d. Does the country's present and future solar electricity utilization pattern correspond with the world trends, considering the current worldwide increasing demand for solar electricity due to downward trend in the investment cost and improved technical efficiency?

e. What is the optimal approach to PVs applications for rural access to energy services considering various energy models available?

f. What are (or likely to be) the country's total CO_2 and other GHG emissions reduction gains for PV applications alternative to fossil fuels?

g. What are the major impediments to policy interventions for promoting PVs applications for energy services provision and rapid economic development in the country?

In an attempt to assist policy makers in their quest for solution to the above issues and research questions the study focuses on the following three specific objectives:

1. Analysis of solar electricity demand for energy services in the rural areas of Nigeria.

2. Simulation the future solar electricity share in the energy demand.

The significance of such a policy oriented study is unquantifiable, some of which are highlighted below:

Firstly, the huge economic, environmental, social cost and the lost associated with the present inefficient energy demand trend in the rural areas of Nigeria could be averted if appropriate quantifiable policy measures are adopted. Sequel to this, a constant critical appraisal of these policy measures and the strategies for their implementation becomes essential. Secondly, results obtained from such exercise usually reveal the nature of the problems upon which scientific solutions are proffered. In view of this, an appraisal study of the impact of these existing solar electricity installations for increasing access to energy services in the country is quite essential to provide the required guide for further policy interventions. Specifically, the outcome of this study is expected to generate comprehensive database of the share of PVs in the Nigerian Energy Mix with particular reference to the rural areas. It will also provide information on roles of PVs in rural electrification projects for poverty alleviation, penetration rate of PVs and its various applications in the household, cottage industry, agricultural and other sectors of the economy. Such information will be useful as veritable tools for attracting investment into the PV industry development in Nigeria. Useful data for policy guide for optimal energy demand and supply balance for overall efficient national energy system will be generated. Baseline data for projection of the future solar energy use and guide for monitoring future rural energy requirements will also be obtained. Input data for environmental analysis of CO_2 and other green-house-gases emissions stock associated with rural energy use will also be made available.

SOLAR ELECTRICITY APPLICATIONS FOR IMPROVED ACCESS TO ENERGY SERVICES IN THE RURAL AREAS OF NIGERIA

The United Nations Millennium Summit of world leaders in September 2000 and the final declaration at the summit, signed by 189 countries emphasized the world's commitment to a specific agenda for reducing global poverty. Eight Millennium Development Goals were listed in this agenda which not only identified the gains needed but quantified them and established yardsticks for measuring improvements in people's lives. The goals, listed below, today guide the efforts of virtually all organizations working in development and have been commonly accepted as a framework for measuring development progress. The Eight point MDG are: Eradicating extreme poverty and hunger; achieving universal primary education; promoting gender equality and empower women, reduction in child mortality; Improving maternal health; combating HIV/AIDS, malaria and other diseases; ensuring environmental sustainability and Global Partnership for Development (MDGs, 2000)

In adopting strategies for the achievement of the MDGs, the world assembly of leaders concluded that these Goals of halving poverty will not be achieved without energy to increase production and income, create jobs and reduce drudgery. They also realized that electricity access in particular is of paramount importance because of it convenience as an energy form for various applications. Clean energy promotion through renewable energy was also given priority. This invariably offers a good opportunity for solar electricity contribution to access increasing electricity access as discuss below with respect to strategies for achieving each of these MDGs in Nigeria (MDGs, 2000).

Eradicate Extreme Poverty and Hunger

The first emphasis under this MDG is to halve, between 1990 and 2015, the proportion of people whose income is less than one dollar a day. The second one also aims at halving, between 1990 and 2015, the proportion of people who suffer from hunger. As a result of insufficient energy to pro-

cess food, people have had to pay more especially from the rural areas out of their meager income. However income generation is necessary to alleviate poverty. Many people in developing countries spend up to a third of their income on energy, most of which is for cooking. As a result, women spend up to three hours day collecting firewood, walking up to ten kilometres and carrying 35kg of wood. Millions depend on bio-mass (wood, charcoal and dung) to meet their household energy needs. A major challenge will be to provide electricity to the rural poor. At the moment, nominal generating capacity is less than 6,000MW which is a far cry from the needed energy supply in the cities alone not to talk of the rural areas. Electricity is needed to power small industry and enterprise. This is bound to strengthen the economy of the country since most Nigerian are resident in the rural areas. The people will ease the government of the large investments made to circumvent their poverty state as the residents will be self sustained. Without this option being considered, rural poverty will not be eradicated. The conventional approach to electrification tends to marginalize rural communities who are located far away from the grid. The cost of energy supply in rural areas is high compared with densely populated areas. Electricity companies–public or private–have little incentive to provide services to these areas. Simple low cost solutions are expedient if the situation must be well tackled. Efficient technologies which will be beneficial to the people in both short- and long-term advances in the quality of life are advised considering the plight of the people and the option of solar photovoltaic energy fits in the process very well as facts show that the cost of solar photovoltaic materials have witnessed a sharp down trend in the last seven years. When the people have an option of solar water heaters, less energy and time is devoted to sourcing for biomass technology. The introduction of solar powered bulbs will encourage the people to stay longer outside working and they are able to generate more funds for themselves and their families (MDGs, 2000).

Recently in the Northern part of Nigeria, solar powered clippers were introduced giving room to more self sustained barbers in the communities. The Federal Government is committed to reaching the MDG targets of reducing by half the number of people living under extreme poverty by the year 2015. This commitment is strongly embedded in the NEEDS (NEEDS 2000).

Achieve Universal Primary Education

This MDG aims at ensuring that, by 2015, children everywhere, boys and girls alike, will be able to complete a full course of primary schooling. Getting children to school in the urban societies in Nigeria is not too difficult because there are laws that make it compulsory and there are agencies available to execute the law like the Nigeria Police and the Local Government marshals but it is quite difficult in the rural areas because of poor monitoring. The heavy workload of mothers and the poverty of the households are two main determinants keeping girls away from school or leading to heavy absenteeism or a high failure rate and dropout rate. In this context, to achieve the universal primary education, girls have to be relieved from their supplementary responsibilities for household and agricultural work. This can be addressed only through increased access to energy services to ease mothers' work burden. Additionally, researches have shown that daughters in poor households have opportunities for education if and only if their mothers have access to income in the rural areas, children dissipate so much energy in basic survival activities like gathering firewood, fetching water etc. This has naturally quenched the zeal for education because you can hardly assimilate when you are tired. Water pumping is another potential source of time savings for women and young girls. The combined amount of time and effort these services can save women—upwards of 5 hours—underscores the critical role that modern energy services have in reducing the gender bias of energy poverty (MDGs, 2000).

Solar PV will make room for the possibility of home study as lighting enhances reading, security. The use of information and communications technologies is also encouraged. Access to energy provides the opportunity to use equipment for teaching (overhead projector, computer, printer, photocopier, science equipment) Modern energy systems and efficient building design reduces heating/cooling costs. Once their basic responsibilities have been taken care of, there will be an aroused interest in education as they will seek a place to divert their energy, and in this case a school readily comes to mind. Considering the falling costs of Solar PV, it remains a most viable option for adoption (MDGs, 2000).

Promote Gender Equality and Empower Women

Cardinal to this component of the MDG is the elimination of gender disparity in primary and secondary education, preferably by 2005, and to all levels of education no later than 2015.

It would be hard to imagine a family in the developed world today spending one or more hours every day gathering biomass such as wood, agricultural residues, and dung, when they could instead buy cooking fuel for the same purpose at a price that reflects income from five or fewer minutes of work. Yet this is the burden of women in some Nigerian cities. Relieving women from some of their work burdens and providing them with income generating opportunities, and ensuring girls' access to education, will contribute significantly to achieving the aforementioned goal. The basic things that take the attention of women and make her lose so much time and energy are cooking, gathering firewood, fetching water and taking care of her house. Solar PV energy interventions will lead to releasing women from their heavy workload and providing them with much desired 'leisure time'. In addition to energy services easing women's work and saving their time, mechanization generally leads to men taking over the tasks, relieving women from some of the drudgery and time-consuming labour. Thus energy intervention not only assists women in performing their established gender roles but also result in 'transformation' of gender roles within the family. Ensuring leisure to women could also lead to women engaging in informal literacy classes so that they are more empowered to sustain and conserve their own and their family member's lives (MDGs, 2000).

In addition, energy interventions can provide access to information and communication technology (ICT) services, which are also important for women's empowerment. Street lighting improves women's safety. The goal of promoting gender equality and women's empowerment could also be through promoting women's control over energy services/enterprises. In line with the government's plan of transferring state-owned enterprises to the private sector, it is recommended that priority be given to women's groups for energy enterprise ownership. The government needs to provide scholarships to female students for acquiring higher education in the solar energy sector; so that women can contribute in a larger way academic experts and professionals in the energy sector. As a critical mass of women is achieved in decision/ policy making positions at all levels (national and local) through 'affirmative actions', the voices of women as a major group of stakeholders in the energy sector will be able to be heard (MDGs, 2000).

Reduce Child Mortality

Reduction by two thirds, between 1990 and 2015, the under-five mortality rate is central to this goal. Attaining this goal depends so much on energy as the lifeblood of human society and economics. Careless avoidable situations that occurred as a result energy based problems has made Nigeria with her dominant her population to record a high rate of child morality in previous years. Improving health and reducing death rates will

not happen without energy for the refrigeration needed for vaccination campaigns. The world's greatest child killer, acute respiratory infection, will not be tackled without dealing with smoke from cooking fires in the home. The heavy work burden, women's lack of awareness about nutrition, women's lack of decision-making power in seeking pre-natal, peri-natal and post-natal care, lack of sexual and reproductive rights and the overall poor status of pregnant women's health lead to high infant and child mortality rates and high maternal mortality rates. A nagging problem of the health centres however is the sustainability of energy. Additionally, in rural households many infants and children die due to getting burned in open fires in the hearth. Access to safe energy devices for cooking and heating like the use Solar PV heaters will save many lives of infants and children. Access to reproductive health care services and raised awareness about nutrition ensures improved maternal health and reduction in infant and child mortality.

Solar PV also helps to reduce malnutrition-related mortality by boosting food production and household incomes. They also help reduce the incidence of waterborne diseases by powering equipment for pumping and treating water.

Improve Maternal Health

Reduction of the maternal mortality ratio by three quarters, between 1990 and 2015 is the main declaration of this Millennium Development Goal. The heavy work burden women carry couple with poor awareness about nutrition and the overall poor status of pregnant women's health lead to not only high infant and child mortality rates but also high maternal mortality rates. Provision of clean energy technology such as solar electricity will help to improve pregnant women's health. When women have access to 'functional literacy'/education which has been lacking as a result of poor energy facilities especially in the rural areas and some other places in the Nigerian cities because of poor power generation, they will naturally observe simple laws of health and live better. Effective communication can be enhanced through solar enabled means. Electricity is essential for many medical instruments, illumination, medical record keeping, communications facilities for reporting medically significant events, and medical training and solar powered means could be a great advantage. This will influence also the kind of kids that will be produced by such mothers, and if the MDGs are to be taken seriously, the issue of energy is a major thing to deal with in Nigeria.

Combat HIV/AIDS, Malaria and Other Diseases

This MDG declaration seeks to half halt by 2015 and begun to reverse the spread of HIV/AIDS. It also quests to half halt by 2015 and begun to reverse the incidence of malaria and other major diseases. Yet, a recent statistics showed that 6% of the Nigerian population is infected with HIV/AIDS. Women and girls become infected with HIV/AIDS due to trafficking for prostitution and due to being infected through their husbands. It was also discovered that most of these occurrences were due to lack of information/orientation. This is quite prominent in the rural areas where the level of information is nothing to write home about. To sensitize women/girls and to make their families aware about combating HIV/AIDS, the role of ICT is very important. Solar PV technology can be adopted to power radio, TV, films and videos becomes very crucial for building awareness among village people and young women/girls. Energy services will ease the work burden of women living with HIV/AIDS and also ease the work burden of women taking care of sick members of their families. Empowerment of young women through educational and entrepreneurial interventions (as mentioned above) decreases trafficking in women/girls and helps them assert their sexual and reproductive rights, possibly saving them from HIV/AIDS infection. Energy is

necessary to preserve and sterilize much-needed medicines and medicine administering systems in the rural areas for combating HIV/AIDS, malaria and other infectious diseases and in the current situation of energy deficiency in the country, it is necessary for the Federal Government of Nigeria to adopt the option of Solar PV to avoid the heavy cost of correction in later years.

Ensure Environmental Sustainability

'Integrating the principles of sustainable development into country policies/programmes, and reversing the losses of environmental resources' form the main target of this MDG declaration. It also aims at 'halving by 2015 the proportion of people without sustainable access to safe drinking water'. And by 2020 it desires to 'have achieved a significant improvement in the lives of at least 100 million slum dwellers'.

The fact still remains that biomass (especially firewood) is the energy of the poor (Adegbulugbe et al, 1995). It is also a source of diminishing valuable forests and not only emitting indoor air pollution, but also accelerating adverse climate changes globally. Indoor pollution impacts on people's health, and climate change has affected farmers' systems and productivity as well as ecological balances. The human disruptions of the twentieth century-driven by a more than twenty fold growth in the use of fossil fuels augmented by a tripling in the use of traditional energy forms such as biomass–have amounted to the emergence of civilization as an ecological and geo-chemical force of global proportions. In other words, the accelerating impact of human life is altering the world at the global level. Poor air quality resulting from solid fuel use for cooking and heating has significant health and environmental impacts at every level. The combustion conditions in small cooking fires are such that a significant amount of unburned hydrocarbon, including some methane, is emitted to the atmosphere. These greenhouses gases are estimated to amount to several percent of the world's total greenhouse gas emissions. The main pollutants emitted in the combustion of fossil fuels are Sulphur and Nitrogen Oxides, Carbon Monoxide, and suspended particulate matter. Acid deposition is a problem because it causes damage to natural and human-made surfaces with which it comes in contact with. Access to and utilization of alternative energy technologies such as Solar PV technology for agro-processing and lighting directly reduces dependency on the forests, resulting in increases in forested areas, which could be protected for maintaining biodiversity.

Similarly, solar energy can reduce imports of fossil fuels and reduce greenhouse gas emissions, as well as saving valuable foreign currency. The saved money through commercial energy substitution and money earned in the market can be used for necessary development activities in the country, especially R&D targeted to people friendly energy technologies.

Develop a Global Partnership for Development

The seven-point declaration of the MDG is as follows:

First, 'develop further an open, rule-based, predictable, non-discriminatory trading and financial system'. Second, 'address the special needs of the least developed countries'. Third, 'address the special needs of landlocked countries and Small Island developing States'. Fourth, 'deal comprehensively with the debt problems of developing countries through national and international measures in order to make debt sustainable in the long term'. Fifth, 'in cooperation with developing countries, develop and implement strategies for decent and productive work for youth'. Sixth, 'in cooperation with pharmaceutical companies, provide access to affordable essential drugs in developing countries'. Seventh, 'in cooperation with the private sector, make available the benefits of new technologies, especially information and communications.'

A recent World Bank report, *Partnerships in Development, Progress in the Fight against Poverty* found that uneven progress was being made in terms of meeting the Millennium Development Goals. The report says if current trends in growth and poverty reduction continue the goal for eradicating extreme income poverty is within reach. But it may well be the only goal to be attained, for many of the other non-income goals–such as universal primary education, promoting gender equality and reducing child mortality–current rates of progress are too slow.

Given the complex and imperfect nature of energy markets, market forces alone cannot be expected to deliver energy services that are sustainable and meet the needs of the most vulnerable communities. Cooperation and partnerships between sectors and regions are needed to enable the use of energy services as a means for meeting all of the MDGs.

Defining commonly shared problems and channeling resources and expertise requires effective partnership among governments, public entities, development agencies, civil society, and the private sector. Without such partnerships it is difficult to secure the financial capital, knowledge and technology necessary to expand energy services at the global, regional, and local levels. It is therefore critical that all sectors, public and private alike, cooperate together to ensure that cleaner, more efficient energy systems—and the markets that are needed to sustain them—are available to the poor.

Partnerships are particularly important in helping countries mainstream energy into broader development strategies and frameworks. Energy is a cross-cutting issue by its very nature and thus requires participation from all development sectors in order to maximize its impact on development. To ensure that the poor benefit fully from greater access to energy, energy planning should be linked to development goals and priorities in other sectors.

Partnerships are also quite important for mobilizing financial resources to expand energy investments and services. Public financing from both domestic resources and official development assistance, combined with private entrepreneurship and investment, are needed to develop energy services for the poor. New forms of risk sharing between the private and the public sectors should be developed under public-private partnerships as a way to attract private sector resources in the area of sustainable energy. Official development assistance should be used strategically to build capacity, assess and prepare projects, and support the creation of enabling policy environments.

ANALYTICAL APPROACH FOR SOLAR ELECTRICITY DEMAND ANALYSIS

The study quantifies the share of solar powered end-use energy services in the energy demand mix of the sector. The specific energy services considered are: lighting, water pumping for drinking and irrigation purposes, and powering of electrical appliances such as TV, Video, Radio, Refrigeration, Personal Computers and Communication Equipment, for the based year 2009 and twenty-five years projected demand and supply trends up-till 2034. CO_2 emissions reduction resulting from solar electricity use coupled with an examination of policy interventions for its promotion and widespread utilization in the country was also calculated. For the demand analysis, analytical approach employed involves calculation of the base year per capita solar electricity use (called energy intensity) for various energy services per dweller. Future energy intensity for various energy services were obtained by modifying the based year energy intensity per period with the projected population share per dweller.

The study uses survey method through questionnaire administration for primary data collection from three categories of sampled stakeholders

drawn from the three ecological regions of the country, on one rural village per region basis. These stakeholders are: firstly, Solar PV users at households sector, cottage industries and public buildings. Secondly, Solar PV Suppliers, thirdly, Energy Commission of Nigeria (ECN) saddled with the mandate for overseeing renewable energy matters at the federal level and other relevant state ministries/agencies with the same mandate.

In selection of sample size for the study, strong consideration was given to the peculiar problems associated with energy use and effects on share of solar electricity on energy mix for energy service provision in Nigeria, especially fuelwood usage by larger percentage of the population in the three ecological regions of the country. This includes: desertification in the North, deforestation in the west and soil erosion in the east. Closely connected to this are environmental pollution caused by gas flaring in the south-south. These selected rural communities are places where solar electricity are already installed and are in use. In each of them the share of solar electricity utilization for increasing access to energy services is quantified. The quantification is done for the base year upon which projection is then made for future trend.

The administered questionnaires for different category of users are shown Table 1 below. For the household sector, 5 individuals living under the same roof as a family constitutes a household, based on National Bureau of Statistics (NBS) definition. Each household received only one questionnaire. Thus, the low, middle and high income groups in the rural areas for the households sector received 10 (questionnaire) each, making a total of 90 questionnaires for all the 3 regions under consideration. 10 questionnaires were administered each for the small and micro agro-based/cottage industries, making a total of 60 for the three regions for this sector. 41 questionnaires were administered in Public buildings such as Hospitals/Clinics, Schools, Barracks, Prisons, Post Offices, Police Stations, Town Halls and others, a total of 123 for the three regions for

Table 1. Questionnaire for different categories of users

Please Indicate the Category you Represent	Household (Estimated Average Monthly Income)			Cottage Industry (Estimated Number of Employees)		Public Buildings					
	Less Than N1000	N1000–N5000	Above N5000	Less Than 10	Above 10	Hospitals/Clinics	Schools	Barracks/Prisons	Churches/Mosques	Police Stations	Palaces/Town Halls

In Meeting the Demand for Lighting, Entertainment, Refrigeration, Water Pumping, Ventilation and Computer/Internet Services, for the Category you represent, please indicate as appropriate your Fuel Sources, End-Use-Appliances per Source and Consumption Rate per Source (KWh/Day) in the table below.

End-Use Appliances for Lighting Services	Grid Supply	Diesel/Gasoline powered Generator	Battery Powered Generator	Wind Powered generator	Solar Panel	Candle	Kerosene	LPG	Others: (Please Specify)
Incandescent Bulbs									
Fluorescent Lambs									
High Eff. Bulbs									
Rechargeable Lambs									
Solar Touches									
Solar Lanterns									
Kerosene Lanterns									
Gas Lambs									
Candle Sticks									

End-Use Appliances for Computing/Internet Services	Grid Supply	Diesel/Gasoline powered Generator	Battery Powered Generator	Wind Powered Generator	Solar Panel	Candle	Kerosene	LPG	Others: (Please Specify)
Computer Systems									
Power Stabilizers									
UPS, Printers									
GSM Hand Sets									
Solar Calculators									
Game Consoles									

continues on following page

Table 1. continued

End-Use Appliances for Entertainment Services	Grid Supply	Diesel/Gasoline powered Generator	Battery Powered Generator	Wind Powered Generator	Solar Panel	Candle	Kerosene	LPG	Others: (Please Specify)
Radio									
TV									
Videos									
End-Use Appliances for Water Pumping Services	Grid Supply	Diesel/Gasoline powered Generator	Battery Powered Generator	Wind Powered Generator	Solar Panel	Candle	Kerosene	LPG	Others: (Please Specify)
Water Pumping Machine (Drinking)									
Water Pumping Machine (Irrigation)									
End-Use Appliances for Refrigeration	Grid Supply	Diesel/Gasoline powered Generator	Battery Powered Generator	Wind Powered Generator	Solar Panel	Candle	Kerosene	LPG	Others: (Please Specify)
Fridge									
End-Use Appliances for Ventilation	Grid Supply	Diesel/Gasoline powered Generator	Battery Powered Generator	Wind Powered Generator	Solar Panel	Candle	Kerosene	LPG	Others: (Please Specify)
Fan									

this sector. 9 questionnaires were administered to administrators in ECN, Solar Energy Research Centres at Sokoto and Nssukka, and administrators in energy based state ministries/agencies of the sampled communities. 20 questionnaires were also administered to Solar PV suppliers dawn from the Solar Energy Society of Nigeria and Solar Energy Association of Nigeria.

Model for **A**nalysis of **E**nergy **D**emand (MAED-II and **M**odel for **E**nergy **S**upply **S**ystem **A**nd their **G**eneral **E**nvironmental impacts (MESSAGE-V) developed at International Atomic Energy Agency (IAEA) Austria were adapted for the study (see International Atomic Energy Agency, Vienna, 2006). As presented in Figure 1, MADE-II is made up of 7 blocks of data in flow for demand analysis. Block 1 deals with general information about energy levels, base year and future time periods while information on development of population is presented in Block 2. Data analysis for useful energy demand in the households sector is treated in Block 3, while block 4 and 5 deal with the Households, Cottage Industries and Community Services sectors. The transport sector is not included in the analysis while block 7 deals with efficiencies, penetration factors and sectoral demand for Solar Electricity Lighting, Water Pumping for Drinking and Irrigation Purposes, Powering of TV, Video, Radio, Refrigeration, Ventilation and Personal Computers.

The basic analytical approach employed in MADE-II which makes it suitable for demand analysis of solar electricity in Nigeria is its flexibility and applicability for demand projection over short or long period range. So also, the analytical techniques adopted in MADE-II rest on the idea that energy is a means to an end and it is used together with other production factors to provide products (in form of goods and services) to the society (Saboohi 1989).

A brief explanation on the useful energy demand for various end-use services in an economy for instance is given below:

(i) Process Heat: Energy Service for processing all forms of heating in the production unit of a manufacturing firm.

(ii) Motive Power: Energy Service for locomotives purposes in all the units of any economic sector. This includes evacuation of products in trucks, trains and also transportation of staff or passengers such sector by vehicles.

(iii) Lighting: Energy Service for generating light for atmospheric illumination at every unit of such sector, including the administrative and staff resident units.

(iv) Non-Substitutable Electricity: Energy Service for driving all electrical appliances, including space cooling or room conditioning in an economic sector, which cannot be driving by any other form of energy like Liquefied Petroleum Gas, Gasoline etc., but only by electricity.

Socio-Economic Variables

These activities are key variables such as population growth, urbanization rate and disposable income distribution pattern of various categories of users of solar electricity to be obtained from annual survey of National Bureau of Statistics (NBS). Reference is particularly made to user's income distribution pattern and not fuel prices because historical evidence in Nigeria already shows strong correlation between income levels and type of energy use, which means that low income earners are more disposable to cheaper and inefficient fuel use while the reverse is the case for high income earners.

The study also adopted scenario analysis technique widely used as scientific method for dealing with the basic inherent uncertainty about future evolution of various variables driving the energy demand. In order to bring out the basic issues relating to population growth, urbanization rate and income levels in the study, three scenarios namely low, medium and high will be used. The basic assumptions behind the three

Figure 1. Adapted data flow in MADE-II (Model for Analysis of Demand of Energy)

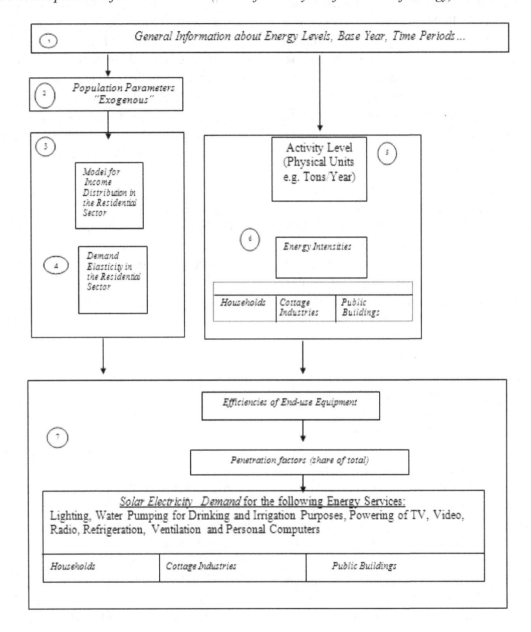

scenarios used pertaining to this study variables are described below.

The Low Scenario

This is characterized by low economic development in the future. Hence, the study assumes a high population growth rate. The underlying assumption is that the economy's dependence crude oil export will increase while government population policy will be ineffective. The share of fossil fuel and fuel wood will be high and the rate of solar electric powered services in the sectoral energy mix will be lowest compared with other two scenarios. The income distribution expected to be more or less the same as at now (i.e. the

Table 2. Evolution of rural population share (in millions) for solar electricity services (low, ref, high scenarios)

1. Solar Lighting.	Scenarios	2009	2014	2019	2024	2029	2034
	Low (2.2% rise)	4.9	5.4	6.0	6.7	7.4	8.3
With 5% Rural-Population Access.	Ref (2.8% rise)	4.9	5.6	6.4	7.3	8.3	9.4
	High (3.2% rise)	4.9	5.7	6.6	7.6	8.9	10.3
2. Solar Powered Radios & Stereos.	Scenarios	2009	2014	2019	2024	2029	2034
	Low (2.2% rise)	0.3	0.3	0.4	0.4	0.4	0.5
With 1% Rural-Population Access.	Ref (2.8% rise)	0.3	0.3	0.4	0.4	0.5	0.6
	High (3.2% rise)	0.3	0.3	0.4	0.5	0.5	0.6
3. Solar Powered Ventilation.	Scenarios	2009	2014	2019	2024	2029	2034
	Low (2.2% rise)	4.9	5.4	6.0	6.7	7.4	8.3
With 5% Rural-Population Access.	Ref (2.8% rise)	4.9	5.6	6.4	7.3	8.3	9.4
	High (3.2% rise)	4.9	5.7	6.6	7.6	8.9	10.3
4. Solar Powered TV.	Scenarios	2009	2014	2019	2024	2029	2034
	Low (2.2% rise)	1.0	1.1	1.2	1.3	1.5	1.7
With 1% Rural-Population Access.	Ref (2.8% rise)	1.0	1.1	1.3	1.5	1.7	1.9
	High (3.2% rise)	1.0	1.1	1.3	1.5	1.8	2.1
5. Solar Powered VCR & CD-DVD Recorder.	Scenarios	2009	2014	2019	2024	2029	2034
	Low (2.2% rise)	1.0	1.1	1.2	1.3	1.5	1.7
With 1% Rural-Population Access.	Ref (2.8% rise)	1.0	1.1	1.3	1.5	1.7	1.9
	High (3.2% rise)	1.0	1.1	1.3	1.5	1.8	2.1
6. Solar Powered Refrigeration.	Scenarios	2009	2014	2019	2024	2029	2034
	Low (2.2% rise)	1.0	1.1	1.2	1.3	1.5	1.7
With 1% Rural-Population Access.	Ref (2.8% rise)	1.0	1.1	1.3	1.5	1.7	1.9
	High (3.2% rise)	1.0	1.1	1.3	1.5	1.8	2.1
7. Solar Powered Water Pumping	Scenarios	2009	2014	2019	2024	2029	2034
	Low (2.2% rise)	0.700	0.777	0.862	0.957	1.063	1.180
With 5% Rural-Population Access.	Ref (2.8% rise)	0.700	0.798	0.910	1.037	1.182	1.348
	High (3.2% rise)	0.700	0.812	0.942	1.093	1.267	1.470
8. Solar Powered PCs & Internet Services.	Scenarios	2009	2014	2019	2024	2029	2034
	Low (2.2% rise)	0.7	0.8	0.8	0.9	1.0	1.2
With 0.7% Rural-Population Access.	Ref (2.8% rise)	0.7	0.8	0.9	1.0	1.2	1.3
	High (3.2% rise)	0.7	0.8	0.9	1.1	1.2	1.4

Table 3. Technical variables for solar electricity demand analysis

Solar Electric Services	Lighting	Radio	Ventilation	TV	VCR	Refrigeration	Water Pumping	PC + Internet
Total Population (Million)	140	140	140	140	140	140	140	140
Total Rural Household (HH) Population (@ 70% of Total Population (Million)	98	98	98	98	98	98	98	98
Sun Shine Hours/day	6	6	6	6	6	6	6	6
Appliance Average Duration of Use per Household per day	6	6	6	6	6	6	0.3	2
Power Rating of Appliance (Wattage)	20	80	20	230	60	600	1000	180
Average Number of Appliances in Use per Household per day	5	1	5	1	1	1	1	1
Total Appliances Usage (KWh/hh/day)	0.6	0.48	0.6	1.38	0.36	3.6	6	1.08

share of population in the low income group will be reduced very slowly). This scenario is more or less a pessimistic view of the Nigerian future but it is structured to demonstrate what could happen if the negative trend in the socio-economic parameters are not reversed.

The Moderate Scenario

This represents the middle of the road case to compare the pessimistic case as represented by low scenario and the very optimistic assumption about the future as represented by the high scenario. The underlying assumption is that policies to turn the economy around work but moderately.

The High Scenario

This scenario assumes a successful population and solar energy policies leading to reduced population growth rate, and increase in solar electric powered energy services in the energy mix for both urban and rural areas. The economy is expected to expand very fast while the income distribution is much better with share of low income groups decreasing substantially. Table 2 shows the rural population growth assumption for Low, Moderate and High Scenarios.

Analytical Results of Solar Electricity Demand for Lighting in Nigeria

As indicated in Table 4, for the base year 2009, at a household daily consumption rate of 100 peak watts of solar electric lighting provided by 5 solar powered bulbs of 20 watts per, a total demand of 109.2 peak Kilowatts solar electric lighting will be required by just 1% of the 14 million households that constitutes the 70% rural Nigerian populace, with virtually no access to the national grid. This is estimated to increase to 4.98, 5.64 and 6.18 peak gigawatt hours per day by 2034 for low, reference and high scenarios at annual growth rate of 2.2%, 2.8% and 3.2% respectively as shown in Figure 2.

The results in Table 3 also shows that the based year solar electricity demand for powering television sets is 1.38 GWh/year for the rural dwellers. By 2032 the demand rose to 2.346, 2.622 and 2.898 GWh/Day for each of the low, ref. and high scenarios respectively as indicated also in Figure 3.

Table 4. Demand analysis for solar electricity services in the rural areas of Nigeria

1. Solar Lighting	Scenarios	2009	2014	2019	2024	2029	2034
	Low	2.94	3.24	3.6	4.02	4.44	4.98
Demand by 5% of Rural-Populace in (GWh/day)	Ref.	2.94	3.36	3.84	4.38	4.98	5.64
	High	2.94	1248.3	3.96	4.56	5.34	6.18
2. Solar Powered Radios & Stereos	Scenarios	2009	2014	2019	2024	2029	2034
	Low	0.144	0.144	0.2	0.192	0.192	0.24
Demand by 1% of Rural-Populace in (GWh/day)	Ref.	0.144	0.144	0.192	0.192	0.24	0.288
	High	0.144	0.144	0.192	0.24	0.24	0.288
3. Solar Powered Ventilation	Scenarios	2009	2014	2019	2024	2029	2034
	Low	2.94	3.24	3.6	4.02	4.44	4.98
Demand by 5% of Rural-Populace in (GWh/day)	Ref.	2.94	3.36	3.84	4.38	4.98	5.64
	High	2.94	3.42	3.96	4.56	5.34	6.18
4. Solar Powered TV	Scenarios	2009	2014	2019	2024	2029	2034
	Low	1.38	1.518	1.7	1.794	2.07	2.346
Demand by 1% of Rural-Populace in (GWh/day)	Ref.	1.38	1.518	1.794	2.07	2.346	2.622
	High	1.38	1.518	1.794	2.07	2.484	2.898
5. Solar Powered VCR & CD-DVD Recorder.	Scenarios	2009	2014	2019	2024	2029	2034
	Low	0.36	0.396	0.4	0.468	0.54	0.612
Demand by 1% of Rural-Populace in (GWh/day)	Ref.	0.36	0.396	0.468	0.54	0.612	0.684
	High	0.36	0.396	0.468	0.54	0.648	0.756
6. Solar Powered Refrigeration.	Scenarios	2009	2014	2019	2024	2029	2034
	Low	3.6	3.96	4.3	4.68	5.4	6.12
Demand by 1% of Rural-Populace in (GWh/day)	Ref.	3.6	3.96	4.68	5.4	6.12	6.84
	High	3.6	3.96	4.68	5.4	6.48	7.56
7. Solar Powered Water Pumping	Scenarios	2009	2014	2019	2024	2029	2034
	Low	4.2	4.662	5.2	5.742	6.378	7.08
Demand by 5% of Rural-Populace in (GWh/day)	Ref.	4.2	4.788	5.46	6.222	7.092	8.088
	High	4.2	4.872	5.652	6.558	7.602	8.82
8. Solar Powered PCs & Internet Services.	Scenarios	2009	2014	2019	2024	2029	2034
	Low	**0.756**	**0.864**	**0.9**	**0.972**	**1.08**	**1.296**
Demand by 0.7% of Rural-Populace in (GWh/day)	Ref.	0.756	0.864	0.972	1.08	1.296	1.404
	High	0.756	0.864	0.972	1.188	1.296	1.512

Figure 2. Solar electricity demand for lighting in rural areas of Nigeria

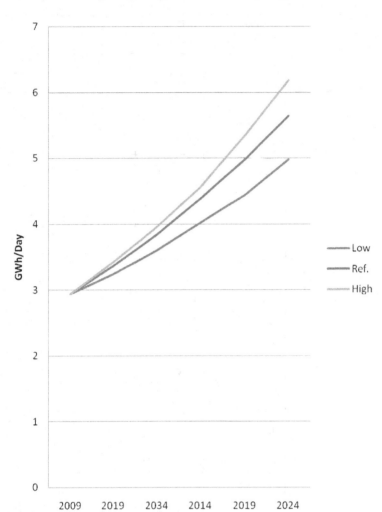

Figure 3. Solar electricity demand for powering television sets in rural Nigeria

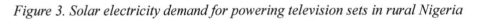
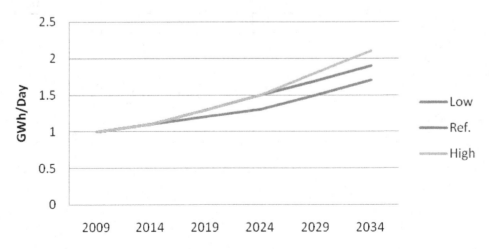

Figure 4. Base year and projected demand for solar electricity services in rural Nigeria (GWh/Day): low scenario

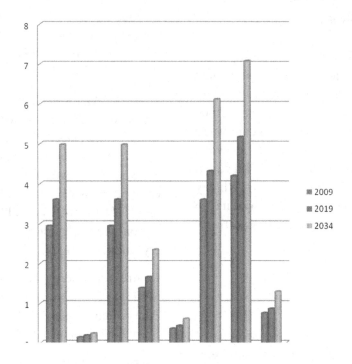

Figure 5. Base year and projected demand for solar electricity services in rural Nigeria (GWh/Day): high scenario

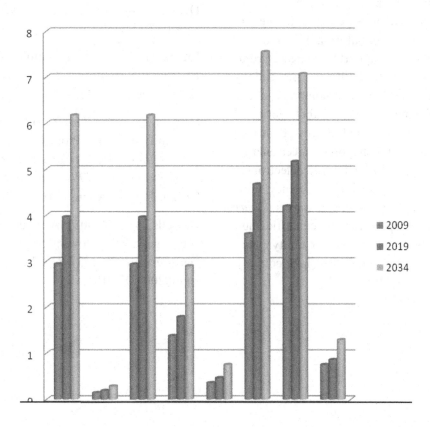

Solar Electricity Demand for Water Pumping in the Rural Areas of Nigeria

This includes demand for irrigation mostly in the rural areas. The calculation is based on the fact that one Solar PV arrays of 1000 watts will pump a gallon per minute, which translates to 2 cubic meters per day to satisfy 10 households daily water requirements of 40 litres per person. As shown in Figure 4, the demand rose from 4.2 GWh/day in 2009 to 5.2 GWh/day in 2019 and 7.08 GWh/day in 2034 for the low. For the high scenarios it rose to 5.65 and 8.82 for 2019 and 2034 respectively as shown in Figure 5.

CONCLUSION

This study articulates a quantitative analysis of solar electricity utilization for lighting, refrigeration, ventilation, powering of radio, video and television sets, personal computers/communication equipment and water pumping by just only 5% of about 100 million Nigerian rural dwellers who lack access to national grid. The study results show that policy support for increased rural access to solar electricity will yield tremendous CO_2 emissions savings and carbon credit under the clean development mechanism for Nigeria and that generating more Solar electricity at cheaper cost will enhance policy support for green energy. This connotes a great future for microelectronics experience in processing high efficiency multijunction solar cells from silicon wafers (*Baert et al, 2009)* at 41.6 percent solar cell efficiency (Spectrolab, 2009) and Nanosolar utility panel being optimized for utility-scale solar electricity systems (Nanosolar, 2009).

REFERENCES

Adegbulugbe, O. A., & Akinbami, J.-F. K. (1995). Urban households energy use pattern in Nigeria. *Natural Resources Forum*, *19*(2), 125–133. doi:10.1111/j.1477-8947.1995.tb00600.x

Baert, K., Van Kerschaver, E., & Poortmans, J. (2009). Crystalline Si solar cells and the microelectronics experience. *Photovoltaics World Magazine for Solar Power.* Renewable Energy World Network. Retrieved from http://www.electroiq.com/index/display/photovoltaics-article-display/4711974914/articles/Photovoltaics-World/silicon-photovoltaics/crystalline-silicon/2009/08/crystalline-si_solar.html

Centre for Energy Economics. (2006). Gas Monetization in Nigeria. In Directory of the U.S. Photovoltaics Industry 1996. Washington, DC: Solar Energy Industries Association.

Energy Commission of Nigeria. ECN (2004), Nigeria Energy Demand and Power Planning Study for the period 2000-2030. Part 1: Energy Demand Projections. Technical Report No ECN/EPA/04/01 Energy Commission of Nigeria, Abuja.

Energy Information Administration. (2004). *Electric Power Annual.* Form EIA-860, Annual Electric Generator Report database.

Hart, H. I. (2000). Gas Turbine Operation in a System Disturbance-Prone Grid. In *Proceedings of the 4th International Conference on Power Systems Operation and Planning (ICPSOP)*, Accra, Ghana, July 31–August 3 (pp. 153-157).

Howells, M. I. (2005). A model of household energy services in a low-income rural African village. *Energy Policy*, *33*, 1833–1851. doi:10.1016/j.enpol.2004.02.019

International Atomic Energy Agency. (2006). *Model for Analysis of Energy Demand (MAED-2) User Manual*. Retrieved from http://www-pub. iaea.org/MTCD/publications/PDF/CMS-18_web. pdf

Jesuleye, O. A., & Siyanbola, W. O. (2008). Solar electricity demand analysis for improved access to energy services in Nigeria. *Nigeria Journal of Solar Energy, 19*(1), 136–152.

Jesuleye, O. A., Siyanbola, W. O., Ilori, M. O., & Sanni, S. A. (2007). Energy demand analysis of Port-Harcourt Refinery, Nigeria and its policy implications. *Energy Policy Journals, 35*, 1338–1345. doi:10.1016/j.enpol.2006.04.004

Kumar, B., Koval, T., Narayanan, S., & Shea, S. (2002). *Commercialization of a silicon Nitride Co-fire through (SINCOT) process for manufacturing high efficient mono-crystalline silicon solar cells*. Pre-print of paper to be presented at 29 IEEE PVSC New Orleans May 20-24, 2002.

Kupolokun, F.(2007, April 1). Nigeria's Gas Reserves Increases. THISDAY NIGERIAN NEWSPAPER.

Lemiti, M., Kaminski, A., Fave, A., & Fourmond, E. (2007). *New solar cell production process may improve efficiency and reduce cost*. SPIE.

Nanosolar. (2009, September 9). Nanosolar Completes Panel Factory, Commences Serial Production. *Nanosolar Communications*.

(2001). *National Council on Privatization (NCP)*. Abuja, Nigeria: National Electric Power Policy.

Nigeria's Electricity Sector. (2006 January). Executive Report, Abuja, Nigeria.

Power Holding Company of Nigeria (PHCN). (2005). *Meeting the Challenges of Power Sector Development: Generation, Transmission and Distribution Expansion*. Paper presented at Presidential Retreat on Power Generation and Supply, 2 December 2005.

Power Holding Company of Nigeria (PHCN). (2006). *Annual Report, 2006*.

Siyanbola, W. O., Adaramola, M. S., Omotoyin, O. O., & Fadare, O. A. (2004). Determination of energy efficiency of non-bio-fuel-household cooking stoves. *Nigerian Journal of Physics, 16*(2), 2004.

The World Bank. (2001). *African Development Indicators*. Washington, DC: The World Bank.

Williams, R. (1960). Becquerel Photovoltaic Effect in Binary Compounds. *The Journal of Chemical Physics, 32*(5), 1505-1514. Retrieved from DOI:10.1063/1.1730950.

World Bank. (2002). *Nigeria Public and Private Electricity Provision as Barrier to Manufacturing Competitiveness*. Retrieved February 10, 2007, from http://www.worldbank.org/afr/findings

Zarma, I. H. (2006). *Hydro Power Resources in Nigeria*. Paper presented at 2nd Hydro Power for Today Conference International Centre on Small Hydro Power (IC-SHP), Hangzhou, China.

Compilation of References

(1994). Hemmed . InCallaghy, T. M., & Ravenhill, J. (Eds.), *Responses to Africa's Economic Decline*. New York: Columbia University Press.

(2001). *National Council on Privatization (NCP)*. Abuja, Nigeria: National Electric Power Policy.

Abalos, J., González, L., & Chervellino, M. (2006). Conceptos, antecedentes históricos y actuales de la institucionalidad nacional y sus instrumentos en ciencia, tecnología e innovación . In *Las regiones de chile ante la ciencia, tecnología e innovación: diagnósticos regionales y lineamientos para sus estrategias* (pp. 24–26). Santiago de Chile, Chile: CONICYT-PBCT.

Abalos, J., González, L., & Chervellino, M. (2006). Diagnóstico y lineamientos estratégicos para el desarrollo de la ciencia, tecnología e innovación en las regiones chilenas . In *Las regiones de chile ante la ciencia, tecnología e innovación: diagnósticos regionales y lineamientos para sus estrategias* (pp. 262–867). Santiago de Chile, Chile: CONICYT-PBCT.

Abeln, G. C. (1998). Approaches to nanofabrication on Si(100) surfaces: Selective area chemical vapor deposition of metals and selective chemisorption of organic molecules. *Journal of Vacuum Science & Technology B Microelectronics and Nanometer Structures, 16*(6), 3874–3878. doi:10.1116/1.590426

Abramovitz, M. (1986). Catching up, forging ahead, and falling behind. *The Journal of Economic History, 46*(1), 385–406. doi:10.1017/S0022050700046209

Acemoglu, D. (2002). Directed technical change. *The Review of Economic Studies, 69*(4), 781–809. doi:10.1111/1467-937X.00226

Ackah, C., & Morrissey, O. (2005). *Trade Policy and performance in Sub-Saharan Africa since the 1980s*. Economic research working paper 78. Tunis. African Development Bank.

Acs, Z. (2002). *Innovation and the Growth of Cities*. Cheltenham, UK: Edward Elgar Publishing Ltd.

Addison, T. (2005). *Agricultural Development for Peace*. Research paper No. 2005/07, United Nations University, World Institute for Development Economics Research.

Adebayo, A. A., & Oladeji, S. I. (2005). *Health Human Capital Condition: An Analysis of the Determinant in Nigeria*. Paper Presented at African Conference 2005 on African Health and Illness, Austin, The University of Texas, March 25-27.

Adegbulugbe, O. A., & Akinbami, J.-F. K. (1995). Urban households energy use pattern in Nigeria. *Natural Resources Forum, 19*(2), 125–133. doi:10.1111/j.1477-8947.1995.tb00600.x

Adegeye, A. J., & Dittoh, J. S. (1985). *Essentials of Agricultural Economics*. Ibadan, Nigeria: Impact Publishers Nigeria Ltd.

Adekoya, E. A., & Tologbonse, E. B. (2005). Adoption and diffusion of innovations . In Adedoyin, S. F. (Ed.), *Agricultural Extension in Nigeria* (pp. 28–37). Ibadan, Nigeria: Federal Agricultural Coordinating Unit.

Adeoti, J. O. (2005). Information Technology Investment in Nigerian Manufacturing Industry: The Progress So Far. In *Selected Papers for the 2004 Annual Conference* (pp. 213-244). Ibadan, Nigeria: Nigerian Economic Society.

Adesina, A. A., & Forson, J. B. (1995). Farmers perception and adoption of new agricultural technology. Evidence from analysis in Burkina-Faso and Guinea. *West Africa, 13*, 1–9.

Adhikarya, R. (1996). Implementing strategic extension campaigns. In Swanson, B. E., Bentz, R. P., & Sofranko, A. J. (Eds.), *Improving Agricultural Extension: A Reference Manual*. Rome: FAO.

Adikwu, M. U. (2007a). *Highlights on Science and Technology Education Post-Basic (STEP-B)*. Presented at the Digital Bridge Institute at a Conference organized by the Nigerian ICT Forum, Abuja, July 24.

Adikwu, M. U. (2007b). *Options for Financing the Change-Management System*. Presented at the Stakeholders' Conference on Engineering and Technology Academic Programmes, Transcorp Hilton, Abuja, November 10.

AEGIS. (2005). *Stocktake of NSW as a Knowledge Hub*. Sydney, Australia: UWS.

African Development Bank- AfDB. (2008). *Selected Statistics on African Countries (Vol. 27)*. Tunis: African Development Bank.

Alcántara, V., & Duro, J. A. (2004). Inequality of energy intensities across OECD countries. *Energy Policy, 32*, 1257–1260. doi:10.1016/S0301-4215(03)00095-8

Alexandre, M. da Silva. (2009). Carrying capacity in agriculture: Environmental significance and some related patents. *Recent Patients on Food . Nutrition and Agriculture, 1*(2), 100–103.

Alivisatos, P. (2007). *Nanotechnology Roadmap for Atomically Precise Nanofabrication and Productive Nanosystems*. Retrieved from http://e-drexler.com/d/07/00/1204TechnologyRoadmap.html

Allende, J., Babul, J., Martínez, S., & Ureta, T. (2005). Publicaciones y patentes. In

Allende, S., Escrig, J., Altbir, D., Salcedo, E., & Bahiana, M. (2008). Angular dependence of the transverse and vortex modes in magnetic nanotubes. *The European Physical Journal B, 66*(1), 37–40. doi:10.1140/epjb/e2008-00385-4

Alliance of Social and Ecological Consumer Organisations (ASECO). (2008). ASECO opinion on nanotechnology, Copenhagen, Denmark, 2008

Allision, J. R., & Lemley, M. A. (2002). The growing complexity of the United States patent system. *Boston University Law Review. Boston University. School of Law, 82*(77).

Altamirano Cabrera, J., Finus, M., & Dellink, R. (2008). Do abatement quotas lead to more successful climate coalitions? *The Manchester School, 76*(104).

Al-Thawaad, R. M. (2008). *Technology transfer and sustainability adapting factors: Culture, physical environment and geographic location*. Paper 152, Session IT 305.

Alting, L., Kimura, F., Hansen, H. N., & Bissacco, G. (2003). Micro engineering. *Annals of the CIRP, 52*(2), 635–657. doi:10.1016/S0007-8506(07)60208-X

Aly, A. H. M. (1994). Economic Cooperation in Africa . In *Search of Direction*. Boulder, CO: Lynne Rienner Publishers.

Amaïzo, Y. E. (1998). From Dependency to Interdependency. Globalization and marginalization: any chance for Africa. Paris, France: L'Harmattan Publisher.

Amaïzo, Y. E. (2004). *Global Value Chains and Production Networks: Promoting Capability formation in South Africa. Discussion Paper*. Dept of Trade and Industry, Ministry of Industry, South Africa: Competitiveness Conference in cooperation with UNIDO. Pretoria, South Africa, 7-11 June. Retrieved July 2, 2009, from http://www.thedti.gov.za/invitations/unidoinvitation.htm

Amaïzo, Y. E., Atieno, R., McCormick, D., & Onjala, J. (2004). *African Productive capacity Initiative: From vision to action*. Vienna, Austria: UNIDO Policy Papers.

Amsden, A. (2001). *The Rise of the Rest: Challenges to the West from Late-Industrializing Economies*. New York: Oxford University Press.

Análisis y proyecciones de la ciencia en Chile (pp. 75-107). Santiago de Chile, Chile: Académia Chilena de Ciencias.

Andersen, B. (2002). *The performance of the IPR system in the new economy: implications for digital inventions and business methods.* Copenhagen, Denmark: The Standing Committees of Research and Industrial Development of the Danish Parliament. Retrieved June 17, 2009, from http://www.druid.dk/conferences/summer2002/Papers/ANDERSEN.pdf

Andersen, M. M. (2005). Path Creation in the Making – the Case of Nanotechnology. In *DRUID 2005 Conference*.

Anderson, K., & Nielsen, C. (2002). *Economic effects of agricultural biotechnology research in the presence of price-distorting policies.* Discussion Paper No. 0232, Centre for International Economic Studies, University of Adelaide.

Ansuategi, A., & Escapa, M. (2002). Economic growth and greenhouse gas emissions. *Ecological Economics, 40*, 23–37. doi:10.1016/S0921-8009(01)00272-5

Anyansi-Archibong, C. (1995). Planning. In *Developing Countries*. Chicago, IL: The Planning Forum.

Anyansi-Archibong, C. (2005). *Role of Women in the Socio-Economic development of the Continent African Forum. Global Leadership Institute, UVI-St.* USA: Thomas.

Anyansi-Archibong, C. (2006). *Entrepreneurship as the Missing Factor in the Economic Development of Developing Nations, Global Entrepreneurship Monitor.* New Zealand: GEM.

Argentine Center for Scientific Information and Technology. (2008 January). *Nanotechnology: Recent trends and scientific research and technological development.* Buenos Aires, Argentina: CAICYT & MINCYT.

Argentine Center for Scientific Information and Technology. (2009 April). *Boletin Estadistico Tecnologico, No. 3.*

Armstrong, T. O. (2008). Nanomics: The Economics Of Nanotechnology And The Pennsylvania Initiative For Nanotechnology. *Pennsylvania Economic Review, 16*(1).

ASTM. (2006). *American Society for Testing and Materials Publications.* Retrieved from http://www.astm.org/

Atkinson, A. B. (1970). On the measurement of inequality. *Journal of Economic Theory, 2*, 244–263. doi:10.1016/0022-0531(70)90039-6

Audebert, F. (2005). *Amorphous and Nanostructured Al-Fe and Al-Ni Based Alloys. NATO Science Series II: Mathematics, Physics and Chemistry, 184. Properties and Applications of Nano-crystalline Alloys from Amorphous Precursors* (Idzikowski, B., Švec, P., & Miglierini, M., Eds.). *Vol. 301*). Dordrecht: Kluwer Acad. Publishers.

Audebert, F., Arcondo, B., Rodríguez, D., & Sirkin, H. (2001). Short range order study in. Al-Fe-X melt spun alloys. *Journal of Metastable and Nanocrystalline Materials, 10*, 155. doi:10.4028/www.scientific.net/JMNM.10.155

Audebert, F., Prima, F., Galano, M., Tomut, M., Warren, P., Stone, I.C., & Cantor, B. (2002). Bulk Amorphous, Nano-Crystalline and Nano-Quasicrystalline *Alloys. Special issue materials Transactions JIM*, 43-8, 2017.

Audretsch, D. (1995). *Innovation and industry evolution.* Cambridge, MA: MIT Press.

Australian Vice-Chancellors Committee-AVCC. (2005 December). *Engagement with Business and Community: Enhancing Universities' Interaction.*

Aydinonat, N. E. (2008). The Invisible Hand in Economics: How Economists Explain Unintended Social Consequences. New-York: Routledge.

Aydoğan, N. (2009). *Innovation Policies, Business Creation and Economic Development: A Comparative Approach.* New York: Springer. doi:10.1007/978-0-387-79976-6

Aydoğan, N., & Chen, Y. P. (2008). *Social Capital and Business Development in High-technology Clusters: An Analysis of Contemporary U.S. Agglomerations.* New York: Springer. doi:10.1007/978-0-387-71911-5

Ayichi, D. (1995). Agricultural Technology Transfer for Sustainable Rural Development in Nigeria . In Eboh, E. C., Okoye, C., & Ayichi, D. (Eds.), *Rural Development in Nigeria: Concepts, Process and Prospects* (pp. 126–134). Enugu, Nigeria: Auto Century Publishing Company.

Ayogu, M. (2007). Infrastructure and economic development in Africa: A review. *Journal of African Economies, 16*(1), 75–126. doi:10.1093/jae/ejm024

Ayoola, G. B. (2001). *Essays on the Agricultural Economy: A Book off Readings on Agricultural Development Policy and Administration in Nigeria*. Ibadan, Nigeria: T. M. A. Publishers and Farm and Infrastructure Foundation.

AZoNano. (2009). *Nanotechnology Led Changes To Manufacturing, Defence, Farming, Human Development and the Possibility of Large Scale Social Disruption As Predicted By The Friends of The Earth*. Retrieved from http://www.azonano.com/Details.asp?ArticleID=1876

Babini, N., & De Asua, M. (2003, Vol. 26, N°56). La Historia de la Ciencia en Argentina en el último cuarto de siglo. *Revista de la Sociedad Española de Historia de las Ciencias y de las Técnicas,* ISSN 0210-8615, 731-738.

Baert, K., Van Kerschaver, E., & Poortmans, J. (2009). Crystalline Si solar cells and the microelectronics experience. *Photovoltaics World Magazine for Solar Power.* Renewable Energy World Network. Retrieved from http://www.electroiq.com/index/display/photovoltaics-article-display/4711974914/articles/Photovoltaics-World/silicon-photovoltaics/crystalline-silicon/2009/08/crystalline-si_solar.html

Balassa, B. (1982). *Development Strategies in Semi-industrial Economies*. Baltimore, MD: Johns Hopkins University Press.

Balassa, B. (1985). Export policy choices and economic growth in developing countries after the oil shock. *Journal of Development Economics, 18*, 23–35. doi:10.1016/0304-3878(85)90004-5

Balbus, J., Denison, R., Florini, K., & Walsh, S. (2005). Getting nanotechnology right the first time. *Issues in Science and Technology*, 65–71.

Baldi, P., & Brunak, S. (2001). *Bioinformatics: The Machine Learning Approach* (2nd ed.). Cambridge, MA: MIT Press.

Barao, S. M. (2005). Behavioural aspects of technology adoption. *Journal of Extension, 43*(2). Retrieved April 14, 2005, from http://www.joe.org/joe/2005april/comm1.shtml

Barnes, D., Openshaw, K., Smith, K., & van der Plas, R. (1993). The design and diffusion of improved cooking stoves. *The World Bank Research Observer, 8*, 119–141. doi:10.1093/wbro/8.2.119

Baron, R., Buchner, B., & Ellis, J. (2009). *Sectoral Approaches and the Carbon Market*. Paris: OECD.

Barro, R. J., & Lee, J.-W. (2000). International Data on Education Attainment: Updates and Implications. *NBER Working Paper 7911*. Cambridge, MA: National Bureau of Economic Research.

Barro, R., & Sala Martin, X. (1995). *Economic growth*. New York: McGraw-Hill.

Bass, F. M. (1969). A new product growth model for consumer durables. *Management Science, 15*, 215–217. doi:10.1287/mnsc.15.5.215

Baum, G., Nemirovsky, A., & Sabelli, N. (2008). La Educación en Ciencia y Tecnología como derecho social en la economía del conocimiento. pp. 133-146 in Propuestas interpretativas para una economía basada en el conocimiento. Argentina, Colombia, México, Estados Unidos, Canadá, Federico Stezano y Gabriel Vélez Cuartas, Editors. 1st Ed. January 2008. ISBN 978-84-96571-58-7

Baum, R., Drexler, K. E., & Smalley, R. (2003). Nanotechnology: Drexler and Smalley make the case for and against molecular assemblers. *Chemical and Engineering News, 81*(48), 37–42.

Beacham, J. (2006). *Succeeding through innovation: 60 minute guide to innovation, turning ideas into profit*. TSO.

Bell, M. (1987). *The acquisition of imported technology for industrial development: Problems of strategies and management in Arab region*. Baghdad: ESCWA, United Nations University.

Bell, T. E. (2007). *Understanding Risk Assessment of Nanotechnology.* Retrieved August 12, 2009, from http://www.nano.gov/Understanding_Risk_Assessment.pdf

Bello, D., Wardle, B. L., Yamamoto, N., deVilloria, R. G., Garcia, E. J., & Hart, A. J. (2009). Exposure to nanoscale particles and fibers during machining of hybrid advanced composites containing carbon nanotubes. *Journal of Nanoparticle Research, 11*, 231–249. doi:10.1007/s11051-008-9499-4

Berkun, S. (2005). Making Things Happen. Mastering Project Management: Mastering Project Management (Theory in Practice) (Rev. Ed.). Sebastopol, CA: O'Reilly Media, Inc.

Bessen, J. (2003). Patent Thickets: Strategic Patenting of Complex Technologies (ROI Working Paper). *Social Science Research Network Database.* Retrieved July 4, 2009, from http://papers.ssrn.com/sol3/papers.cfm?abstract_id=327760

Bessen, J., & Maskin, E. S. (2005). Geistiges Eigentum im Internet: Ist alte weishit ewig gültig? [Intellectual property on the Internet: What's wrong with the conventional wisdom?] In Bernd, L., Gehring, R. A., & Bärwolff, M. (Eds.), *Open Source Jahrbuch 2005: Zwischen Softwareentwicklung und Gesellschaftsmodell.* Berlin: Lehmanns Media.

Bhagwati, J. N. (1978). *Foreign Trade Regimes and Economic Development: Anatomy of Consequences of Exchange Control Regimes.* Cambridge, MA: Ballinger.

Bhagwati, J., & Srinivasan, T. N. (2002, May). Trade and poverty in the poor countries. *The American Economic Review, 92*(2), 180–183. doi:10.1257/000282802320189212

Bhat, P. R., Chanakya, H. N., & Ravindranath, N. H. (2001). Biogas plant dissemination: success story of Sirsi, India. *Energy for Sustainable Development, 5*(1), 39–46. doi:10.1016/S0973-0826(09)60019-3

Bikson, T. K., Law, S. A., Markovich, M., & Harder, B. T. (1996). *NCHRP Report 382: Facilitating the Implementation of Research Findings: A Summary Report.* Washington, DC: Transportation Research Board, National Research Council.

Binswanger, H. P., & Ruttan, V. W. (1978). *Induced innovation. Technology, institutions and development.* Baltimore: The Johns Hopkins University Press.

Blalock, D. J. (2004). Commercializing USDA innovations via public-private partnerships. Presentation to the agricultural Biotechnology Research Advisory Committee, Washington, DC.

Blanchard, O., Criqui, P., & Kitous, A. (2002). *After The Hague, Bonn and Marrakech: The Future International Market for Emissions Permits and the Issue of Hot Air. Cahier de Recherche no. 27 bis.* Grenoble: Institut d'Economie et de Politique de l'Energie.

Böhringer, C., & Löschel, A. (2003). Climate policy beyond Kyoto: Quo vadis? A computable general equilibrium analysis based on expert judgements. *Kyklos, 58*, 467–493. doi:10.1111/j.0023-5962.2005.00298.x

Boldrin, M., & Levine, D. K. (2008). Perfectly competitive innovation. *Journal of Monetary Economics, 55*(3), 435–453. doi:10.1016/j.jmoneco.2008.01.008

Bonaccorsi, A., & Daraio, C. (Eds.). (2007). *Universities and Strategic Knowledge Creation: Specialization and Performance in Europe.* PRIME Series on Research and Innovation Policy in Europe.

Bonny, B. P., & Vijayaragavan, K. (2001). Adoption of sustainable agricultural practices by traditional rice growers. *Journal of Tropical Agriculture, 39*, 151–156.

Borges da Cunha, K., Walter, A., & Rei, F. (2007). CDM implementation in Brazil's rural and isolated regions: the Amazonian case. *Climatic Change, 84*(1–2), 111–129. doi:10.1007/s10584-007-9272-1

Bosetti, V., Carraro, C., & Galeotti, M. (2006). The Dynamics of Carbon and Energy Intensity in a Model of Endogenous Technical Change. The Energy Journal, Special Ed.: Endogenous Technological Change and the Economics of Atmospheric Stabilisation.

Bosetti, V., Carraro, C., Massetti, E., & Tavoni, M. (2007). International energy R&D spillovers and the economics of greenhouse gas atmospheric stabilisation. *Energy Economics,* 2912–2929.

Boston Consulting Group and National Association of Manufacturers. (2009 March). *The Innovation Imperative in Manufacturing: How the United States Can Restore Its Edge*. Boston, USA.

Bowman, D. M., & Hodge, G. A. (2007). A small matter of regulation: An international review of nanotechnology regulation. *Columbia Science & Technology Law Review, 8*, 1–32.

Bozeman, B., Hardin, J., & Link, A. N. (2008). Barriers to the Diffusion of Nanotechnology. *Economics of Innovation and New Technology, 17*(7&8), 749–761. doi:10.1080/10438590701785819

Bozeman, B., Papadakis, M., & Coker, K. (1995). *Industry perspectives on commercial Interactions with Federal Laboratories*. Report to the National Science Foundation, Research on Science and Technology Program, Contract No. 9220125.

Bresnahan, T. F., & Tajtenberg, M. (1995). General purpose technologies: Engines of growth. *Journal of Econometrics, 65*, 83–108. doi:10.1016/0304-4076(94)01598-T

Brinkerhoff, J. M. (2006a). Diasporas, Mobilization Factors, and Policy Options . In Wescott, C., & Brinkerhoff, J. (Eds.), *Converting Migration Drains into Gains: Harnessing the Resources of Overseas Professionals* (pp. 127–153). Manila, Philippines: Asian Development Bank.

Brinkerhoff, J. M. (2006b). Diasporas, Skills Transfer, and Remittances: Evolving Perceptions and Potential . In Wescott, C., & Brinkerhoff, J. (Eds.), *Converting Migration Drains into Gains: Harnessing the Resources of Overseas Professionals* (pp. 1–32). Manila, Philippines: Asian Development Bank.

Brinkerhoff, J. M., Smith, S. C., & Teegen, H. (2007). Beyond the Non: The Strategic Space for NGOs in Development . In Brinkerhoff, J. M., Smith, S. C., & Teegen, H. (Eds.), *NGOs and the Millennium Development Goals, Citizen Action to Reduce Poverty* (pp. 53–79). New York: Palgrave - Macmillan. doi:10.1057/9780230604933

Brinkman, H.-J. (1992). *Economic Performance, Exogenous Factors, and Adjustment Policies in Africa*. Paper presented at the African Finance and Economics Association at Howard University, Washington, D.C.

Brinksmeier, E., Riemer, O., & Stern, R. (2001). Machining of precision parts and microstructures. In *Proceedings of the 10th International Conference on Precision Engineering*, Yokohama, Japan (pp. 3-11).

Brown, H. I. (1977). *Perception, Theory and Commitment: The New Philosophy of Science*. Chicago: Precedent Publishing.

Brown, J., Vigneri, M., & Sosis, K. (2009). *Innovative Carbon-Based Funding for Adaptation*. Overseas Development Institute. Retrieved September 9, 2009, from http://www.odi.org.uk/ccef/resources/reports/s0198_oecd_adaptation.pdf

Brunner, J. (2001). Factores comunes o transversales . In *Chile: Informe sobre capacidad tecnológica* (pp. 6–13). Santiago de Chile, Chile: Universidad Adolfo Ibáñez.

Bunders, J., Haverkort, W., & Hiemstra, W. (1996). *Biotechnology: Building on Farmer's Knowledge*. London: Macmillan Education Ltd.

Buonanno, P., Carraro, C., & Galeotti, M. (2003). Endogenous induced technical change and the costs of Kyoto. *Resource and Energy Economics, 25*, 11–34. doi:10.1016/S0928-7655(02)00015-5

Burns, T. (1969). Models, images and myths . In Gruber, W. H., & Marquis, D. G. (Eds.), *Factors in Transfer of Technology* (pp. 11–23). Cambridge, MA: MIT Press.

Business Review Weekly (BRW). (2005, August 4). Big Hopes for Tiny Floats. *BRW*, 24.

Buttel, F. H., Kenney, M., & Kloppenburg, J. Jr. (1985). From green revolution to biorevolution: some observations on the changing technological bases of economic transformation in the Third World. *Economic Development and Cultural Change, 34*(1), 31–55. doi:10.1086/451508

Buzea, C. (2007). Nanomaterials and nanoparticles: Sources and toxicity. *Biointerphases, 2*(4), MR17–MR71. doi:10.1116/1.2815690

Byerlee, D., & Fischer, T. (2001). Accessing modern science: policy and institutional options for agricultural biotechnology in developing countries. *IP Strategy Today,1*. Retrieved June 12, 2009, from http://www.biodevelopments.org

Callaghy, T. M. (1990). Lost Between State and Market: The Politics of Economic Adjustment in Ghana, Zambia and Nigeria. In Nelson, J. M. (Ed.), *Economic Crisis and Policy Choices* (pp. 257–319). Princeton, NJ: Princeton University Press.

Callaghy, T. M., & Ravenhill, J. (1993). Towards State Capability and Unbedded Liberalism in the Third World: Lessons for Adjustment. In Nelson, J. M. (Ed.), *Fragile Coalitions: The Policies of Economic Adjustment*. Oxford, UK: Transaction Books.

Callon, M. (1992). The Dynamics of Techno-Economic Networks. In Coombs, R., Saviotti, P., & Walsh, V. (Eds.), *Technical Change and Company Strategies* (pp. 72–102). London: Academic Press.

Calvert, J., & Patel, P. (2002). *University-Industry Research Collaborations in the UK*. SPRU working paper, University of Sussex, Brighton, UK.

Caneschi, A., Gatteschi, D., Lalioti, N., Sangregorio, C., Sessoli, R., Venturi, G., et al. (2001). Cobalt (II)-nitronyl nitroxide chains as molecular magnetic nanowires. Angewandte Chemie-International Ed., 40(9), 1760-1763.

Caniëls, M., & Verspagen, B. (2001). Barriers to knowledge spillovers and regional convergence in an evolutionary model. *Journal of Evolutionary Economics, 11*, 307–329. doi:10.1007/s001910100085

Cannady, C. (2009). *Access to Climate Change Technology by Developing Countries: A Practical Strategy. ICTSD's Programme on IPRs and Sustainable Development, Issue Paper No. 25*. Geneva, Switzerland: International Centre for Trade and Sustainable Development.

Cantore, N. (2009). International spillovers and learning by doing in a regionalised model of climate change: a post-Kyoto analysis. In Marques, H., Soukiazis, E., & Cerqueira, P. (Eds.), *Integration and globalisation: challenges for developed and developing countries* (pp. 125–142). Cheltenham, UK: Edward Elgar Publishing.

Cantore, N., & Canavari, M. (in press). Reconsidering the Environmental Kuznets Curve Hypothesis: the trade off between environment and welfare. In Montini, A., & Mazzanti, M. (Eds.), *Environmental efficiency, innovation and economic performance*. New York: Routledge.

Cantore, N., & Padilla, E. (2007). *Equity and emissions concentration in climate change integrated assessment modelling*. Dei Agra WP-03-2007 and Working Paper 07-05, Department of Applied Economics, Univ. Autónoma de Barcelona.

Cantore, N., Canavari, M., & Pignatti, E. (2008). *International distribution of CO_2 emissions according to the climate change integrated assessment models*. Paper presented at the AISSA (Congress of the Italian Societies of the Scientific Agricultural Associations), 26–28 November, Imola, Italy.

Cantwell, J., & Molero, J. (Eds.). (2003). *Multinational Enterprises, Innovative Strategies and Systems of Innovation*. Cheltenham, UK: Edward Elgar Publishing Limited.

Caram, J., Sandoval, C., Tirado, M., Comedi, Czaban, J., Thompson, D.A., et al. (2009). Electrical characteristics of core-shell p-n GaAs nanowire structures with Te as the n-dopant. Submitted to *Nanotechnology*.

Carlsson, B. (2006). Internationalization of innovation systems: A survey of the literature. *Research Policy, 35*, 56–67. doi:10.1016/j.respol.2005.08.003

Carraro, C., Eyckmans, J., & Finus, M. (2006). Optimal transfers and participation decisions in international environmental agreements. *The Review of International Organizations, 1*, 379–396. doi:10.1007/s11558-006-0162-5

Catalog, I. N. N. O. V. A. R. (2008). 4rd Ed. of the National Innovation Competition. Ministry of Science, Technology and Productive Innovation, Argentina.

Centre for Energy Economics. (2006). Gas Monetization in Nigeria. In Directory of the U.S. Photovoltaics Industry 1996. Washington, DC: Solar Energy Industries Association.

Centre for Nanotechnology and Society. (2008). *Nanotechnology and Society*. Santa Barbara, CA: University of California. Retrieved from http://www.cns.ucsb.edu/nanotechnology-society

Chandra, V., & Kolavalli, S. (2006). Technology, Adaptation, and Exports: How Some Developing Countries Got It Right: How Some Countries Got It Right . In Chandra, V. (Ed.), *Technology, Adaptation, and Exports* (pp. 1–47). Washington, DC: World Bank Publications. doi:10.1596/978-0-8213-6507-6

Chang, H.-J. (2002). *Kicking away the ladder. Development Strategy in Historical Perspective*. London: Anthem Press.

Chang, H.-J. (2010). *Market as means rather than master: Towards new developmentalism*. London: Routledge Ed.

Chazan, N. (1992). *Politics and Society in Contemporary Africa*. Boulder, CO: Lynne Rienner Publishers.

Chen, H., Roco, M. C., Li, X., & Lin, Y. (2008). Trends in nanotechnology patents. *Nature Nanotechnology, 3*. Retrieved June 20, 2009 from http://mis.eller.arizona.edu/docs/news/2008/nature_nano_42008.pdf

Chesbrough, H. (2003). *Open Innovation: the new imperative for creating and profiting from technology*. Cambridge, MA: Harvard Business School Press.

Chesbrough, H., & Teece, D. J. (1996). When is virtual virtuous? Organizing for innovation. *Harvard Business Review, 74*(1), 65–73.

Chinn, M. D., & Fairlie, R. W. (2006 December). The determinants of the global digital divide: a cross-country analysis of computer and internet penetration. *Oxford Economic Papers*.

Coe, D. T. (1997). North-South R&D spillovers. The Economic Journal, 107(134).

Cohen, G. (2004). *Technology Transfer. Strategic management in Developing countries* (1st ed.). New Delhi, India: SAGE Publications Pvt. Ltd.

Cohen, W. M., & Levinthal, D. A. (1990). Absorptive capacity: A new perspective on learning and innovation. *Administrative Science Quarterly, 35*, 128–152. doi:10.2307/2393553

Collier, P. (2007). The Bottom Billion: Why the Poorest Countries are Failing and What Can Be Done About It (1st Ed.). New-York: Oxford University Press.

Colvin, V. (2003). The potential environmental impact of engineered nanomaterials. *Nature Biotechnology, 21*(10), 1166–1170. doi:10.1038/nbt875

Combs, K. L. (1992). Cost sharing vs. multiple research projects in cooperative R&D. *Economics Letters, 39*, 353–371. doi:10.1016/0165-1765(92)90273-2

Comedi, D., Tirado, M., Zapata, C., Heluani, S. P., Villafuerte, M., Mohseni, P., et al. (2009). Randomly Oriented ZnO Nanowires Grown on Amorphous SiO_2 by Metal-Catalized Vapour Deposition. Submitted to *Journal of Alloys and Compounds*.

Consejo Nacional de Ciencia y Tecnología México. (2008, December 16). PECyTI 2007-2012. *Official periodical*. Retrieved from http://74.125.155.132/search?q=cache:9ggPNl8FioMJ:www.siicyt.gob.mx/siicyt/docs/contenido/PECYT.pdf+pecyt+2007-2012&cd=2&hl=es&ct=clnk&gl=mx

Consejo Nacional de Investigaciones Científicas y Técnicas Argentina. (2009). II Nanomercosur Meeting: Opportunities for micro- and Nanotechnology. *CONICET's catalog on Nanoscience & Nanotechnology Institutes*.

Conway, G. (1997). *The Doubly Green Revolution*. London: Penguin Books.

Conway, G. (2000). *Crop biotechnology: benefits, risks and ownership*. Rockefeller Foundation, New York (. *Foundation News*, 03062000.

Cook, P. (2005). Regionally asymmetric knowledge capabilities and open innovation . *Research Policy, 34*, 1128–1149. doi:10.1016/j.respol.2004.12.005

Coombs, R., Harvey, M., & Tether, B. (2003). Analysing distributed processes of provision and innovation. *Industrial and Corporate Change, 12*(6), 1125–1155. doi:10.1093/icc/12.6.1125

Coondoo, D., & Dinda, S. (2007). Carbon dioxide emission and income: a temporal analysis of cross-sectional distributional patterns. *Ecological Economics, 65*, 375–385. doi:10.1016/j.ecolecon.2007.07.001

Cooper, R. G. (2001). *Winning at New Products*. Cambridge, MA: Perseus Books.

Córdova, V. S. (2009). Multicultural and Creative On-Line Learning. *International Journal of Education and Development using ICT, 5*(2), 1-10.

Corley, E. A., Scheufele, D. A., & Hu, Q. (2009). Of Risks and Regulations: How Leading U.S. Nanoscientists Form Policy Stances about Nanotechnology. *Journal of Nanoparticle Research.* Retrieved from DOI 10.1007/s11051-009-9671-5

CortesA.RiverosG.PalmaJ.DenardinJ.MarottiR.DalchieleE. (2009).

Cosbey, A., Parry, J., & Browne, J. Babu, Y., Bhandari, P., Drexhage, J., & Murphy, D. (2005). *Realising the Development Dividend: Making the CDM Work for Developing Countries.* Report of the International Institute for Sustainable Development IISD, Canada. Retrieved September 3, 2009, from http://www.iisd.org/ publications/ pub.aspx? id=694

Crespilho, F. N., Pavinatto, F. J., Zucolotto, V., Avansi, W., Barioto, V., Gasparotto, L. H. et al. (2008). *Processo de obtenção de um produto a base nanopartículas metálicas e polímeros para tecidos autolimpantes e auto-esterilizantes e produtos resultantes.* Patented in Brazil, PI 0.802.649-1.

Criscuolo, P., & Narula, R. (2001). *A novel approach to national technological accumulation and absorptive capacity: Aggregating Cohen and Levinthal.* Oslo, Norway: University of Oslo.

CRISP/OST Foresight Briefing Paper. (2001). *Nanotechnology: engineering with atoms. Presented in Foresight/CRISP workshop on nanotechnology: What is nanotechnology? What are its implications for construction.* London: The Royal Society of Arts.

Custer, M. (2005 June). Investing in Nanotechnology is its Own Science. *Bulletin Special, Credit Suisse*, 32-33.

Cypher, J. M., & Dietz, J. L. (2004). *The Process of Economic Development* (2nd ed.). London: Routledge.

d'Aspremont, C., & Jacquemin, A. (1988). Cooperative and non-cooperative R&D in duopoly with spillovers. *The American Economic Review, 78*, 1133–1137.

D'Silva, J. (2008). Pools, Thickets and Open Source Nanotechnology. *Social Science Research Network Database.* Retrieved June 4, 2009, from http://ssrn.com/abstract=1368389

Da Róz, A. L., Pereiro, L. V., Pavinatto, F. J., Crespilho, F. N., Zucolotto, V., Carvalho, A. J. F., et al. (2008). *Produto a base de quitosana e processo de impregnação do mesmo em têxteis.* Patented in Brazil, PI 0.802.290-9.

Dabic, M. (2007). *Gaining from Partnership: Transfer Technology - Issues and Challenges in Transitional Economies.* Retrieved from http://inderscience.blogspot.com/2007/10/call-for-papers-gaining-from.html

Daily, S. (2006). *Nanotechnology? What's That?! Engineers Create Exhibits on Achievements, Promise.* Retrieved August 20, 2009, from http://www.sciencedaily.com/videos/2006/0611-nanotechnology_whats_that.htm

Dandolopartners, 2005a. *Nanotechnology and the Business Community: A study of business' understanding of and attitudes towards nanotechnology.* Report for Nanotechnology Victoria. 11 July 2005.

Dandolopartners. (2005b). *Nanotechnology: National Business Interviews.* Detailed Report. Report commissioned by Nanovic and DITR.

Dandolopartners. (2005c). *Nanotechnology: A National Survey of Consumers.* Detailed Report. 11 July 2005.

Daneke, G. A. (1998). Beyond schumpeter: Nonlinear economics and the evolution of the U.S. innovation system. *Journal of Socio-Economics, 27*(1), 97–115. doi:10.1016/S1053-5357(99)80079-1

David, P. A. (2003). The economic logic of 'open science' and the balance between private property rights and the public domain in scientific data and information: a primer. In J. Esanu & P. F. Uhlir (Eds.), *The Role of the Public Domain in Scientific and Technical Data and Information: Proceedings of a Symposium.* Washington, DC: National Academies Press.

Day, R. K. (2003). Transferring public research: The patent licensing mechanism in agriculture. *The Journal of Technology Transfer, 28*(2), 111–130. doi:10.1023/A:1022934330322 de Gouvello, C., Dayo, F., & Thioye, M. (2008). *Low carbon energy projects for development in Sub-Saharan Africa*. Washington, DC: World Bank. Retrieved September 9, 2009, from http://wbcarbonfinance.org/docs/Main_Report_Low_Carbon_Energy_projects_for_Development_of_Sub_Saharan_Africa_8-18-08.pdf

De Janvry, A. (1973). A socioeconomic model of induced innovations for Argentine agricultural development. *The Quarterly Journal of Economics*, *87*(3), 410–435. doi:10.2307/1882013

de Janvry, A., Graff, G., Sadoulet, E., & Zilberman, D. (2000). *Technological Change in Agriculture and Poverty Reduction*. Concept paper for the World Development Report on Poverty and Development 2000/01University of California at Berkeley.

De Reca, W. (2007). Anales de la Academia Nacional de Ciencias Exactas, Físicas y Naturales. *Nanostructured Materials: Synthesis, Characterization . Properties and Applications*, *59*, 59–94.

De Walt, B. R. (1978). Appropriate technology in rural Mexico: Antecedents and consequences of an indigenous peasant innovation. *Technology and Culture*, *19*(1), 32–52. doi:10.2307/3103307

DeCourcy, J. (2001). *International trade policy and the National Cooperative Research Act*. PhD dissertation, Michigan State University.

Deheer, W. A., Chatelain, A., & Ugarte, D. (1995). A carbon nanotube field-emission electron. *Science*, *270*(5239), 1179–1180. doi:10.1126/science.270.5239.1179

Delgado, L. C., Hopkins, J., & Kelly, V. A. (1998). *Agricultural growth linkages in Sub-Saharan Africa*. IFPRI Research Report No. 107. Washington, DC: International Food Policy Research Institute.

Demou, E., Peter, P., & Hellweg, S. (2008). Exposure to manufactured nanostructured particles in an industrial pilot plant. *The Annals of Occupational Hygiene*, *52*(8), 695–706. doi:10.1093/annhyg/men058

DESERTEC. (2009). *Clean Power from Deserts, The DESERTEC Concept for Energy, Water and Climate Security* (4th Ed.). Retrieved from http://www.desertec.org

Dienwiebel, M. (2004). Superlubricity of Graphite. *Physical Review Letters*, *92*(12). doi:10.1103/PhysRevLett.92.126101

Dimov, S. S., Matthews, C. W., Glanfield, A., & Dorrington, P. (2006). A roadmapping study in multi-material micro manufacture. In *Proceedings of the Second International Conference on Multi-Material Micro Manufacture*, Grenoble, France (pp. 11-25).

Dinar, A., & Yaron, D. (1992). Adoption and abandonment of irrigation technologies. *Agricultural Economics*, *6*, 315–332. doi:10.1016/0169-5150(92)90008-M

Dixon, R. (1980). Hybrid corn revisited. *Econometrica*, *48*, 1451–1461. doi:10.2307/1912817

Doessel, D. P., & Strong, S. M. (1991). A neglected problem in the analysis of the diffusion process. *Applied Economics*, *23*, 1335–1340. doi:10.1080/00036849100000054

Donald, A. (1999). Political economy of technology transfer. *British Medical Journal*, *319*(7220).

Dőry, T. (2005). *Regionális innováció-politika. Kihívások az Európai Unióban és Magyarországon*. Budapest, Pécs: Dialóg Campus.

Dosi, G. (1982). Technological paradigms and technological trajectories. *Research Policy*, *11*, 147–162. doi:10.1016/0048-7333(82)90016-6

Dosi, G. (1988). The Nature of the innovation process . In Dosi, G., Freeman, C., Nelson, R., Silverberg, G., & Soete, L. (Eds.), *Technical Change and Economic Theory*. London: Pinter.

Doss, C. R., & Morris, M. L. (2001). How does gender affect the adoption of agricultural innovations? The case of improved maize technology in Ghana. *Agricultural Economics*, *25*(1), 32.

Douglas, S. M. (2009, May 21). Self-assembly of DNA into nanoscale three-dimensional shapes. *Nature*, *459*, 414–418. doi:10.1038/nature08016

Douthwaite, B., Keatinge, J. D. H., & Park, J. R. (2001). *Learning Selection: An Evolutionary Model for Understanding, Implementing and Evaluating Participatory Technology Development.* UK: Department of Agriculture, University of Reading.

Dresselhaus, M. S., Dresselhaus, G., Jorio, A., Souza, A. G., & Saito, R. (2002). Raman spectroscopy on isolated single wall carbon nanotubes. *Carbon, 40*(12), 2043–2061. doi:10.1016/S0008-6223(02)00066-0

Drexler, E. (1992). *Nanosystems: molecular machinery, manufacturing, and computation.* Hoboken, NJ: Wiley.

Drexler, E. (2009). *Toward Advanced Nanotechnology.* Retrieved from http://e-drexler.com/

Drexler, K. E. (1986). *Engines of Creation: The Coming Era of Nanotecnology.* New York: Anchor Books, Doubleday.

Drexler, K. E. (2001). Machine-phase nanotechnology. *Scientific American, 285*(3), 74–75. doi:10.1038/scientificamerican0901-74

Driouchi, A., Azelmad, E., & Anders, G. (2006). An econometric analysis of the role of knowledge in economic performance. *The Journal of Technology Transfer, 31*(2), 241–255. doi:10.1007/s10961-005-6109-9

Dumond, J. E. (1994). Making best use of performance measures and information. *International Journal of Operations & Production Management, 14*(9), 16–31. doi:10.1108/01443579410066712

Dunning, J. H. (2005). The evolving world scenario . In Passow, S., & Runnbeck, M. (Eds.), *What's Next? Strategic Views on Foreign Direct Investment* (pp. 12–17). Stockholm, Sweden: Invest in Sweden Agency.

Dupont, J., Fonseca, G. S., Umpierre, A. P., Fichtner, P. F. P., & Teixeira, S. R. (2002). Transition-metal nanoparticles in imidazolium ionic liquids: recyclable catalysts for biphasic hydrogenation reactions. *Journal of the American Chemical Society, 124*(16), 4228–4229. doi:10.1021/ja025818u

Duro, J. A., & Padilla, E. (2006). International inequalities in per capita CO_2 emissions: a decomposition methodology by Kaya factors. *Energy Economics, 28,* 170–187. doi:10.1016/j.eneco.2005.12.004

Duro, J. A., & Padilla, E. (2008). Analysis of the international distribution of per capita CO_2 emissions using the polarisation concept. *Energy Policy, 36,* 456–466. doi:10.1016/j.enpol.2007.10.002

Duro, J. A., Alcántara, V., & Padilla, E. (2009) *La desigualdad en las intensidades energéticas y la composición de la producción.* Un análisis para los países de la OCDE, Working Paper 09.05, Departamento de Economía Aplicada, Universidad Autónoma de Barcelona.

EBRD. (2005). Transition Report: Business in Transition. London.

Echevarría, R. (1998). Agricultural research policy issues in Latin America: an overview. *World Development, 26*(6), 107.

Economic Commission for Africa and African Union. (2007). *Economic Report on Africa 2007: Accelerating Africa's Development through Diversification.* Addis Ababa, Ethiopia: United Nations Economic Commission for Africa.

Economic Expert. (2009). *Microelectronics.* Retrieved from http://www.economicexpert.com/a/Microelectronics.htm

Edler, J. (2008). Creative internationalization: Widening the perspectives on analysis and policy regarding international R&D activities. *The Journal of Technology Transfer, 33*(4). doi:10.1007/s10961-007-9051-1

Edler, J., Mayer-Krahmer, F., & Reger, G. (2002). Changes in the strategic management of technology: Results of a global benchmarking study. *R & D Management, 32*(2), 149–164. doi:10.1111/1467-9310.00247

Edoho, F. M. (1997). International Technology Transfer in the Emerging Global Order . In Edoho, F. M. (Ed.), *Globalization And the New World Order, Promises, Problems, and Prospects for Africa in the Twenty-First Century* (pp. 99–126). Westport, CT: Praeger Publishers.

Edwards, S. (1993, September). Openness, trade liberalization, and growth in developing countries. *Journal of Economic Literature, 31*, 1358–1393.

Ekanem, N. F. (1997). Economic Instability and Africa's Marginal Role in the New World Order: A Study of Selected African Countries . In Edoho, F. M. (Ed.), *Globalization And the New World Order* (pp. 47–60). Westport, CT: Praeger.

Ekekwe, N. (2002). *Telematics and Internet Banking: Implications, SWOT analyses and Strategies-A case study of Diamond Bank Limited*. MBA Dissertation, University of Calabar, Nigeria.

Ekekwe, N. (2008). Neuromorphs: Replaceable human organs of the future? *IEEE Potentials, 27*(1), 8–25. doi:10.1109/MPOT.2007.913683

Ekekwe, N. (2009). *Towards Competitiveness and Global Outsourcing: Practical Model for Microelectronics Diffusion in Africa*. Paper Presented at International Conference on Industry Growth, Investment, Competitiveness in Africa (IGICA), Abuja, Nigeria, 8-10ᵗʰ June.

Ekekwe, N. (2009b). *Reconfigurable Application-Specific Instrumentation and Control Integrated Systems*. PhD Dissertation, Electrical & Computer Engineering, Johns Hopkins University, Baltimore.

Ekekwe, N. (2009c). Adaptive Application-Specific Instrumentation and Control Integrated Microsystems. Koln, Germany: LAP Academic Publishing.

Ekekwe, N., & Ekenedu, C. (2007), Challenges and Innovations in Microelectronics Education in Developing Nations. In *IEEE International Conference on Microelectronic Systems Education*, San Diego, CA, USA.

Ekekwe, N., & Etienne-Cummings, R. (2006). Power dissipation sources and possible control techniques in ultra deep submicron CMOS technologies. Microelectronics Journal, 37(9), 851–860. doi:10.1016/j.mejo.2006.03.008doi:10.1016/j.mejo.2006.03.008

Ekong, E. E. (2003). *Rural Sociology an Introduction and Analysis of Rural Nigeria*. Uyo, Nigeria: Dove Educational Publishers.

Ellis, F. (2007). *Small-Farms, Livelihood diversification, and Rural-Urban Transitions: Strategic Issues in Sub-Saharan Africa*. Paper presented at the future of small farms workshop, June 26, Wye, Kent, UK.

Ellis, J. (2006). *Issues Related to a Programme of Activities Under the CDM*. Paris: OECD. Retrieved September 9, 2009, from http://www.olis.oecd.org/olis/2006doc.nsf/FREDIRCORPLOOK/NT00000A2E/$FILE/JT03208489.PDF

Energy Commission of Nigeria. ECN (2004), Nigeria Energy Demand and Power Planning Study for the period 2000-2030. Part 1: Energy Demand Projections. Technical Report No ECN/EPA/04/01 Energy Commission of Nigeria, Abuja.

Energy Information Administration. (2004). *Electric Power Annual*. Form EIA-860, Annual Electric Generator Report database.

Engle, R., & Granger, C. (1987). Co-integration and error-correction: Representation, estimation and testing. *Econometrica, 55*, 251–276. doi:10.2307/1913236

Environmental Defense Fund (EDF). (2009). *Nanotechnology and Health: Getting nanotechnology Right*. Unregulated Nanotechnology - Are Your Products Safe.

EPA's Science Policy Council (EPASPC). (2007 February). *Nanotechnology Workgroup, US Environmental Protection Agency Nanotechnology White Paper*. EPA 100/B-07/001.

Erlebach, R. W. (2006). *The Importance of Wage Labor in the Struggle to Escape Poverty: Evidence from Rwanda*. UK: University of London.

Escoba, J., Reardon, T., & Agreda, V. (2000). Endogenous institutional innovation and agro-industrialization on the Peruvian coast. *Agricultural Economics, 23*(3), 267–277. doi:10.1111/j.1574-0862.2000.tb00278.x

Escrig, J., Lavin, R., Palma, J., Denardin, J., Altbir, D., Cortes, A., et al. (2008). Geometry dependence of coercivity in Ni nanowire arrays. *Nanotechnology, 19*(7). Retrieved from DOI: 10.1088/0957-4484/19/7/075713.

Esteban, J., Gradin, C., & Ray, D. (1999). *Extension of a Measure of Polarization, with an Application to the Income Distribution of Five OECD Countries. Papers 24.* El Instituto de Estudios Economicos de Galicia Pedro Barrie de la Maza.

ETC Group. (2004 November). *Down on the Farm. The Impacts of Nano-Scale Technologies on Food and Agriculture.*

Eto, H. (2003). Interdisciplinary information input and output of nano-technology project. *Scientometrics, 58*(1), 5–33. doi:10.1023/A:1025423406643

Etzkowitz, H. (2008). *The Triple Helix: University-Industry-Government Innovation in Action.* London: Routledge. doi:10.4324/9780203929605

Etzkowitz, H., & Leydesdorff, L. (2000). The dynamics of innovation: from national systems and 'mode 2' to a Triple Helix of university-industry-government relations. *Research Policy, 29*(2), 109–123. doi:10.1016/S0048-7333(99)00055-4

Etzkowitz, H., & Leydesdorff, L. (Eds.). (1997). *Universities in the Global Economy: A Triple Helix of University–Industry–Government Relations.* London: Cassell Academic.

EU. (2005). *The Handbook on Responsible Partnering—Joining forces in a world of open innovation. A guide to better practices for collaborative research and knowledge transfer between science and industry.* EUA, ProTon Europe, EARTO and EIRMA. Retrieved from http://www.responsible-partnering.org/library/rp-2005-v1.pdf

European Commission. (2004). *Nanotechnology - Innovation for tomorrow's world.* Luxembourg: European Commission Community Research.

European Union (EU). (2004). Towards a European strategy for nanotechnology. In EU Policy for Nanosciences and Nanotechnologies, Brussels, Belgium.

Evenson, R. E. (2002). From the Green Revolution to the Gene Revolution . In Evenson, R. E., Santaniello, V., & Zilberman, D. (Eds.), *Economic and Social Issues in Agricultural Biotechnology* (pp. 1–16). London: CAB International. doi:10.1079/9780851996189.0001

Evenson, R. E. (2003). *GMOs: Prospects for increased crop productivity in developing countries.* Center Discussion Paper No. 878, Economic Growth Center, Yale University. Retrieved June 12, 2009, from http://ssrn.com/abstract=487503

Evenson, R. E., & Gollin, D. (2003). Assessing the impact of the Green Revolution: 1960-1980. *Science, 300,* 758–762. doi:10.1126/science.1078710

Evenson, R. E., & Rosegrant, M. (2003). The economic consequences of crop genetic improvement programs . In Evenson, R. E., & Gollin, D. (Eds.), *Crop Variety Improvement and Its Effect on Productivity: The Impact of International Agricultural Research* (pp. 473–498). Wallingford, UK: CAB International. doi:10.1079/9780851995496.0473

Eyckmans, J., van Regemorter, D., & van Steenberghe, V. (2001). *Is Kyoto Fatally Flawed?* Working Paper 2001-18, Center for Economic Studies, Katholieke Universiteit Leuven, Leuven, November.

Eze, S. O. (2005). Diffusion and Adoption of Innovation . In Nwachukwu, I., & Onuekwusi, G. C. (Eds.), *Agricultural Extension and Rural Sociology* (p. 249). Enugu, Nigeria: Snaap Press Ltd.

Falk, J. (2007). Transitioning to new technologies: Challenges and choices in a changing world. *Journal of Futures Studies, 12*(2), 69–90.

Faulkner, W., & Senker, J. (1995). *Knowledge Frontiers.* Oxford, UK: Oxford University Press.

Feder, G., & Umali, D. L. (1993). The adoption of agricultural innovations: A review. *Technological Forecasting and Social Change, 43,* 215–239. doi:10.1016/0040-1625(93)90053-A

Felcher, E. M. (2008). *The Consumer Product Safety Commission and Nanotechnology. Project on Emerging Nanotechnologies (PEN) Report.* Retrieved July 10, 2009, from http://www.nanotechproject.org/process/assets/files/7033/pen14.pdf

Ferreira, E. J., Pereira, R. C. T., Delbem, A. C. B., Oliveira, O. N. Jr, & Mattoso, L. H. C. (2007). Random subspace method for analyzing coffee with electronic tongue. *Electronics Letters*, *43*(21), 1138–1140. doi:10.1049/el:20071182

Ferreira, M., Riul, A. Jr, Wohnrath, K., Fonseca, F. J., Oliveira, O. N. Jr, & Mattoso, L. H. C. (2003). High-performance taste sensor made from Langmuir-Blodgett films of conducting polymers and a ruthenium complex. *Analytical Chemistry*, *75*(4), 953–955. doi:10.1021/ac026031p

Feynman, R. (1959). *There's Plenty of Room at the Bottom*. Retrieved from http://www.zyvex.com/nanotech/feynman.html

Feynman, R. P. (1960). *There's Plenty of Room at the Bottom: An Invitation to Enter a New Field of Physics*. Retrieved August 20, 2009, from http://www.zyvex.com/nanotech/feynman.html

Figueres, C. (2006). Sectoral CDM: Opening the CDM to the yet unrealized goal of sustainable development. *International Journal of Sustainable Development Law and Policy*, *2*(1).

Financial Times. (2009, April 29). Climate talks: What India, China and Brazil want. Retrieved September 9, 2009, from http://blogs.ft.com/energy-source/2009/04/29/developing-countries-and-the-climate-negotiations/

Fliegel, F. C. (1984). Extension communication and the adoption process . In Swanson, B. E., Bentz, R. P., & Sofranko, A. J. (Eds.), *Improving Agricultural Extension: A Reference Manual*. Rome: FAO.

FMNT. (2009). *What is microelectronics*. Retrieved from http://www.fmnt.fi/index.pl?id=2408

Foray, D. (2006). *Globalization of R&D: linking better the European economy to 'foreign' sources of knowledge and making EU a more attractive place for R&D investment*. Technical Report, Expert Group 'Knowledge for Growth.'

Förster, W. (2007). Rate, direction, and lifecycle of patenting investments. In *Proceedings from the HEC Paris Symposium Intellectual Property: Les Entretiens de Paris 2007*.

Foster, R. (1986). *Innovation: The attacker's advantage*. London: McMillan.

Fosu, A. K. (1996). Primary exports and economic growth in developing countries. *World Economy*, *19*(4), 465–475. doi:10.1111/j.1467-9701.1996.tb00690.x

Franklin, U. (1989). *Real World of Technology*. Toronto, Canada: House of Anansi Press.

Freeman, C. (1982). Schumpeter or Schmookler? In Freeman, C., Clark, J., & Soete, L. (Eds.), *Unemployment and Technical Innovation*. London: Pinter.

Freeman, C., & Perez, C. (1986). *The Diffusion of Technical Innovations and Changes of Techno-economic Paradigm*. Paper prepared for the Venice Conference, March 1986. Science Policy Research Unit, University of Sussex.

Freeman, C., & Perez, C. (1988). Structural crises of adjustment, business cycles and investment behaviour . In Dosi, G., & Freeman, C. (Eds.), *Technical Change and Economic Theory*. London: Pinter.

Freeman, C., & Soete, L. (1997). *The Economics of Industrial Innovation* (3rd ed.). London: Pinter.

Frenzel, C. W. (1996). *Information Technology Management*. Cambridge, UK: Thomson Publishing Company.

Fukuyama, F. (2002). *Our Post Human Future: Consequences of Biotechnology Revolution*. New York: Farrar, Straus, and Giroux.

Galano, M., Audebert, F., García Escorial, A., Stone, I. C., & Cantor, B. (2009). Nanoquasicrystalline Al-Fe-Cr-based alloys: Part II mechanical properties. *Acta Materialia*, *57*(17), 5120–5130. doi:10.1016/j.actamat.2009.07.009

Galano, M., Audebert, F., Stone, I. C., & Cantor, B. (2008). Effect of Nb on nanoquasicrystalline Al-based alloys. *Philosophical Magazine Letters*, *88*(4), 269. doi:10.1080/09500830801935277

Galano, M., Audebert, F., Stone, I. C., & Cantor, B. (2009). Nanoquasicrystalline Al-Fe-Cr-based alloys: Part I phase transformation process. *Acta Materialia*, *57*(17), 5107–5119. doi:10.1016/j.actamat.2009.07.011

Galembeck, F., & De Brito, J. (2006). *Aluminum phosphate or polyphosphate particles for use as pigments in paints and method of making same.* U.S. Patent 2006045831.

Gallini, N. T., & Winter, R. A. (1985). Licensing in the theory of innovation. *The Rand Journal of Economics, 16*, 237–252. doi:10.2307/2555412

Gans, J., Hsu, D. H., & Stern, S. (2002). When does start-up innovation spur the gale of creative destruction? *The Rand Journal of Economics, 33*(4), 571–586. doi:10.2307/3087475

Gao, X. (2001). A flexible light-directed DNA chip synthesis gated by deprotection using solution photogenerated acids. *Nucleic Acids Research, 29*(22). doi:10.1093/nar/29.22.4744

Garlic, S. (2000). *Engaging Universities and Regions: Knowledge contribution to regional economic development in Australia.* Occasional Paper Series 00/15, Higher Education Division, Department of Education, Training and Youth Affairs, Canberra. Retrieved from http://www.detya.gov.au/highered/eippubs/eip00_15/00_15.pdf

Gassmann, O. (2006). Opening up the innovation process: Towards and agenda. *R & D Management, 36*(3), 223–228. doi:10.1111/j.1467-9310.2006.00437.x

Gassmann, O., & Enkel, E. (2004). Towards a Theory of Open Innovation: Three Core Process Archetypes, In *Proceedings of the R&D Management Conference (RADMA), Sessimbra, Portugal July 8-9, 2004.*

GEF. (2008). Transfer of Environmentally Sound Technologies: The GEF Experience. *Global Environmental Facility.* Retrieved September 9, 2009, from http://thegef.org/uploadedFiles/Publications/GEF_TTbrochure_final-lores.pdf

Genaidy, A., & Waldemar, K. (2008). A roadmap for a methodology to assess, improve and sustain intra-and inter-enterprise system performance with respect to technology-product life cycle in small and medium manufacturers. *Human Factors and Ergonomics in Manufacturing, 18*(1), 70–84. doi:10.1002/hfm.20097

Gerlagh, R. (2006). ITC in a Global Growth-Climate Model with CCS. The Value of Induced Technical Change for Climate Stabilization. *The Energy Journal, Special issue on Induced Technological Change and Climate Change,* 55-72.

Gerlagh, R., & Kuik, O. (2007). *Carbon leakage with international technology spillovers.* Nota di Lavoro FEEM 33.2007.

Gerschenkron, A. (1962). *Economic Backwardness in Historical Perspectives.* Cambridge, MA: Belknap Press.

Ghatak, S. (2001). *An Introduction to Development Economics* (4th ed.). London: Routledge.

Giddens, A. (2003). *The Runaway World- How Globalization is Reshaping Our Lives.* New York: Routledge.

Glänzel, W., Debackere, K., & Meyer, M. (2006). *Triad or Tetrad? On Global Changes in a Dynamic World.* Paper presented at the 9th International Conference on S&T Indicators Leuven (Belgium), September.

Glass, J. C., & Johnson, W. (1989). *Economics: Progression, Stagnation or Degeneration?* Hemel Hempstead, UK: Harvester Wheatsheaf.

Golding, D. (2006). *United Kingdom Technology Strategy Board Annual Report.*

Gore, A. P., & Lavaraj, U. A. (1987). Innovation diffusion in a heterogeneous population. *Technological Forecasting and Social Change, 32*, 163–168. doi:10.1016/0040-1625(87)90037-0

Green, C. J., Kimuyu, P., Manos, R., & Murinde, V. (2007). How do small firms in developing countries raise capital? Evidence from a large-scale survey of Kenyan micro and small-scale enterprises. *Advances in Financial Economics, 12*, 379–404. doi:10.1016/S1569-3732(07)12015-6

Gregg, J. V., Hassell, C. H., & Richardson, J. T. (1964). *Mathematical Trend Curves: An Aid to Forecasting.* Edinburgh, UK: Oliver and Boyd.

Griliches, Z. (1957). Hybrid corn: An exploration in the economics of technological change. *Econometrica, 25*, 501–522. doi:10.2307/1905380

Griliches, Z. (1980). Hybrid corn revisited: A reply. *Econometrica*, *48*, 1451–1461. doi:10.2307/1912818

Grossman, G. M., & Krueger, A. B. (1991). *Environmental impacts of the North American Free Trade Agreement*. NBER working paper 3914.

Grove-White, R., Keraned, M., Miller, P., Machnaghten, P., Wilsdon, J., & Wynne, B. (2004). *Bio-to-Nano? Learning the Lessons, Interrogating the Comparison*. Working Paper. Institute for Environment, Philosophy and Public Policy and Demos.

Grütter, J. (2001). The GHG market after Bonn. *Grütter Consulting and Joint Implementation Quarterly*, *7*(3), 9.

Guellec, D., & van Pottelsberghe de la Potterie. (2001). The internationalisation of technology analysed with patent data. *Research Policy*, *30*(8), 1253–1266. doi:10.1016/S0048-7333(00)00149-9

Guimón, J. (2009). Government strategies to attract R&D-intensive FDI. *Journal of Technology Transfer*.

Gulbrandsen, M., & Slipersaeter, S. (2007). The 3rd Mission and the Entrepreneurial University Model. In Bonaccorsi, A., & Dario, C. (Eds.), *Universities and Strategic Knowledge Creation. Specialization and Performance in Europe* (pp. 112–143). Cheltenham, UK: Edward Elgar.

Gupta, S., Macala, M., & Schafmeister, C. E. (2006). Synthesis of structurally diverse bis-peptide oligomers. *The Journal of Organic Chemistry*, (71): 8691–8695. doi:10.1021/jo0609125

Gupta, S., Tirpak, D., Burger, N., Gupta, J., Höhne, N., Boncheva, A., et al. (2007). Policies, Instruments and Co-operative Arrangements. In *Climate Change 2007: Mitigation*. Cambridge, UK: Cambridge University Press. Retrieved September 9, 2009, from http://www.ipcc.ch/pdf/assessment-report/ar4/wg3/ar4-wg3-chapter13.pdf

Hagerstrand, T. (1972). Diffusion of Innovations. In D. C. Sills (Ed.), International Encyclopedia of Social Sciences (Vol. 3 & 4). New York: The Free Press.

Haggable, S. (2005). *The Rural Nonfarm Economy: Pathway out of Poverty or Pathway In*. Paper presented at the Future of Small Farms conferences, June 25, Wye, UK.

Haggblade, S., Hazell, P., & Brown, J. (1989). Farm-non-farm linkages in rural Sub-Saharan Africa. *World Development*, *17*(18), 22–31.

Haites, E. (2004). Estimating the Market Potential for the Clean Development Mechanism: Review of Models and Lessons Learned. PCFplus Report 19, The World Bank Carbon Finance Business PCFplus Research Program, Washington, DC.

Hall, G. E., & Loucks, S. (1998). Teachers concerns as a basis for facilitating and personalizing staff development. *Teachers College Record*, *80*(1), 36–53.

Halsnæs, K. (2000). Estimation of the Global Market Potential for Cooperative Implementation Mechanisms under the Kyoto Protocol. In Ghosh, P. (Ed.), *Implementation of the Kyoto Protocol: Opportunities and Pitfalls for Developing Countries*. Manila, Philippines: Asian Development Bank.

Hamilton, K., Sjardin, M., Shapiro, A., & Marcello, T. (2009). Fortifying the Foundation: State of the Voluntary Carbon Markets. *Ecosystem Marketplace and New Carbon Finance*. Retrieved September 9, 2009, from http://ecosystemmarketplace.com/documents/cms_documents/StateOfTheVoluntaryCarbonMarkets_2009.pdf

Harder, B. T. (2003). *NCHRP Synthesis of Highway Practice 312: Facilitating Partnerships in Transportation Research*. Washington, DC: Transportation Research Board, National Research Council.

Harder, B. T., & Benke, R. (2005). *Transportation technology transfer: Successes, challenges, and needs: A synthesis of highway practice*. Washington, DC: Transportation Research Board.

Hardin, G. (1968). The tragedy of the commons. *Science*, *162*(3859), 1243–1248. doi:10.1126/science.162.3859.1243

Hardy, C. (1992). The Prospects for Intra-Regional Trade Growth in Africa. In Stewart, F. (Eds.), *Alternative Development Strategies in Sub-Saharan Africa* (pp. 426–442). London: Palgrave Macmillan.

Hare, B., & Meinshausen, M. (2006). How much warming are we committed to and how much can be avoided? *Climatic Change, 75,* 111–149. doi:10.1007/s10584-005-9027-9

Harper, T. (2009). Nanotechnologies In 2009:Creative Destruction or Credit Crunch? Cientifica.

Hart, H. I. (2000). Gas Turbine Operation in a System Disturbance-Prone Grid. In *Proceedings of the 4th International Conference on Power Systems Operation and Planning* (*ICPSOP*), Accra, Ghana, July 31–August 3 (pp. 153-157).

Hassan, M. H. A. (2007). Reforming universities is key to technology transfer. Trieste, Italy: Academy of Sciences for the Developing World.

Hayami, Y., & Godo, Y. (2005). *Development Economics: From the Poverty to the Wealth of Nations* (3rd ed.). Oxford, UK: Oxford University Press.

Hayami, Y., & Ruttan, V. M. (1985). *Agricultural Development: An International Perspective*. Baltimore, MD: Johns Hopkins University Press.

Hayami, Y., et al. (1978). Anatomy of a peasant economy: a rice village in the Philippines. Los Baños, The Philippines: International Rice Research Institute (IRRI).

Hazell, P., & Ramaswamy, G. (Eds.). (1991). *The Green Revolution reconsidered. The impact of high-yielding rice varieties in South India*. Baltimore, MD: Johns Hopkins University Press.

Hazlett, T. (2003, April 18). Tragedies of the tele-commons. *Financial Times*.

Hedenus, F., & Azar, C. (2005). Estimates of trends in global income and resource inequalities. *Ecological Economics, 55*(3), 351–364. doi:10.1016/j.ecolecon.2004.10.004

Heil, M., & Wodon, Q. (1997). Inequality in CO2 emissions between poor and rich countries. *Journal of Environment & Development, 6,* 426–452. doi:10.1177/107049659700600404

Heil, M., & Wodon, Q. (2000). Future inequality in CO2 emissions and the impact of abatement proposals. *Environmental and Resource Economics, 17,* 163–181. doi:10.1023/A:1008326515058

Heller, M. A. (1998). The tragedy of the anticommons: property in the transition from Marx to markets. *Harvard Law Review, 111*(3), 621–688. doi:10.2307/1342203

Helpman, E. (1998). *General Purpose Technologies and Economic Growth*. Cambridge, MA: MIT Press.

Herdt, R. W., Castillo, L., & Jayasuriya, S. (1984). The economics of insect control in the Philippines. In *Judicious and Efficient Use of Pesticides on Rice, Proceedings of the FAO/IRRI Workshop, International Rice Research Institute*, Los Baños, Laguna, Philippines.

Hicks, D., & Hamilton, K. (1999). Does University-Industry Collaboration Adversely Affect University Research? *Issues in Science and Technology Online*. Retrieved from http://www.nap.edu/issues/15.4/realnumbers.htm

Hogeweg, P. (1978). Simulation of cellular forms. *Frontiers in Systems Modeling Simulation, 31,* 90–95. doi:10.1177/003754977803100305

Holtsmark, B. (2003). Russian behaviour in the market for permits under the Kyoto Protocol. *Climate Policy, 3*(4), 399–415. doi:10.1016/j.clipol.2003.08.004

Hosni, Y. A., & Khalil, T. M. (Eds.). (2004). *Management of technology. Internet Economy: Opportunities and Challenges for Developed and Developing Regions of the World*. Selected Papers from the 11th International Conference on Management of Technology, Oxford, UK: Elsevier Ltd.

Howells, M. I. (2005). A model of household energy services in a low-income rural African village. *Energy Policy, 33,* 1833–1851. doi:10.1016/j.enpol.2004.02.019

Human Development Report. (2006). *Beyond Scarcity: Power, Poverty and the Global Water Crisis*. New York: United Nations Development Programme.

Humphrey, J. (2004). The Clean Development Mechanism: How to Increase Benefits for Developing Countries. *IDS Bulletin 35.3: Climate Change and Development, 88.*

Hung, S.-Ch., & Chu, Y.-Y. (2006). Stimulating new industries from emerging technologies: challenges for the public sector. *Technovation*, *26*, 104–110. doi:10.1016/j.technovation.2004.07.018

Hunter, D. (2003). Cyberspace as place, and the tragedy of the digital anticommons. *California Law Review*, *91*(2), 442–519. doi:10.2307/3481336

Huq, S. (2002). Applying Sustainable Development Criteria to CDM Projects: PCF Experience. PCFplus Report 10, Washington DC, April 2002.

Iammarino, S., & McCann, P. (2006). The structure and evolution of industrial clusters: Transactions, technology and knowledge spillovers. *Research Policy*, *35*(7), 1018–1036. doi:10.1016/j.respol.2006.05.004

IARU. (2009). *Climate Change–Global Risks, Challenges and Decisions*. Scientific Conference, Copenhagen 2009, 10-12- March, Synthesis Report, International Alliance of Research Universities IARU. Retrieved September 9, 2009, from http://www.climatecongress.ku.dk

Idachaba, F. S. (2006). Good Intensions are not Enough: collected Essays on Government and Nigerian Agriculture: *Vol. 3. Agricultural Research and Uncertainty and Diversification*. Ibadan, Nigeria: University Press.

IEA. (2008). *International Energy Annual 2006*. Retrieved from http://www.eia.doe.gov/iea/

IETA. (2009). State of the CDM 2009: Reforming for the Present and Preparing for the Future. *International Emissions Trading Association (IETA)*. Retrieved April 12, 2010, from http://www.ieta.org/ieta/www/pages/getfile.php?docID=3363

IFPRI. (2001). *Good News from Africa - Farmers, Agricultural Research and Food in the Pantry* (Schioler, E., Ed.). Washington, DC: Author.

Igami, M., & Okazaki, T. (2007). *Capturing Nanotechnology's Current State of Development via Analysis of Patents*. STI Working Paper 2007/4, OECD Directorate for Science, Technology and Industry.

Igbokwe, E. M. (2004). Rationalizing and streamlining agricultural research institutions in Nigeria. In Legislative and Policy Agenda for Nigerian Agriculture (Vol. 1). Technical Assistance to the House Committee on Agriculture, Briefing Paper No. 4. Enugu, Nigeria: African Institute for Applied Economics.

Igbokwe, E. M. (2005). Strengthening the linkage between agricultural research, extension and the farmer. In Legislative and Policy Agenda for Nigerian Agriculture (Vol. 2). Technical Assistance to the House Committee on Agriculture, Briefing Paper Number 4. Enugu, Nigeria: African Institute for Applied Economics.

Ikezawa, N. (2001). Nanotechnology: Encounters of Atoms, Bits and Genomes. *NRI Papers 37*.

IMF. (2005). *Review of the IMF's trade restrictiveness index. Background paper to the Review of Fund Work on Trade*. Washington, DC: International Monetary Fund.

International Atomic Energy Agency. (2006). *Model for Analysis of Energy Demand (MAED-2) User Manual*. Retrieved from http://www-pub.iaea.org/MTCD/publications/PDF/CMS-18_web.pdf

International Monetary Fund (IMF). (2006). Integrating Poor Countries into the World Trading System. *Economic Issues, 37*.

International Monetary Fund (IMF). (2008). *International Financial Statistics*. Washington, DC: IMF.

International Technology Education Association. (1996). *Technology for all Americans: A Rational and Structure for the Study of Technology*. Reston, VA: Author.

International Telecommunication Union (ITU). (2008). *Information and Communication Technology Statistics*. Geneva: Author.

Inzelt, A. (2004). The evolution of university-industry-government relationships during transition. *Research Policy*, *33*, 975–995. doi:10.1016/j.respol.2004.03.002

Inzelt, A. (2008a). Strengthen and upgrade regional capabilities (Regional University Knowledge Centre Programme in Hungary). *Romanian Journal of Economics*, *26*(1), 133–154.

Inzelt, A. (2008b). The inflow of highly skilled workers into Hungary: a by-product of FDI. *The Journal of Technology Transfer, 33*, 422–438. doi:10.1007/s10961-007-9053-z

Inzelt, A., & Csonka, L. (2008). Strengthening and Upgrading Regional Knowledge Capabilities in Hungary . In Filho, W. L., & Weresa, M. (Eds.), *Fostering Innovation and Knowledge Transfer in European Regions* (pp. 109–132). Frankfurt, Germany: Peter Lang.

Inzelt, A., & Schubert, A. (2009). *Collaboration between Professionals in Academia and in Practice (in the light of scientometric indicators for 12 universities).* Minerva.

Inzelt, A., Laredo, P., Sanchez, P., Marian, M., Vigano, F., & Carayol, N. (2006). Third mission in Methodological Guide, Observatory of European University. *PRIME NoE.* Retrieved from http://www.prime-noe.org

Inzelt, A., Schubert, A., & Schubert, M. (2009). Incremental citation impact due to international co-authorship in Hungarian higher education institutions. *Scientometrics, 78*(1), 37–43. doi:10.1007/s11192-007-1957-8

IPCC. (2007). *Contribution of Working Group II to the Fourth Assessment Report of the Intergovernmental Panel on Climate Change.* (M. L. Parry, O. F. Canziani, J. P. Palutikof, P. J. van der Linden & C. E. Hanson, Eds.). Cambridge, UK: Cambridge University Press. Retrieved September 9, 2009, from http://www.ipcc.ch/publications_and_data/publications_ipcc_fourth_assessment_report_wg2_report_impacts_adaptation_and_vulnerability.htm

IPCC. 2000. Methodological and Technological Issues in Technology Transfer. Intergovernmental Panel on Climate Change (IPCC). Cambridge: B. Metz, O.R. Davidson, J-W. Martens, S.N.M. van Rooijen and L Van Wie McGrory, eds. Cambridge University Press.

ISAAA. (2008). *Global Status of Commercialized Biotech/GM Crops: 2008. The First Thirteen Years, 1996 to 2008.* ISAAA Brief 39-2008.

Islam, N. (2009). Innovative manufacturing readiness levels (IMRLs): A new readiness matrix. *International Journal of Nanomanufacturing.*

Islam, N., & Miyazaki, K. (2009). Nanotechnology innovation system: Understanding hidden dynamics of nanoscience fusion trajectories. *Technological Forecasting and Social Change, 76*(1), 128–140. doi:10.1016/j.techfore.2008.03.021

ISR. (2001). *Nanotechnology in Australian Industry. Proceedings and outcomes report.* Canberra, Australia: ISR.

Istanbul Technical University. (n.d.). Retrieved from http://www.itu.edu.tr/

Janssen, R., & Munda, G. (1999). Multi-criteria methods for quantitative, qualitative and fuzzy evaluation problems . In Van den Bergh, J. (Ed.), *Handbook of environmental and resource economics.* Cheltenham, UK: Edward Elgar.

Jensen, K. (2007). Nanotube Radio. *Nano Letters, 7*(11), 3508–3511. Retrieved from http://dx.doi.org/10.1021/nl0721113. doi:10.1021/nl0721113

Jensen, P. H., & Webster, E. (2004). Achieving the optimal power of patent rights. *The Australian Economic Review, 37*(4), 419–426. doi:10.1111/j.1467-8462.2004.00343.x

Jesuleye, O. A., & Siyanbola, W. O. (2008). Solar electricity demand analysis for improved access to energy services in Nigeria. *Nigeria Journal of Solar Energy, 19*(1), 136–152.

Jesuleye, O. A., Siyanbola, W. O., Ilori, M. O., & Sanni, S. A. (2007). Energy demand analysis of Port-Harcourt Refinery, Nigeria and its policy implications. *Energy Policy Journals, 35*, 1338–1345. doi:10.1016/j.enpol.2006.04.004

Johnson, S., Gatz, E. T., & Hicks, T. (1999). Expanding the content base of technology and education transfers: A topic of study. *Journal of Technology Education, 8*(2).

Jolly, J. A., & Creighton, J. W. (1977). The technology transfer process: Concepts, framework and methodology. *The Journal of Technology Transfer, 1*(2), 77–91. doi:10.1007/BF02622191

Joseph, T., & Morrison, M. (2006 May). *Nanotechnology in agriculture and food.* Institute of Nanotechnology. Retrieved June 12, 2009, from http://www.nanoforum.org

Jotzo, F., & Michaelowa, A. (2002). Estimating the CDM market under the Marrakech Accords. *Climate Policy*, *2*(1), 179–196.

Jotzo, F., & Pezzey, J. C. V. (2007). Optimal intensity targets for greenhouse gas emissions trading under uncertainty. *Environmental and Resource Economics*, *38*, 259–284. doi:10.1007/s10640-006-9078-z

Jovanovic, M. N. (1992). *International Economic Integration*. London: Routlege Publishers.

Joy, B. (1992). *Why the future doesn't need us*. Retrieved from http://www.wired.com/wired/archive/8.04/joy.html

Joy, B. (2000). Why the future doesn't need us. *Wired*, *8*(4), 238–262.

Just, R. E., & Huffman, W. E. (2004). *The role of patents, royalties, and public-private in university funding. Mimeo*. College Park, MD: University of Maryland.

Kabiraj, T. (2006). On the incentive for cooperative and non-cooperative R&D in duopoly. *Arthaniti*, *5*, 24–33.

Kabiraj, T. (2007). On the incentives for cooperative research. *Research in Economics*, *61*, 17–23. doi:10.1016/j.rie.2006.12.003

Kabiraj, T., & Mukherjee, A. (2000). Cooperation in R&D and production: a three firm analysis. [Zeitschrift fur Nationalokonomie]. *Journal of Economics*, *71*, 281–304. doi:10.1007/BF01228744

Kagame, P. (2009, May 7). Africa has to find its own road to prosperity. *Financial Times*.

Kállay, L., & Lengyel, I. (2008). The Internationalisation of Hungarian SMEs . In Dana, L., Han, M., Ratten, V., & Welpe, I. (Eds.), *A Theory of Internationalisation for European Entrepreneurship* (pp. 22–36). Cheltenham, UK: Edward Elgar.

Kalotay, K., & Filippov, S. (2009). *Foreign Direct Investment in Times of Global Economic Crisis: Spotlight on New Europe*. UNU-MERIT Working Paper, 2009-021.

Kamien, M. (1992). Patent licensing . In Aumann, R. J., & Hart, S. (Eds.), *Handbook of game theory*. Amsterdam: Elsevier.

Kamien, M. I., Muller, E., & Zang, I. (1992). Research joint ventures and R&D cartels. *The American Economic Review*, *82*, 1293–1306.

Kamra, A. (2009). Patenting Nanotechnology: In Pursuit of a Proactive Approach. *Social Science Research Network Database*. Retrieved June 12, 2009, from http://ssrn.com/abstract=1399332

Katz, E., & Lazarsfeld, P. (1955). *Personal influence: The part played by people in the flow of mass communications*. Glencoe, IL: Free Press.

Katz, M. L., & Shapiro, C. (1985). On the licensing of innovations. *The Rand Journal of Economics*, *16*, 504–520. doi:10.2307/2555509

Kaushik, P. D. (2000). *TRIPS: IPR regime for the digital medium*. Retrieved July 6, 2009, from http://unpan1.un.org/intradoc/groups/public/documents/APCITY/UNPAN006305.pdf

KAWAX. (2009). *Observatorio Chileno de CTI*. CONICYT.

Kay, L., & Shapira, P. (2009). Developing nanotechnology in Latin America. *Journal of Nanoparticle Research*, *11*, 259–278. doi:10.1007/s11051-008-9503-z

Kaya, Y. (1989). *Impact of Carbon Dioxide Emission Control on GNP Growth: Interpretation of Proposed Scenarios*. Paper presented to the Energy and Industry Subgroup, Response Strategies Working Group, Intergovernmental Panel on Climate Change, Paris, France.

Kershner, R. J. (2009). Placement and orientation of individual DNA shapes on lithographically patterned surfaces. *Nature Nanotechnology*, *4*, 557–561. doi:10.1038/nnano.2009.220

Khanijou, S. (2006). Patent inequality?: Rethinking the application of strict liability to patent law in the nanotechnology era. *Journal of Technology . Law & Policy*, *12*, 179–181.

Khush, G. S. (1992). Selecting rice for simple inherited resistances . In Stalker, H. T., & Murphy, J. P. (Eds.), *Plant breeding in the 1990s* (pp. 303–322). Wallingford, UK: Commonwealth Agricultural Bureaux International.

Khush, G. S. (1995). Modern varieties: Their real contribution to food supply and equity. *GeoJournal, 35*, 275–284. doi:10.1007/BF00989135

Khush, G. S. (1999). Green Revolution: preparing for the 21st century. *Genome, 42*, 646–655. doi:10.1139/gen-42-4-646

Kline, S. J., & Rosenberg, N. (1986). *The Positive Sum Strategy, Harnessing Technology for Economic Growth An Overview of Innovation.* Washington, DC: National Academy Press.

Kloppenburg, J. R. (2004). *First the Seed: The Political Economy of Plant Biotechnology 1492-2000* (2nd ed.). Madison, WI: University of Wisconsin Press.

Knop, K. (2005 June). Nanotechnology: A Big Future for Small Things. *Bulletin Special, Credit Suisse,* 9-11.

Knudson, M. K. (1991). Incorporating technological change in diffusion models. *American Journal of Agricultural Economics, 73*, 724–733. doi:10.2307/1242824

Koç University. (n.d.). Retrieved from http://www.ku.edu.tr/

Kogan, M., Bastus, N., & Amigo, R. (2006). Nanoparticle-mediated local and remote manipulation of protein aggregation. *Nano Letters, 6*(1), 110–115. doi:10.1021/nl0516862

Kogan, M., Olmedo, I., Hosta, L., Guerrero, A., Cruz, L., & Albericio, F. (2007). Peptides and metallic nanoparticles for biomedical applications. *Nanomedicine; Nanotechnology, Biology, and Medicine, 2*(3), 287–306.

Kostoff, R. N., Murday, J. S., Lau, C. G. Y., & Tolles, W. M. (2006b). The seminal literature of nanotechnology research. *Journal of Nanoparticle Research, 8*(1).

Kostoff, R. N., Murday, J. S., Stump, J. A., Johnson, D., Lau, C. G. Y., & Tolles, W. M. (2006a). The structure and infrastructure of the global nanotechnology literature. *Journal of Nanoparticle Research, 8*(1).

Kozul-Wright, R., & Rayment, P. (2007). *The Resistible Rise of Market Fundamentalism: The Struggle for Economic Development in a Global Economy.* London: Zed Books.

Krauss, L. M., & Starkman, G. D. (2004 May). Universal Limits on Computation. *Physical Review Letters.*

Krueger, A. (1978). *Foreign Trade Regimes and Economic Development: Liberalization Attempts and Consequences.* Cambridge, MA: Ballinger Publishing Company.

Krueger, A. (1990). Import Substitution Versus Export Promotion . In King, P. (Ed.), *International Economics and International Economic Policy: A Reader* (pp. 155–165). New York: McGraw Hill.

Krueger, A. (1997, March). Trade policy and economic development how we learn. *The American Economic Review, 87.*

Kuhlman, B. (2003, November 21). Design of a novel globular protein fold with atomic-level accuracy. *Science, 302*(5649), 1364–1368. doi:10.1126/science.1089427

Kuhn, T. S. (1970). *The Structure of Scientific Revolutions.* Chicago: University of Chicago Press.

Kuiper, M., Meijerink, G., & Eaton, D. (2007). Rural Livelihood interplay between farm activities, non-farm activities and the resource base . In *Science for agriculture and rural development in low-income countries* (pp. 77–95). Dordrecht, The Netherlands: Springer. doi:10.1007/978-1-4020-6617-7_5

Kumar, B., Koval, T., Narayanan, S., & Shea, S. (2002). *Commercialization of a silicon Nitride Co-fire through (SINCOT) process for manufacturing high efficient mono-crystalline silicon solar cells.* Pre-print of paper to be presented at 29 IEEE PVSC New Orleans May 20-24, 2002.

Kupolokun, F.(2007, April 1). Nigeria's Gas Reserves Increases. THISDAY NIGERIAN NEWSPAPER.

Kurzweil, R. (1999). *The Age of Spiritual Machines.* New York: Penguin Books.

Kurzweil, R. (2005). The Singularity is Near. New York: Penguin Books.

Lall, S. (2000). The technological structure and performance of developing country manufacturing exports, 1985–98. *Oxford Development Studies, 28*(3), 337–369. doi:10.1080/713688318

Laredo, P. (2007). Revisiting the third mission of universities: toward a renewed categorisation of university activities? *Higher Education Policy, 20*(4), 441–456. doi:10.1057/palgrave.hep.8300169

Larson, D. F., & Breustedt, G. (in press). Will markets direct investments under the Kyoto Protocol? Lessons from the activities implemented jointly pilots. *Environmental and Resource Economics.*

Larson, D. F., Ambrosi, P., Dinar, A., Rahman, S. M., & Entler, R. (2008). A review of carbon market policies and research. *International Review of Environmental and Resource Economics, 2*(3), 177–236. doi:10.1561/101.00000016

Laudel, G. (2006). The art of getting funded: how scientists adapt to their funding conditions. *Science & Public Policy, 33*(7), 489–504. doi:10.3152/147154306781778777

Layne-Farrar, A., & Evans, D. S. (2004) Software patents and open source: The battle over intellectual property rights. *Virginia Journal of Law and Technology, 9*(10).

LEAD. (2009). Leading the way: A role for regional institutions. African leadership on climate change in Africa. Based on the workshop 'African Leadership on Climate Change: Challenges and Opportunities', Tunis, January 2009. LEAD Africa, Enda Energy, LEAD International, African Development Bank.

Lecocq, F., & Ambrosi, P. (2007). The clean development mechanism: history, status and prospects. *Review of Environmental Economics and Policy, 1*(1), 134–151. doi:10.1093/reep/rem004

Leimbach, M. (2003). Equity and carbon emissions trading: a model analysis. *Energy Policy, 31*, 1033–1044. doi:10.1016/S0301-4215(02)00180-5

Leipziger, D., Motta, M., & Dahlman, C. (2004). Chile new economy study (Report No. 25666-CL). In *Documents of the World Bank (Vol. 1): Executive Summary and Policy Recommendations* (pp. 1-31).

Lemiti, M., Kaminski, A., Fave, A., & Fourmond, E. (2007). *New solar cell production process may improve efficiency and reduce cost.* SPIE.

Lemley, M. A., & Shapiro, C. (2007). Patent holdup and royalty stacking. *Texas Law Review*, 85.

Lengyel, B., & Leydesdorff, L. (2007). *Measuring the knowledge base in Hungary: Triple Helix dynamics in a transition economy.* Paper presented at the 6th Triple Helix Conference, 16-19 May 2007, Singapore.

Leon, A., Barticevic, Z., & Pacheco, M. (2009). Coupling and chemical shifts in carbon nanostructures for quantum computing. *Chemical Physics Letters, 470*(4-6), 249–254. doi:10.1016/j.cplett.2009.01.052

Lévêque, F., & Ménière, Y. (2004). *The economics of patents and copyright.* Berkeley, CA: Berkeley Electronic Press.

Liao, S., et al. (2004, December 17). Translation of DNA Signals into Polymer Assembly Instructions. *Science, 306*(5704), 2072–2074. Retrieved from DOI: 10.1126/science.1104299

Libaers, D., Meyer, M., & Geuna, A. (2006). The Role of University Spinout Companies in an Emerging Technology: The Case of Nanotechnology. *The Journal of Technology Transfer, 31*, 443–450. doi:10.1007/s10961-006-0005-9

Libecap, G. (Ed.). (2005). *University Entrepreneurship and Technology Transfer: Process, Design, and Intellectual Property. Advances in the Study of Entrepreneurship, Innovation. Innovation and Economic Growth.* London: Jay Press.

LizardFire Studios. (2009). Retrieved from http://www.lizardfire.com/html_nano/themovies.html

Longo, E., Paskocimas, C. A., Leite, E. R., Magnani, R. A., Pontes, F. M. L., & Barroso, V. (2003). *Método para transformar carbono amorfo em grafite cristalino em minas de grafite utilizando nanopartículas de metais de transição.* Pending patent in Brazil.

Lundvall, B.-A. (Ed.). (1992). *National Systems of Innovation–Towards a theory of innovation and interactive learning.* London: Pinter Publishers.

Luther, W. (2004). *Industrial Applications of Nanomaterials – chances and risks. Technology Analysis.* VDI Technologiezentrum.

Luther, W. (2004). *International Strategy and Foresight Report on Nanoscience and Nanotechnology.* VDI Technologiezentrum.

Lux Research. (2005). *Ranking the nations: Nanotech's Shifting Global Leaders.*

Lux Research. (2006). *The Nanotech Report* (4th ed.). New York: Lux Research.

Lux. (2008). *Nanomaterials State of the Market Q3 2008.* Lux Research.

Lyakurwa, W. (1997). Regional Integration in Sub-Saharan Africa: A Review of Experiences and Issues. In Oyejide, A., Elbadawi, I., & Collier, P. (Eds.), *Regional Integration and Trade Liberalization in Sub-Saharan Africa.* New York: St. Martin's Press Inc.

Maclurcan, D. C. (2005 October). Nanotechnology and Developing Countries Part 2: What Realities? *Journal of Nanotechnology Online.*

Maclurcan, D. C. (2005 September). Nanotechnology and Developing Countries Part 1: What Possibilities? *Journal of Nanotechnology Online.*

Macnaghten, P., Kearnes, M., & Wynne, B. (2005). Nanotechnology, governance, and public deliberation: what role for the social sciences? *Science Communication, 27*(2), 1–25. doi:10.1177/1075547005281531

Madou, M. J. (2002). *Fundamentals of microfabrication: The Science of miniaturization* (2nd ed.). Boca Raton, FL: CRC press.

Mafimisebi, T. E. (2007). Long-run price integration in the Nigerian fresh fish market: Implications for marketing and development. In S. M. Baker & D. Westbrook (Eds.), *Proceedings of a Joint Conference of The International Society of Marketing and Development and the Macromarketing Society,* Washington, DC (pp.149-158).

Mafimisebi, T. E., & Fasina, O. O. (2009). Rural women's productivity and welfare issues: A cause for concern. In Agbamu, J. U. (Ed.), *Perspectives in Agricultural Extension and Rural Development* (pp. 361–386). Owerri, Nigeria: Springfield Publishers Ltd.

Mafimisebi, T. E., Onyeka, U. P., Ayinde, I. A., & Ashaolu, O. F. (2006b). Analysis of farmer-specific socio-economic determinants of adoption of modern livestock management technologies by farmers in Southwest Nigeria. [Retrieved from http://www.World-foodnet]. *Journal of Food Agriculture and Environment, 4*(1), 183–186.

Mahajan, V., & Schoeman, M. E. F. (1977). Generalized model for the time pattern of the diffusion process. *IEEE Transactions on Engineering Management, 24,* 12–18.

Mairesse, J., & Mohnen, P. (2003). Intellectual Property in Services. What do we learn from innovation surveys? In *Proceedings from the OECD conference.* IPR, Innovation and Economic Performance.

Malanowski, N., Heimer, T., Luther, W., & Werner, M. (Eds.). (2006). *Growth Market Nanotechnology – An Analysis of Technology and Innovation.* Weinheim, Germany: Wiley-VCH Verlag.

Malsch, I. (2008). *Nanotechnology in Argentina.* Utrecht, Netherlands: Malsch TechnoValuation.

Mankins, J. C. (1995). *Technology Readiness Levels. Advanced Concepts Office, Office of Space Access and Technology.* NASA.

Mansfield, E. (1961). Technical change and the rate of imitation. *Econometrica, 29,* 741–765. doi:10.2307/1911817

Marjit, S. (1991). Incentives for cooperative and non-cooperative R&D in duopoly. *Economics Letters, 37,* 187–191. doi:10.1016/0165-1765(91)90129-9

Marmara University. (n.d.). Retrieved from http://www.marmara.edu.tr/

Martin, R. B. (2003). The changing social contract for science and the evolution of the university. In Geuna, A., Salter, A., & Steinmueller, W. E. (Eds.), *Science and innovation: Rethinking the rationales for funding and governance* (pp. 1–29). Cheltenham, UK: Edward Elgar.

Martinez-Fernandez, M. C. (2004). Regional Collaboration Infrastructure: Effects in the Hunter Valley of NSW. *Australian Planner, 41*(4).

Martinez-Fernandez, M. C. (2005b). *Knowledge Intensive Service Activities (KISA) in Innovation of Mining Technology Services in Australia.* Sydney, Australia: University of Western Sydney.

Martinez-Fernandez, M. C., & Leevers, K. (2004). Knowledge creation, sharing and transfer as an innovation strategy: The discovery of nano-technology by South-West Sydney. [IJTM]. *International Journal of Technology Management, 28*(3-6), 560–581.

Martinez-Fernandez, M. C., & Potts, T. (2006). *Innovation at the Edges of the Metropolis: An Analysis of Innovation Drivers in Peripheral Suburbs of Sydney.* Opolis.

Martinez-Fernandez, M. C., & Potts, T. (2007 March). Innovation at the Edges of the Metropolis: An Analysis of Innovation Drivers in Peripheral Suburbs of Sydney. *Opolis.*

Martinez-Fernandez, M. C., & Rerceretnam, M. (2006). *The Role of UWS as a Knowledge Hub.* Sydney, Australia: UWS.

Martinez-Fernandez, M. C., Potts, T., Receretnam, M., & Bjorkli, M. (2005a). *Innovation at the Edges: An analysis of Innovation Drivers in South West Sydney.* Sydney, Australia: University of Western Sydney. Retrieved April 28, 2005, from http://aegis.uws.edu.au/innovationedges/main.html

Martinez-Fernandez, M. C., Soosay, C., & Tremayne, K. (2005e). *Learning Spaces, Co-production of Knowledge and Capacity Building in the Service Firm.* AEGIS Working Paper 2005-06. Sydney, Australia: AEGIS, UWS.

Martinez-Fernandez, M. C., Soosay, C., Krishna, V. V., Turpin, T., & Bjorkli, M. (2005d). *Knowledge Intensive Service Activities (KISA) in Innovation of the Software Industry in Australia.* Sydney, Australia: University of Western Sydney.

Martinez-Fernandez, M. C., Soosay, C., Krishna, V. V., Turpin, T., Bjorkli, M., & Doloswala, K. (2005c). *Knowledge Intensive Service Activities (KISA) in Innovation of the Tourism Industry in Australia.* Sydney, Australia: University of Western Sydney.

Maskell, P. (2001). Towards a knowledge-based theory of the geographical cluster. *Industrial and Corporate Change, 10*(4), 921–943. doi:10.1093/icc/10.4.921

Masuzawa, T. (2000). State of the art of micro-machining. *Annals of the CIRP, 49*(2), 473–488. doi:10.1016/S0007-8506(07)63451-9

McCalla, A. F., & Nash, J. (2007). *Reforming Agricultural Trade for Developing Countries, Key Issues for a Pro-development Outcome of the Doha Round (Vol. 1).* Washington, DC: The World Bank.

McConnell, H. (2008). Leapfrog technologies for health and development: Technological innovations. *Global Forum Update on Research for Health, 5.*

McCraw, T. K. (2007). *Prophet of Innovation: Joseph Schumpeter and Creative Destruction.* Cambridge, MA: Harvard University Press.

McKibben, B. (2003). *Enough: Staying Human in an Engineered Age.* New York: Henry Holt & Co.

McKinley, B. (2000). Statement to the Migration in Africa conference, Addis Ababa, October 18, 2000. International Organization for Migration (IOM).

Meléndrez, D., Motta, M., & Dahlman, C. (2004). Synthesis and aggregation study of tin nanoparticles and colloids obtained by chemical liquid deposition. *Colloid & Polymer Science, 287*(1), 13–22. doi:10.1007/s00396-008-1950-7

Mendoza Zélis, P., Sánchez, M., & Vázquez, J. (2007). Magnetostrictive bimagnetic trilayer ribbons for temperature sensing. *Journal of Applied Physics, 101*, 034507. doi:10.1063/1.2422905

Mengistae, T., & Pattillo, C. (2004). Export orientation and productivity in Sub-Saharan Africa. *IMF Staff Papers, 51*(2), 327–353.

Merkle, R. C. (1996). Design considerations for an assembler. *Nanotechnology, 7*, 210–215. doi:10.1088/0957-4484/7/3/008

Merkle, R. C. (1999). Biotechnology as a route to nanotechnology. *Trends in Biotechnology, 17*(7), 271–274. doi:10.1016/S0167-7799(99)01335-9

Messer, E., & Heywood, P. (1990). Trying technology: Neither sure nor soon. *Food Policy*, *15*(4), 336–345. doi:10.1016/0306-9192(90)90076-C

Metcalfe, J. S. (1981). Impulse and diffusion in the study of technical change. *Futures*, *13*, 347–359. doi:10.1016/0016-3287(81)90120-8

Meyer, M. (2006). *What do we know about innovation in nanotechnology? Some propositions about an emerging field between hype and path-dependency.* Paper presented at the 2006 technology transfer society conference, Atlanta, Georgia, 27–29 September.

Meyer, M., & Persson, O. (1998). Nanotechnology: Interdisciplinarity, patterns of collaboration and differences in application. *Scientometrics*, *42*(2), 195–205. doi:10.1007/BF02458355

Michaelowa, A., & Michaelowa, K. (2007a). Does climate policy promote development? *Climatic Change*, *84*(1-2), 1–4. doi:10.1007/s10584-007-9266-z

Michaelowa, A., & Michaelowa, K. (2007b). Climate or development: is ODA diverted from its original purpose? *Climatic Change*, *84*(1–2), 5–21. doi:10.1007/s10584-007-9270-3

Michaelowa, A., & Müller, B. (2009). The Clean Development Mechanism in the Future Climate Change Regime, Summary for Policy Makers. *Climate Strategies*. Retrieved September 9, 2009, from http://www.climatestrategies.org/our-reports/category/39/149.html

Michelson, E. S. (2006 October). *Nanotechnology Policy: An Analysis of Transnational Governance Issues Facing the United States and China.* Project on Emerging Nanotechnologies, Woodrow Wilson International Center for Scholars, Washington, DC.

Michelson, E. S. (2008). Globalization at the nano frontier: The future of nanotechnology policy in the United States, China, and India. *Technology in Society*, *30*, 405–410. doi:10.1016/j.techsoc.2008.04.018

Michelson, E. S., Sandler, R., & Rejeski, D. (2008). Nanotechnology. In M. Crowley (Ed.), From Birth to Death and Bench to Clinic: The Hastings Center Bioethics Briefing Book for Journalists, policymakers, and Campaigns (pp. 111–116). Garrison, NY: The Hastings Center.

Miketa, A., & Schrattenholzer, L. (2006). Equity implications of two burden-sharing rules for stabilizing greenhouse-gas concentrations. *Energy Policy*, *34*, 877–891. doi:10.1016/j.enpol.2004.08.050

Miller, G. (2008). Contemplating the social implications of a nanotechnology revolution . In Fisher, E., Selin, C., & Wetmore, J. (Eds.), *Yearbook of Nanotechnology in Society* (pp. 215–225). Berlin: Springer. doi:10.1007/978-1-4020-8416-4_19

Millock, K. (2001). Technology transfers in the clean development mechanism: an incentive issue. *Environment and Development Economics*, *7*, 449–466.

Ministry of Education and Science. (2004). *PAV Projects Report. SECYT.* Argentina: ANPCYT.

Ministry of Education and Science. (2006). *Science and Technology Indicators.* Argentina: SECYT.

Ministry of Science, Technology and Productive Innovation, Argentina. (2009 April/June). *Statistic Bulletin on Technology No. 3.*

Mnyusiwalla, A., Daar, A. S., & Singer, P. A. (2003). Mind the gap: science and ethics in nanotechnology. *Nanotechnology*, *14*, R9–R13. doi:10.1088/0957-4484/14/3/201

Mock, J. E., Knenkeremath, D. C., & Janis, F. T. (1993). *Moving R&D to the Marketplace: A Guidebook for Technology Transfer Managers.* Washington, DC: J. E. Mock.

Molas-Gallart, J., Salter, A., Patel, P., & Scott, A. (2002). *Measuring Third Stream Activities – Final Report to the Russell Group of Universities.* University of Sussex, Science & Technology Policy Research.

Molas-Gallart, J., Salter, A., Patel, P., Scott, A., & Duran, X. (2002). *Final Report to the Russell Group of Universities, SPRU.* Unpublished paper.

Montague, Y., Ratta, A., & Nygaard, D. (1998). *Pest Management and Food Production: Looking to the Future*. Food, Agriculture and the Environment Discussion Paper 25. Washington, DC: International Food Policy Research Institute.

Mooney, P. (1999). The ETC century erosion, technological transformation and corporate concentration in the 21st century. Development Dialogue, (1-2): 1–128.

Moore, G. A. (1999). *Crossing the Chasm*. New York: Harper Business.

Moraru, C., Panchapakesan, C., Huang, Q., Takhistov, P., Liu, S., & Kokini, J. (2003). Nanotechnology: A new frontier in food science. *Food Technology, 57*(12), 24–29.

Morris, M. L., & Doss, C. R. (2002). *How does gender affect adoption of agricultural innovations? The case of improved maize technology in Ghana.*

Mosley, P., & Sulieman, A. (2007). Aid, agriculture and poverty in developing countries. *Review of Development Economics, 11*(1), 139–158. doi:10.1111/j.1467-9361.2006.00354.x

Moyo, D. (2009). *Dead Aid: Why Aid Is Not Working and How There Is a Better Way for Africa*. London: Penguin Group.

MSU. (2009). *Diffusion of innovations*. Retrieved from http://www.educ.msu.edu/epfp/meet/02-06-06files/Diffusion_Innovations.pdf

Mukherjee, A. (2005). Innovation, licensing and welfare. *The Manchester School, 73*, 29–39. doi:10.1111/j.1467-9957.2005.00422.x

Mukherjee, A. (2008). Technology licensing . In Rajan, R. S., & Reinert, K. A. (Eds.), *Princeton Encyclopaedia of the World Economy*. Princeton, NJ: Princeton University Press.

Mukherjee, A., & Balasubramanian, N. (2001). Technology transfer in horizontally differentiated product-market. [Richerche Economiche]. *Research in Economics, 55*, 257–274. doi:10.1006/reec.2001.0254

Mukherjee, A., & Marjit, S. (2004). R&D organization and technology transfer. *Group Decision and Negotiation, 13*, 243–258. doi:10.1023/B:GRUP.0000031079.32373.a4

Mukherjee, A., & Ray, A. (2009). Unsuccessful patent application and cooperative R&D. *Journal of Economics, 97*, 251–263. doi:10.1007/s00712-009-0071-1

Muller, A. (2007). How to make the CDM more sustainable: The potential of rent extraction. *Energy Policy, 35*(6), 3203–3212. doi:10.1016/j.enpol.2006.11.016

Muller, A. (2009a). Sustainable agriculture and the production of biomass for energy use. *Climatic Change, 94*(3-4), 319–331. doi:10.1007/s10584-008-9501-2

Muller, A. (2009b). *Benefits of Organic Agriculture as a Climate Change Adaptation and Mitigation Strategy in Developing Countries*. Scandinavian Working Papers in Economics 343 and EfD Discussion paper 09-09, Environment for Development Initiative and Resources for the Future, Washington DC.

Murashov, V., Engel, S., Savolainen, K., Fullam, B., Lee, M., & Kearns, P. (2009). Occupational safety and health in nanotechnology and Organisation for Economic Cooperation and Development. *Journal of Nanoparticle Research - Special Issue: Environmental and Human Exposure of Nanomaterials*. Retrieved April 29, 2009 from DOI: 10.1007/s11051-009-9637-7

Nagy, J. G., & Sanders, H. (1990). Agricultural technology development and dissemination within a farming systems perspective. *Agricultural Systems, 32*, 305–320. doi:10.1016/0308-521X(90)90097-A

Nanosolar. (2009, September 9). Nanosolar Completes Panel Factory, Commences Serial Production. *Nanosolar Communications*.

Nanotechnology Now. (2009). *Nanotechnology Introduction*. Retrieved from http://www.nanotech-now.com/introduction.htm

Narain, U., & van t'Veld, K. (2008). The clean development mechanism's low-hanging fruit problem: When might it arise, and how might it be solved? *Environmental and Resource Economics, 40*(3), 445–465. doi:10.1007/s10640-007-9164-x

Narula, R., & Zanfei, A. (2005). Globalization of Innovation: The Role of Multinational Enterprises . In Fagerberg, J., Mowery, D., & Nelson, R. (Eds.), *The Oxford Handbook of Innovation* (pp. 318–345). New York: Oxford University Press.

Nassauer, J. I. (2002). Agricultural Landscapes in Harmony with Nature . In Kimbrell, A. (Ed.), *The Fatal Harvest Reader: The Tragedy of Industrial Agriculture*. Washington, DC: Island Press.

Nastas, T. (2007). *Scaling up Innovation: The Go-Forward Plan to Prosperity*. Washington, DC: World Bank Institute.

National Cancer Institute (NCI) Alliance for Nanotechnology in Cancer. (2009). *Nanotech News: Tumors Feel the Deadly Sting of Nanobees*. Retrieved August 17, 2009, from http://nano.cancer.gov/news_center/2009/aug/nanotech_news_2009-08-27a.asp

National Science and Technology Council (NSTC). (2006). *The National Nanotechnology Initiative: Research and Development Leading to a Revolution in Technology and Industry*. NSTC Report, July 2006.

National Science Foundation. (2002). *Science and Engineering Indicators* (*Vol. 2*). Washington, DC: Author.

Ndikumana, L. (2000). Financial determinants of domestic investment in Sub-Saharan Africa: evidence from panel data. *World Development*, *28*(2), 381–400. doi:10.1016/S0305-750X(99)00129-1

Nedeva, M. (2008). New tricks and old dogs? The 'third mission' and the re-production of the university . In Epstein, D., Boden, R., Rizvi, F., Deem, R., & Wright, S. (Eds.), *The World Yearbook of Education 2008: Geographies of Knowledge/Geometries of Power–Higher Education in the 21st Century*. New York: Routledge.

Nelson, R. (Ed.). (1993). *National Innovation Systems*. New York: Oxford University Press.

New South Wales Department of State and Regional Development. (2002 December). Retrieved from http://ats.business.gov.au

New World Encyclopedia. (2008). *Agricultural technology*. Retrieved from http://www.newworldencyclopedia.org/entry/Agricultural_technology

New, P. A. (2009). *Pennsylvania Initiative for Nanotechnology*. Retrieved from http://www.newPA.com

Niemeyer, J. K., & Whitney, D. E. (2002). Risk reduction of jet engine product development using technology readiness metrics. In . *Proceedings of the ASME Design Engineering Technical Conference, 4*, 3–13.

Nigeria's Electricity Sector. (2006 January). Executive Report, Abuja, Nigeria.

Niosi, J., & Reid, S. E. (2007). Biotechnology and Nanotechnology: Science-based Enabling Technologies as Windows of Opportunity for LDCs? *World Development, 35*(3), 426–438. doi:10.1016/j.worlddev.2006.11.004

Nogueira, A. F., Durrant, J. R., & De Paoli, M. A. (2001). Dye-sensitized nanocrystalline solar cells employing a polymer electrolyte. *Advanced Materials (Deerfield Beach, Fla.), 13*(11), 826. doi:10.1002/1521-4095(200106)13:11<826::AID-ADMA826>3.0.CO;2-L

Nordhaus, W. (1994). *Managing the global commons, The Economics of Climate Change*. Cambridge, MA: MIT Press.

Nordhaus, W. D. (1962). *Invention, growth, and welfare: A theoretical treatment of technological change*. Cambridge, MA: MIT Press.

Nordmann, A. (2004). *Converging Technologies: Shaping the Future of European Societies. Report of the High Level Expert Group: Foresighting the New Technology Wave*. Brussels, Belgium: European Commission Research.

North, D. C. (2000). Building Institutions: Complement, Innovate, Connect, and Compete . In *World Development Report 2002: Building Institutions for Markets*. Washington, DC: Oxford University Press.

NSTC (National Science and Technology Council). (1999). *Nanotechnology: Shaping the World Atom by Atom*. Washington, DC: National Science and Technology Council. (NTP) Nanotech Project. (2008). *Nanoscale Silver: No Silver Lining?* Retrieved July 10, 2009, from http://www.nanotechproject.org/news/archive/silver/

O'Dwyer, C., Lavayen, V., Fuenzalida, D., Lozano, H., Santa Ana, M., & Benavente, E. (2008). Low-dimensional, hinged bar-code metal oxide layers and free-standing, ordered organic nanostructures from turbostratic vanadium oxide. *Small*, *4*(7), 990–1000. doi:10.1002/smll.200701014

O'Dwyer, C., Navas, D., Lavayen, V., Benavente, E., Santa Ana, M., & González, G. (2006). Nano-urchin: The formation and structure of high-density spherical clusters of vanadium oxide nanotubes. *Chemistry of Materials*, *18*(13), 3016–3022. doi:10.1021/cm0603809

Obioma, G. (2007). *The 9 year Basic Education Curriculum (Structure, Content and Strategy for Implementation): Nigeria's Experience in Educational reform.* NERDC.

Ocampo, J. A., & Parra, M. A. (2003). The terms of trade for commodities in the twentieth century. [Santiago, Chile: United Nations Economic Commission for Latin America.]. *CEPAL Review*, 79.

Ocampo, J. A., Jomo, K. S., & Vos, R. (2007). *Growth Divergences: Explaining Differences in Economic Performance.* London: Zed Books.

Odigie, H. A., & Li-Hua, R. (2008). *Unlocking the channel of tacit Knowledge transfer.* Retrieved from http://www.mostc.coms

OECD. (1999). *Benchmarking Knowledge-based Economies.* Paris: OECD.

OECD. (2001). *Cities and Regions in the New Learning Economy.* Paris: OECD.

OECD. (2003). *OECD Science, Technology and Industry Scoreboard.* Paris: OECD.

OECD. (2004). *Patents and innovation: trends and policy challenges.* Retrieved May 16, 2009, from http://www.oecd.org/dataoecd/48/12/24508541.pdf

OECD. (2004, May). Nanotech is not small. *The OECD Observer. Organisation for Economic Co-Operation and Development*, 243.

OECD. (2005). *University Research Management. Developing Research in New Institutions.* Paris: OECD.

OECD. (2006a). *Science, Technology and Innovation Outlook.* Paris: OECD Publishing.

OECD. (2007). *OECD Science, Technology and Industry Scoreboard 2007, Innovation and Performance in the Global Economy.* Paris: OECD Publishing.

OECD. (2008a). *Open Innovation in Global Networks.* Paris: OECD Publishing.

OECD. (2008b). *The Internationalisation of Business R&D. Evidence, impacts and implications* (Guinet, J., & De Backer, K., Eds.). Paris: OECD Publishing.

OECD. (2008c). *Review of Innovation Policy, Hungary.* Paris: OECD Publishing.

Okojie, J. A. (1999). The role of government and Universities of Agriculture in improving animal production and consumption in Nigeria. *Tropical Journal of Animal Science*, *2*(2), 1–7.

Okunmadewa, F. Y. (1999). Livestock industry as a tool for poverty alleviation. *Tropical Journal of Animal Science*, *2*(2), 21–30.

Okunmadewa, F. Y., Mafimisebi, T. E., & Fateru, O. O. (2002). Resource use efficiency of commercial livestock farmers in Oyo State, Nigeria. *Tropical Animal Production Investigations*, *5*(1), 47–57.

Olalokun, E. A. (1998). Sustainable animal production for food self-sufficiency in the 21st century. In Animal Science at the University of Ibadan: A Commemorative Brochure of Anniversary Home-coming Celebrations.

Olayiwola, K., Aderinkomi, A., & Atunwa, O. (2004). Information Technology and Business Development in the Nigerian Banking Sector. In R. J. Bauerly, P. C. Thistlethwaite, & D. W. Schofield (Eds.), Emerging Issues in Business and Technology (pp. 214-223). Macomb, IL: Western Illinois University.

Olopoenia, R. A. (1983). On the meaning of economic development . In Osayimwese, I. (Ed.), *Development Economics and Planning: Essays in Honour of Ojetunji Aboyade* (pp. 13–29).

Olsen, K. H. (2007). The clean development mechanism's contribution to sustainable development: a review of the literature. *Climatic Change, 84*(1-2), 59–73. doi:10.1007/s10584-007-9267-y

Omabegho, T., et al. (2009). A Bipedal DNA Brownian Motor with Coordinated Legs. *Science, 324*(5923), 67. Retrieved from DOI: 10.1126/science.1170336

Orszag, J. (2000). *Statement before the U.S. Senate on Indian Affairs*. Retrieved from http///www.education.elibrary.com.au

Osabuohien, E. S. (2007). Trade openness and economic performance of ECOWAS members: Reflections from Ghana and Nigeria. *African Journal of Business and Economic Research, 2*(2-3), 57–73.

Osabuohien, E. S. (2008). ICT and Nigerian banks' reforms: Analysis of anticipated impacts in selected banks. *Global Journal of Business Research, 2*(2), 67–76.

Osabuohien, E. S. (2009). *Industrial Conflicts and Health Care Provision in Nigeria*. Paper Presented at the Faculty of Social Sciences Discourse Series in Conjunction with Sociology Dept, University of Lagos on Human Resources, the Moral Domain and the Human Condition, 14th -16th April.

Oshikoya, T. W., & Hussain, M. N. (1998). Information Technology and the Challenge of Economic Development in Africa. *Economic Research Papers, 36*.

Ostergaard, T. (1993). *Classical Models of Regional Integration- What relevance for Southern Africa?* Uppsala, Sweden: ODEN.

Osterloh, M., & Rota, S. (2007). Open source software development- Just another case of collective invention? *Research Policy, 36*(2), 157–171. doi:10.1016/j.respol.2006.10.004

Ostrowski, A. D., Martin, T., Conti, J., Hurt, I., & Harthorn, B. H. (2009). Nanotoxicology: characterizing the scientific literature, 2000–2007. *Journal of Nanoparticle Research, 11*(2), 251-257. Retrieved from DOI 10.1007/s11051-008-9579-5.

Osuji, L. O. (1997). Regional Trade Blocs and Economic Development: A Study of the West African Subregion . In Edoho, F. M. (Ed.), *Globalization And The New World Order* (pp. 171–182). Westport, CT: Praeger Publishing Company.

Osuji, L. O., & Chukwuanu, M. (2009). *Regional Trade Agreements and South-South Trade: Analysis of the experience of the major Sub-Saharan African Groupings*. Paper delivered at the MBAA Conference, Drake Hotel, Chicago Illinois.

Owens, R. (2008). Intellectual Property and Software: Challenges and Prospects. *WIPO publications*. Retrieved June 3, 2009, from http://www.tecpar.br/appi/SeminarioTI/Richard%20Owens.pdf

Ozor, N., & Igbokwe, E. M. (2007). Roles of biotechnology in ensuring adequate food security in developing societies. *African Journal of Biotechnology, 6*(14), 1597–1602.

Pacey, A. (1986). *The Culture of Technology*. Cambridge, MA: MIT Press.

Paddock, W., & Paddock, P. (1967). *Time of Famines*. Boston: Little Brown & Company.

Padilla, E., & Serrano, A. (2006). Inequality in CO_2 emissions across countries and its relationship with income inequality: a distributive approach. *Energy Policy, 34*, 1762–1772. doi:10.1016/j.enpol.2004.12.014

Panitchpakdi, S. (2009, June 24). *Economic Development in Africa*. Report presented in New York to UNCTAD.

Parayil, G. L. (1992). The green revolution in India: A case study of technological change. *Technology and Culture, 33*(4), 737–756. doi:10.2307/3106588

Parchmovsky, G., & Wagner, R. P. (2005). *Patent portfolios* (Public Law Working Paper 56). Philadelphia, PA: University of Pennsylvania Law School Rochester Institute of Technology. (2008). *About MicroE: What is Microelectronic Engineering?* Retrieved June 30, 2009, from http://www.rit.edu/kgcoe/ue/about.php

Park, J., Kwak, B. K., Bae, E., Lee, J., Kim, Y., Choi, K., & Yi, J. (2009). Characterization of exposure to silver nanoparticles in a manufacturing facility. *Journal of Nanoparticle Research*. Retrieved August 2, 2009, from DOI 10.1007/s11051-009-9725-8.

Parker, R. (2008). *Leapfrogging Development through Nanotechnology Investment: Chinese and Indian Science and Technology Policy Strategies*. Presented at China-India-US Workshop on Science, Technology and Innovation Policy.

Parry, M., Arnell, N., Berry, P., Dodman, D., Fankhauser, S., & Hope, C. (2009). *Assessing the Costs of Adaptation to Climate Change: A Review of the UNFCCC and Other Recent Estimates*. London: International Institute for Environment and Development and Grantham Institute for Climate Change.

Partners, B. C. T. (2009). *S-curve*. Retrieved from http://www.bctpartners.com/

Pavitt, K. (1997). National Policies for Technological Change: Where are the Increasing Returns to Economic Research? In *Proceedings of the National Academy of Sciences*, Washington DC.

Pavitt, K. (2002). Public policies to support basic research: What can the rest of the world learn from US theory and practice? (And what they should not learn). *Industrial and Corporate Change, 11*, 117–133. doi:10.1093/icc/11.1.117

PBL. (2009). *Meeting the 2 °C target. From climate objective to emission reduction measures*. Netherlands Environmental Assessment Agency (PBL). Retrieved April 12, 2010, from http://www.rivm.nl/bibliotheek/rapporten/500114012.pdf.

Pehu, E., & Ragasa, C. (2007). *Agricultural Biotechnology Transgenics in Agriculture and their Implications for Developing Countries*. Background paper for the World Development Report, 2008.

Pemberton, H. E. (1936). The curve of culture diffusion rate. *American Sociological Review, 1*(4), 547–556. doi:10.2307/2084831

Perreault, W., Cannon, J., & McCarthy, E. J. (2009). *Basic Marketing: A Marketing Strategy Planning Approach*. New York: McGraw-Hill Publishing.

Persley, G. J., & George, P. (1999). *Banana, breeding and biotechnology: Commodity advancement through improvement project research, 1994-1998. Banana Improvement Project Report 2*. Washington, DC: World Bank.

Peters, L. S. (2001). Radical innovation and global patterns of breakthrough technology development: an analysis of biotechnology and nanotechnology. In *Proceedings of IEEE Engineering Management Society Proceedings* (pp. 206-212).

Phoenix, C. (2003 October). Design of a Primitive Nanofactory. *Journal of Evolution and Technology, 13.*

Phoenix, C. (2005). Large-Product General -Purpose Design and Manufacturing Using Nanoscale Modules. *NASA Institute for Advanced Concepts*. CP-04-01 Phase I Advanced Aeronautical/Space Concept Studies. Retrieved from http://www.niac.usra.edu/files/studies/final_report/1030Phoenix.pdf

Pingali, P. (2007). *Will the Gene Revolution reach the poor? Lessons from the Green Revolution*. Mansholt Lecture, Wageningen University, 26 January 2007.

Pingali, P. L., & Heisey, P. W. (2001). Cereal-crop productivity in developing countries. Past trends and future prospects . In Alston, J. M., Pardey, P. G., & Taylor, M. (Eds.), *Agricultural science policy*. Washington, DC: IFPRI & Johns Hopkins University Press.

Pingali, P. L., & Traxler, G. (2002). Changing the locus of agricultural research: will the poor benefit from biotechnology and privatization trends? *Food Policy, 27*, 223–238. doi:10.1016/S0306-9192(02)00012-X

Polson, R. A., & Spencer, D. S. C. (1991). The technology adoption process in subsistence agriculture: The case of cassava in Southwestern Nigeria. *Agricultural Systems, 36*, 65–78. doi:10.1016/0308-521X(91)90108-M

Poncharal, P., Wang, Z. L., Ugarte, D., & de Heer, W. A. (1999). Electrostatic deflections and electromechanical resonances of carbon nanotubes. *Science, 283*(5407), 1513–1516. doi:10.1126/science.283.5407.1513

Porter, A. L., & Cunningham, S. W. (2005). *Tech Mining. Exploiting New Technologies for Competitive Advantage.* Hoboken, NJ: Wiley-Interscience.

Porter, M. (1990). *The Competitive Advantage of Nations.* London: Macmillan Publications.

Pourrezaei, K., Carpick, R., & Anthony, P. G. (2007). The Nanotechnology Institute: A Comprehensive Model for Nano-Based Development (pp. 1–148). Proposal to the Ben Franklin Technology Development Authority, Drexel University, University of Pennsylvania, Ben Franklin Technology Partners of Southeastern Pennsylvania.

Power Holding Company of Nigeria (PHCN). (2005). *Meeting the Challenges of Power Sector Development: Generation, Transmission and Distribution Expansion.* Paper presented at Presidential Retreat on Power Generation and Supply, 2 December 2005.

Power Holding Company of Nigeria (PHCN). (2006). *Annual Report, 2006.*

Prentiss, M. C. (2006). Protein structure prediction: The next generation. *Journal of Chemical Theory and Computation, 2*(3), 705–716. doi:10.1021/ct0600058

Prime Minister's Science, Engineering and Innovation Council (PMSEIC). (2005). *Nanotechnology: Enabling Technologies for Australian Innovative Industries.* Canberra, Australia: DEST.

Pursell, C. (1993). Knowledge innovation system: The common language. *Journal of Technology Studies, 19*(2), 2–8.

Qaim, M., & Traxler, G. (2005). Roundup ready soybeans in Argentina: Farm level and aggregate welfare effects. *Agricultural Economics, 32*(1), 73–86.

Radwan, I., & Pellegrini, G. (2009 February). *The Knowledge Economy Gateway to Nigeria's Future.* Abuja, Nigeria: World Bank.

Raney, T. (2006). Economic impact of transgenic crops in developing countries. *Current Opinion in Biotechnology, 17,* 1–5.

Raney, T., & Pingali, P. (2007 September). Sowing a gene revolution. *Scientific American.*

Rao, S., Keppo, I., & Rihai, K. (2006). Importance of technological change and spillovers in long term climate policy. The Energy Journal, Special Ed.: Endogenous Technological Change and the Economics of Atmospheric Stabilisation.

Rass, N. (2006). *Policies and Strategies to Address the Vulnerability of Pastoralist In Sub- Saharan Africa.* Rome: FAOP, Pro-poor Livestock Policy Initiative. Working paper series.

Ravenhill, J. (1990). Overcoming Constraints to Regional Cooperation in Africa: Coordination rather than Integration. In *Long-Term Perspective Study of Sub-Saharan Africa, Vol. 4.: Proceedings of a Workshop on Regional Integration and Cooperation* (pp. 81-85). Washington DC: World Bank.

Raymond, S., & Taggart, J. H. (1998). Strategy shifts in MNC subsidiaries. *Strategic Management Journal, 19*(7), 663–681. doi:10.1002/(SICI)1097-0266(199807)19:7<663::AID-SMJ964>3.0.CO;2-Y

Red de Nanociencias y Nanotecnologioa. (2008). *Diagnóstico y Perspectivas de la Nanotecnología en México.* Retrieved from http://www.nanored.org.mx/documentos.aspx

Reinert, E. S. (2007). *How Rich Countries Got Rich... and Why Poor Countries Stay Poor.* London: Constable & Robinson Ltd.

Reinert, E., Amaïzo, Y. E., & Kattel, R. (2010). The Economics of Failed, Failing and Fragile States: Productive Structure as the Missing Link . In Chang, H.-J. (Eds.), *Market as means rather than master: Towards new developmentalism.* London: Routledge Ed.

Renkow, M. (2005). Poverty, productivity, and production environment: A review of the evidence. *Food Policy, 25*(4), 463–478. doi:10.1016/S0306-9192(00)00020-8

Riul, A. Jr, De Sousa, H. C., Malmegrim, R. R., Dos Santos, D. S. Jr, Carvalho, A. C. P. L. F., & Fonseca, F. J. (2004). Wine classification by taste sensors made from ultra-thin films and using neural networks. *Sensors and Actuators. B, Chemical*, *98*(1), 77–82. doi:10.1016/j.snb.2003.09.025

Riul, A. Jr, Dos Santos, D. S. Jr, Wohnrath, K., Di Tommazo, R., Carvalho, A. C. P. L. F., & Fonseca, F. J. (2002). Artificial taste sensor: efficient combination of sensors made from Langmuir-Blodgett films of conducting polymers and a ruthenium complex and self-assembled films of an azobenzene-containing polymer. *Langmuir*, *18*(1), 239–245. doi:10.1021/la011017d

Riveros, G., Vasquez, J., Gomez, H., Makarova, T., Silva, D., Marotti, R., & Dalchiele, E. (2008). Single-step electrodeposition of polycrystalline CdSe microwire arrays: structural and optical properties. *Applied Physics. A, Materials Science & Processing*, *90*(3), 423–430. doi:10.1007/s00339-007-4318-9

RNCOS. (2006 August). *The World Nanotechnology Market*. Noida, India: Author.

Rocco, M. C. (2002). *International Strategy for Nanotechnology Research and Development Report*. Washington, DC: National Science Foundation.

Rockett, K. (1990). The quality of licensed technology. *International Journal of Industrial Organization*, *8*, 559–574. doi:10.1016/0167-7187(90)90030-5

Roco, M. C. (2001). International strategy for nanotechnology research and development. *Journal of Nanoparticle Research*, *3*, 353–360. doi:10.1023/A:1013248621015

Roco, M. C. (2003). Nanotechnology: convergence with modern biology and medicine. Current Opinion in Biotechnology, 14, 337–346. PubMeddoi:10.1016/S0958-1669(03)00068-5doi:10.1016/S0958-1669(03)00068-5

Roco, M. C., & Bainbridge, W. S. (2002). Converging technologies for improving human performance: Integrating from the nanoscale. *Journal of Nanoparticle Research*, *4*(4), 281–295. doi:10.1023/A:1021152023349

Roco, M. C., & Bainbridge, W. S. (2003). *Converging technologies for improving human performance: nanotechnology, biotechnology information technology and cognitive science*. Dordrecht, The Netherlands: Kluwer Academic Publishers.

Rodrigues, V., Fuhrer, T., & Ugarte, D. (2000). Signature of atomic structure in the quantum conductance of gold nanowires. *Physical Review Letters*, *85*(19), 4124–4127. doi:10.1103/PhysRevLett.85.4124

Rodrik, D. (2007). *One Economics, Many Recipes*. Princeton, NJ: Princeton University Press.

Roetter, R. P., Van Keulen, H., Hengsdijk, H., & Laar, H. H. (2007). Editorial sustainable resource management and policy option for rice ecosystem. *Agricultural Systems*, *94*(Special Issue), 763–765. doi:10.1016/j.agsy.2006.11.003

Rogelj, J., Hare, B., Nabel, J., Macey, K., Schaeffer, M., Markmann, K., & Meinshausen, M. (2009). Halfway to Copenhagen, no way to 2°C. *Nature Reports Climate Change*. Retrieved September 9, 2009, from http://sites.google.com/a/climateanalytics.org/test/welcome/briefing-papers

Rogers, E. M. (1995). *Diffusion of Innovations* (4th ed.). New York: The Free Press.

Rohit, S., Hwang, V., Sood, K., Klein, J., & Cohn, K. (2003). *Nanotechnology What to Expect*. A Larta White Paper. Los Angeles, CA: Larta

Roling, N., & Pretty, J. (1998). Extension's role in sustainable agricultural development. In B. Swanson, R. Bentz, & A. Sofranko (Eds.), Improving Agricultural Extension. A Reference Manual. Rome, Italy: Food and Agriculture Organization of the United Nations (FAO).

Rooks, B. (2004). .. *Assembly Automation*, *24*(4), 352–356. doi:10.1108/01445150410562534

Roson, R., & Bosello, F. (2002). Carbon emissions trading and equity in international agreements. *Environmental Modeling and Assessment*, *7*, 29–37. doi:10.1023/A:1015218031905

Rostoker, M. (1984). A survey of corporate licensing. *IDEA*, *24*, 59–92.

Roth, C. (2007). New technologies and the ergonomic risk to users: Are new technologies part of a technology nirvana or newly identified ergonomics risk factors? *EHS Today, The Magazine of Environment, Health, and Safety Leaders*. Retrieved June 3, 2009, from http://ehstoday. com/health/ergonomics/ehs_imp_70688/

Rothaermel, F. T., & Thursby, M. (2007). The Nanotech versus The Biotech Revolution: Sources of Productivity in Incumbent Firm Research. *Research Policy*, *36*, 832–849. doi:10.1016/j.respol.2007.02.008

Rothemund, P. W. K. (2006, March 16). Folding DNA to create nanoscale shapes and patterns. *Nature, 440*. Retrieved from doi:10.1038/nature04586

Ruttan, V. W. (1999). Biotechnology and agriculture: a skeptical perspective. *AgBioForum*, *2*(1), 54–60.

Ruttan, V. W., & Binswanger, H. P. (1978). Induced innovation and the Green Revolution. In Binswanger, H. P., & Ruttan, V. W. (Eds.), *Induced innovation: Technology, institutions and development* (pp. 358–408). Baltimore, MD: Johns Hopkins University Press.

Ruttan, V. W., & Hayami, Y. (1975). Technology transfer and agricultural development. *Technology and Culture*, *14*(2), 151–199.

Sabancı University. (n.d.). Retrieved from http://www. sabanciuniv.edu.tr/

Sachs, J. (2003). *The Case for Fertilizer subsidies for Subsistence farmers*. New York: Colombia University.

Sala-i-Martin, X., & Subramanian, A. (2003). *Addressing the Natural Resource Curse: An Illustration from Nigeria*. Mimeo, Columbia University.

Sanchez, M. P., & Elena, S. (2006). Intellectual capital in universities: Improving transparency and internal management. *Journal of Intellectual Capital*, *7*(4), 529–548. doi:10.1108/14691930610709158

Santos-Paulino, A., & Thirlwall, A. (2004). The impact of trade liberalization on exports, imports and the balance of payments of developing countries. *The Economic Journal*, *114*, 50–72. doi:10.1111/j.0013-0133.2004.00187.x

Sanyang, S. E., Te-Cheng, K., & Wen-Chi, H. (2008). Comparative study of sustainable and non-sustainable interventions in technology and transfer to the women's vegetable gardens in the Gambia. *J. Technology Transfer*, *34*, 59-75. Retrieved from DOI 10.1007/s10691-008-9084-0.

Sanyang, S. E., Te-Cheng, K., & Wen-Chi, H. (2009). The impact of agricultural technology transfer to women vegetable production and marketing groups in the Gambia. *World Journal of Agricultural Sciences*, *5*(2), 169–179.

Saporiti, F., Boudard, M., & Audebert, F. (2009). Short range order in Al-Fe-Nb, Al-Fe-Ce and Al-Ni-Ce metallic glasses. Submitted to *J. Alloys & Compounds*.

Sardar, Z. (1999). Development and the Locations of Eurocentrism . In Munck, R., & O'Hearn, D. (Eds.), *Critical Development Theory*. New York: Zed Books.

Sargent Jr., J. F. (2009 February). *Nanotechnology: a policy primer*. Washington, DC: Congressional Research Service.

Sarr, M. D. (2008 September). Africa's Search for Collective Development Strategies. *The African Executive*, *179*.

Saxenian, A. (2006). *The New Argonauts: Regional Advantage in a Global Economy*. Cambridge, MA: Harvard University Press.

Saxton, J. (2007 March). *Nanotechnology: The Future is Coming Sooner Than You Think*. Washington, DC: Joint Economic Committee.

SBSTA. (2006). *Synthesis report on technology needs identified by Parties not included in Annex I to the Convention, FCCC/SBSTA/2006/INF.1. Subsidiary Body for Scientific and Technological Advice to the United Nations Framework Convention on Climate Change*. New York: United Nations.

Schacht, W. H. (2008). *The Bayh-Dole Act: Selected Issues in Patent Policy and the Commercialization of Technology* (*CRS Report RL32076*). Retrieved July 7, 2009, from http://www.usembassy.it/pdf/other/RL32076.pdf

Schiff, M., & Yanling, W. (2006). North–South and South–South trade-related technology diffusion: An industry-level analysis of direct and indirect effects. *The Canadian Journal of Economics. Revue Canadienne d'Economique*, *39*(3), 831–844. doi:10.1111/j.1540-5982.2006.00372.x

Schneider, L., & Cames, M. (2009). *A framework for a sectoral crediting mechanism in a post-2012 climate regime*. Report for the Global Wind Energy Council Berlin, May 2009.

Schneider, O., & Sanchez, R. (1980). Evolución de las ciencias en la Republica Argentina 1923-1973: *Vol. 8. Geofísica y Geodesia*. Buenos Aires, Argentina: Sociedad Científica Argentina.

Schummer, J. (2004). Multidisciplinarity, interdisciplinarity, and patterns of research collaboration in nanoscience and nanotechnology. *Scientometrics*, *59*, 425–465. doi:10.1023/B:SCIE.0000018542.71314.38

Schumpeter, J. A. (1939). *Business cycles: A Theoretical, Historical and Statistical Analysis of the Capitalist Process (2 Vols.)*. New York: McGraw-Hill.

Schumpeter, J. A. (1967). *The Theory of Economic Development* (5th ed.). New York: Oxford University Press.

Scocco, D. (2006). *Innovation Zen*. Retrieved from http://innovationzen.com/

Scott, I. (2007 December). *Revitalizing Ontario's Microelectronics Industry*. Information Technology Association of Canada.

Scrinis, G. (2006). Nanotechnology and the Environment: The Nano-atomic reconstruction of nature. *Chain Reaction*, *97*, 23–26.

Scrinis, G., & Lyons, K. (2007). The emerging nano-corporate paradigm: nanotechnology and the transformation of Nature, food and agri-food systems. *International Journal of Sociology of Food and Agriculture*, *15*(2), 22–44.

Seaton, A. L. T., Aitken, R., & Donaldson, K. (2009). Nanoparticles, human health hazard and regulation . *Journal of the Royal Society, Interface*. Published online 2 September 2009. doi:.doi:10.1098/rsif.2009.0252.focus

Seers, D. (1972). What are we trying to measure? *The Journal of Development Studies*, *8*(3), 36–47. doi:10.1080/00220387208421410

Sen, A. (1974). Information bases of alternative welfare approaches: Aggregation and income distribution. *Journal of Public Economics*, *3*, 387–403. doi:10.1016/0047-2727(74)90006-1

Senker, J., Faulkner, W., & Velho, L. (1998). Science and technology knowledge flows between industrial and academic research: a comparative study . In Etzkowitz, H., Webster, A., & Healey, P. (Eds.), *Capitalizing Knowledge: New Intersections of Industry and Academia* (pp. 111–132). New York: State University of New York Press.

Shapira, P., & Youtie, J. (2008). Emergence of Nano-districts in the United States: Path Dependency or New Opportunities? *Economic Development Quarterly*, *22*, 187–199. doi:10.1177/0891242408320968

Shapiro, C. (2000). Navigating the patent thicket: Cross licenses, patent pools, and standard setting. *Innovation Policy and the Economy, National Bureau of Economic Research*, 1. Retrieved June 12, 2009, from http://faculty.haas.berkeley.edu/shapiro/thicket.pdf

Shea, C. M. (2005). Future management research direction in nanotechnology: A case study. *Journal of Engineering and Technology Management*, *22*, 185–200. doi:10.1016/j.jengtecman.2005.06.002

Shetty, S. (2006). *Water, Food Security and Agricultural Policy in the Middle East and North Africa Region*. World Bank: Middle East and North Africa Working paper 47.

Sicard, E. (2009). Electromagnetic Compatibility of IC's. Toulouse, France: Advances and Issues, Past, Present and Future.

Sicard, E., & Bendhia, S. (2006). Basics of CMOS Design. New York: McGraw Hill.

Simons, J., Zimmer, R., Vierboom, C., Harlen, I., Hertel, R., & Bol, G-F. (2009).The slings and arrows of communication on nanotechnology. *Journal of Nanoparticle Research*. Retrieved May 20, 2009, from DOI 10.1007/s11051-009-9653-7.

Single-Crystal Growth of Nickel Nanowires.. Influence of deposition conditions on structural and magnetic properties. *Journal of Nanoscience and Nanotechnology*, *9*(3), 1992–2000. doi:10.1166/jnn.2009.374

Sirohi, S. (2007). CDM: is it a win–win strategy for rural poverty alleviation in India? *Climatic Change*, *84*(1–2), 91–110. doi:10.1007/s10584-007-9271-2

Siyanbola, W. O., Adaramola, M. S., Omotoyin, O. O., & Fadare, O. A. (2004). Determination of energy efficiency of non-bio-fuel-household cooking stoves. *Nigerian Journal of Physics*, *16*(2), 2004.

Smale, M., & Zambrano, P. Falck-Zepeda, J., & Gruère, G. (2006). *Parables: Applied economics literature about the impact of genetically engineered crop varieties in developing economies*. EPTD Discussion Paper 158, International Food Policy Research Institute, Washington, DC.

Smale, M., Zambrano, P., Gruère, G., & Falck-Zepeda, J. Matuschke, I., Horna, D., Nagarajan, L., Yerramareddy, I., & Jones, H. (2009). Measuring the Economic Impacts of Transgenic Crops in Developing Agriculture during the First Decade: Approaches, Findings, and Future Directions. Food Policy Review 10. Washington, DC: International Food Policy Research Institute.

Smith, A. (1998). *An Inquiry into the Nature and causes of the Wealth of Nations*. Washington, DC: Regnery Publishing. (Original work published 1776)

Smith, G. W., Audebert, F., Galano, M., & Grant, P. (n.d.). *Metal Matrix Composite Material*. Patent Application N[er]: PCT/GB2007/004004.

Smith, K. (2000). Innovation as a systemic phenomenon: Rethinking the role of policy. *Enterprise and Innovation Management Studies*, *1*(1), 73–102. doi:10.1080/146324400363536

Soludo, C. C., Ogbu, O., & Chang, H. J. (2004). The Politics of Trade and Industrial Policy in Africa: Forced Consensus? Ottawa, Canada: Africa World Press and the International Development Research Centre, Speser, P. L. (2006). The Art and Science of Technology Transfer (1st Ed.). Hoboken, NJ: John Wiley and Sons.

Spotlight. (1999). *Biotechnology in agriculture*. Rome: FAO

Sproats, K. (2003). *The role of universities as economic drivers in developing their local environment*. Presented at ACU General Conference, Belfast.

Sterk, W., & Wittneben, B. (2006). Enhancing the clean development mechanism through sectoral approaches: Definitions, applications and ways forward. *International Environmental Agreement: Politics, Law and Economics*, *6*(3), 271–287. doi:10.1007/s10784-006-9009-z

Sumner, J. (2005). *Sustainability and the Civil Commons: Rural Communities in the Age of Globalization*. Toronto, Canada: University of Toronto Press.

Sun, J. W. (2002). The decrease in the difference of energy intensities between OECD countries from 1971 to 1998. *Energy Policy*, *30*, 631–635. doi:10.1016/S0301-4215(02)00026-5

Sutter, C. (2003). *Sustainability Check-Up for CDM Projects. How to assess the sustainability under the Kyoto Protocol*. Berlin: Wissenschaftlicher Verlag.

Sutter, C., & Parreno, J. C. (2007). Does the current Clean Development Mechanism (CDM) deliver its sustainable development claim? An analysis of officially registered CDM projects. *Climatic Change*, *84*(1-2), 75–90. doi:10.1007/s10584-007-9269-9

Suzumura, K. (1992). Cooperative and non-cooperative R&D in an oligopoly with spillovers . *The American Economic Review*, *82*, 1307–1320.

Swanson, B. E. (1996). Strengthening research-extension-farmer linkages . In Swanson, B. E., Bentz, R. P., & Sofranko, A. J. (Eds.), *Improving Agricultural Extension: A Reference Manual*. Rome: FAO.

Swanson, B. E., Sands, C. M., & Peterson, V. (1990). *Analyzing Agricultural Technology Systems: Some Methodological Tools.* The Hague: ISNAR.

Taggart, J. H. (1998). Determinants of increasing R&D complexity in affiliates of manufacturing multinational corporations in the UK. *R & D Management, 28*(2), 101–110. doi:10.1111/1467-9310.00086

Tao, L., Probert, D., & Phaal, R. (2008). Towards an integrated framework for managing the process of innovation. In *Proceedings of the R&D Management Conference 2008,* Ottawa, Canada.

Taylor, R., Govindarajalu, C., Levin, J., Meyer, A. S., & Ward, W. A. (2008). *Financing Energy Efficiency: Lessons from Brazil, China, India and Beyond.* Washington, DC: The World Bank Group.

TechnologyReadiness Assessment Desk Book DUSD(S&T). (2005). Department of Defense (DOD) report, United States, May.

Teece, D. (1987). Profiting from technological innovation: Implications for integration, collaboration, licensing and public policy . In Teece, D. (Ed.), *The competitive challenge: Strategies for industrial innovation and renewal* (pp. 185–219). Cambridge, MA: Ballinger.

Tello, A., Cardenas, G., Haberle, P., & Segura, R. (2008). The synthesis of hybrid nanostructures of gold nanoparticles and carbon nanotubes and their transformation to solid carbon nanorods. *Carbon, 46*(6), 884–889. doi:10.1016/j.carbon.2008.02.024

Tenkasi, R. V., & Mohram, S. A. (1995). Technological transfer as collaborative learning . In Backer, T. E., David, S. L., & Soucy, O. (Eds.), *Reviewing the Behavioral Science Knowledge Base on Technology Transfer* (pp. 147–167). Rockville, MD: U.S. Department of Health and Human Sciences, Public Health Service, National Institute of Health.

Thakur, D. (2008, November), The Implications of Nano-technologies for Developing Countries - Lessons from Open Source Software, workshop on Nanotechnology, Equity, and Equality Center for Nanotechnology in Society, Arizona State University The Economist. (2008). *Pocket World in Figures* (2009 Ed.). New York: Profile Books Ltd.

The Chronicle of Higher Education. (2009). *Microsoft's Encarta, rendered obsolete by Wikipedia, will shut down.* Retrieved May 27, 2009, from http://chronicle.com/wiredcampus/article/3715/microsofts-encarta-rendered-obsolete-by-wikipedia-will-shut-down

The Interagency Working Group on Nanoscience (IWGN). (1999). *Nanotechnology: Shaping the World Atom by Atom.* Washington, DC: National Science and Technology Council Committee on Technology. Retrieved from http://www.wtec.org/loyola/nano/IWGN.Public.Brochure/IWGN.Nanotechnology.Brochure.pdf

The National Council of Science and Technology in Mexico. (2006). *Programa Especial de Ciencia y Tecnología 2001-2006.* Retrieved from http://www.siicyt.gob.mx/siicyt/docs/Programa_Nacional_de_C_y_T_1970-2006/documentos/PECYT.pdf

The Royal Society & the Royal Academy of Engineering. (2004). *Nanoscience and nanotechnologies: opportunities and uncertainties.* London, UK. Retrieved from http://www.royalsoc.ac.uk/policy

The Times of India. (2008, November 11). *Maldives plans to buy 'new homeland.'* Retrieved from http://timesofindia.indiatimes.com/World/Maldives_plans_to_buy_new_homeland/articleshow/3696018.cms

The World Bank. (2001). *African Development Indicators.* Washington, DC: The World Bank.

The World Conservation Union. (2007). *The World Conservation Union - About IUCN.* Retrieved from http://www.iucn.org/eng/about

Thinking Open. (2008). *The Decision All Of Open Source Has Been Waiting For.* Retrieved June 3, 2009, from http://thinkingopen.wordpress.com/2008/08/15/the-decision-all-of-open-source-has-been-waiting-for/

TIAC. (2002). *The Organisation of Knowledge: Optimising the Role of Universities in a Western Australia 'Knowledge Hub.* Australia: TIAC.

Tilt, B. (2006). Perceptions of risk from industrial pollution in China: A comparison of occupational groups. *Human Organization, 65,* 115–127.

Tol, R., Downing, T., Kuik, O., & Smith, J. (2004). Distributional aspects of climate change impacts. *Global Environmental Change, 14,* 259–272. doi:10.1016/j. gloenvcha.2004.04.007

Toma, H. E. (2008). '08 Workshop of the Thematic Net of Petrobras: Research on molecular interfaces applied to oil and gas nanotechnology.

Tomalia, D. A. (2009). In Quest of a Systematic Framework for Unifying and Defining Nanoscience. Journal of Nanoparticle Research, 11, 1251–1310. Retrieved from DOI 10.1007/s11051-009-9632-z

Torimiro, D. O., & Kolawole, O. D. (2005). New Partnership for Africa's Development (NEPAD) vision for agricultural technology transfer . In Adedoyin, S. F. (Ed.), *Agricultural Extension in Nigeria* (pp. 170–176).

Tornatzky, L., Gray, O., & Waugaman, P. (2001). *Making the Future: Universities, Their States and the Knowledge Economy.* Raleigh, NC: Southern Growth Policies Board.

Tran, L., Aitken, R., Ayres, J., Donaldson, K., & Hurley, F. (2009). Human Effects of Nanoparticle Exposure . In Hester, R. E., & Harrison, R. M. (Eds.), *Nanotechnology: Consequences for Human Health and the Environment* (pp. 102–113). Cambridge, UK: The Royal Society of Chemistry.

Traxler, G. (2004). *The Economic Impacts of Biotechnology-Based Technological Innovations.* ESA Working Paper No. 04-08 Agricultural and Development Economics Division. The Food and Agriculture Organization of the United Nations. Retrieved June 12, 2009 from http://www.fao.org/es/esa

Turpin, T., & Martinez-Fernandez, C. (2003). *Riding the Waves of Policy.* AEGIS Working Paper Series 2003-02. Sydney, Australia: AEGIS.

Tushman, M., & Anderson, P. (1987). Technological Discontinuities and Organization Environments . In Pettigrew, A. (Ed.), *The Management of Strategic Change.* Oxford, UK: Blackwell.

U.S. Department of Commerce (USDO). (2006 August). *Resource Guide for Technology-Based Economic Development: Positioning Universities as Drivers, Fostering Entrepreneurship, Increasing Access to Capital.* Washington, DC: Economic Development Administration, U.S. Department of Commerce.

Umali, D. D. (1997). Public and private agricultural extension: partners or rivals? *The World Bank Research Observer, 12*(2), 203.

UN ICT Task Force. (2002). *Information and Communication Technologies in Africa – A Status report.* Third Task Force Meeting, United Nations headquarters, 30 September – 1 October 2002.

UN. (2004). *The role of science and technology in the achievement of the MDGs.* Note by the UNCTAD secretariat, United Nations Conference on Trade and Development (UNCTAD). Retrieved September 9, 2009, from http://www.unctad.org/en/docs/tdxibpd4_en.pdf

UNCTAD. (1995). *Bridging the technology gap.* CSTD Issues Paper, Inter-sessional panel, UNCTAD Commission on Science and Technology for Development, 10-12 November. Retrieved April 16, 2009, from http://www. unctad.org/en/docs/ecn162006crp1_en.pdf

UNCTAD. (2001). *World Investment Report: Promoting Linkages.* New York: United Nations.

UNCTAD. (2003a). *Economic Development in Africa: Trade Performance and Commodity Dependence. United Nations publication No. E. 03. 11. D. 34.* New York: United Nations.

UNCTAD. (2004). Round Table on Harnessing Emerging Technologies to Meet the Development Goals Contained in the Millennium Declaration. In *United Nations Conference on Trade and Development.* Retrieved September 9, 2009, from http://www.unctad.org/en/docs/tdl383_en.pdf

UNCTAD. (2005). *World Investment Report, Transnational Corporations and the Internationalisation of R&D.* New York: United Nations.

UNCTAD. (2007). *Economic Development in Africa: Reclaiming Policy Space-Domestic Resource Mobilization and Developmental States.* New York: United Nations.

UNCTAD. (2007a). International Investment Rule-setting: Trends, Emerging Issues and Implications. Note by the UNCTAD Secretariat, Commission on Investment, Technology and Related Financial Issues, Geneva, 8-14 March 2007, TD/B/COM.2/73. New York: United Nations.

UNCTAD. (2007b). *The Least Developed Countries 2007: Knowledge, Technological Learning and Innovation for Development.* Geneva, Switzerland: UNCTAD.

UNCTAD. (2008). *Handbook of Statistics.* Washington, DC: United Nations.

UNCTAD. (2008b). *The Changing face of commodities in the twenty-first century.* TD/428. Note prepared by the UNCTAD secretariat, UNCTAD X11, Accra, Ghana, 20-25 April.

UNCTAD. (2009). *The Least Developed Countries 2009: The State and Development Governance.* Geneva, Switzerland: UNCTAD.

UNDP. (2007). *Human Development Report 2007/2008 - Fighting Climate Change: Human Solidarity in a Divided World. United Nations Development Programme.* New York: Palgrave Macmillan.

UNDP. (2007). *Making Globalization Work for All.* United Nations Development Programme Annual Report. Retrieved from http://www.undp.org/publications/annualreport2007/IAR07-ENG.pdf

UNDP. (2008). *Millennium Development Goals (MDGs).* Retrieved March 15, 2009, from http://www.undp.org/publications/MDG_Report_2008_En.pdf

UNEP. (2009, September 1). *UNEP Risoe CDM/JI Pipeline Analysis and Database.* Retrieved September 9, 2009, from http://www.CDMpipeline.org/

UNFCCC. (1992). *United Nations Framework Convention on Climate Change.* Retrieved September 9, 2009, from http://unfccc.int/resource/docs/convkp/conveng.pdf

UNFCCC. (1998). *Kyoto Protocol to the United Nations Framework Convention on Climate Change.* Retrieved September 9, 2009, from http://unfccc.int/resource/docs/convkp/kpeng.pdf

UNFCCC. (2001). *Report of the COP 7 - Marrakesh-Accords.* Retrieved September 9, 2009, from http://unfccc.int/resource/docs/cop7/13a01.pdf and .../13a02.pdf

UNFCCC. (2006). *Preparing and presenting proposals - A guidebook on preparing technology transfer projects for financing.* Retrieved September 9, 2009, from http://unfccc.int/ttclear/pdf/PG/EN/unfccc_guidebook.pdf

UNFCCC. (2007). *Investment and financial flows relevant to the development of an effective and appropriate international response to Climate Change.* Retrieved September 9, 2009, from http://unfccc.int/cooperation_and_support/financial_mechanism/items/4053.php

UNFCCC. (2007). *The Bali Action Plan.* Retrieved September 9, 2009, from http://unfccc.int/files/meetings/cop_13/application/pdf/cp_bali_action.pdf

UNFCCC. (2008). *Ad Hoc Working Group On Long-Term Cooperative Action Under The Convention, Fourth Session.* Retrieved September 9, 2009, from http://unfccc.int/resource/docs/2008/awglca4/eng/misc05.pdf

UNFCCC. (2009). Further elaboration of possible improvements to emissions trading and the project-based mechanisms under the Kyoto Protocol. *Ad Hoc Working Group On Further Commitments For Annex I Parties Under The Kyoto Protocol, Seventh Session.* Retrieved September 9, 2009, from http://unfccc.int/resource/docs/2009/awg7/eng/inf02.pdf

UNFCCC. (2010). *The Copenhagen Accord, Appendix II - Nationally appropriate mitigation actions of developing country Parties.* Retrieved April 12, 2010, from http://unfccc.int/home/items/5265.php

UNICEF. (1990). Strategy for Improved Nutrition of Children and Women in Developing Countries. UNICEF Policy Review Paper 1990/71.

United Nations Conference on Trade and Development (UNCTD). (1986). *Periodic Report 1986: Policies, Laws, and Regulations on Transfer, Application, and Development of Technology* (TD/B/C.6/133). New York: United Nations.

United Nations Convention on Biological Diversity. (2002). *Convention on Biological Diversity.* Montreal, Canada: Secretariat of the Convention on Biological Diversity.

United Nations Statistics Division (UNTSAT). (2008). *Country Table.* Retrieved from http://unstats.un.org/unsd/snaama/SelectionCountry.asp

United Nations. (2006 May). *Bridging the Technology Gap between and within nations.* Commission on science and Technology for development Report. Geneva: Author

United States Department of Agriculture (USDA). (2008). *Biotechnology.* Washington, DC: Author.

Uranga, M. G., Kerexeta, G. E., & Campas-Velasco, J. (2007). The Dynamics of Commercialization of Scientific Knowledge in Biotechnology and Nanotechnology. *European Planning Studies, 15*(9), 1199–1214. doi:10.1080/09654310701529136

Vago, M., Tagliazucch, M., Williams, F. J., & Calvo, E. J. (2008). Electrodeposition of a palladium nanocatalyst by ion confinement in polyelectrolyte multilayers. *Chemical Communications, 44,* 5746. doi:10.1039/b812181h

Vaidhyanathan, S. (2006). Nanotechnologies and the law of patents: a collision course . In Hunt, G., & Mehta, M. (Eds.), *Nanotechnology: Risk, Ethics and Law.* London: Earthscan.

Vaillancourt, K., & Waaub, J.-P. (2004). Equity in international greenhouse gases abatement scenarios: A multicriteria approach. *European Journal of Operational Research, 153,* 489–505. doi:10.1016/S0377-2217(03)00170-X

Vaillancourt, K., & Waaub, J.-P. (2006). A decision aid tool for equity issues analysis in emission permit allocations. *Climate Policy, 5,* 487–501. doi:10.3763/cpol.2005.0538

Van de Ban, A. W., & Hawkins, H. S. (1996). *Agricultural Extension* (2nd ed.). London: Blackwell Science Ltd.

Van de Ven. A. H., Polley, D. E., Garud, R., & Venkataraman, S. (1999). Innovation and agricultural technology transfer and adoption. New York: Oxford University Press.

van den Ban, A. W., & Hawkins, H. S. (1996). Agricultural Extension. Cambridge, MA: Black-well Sciences Inc.

Van Dijk, J., & Hacker, K. (2003). The digital divide as a complex and dynamic phenomenon. *The Information Society, 19,* 315–327. doi:10.1080/01972240309487

Varga, A. (2000). Local academic knowledge spillovers and the concentration of economic activity. *Journal of Regional Science, 40,* 289–309. doi:10.1111/0022-4146.00175

Varga, A. (2005). Localized knowledge inputs and innovation: The role of spatially mediated knowledge spillovers in the new EU member countries from Central Europe: The case of Hungary . In *The impact of European integration on the national economy* (pp. 118–133). Cluj-Napoca, Romania: Babes Bolyai University Press.

Veneris, Y. (1990). Modeling the transition from the industrial to the informational revolution. *Environment & Planning A, 22*(3), 399–416. doi:10.1068/a220399

Vonortas, N. (1997). Research joint ventures in the U.S. *Research Policy, 26,* 577–595. doi:10.1016/S0048-7333(97)00032-2

Vrolijk, C. (2000). Quantifying the Kyoto Commitments. *Review of European Community & International Environmental Law, 9*(3), 285–295. doi:10.1111/1467-9388.t01-1-00277

Walker, T. (2007). *Participatory Varietal Selection, Participatory Plant Breeding, and Varietal Change.* Background paper for the WDR, 2008.

Walker, W. (2000). Entrapment in large technology systems: institutional commitments and power relations. *Research Policy, 29*(7-8), 833–846. doi:10.1016/S0048-7333(00)00108-6

Wambugu, F. M., & Romano, M. K. (2001). The Benefits of Biotechnology for Small-scale Banana Producers in Kenya. *ISAA Briefs, 22*.

WCED. (1987). *Our common future.* New York: United Nations.

Wilkinson, J. (2002). Genetically modified organisms, organics and the contested construction of demand in the agrofood system. *International Journal of Sociology of Agriculture and Food, 10*(2), 3–11.

William, M. R. (2001). *Agricultural and Rural Extension Worldwide: Options for Institutional Reform in Developing Countries. Paper prepared for the Extension.* Rome: Education and Communication Service, Draft, Food and Agriculture Organization of the United Nations.

Williams, A. (2005). *The patent explosion* (Working paper). New Paradigm Learning Corporation. Retrieved June 25, 2009, from http://anthonydwilliams.com/wp-content/uploads/2006/08/The_Patent_Explosion.pdf

Williams, R. (1960). Becquerel Photovoltaic Effect in Binary Compounds. *The Journal of Chemical Physics, 32*(5), 1505-1514. Retrieved from DOI:10.1063/1.1730950.

Williamson, G., & Payne, W. J. A. (1978). *An Introduction to Animal Husbandry in the Tropics* (3rd ed.). New York: ELBS/Longman.

Wilson, I. (1986). The strategies management of technology: corporate fad or strategic necessity? *Long Range Planning, 19*(2), 45–80. doi:10.1016/0024-6301(86)90216-5

Wilson, M. (2002). *Nanotechnology: Basic Science and Emerging Technologies.* Boca Raton, FL: Chapman & Hall/CRC. doi:10.1201/9781420035230

Wood, S., Jones, R., & Geldart, A. (2003). *The Social and Economic Challenges of Nanotechnology.* Swindon, UK: Economic & Social Research Council.

World Bank. (1981). *Accelerated Development in Sub-Saharan Africa: An Agenda for Action.* Washington, DC: World Bank.

World Bank. (1987). *World Development Report 1987.* New York: Oxford University Press.

World Bank. (1989). *Sub-Saharan Africa, From Crisis to Sustainable Growth: A Long-term Perspective Study.* Washington, DC: World Bank.

World Bank. (1990). Household Food Security and the Role of Women. (J. P. Gittinger, Ed.). World Bank Discussion papers.

World Bank. (1993). A strategy to develop agriculture in Sub-Saharan Africa and a focus for the World Bank. *Africa Technical Department Series, 203*, 83–90.

World Bank. (1993). *World Development Report.* New York: Oxford University Press.

World Bank. (1994). *Adjustment in Africa: Reforms, Results, and the Road Ahead.* Washington, DC: World Bank.

World Bank. (1996). Nigeria, Poverty in the Midst of Plenty: The Challenge of Growth with Inclusion. Report No 13053 UNI Western Africa Department, Country Operations Division, May 13.

World Bank. (2002). *A Sourcebook for Poverty Reduction Strategies.* Washington, DC: World Bank.

World Bank. (2002). *Nigeria Public and Private Electricity Provision as Barrier to Manufacturing Competitiveness.* Retrieved February 10, 2007, from http://www.worldbank.org/afr/findings

World Bank. (2005). *Agriculture Investment Sourcebook.* Washington, DC: Agriculture and Rural Development Department, World Bank.

World Bank. (2006). Increasing Fertilizer Use in Africa: What Have we Learned? In C. Poulton, J. Kydd & A. Doward (Eds.), *Agriculture and Rural Development Discussion Paper 25.* Washington, DC: World Bank.

World Bank. (2007). Project Appraisal Document (PAD) for Science and Technology Education Post-Basic Program.

World Bank. (2008). *Country Assistance Evaluation*. Nigeria: Independent Evaluation Group Approach Paper.

World Bank. (2008a). Global Economic Prospects- Technology Diffusion in the Developing World. Washington, DC: Author.

World Bank. (2008a). *World Development Indicators*. Washington, DC: The World Bank.

World Bank. (2008a). *World Trade Indicators*. Washington, DC: World Bank.

World Bank. (2008b). *Africa Development Indicators 2007*. Washington, DC: World Bank World Economic Forum. (2008). *The Global Competitiveness Report*. Retrieved February 19, 2008, from http://www.weforum.org/en/initiatives/gcp/Global%20Competitiveness%20Report/index.htm

World Bank. (2008b). *Knowledge Assessment Methodology (KAM)*. Retrieved from http://www.worldbank.org/kam

World Bank. (2009). *Annual Report 2008: Carbon Finance for Sustainable Development 2008*. The World Bank Carbon Finance Unit. Retrieved September 9, 2009, from http://wbcarbonfinance.org/Router.cfm?Page=DocLib&CatalogID=47061

World Bank. (2009). *World Development Report 2009: Reshaping Economic Geography*.

World Economic Forum. (WEF). (2008). *The Competitiveness Report 2008-2009*. Geneva, Switzerland: Author.

World Economic Forum. (WEF). (2009a). The Africa Competitiveness Report 2009. Geneva, Switzerland: Author.

World Economic Forum. (WEF). (2009b). The Global Enabling Trade Report 2009. Geneva, Switzerland: Author.

Wright, R. (2001). NonZero: The Logic of Human Destiny. New-York: Vintage Books.

Wu, Y., & Zeng, L. (2008). *The impact of trade liberalization on the trade balance in developing countries*. IMF working paper WP/08/14. Washington, DC: IMF.

Yamaji, K., Matsuhashi, R., Nagata, Y., & Kaya, Y. (1991). An Integrated Systems for CO2/Energy/GNP Analysis: Case Studies on Economic Measures for CO_2 Reduction in Japan. In *Workshop on CO_2 Reduction and Removal: Measures for the Next Century*, 19–21 March 1991. Laxenburg, Austria: International Institute for Applied Systems Analysis.

Zanetti-Ramos, B. G., & Creczynski-Pasa, T. B. (2008). Nanotechnology development: world-wide and national investments. *Rev. Bras. Farm., 89*(2), 95–101.

Zarma, I. H. (2006). *Hydro Power Resources in Nigeria*. Paper presented at 2nd Hydro Power for Today Conference International Centre on Small Hydro Power (IC-SHP), Hangzhou, China.

Zerfass, A. (2005). Innovation readiness: A framework for enhancing corporations and regions by innovation communications. *Innovation Journalism, 2*(8), 1–27.

Zhang, F., Wang, R., Xiao, Q., Wang, Y., & Zhang, J. (2006). Effects of slow/controlled release fertilizer cemented and coated by nano-materials on biology. *Nanoscience, 11*(1), 18–26.

Zhang, Z. (1999). *Estimating the Size of the Potential Market for the Kyoto Flexibility Mechanisms*. FEEM Working Paper No. 8.2000. Milano, Italy: Fondazione Eni Enrico Mattei.

Zheng, M., Jagota, A., Strano, M. S., Santos, A. P., Barone, P., & Chou, S. G. (2003). Structure- based carbon nanotube sorting by sequence-dependent DNA assembly. *Science, 302*(5650), 1545–1548. doi:10.1126/science.1091911

Zilberman, D., Yarkin, C., & Heiman, A. (1997 August). *Agricultural biotechnology: economic and international implications*. Invited Paper Presented at the meeting of the International Agricultural Economics Association. Sacramento, CA.

Zucker, L. G., Darby, M. R., Furner, J., Liu, R. C., & Ma, H. (2007). Minerva Unbound: Knowledge Stocks, Knowledge Flows and New Knowledge Production. *Research Policy, 36*, 850–863. doi:10.1016/j.respol.2007.02.007

Zucolotto, V., Daghastanli, K. R. P., Hayasaka, C. O., Riul, A. Jr, Ciancaglini, P., & Oliveira, O. N. Jr. (2007). Using capacitance measurements as the detection method in antigen-containing layer-by-layer films for biosensing. *Analytical Chemistry*, *79*(5), 2163–2167. doi:10.1021/ac0616153

About the Contributors

Ndubuisi Ekekwe holds two doctoral and four master's degrees, including a PhD in electrical and computer engineering from the Johns Hopkins University, Baltimore and MBA from University of Calabar, Nigeria. During this MBA and Doctor of Management program, he specialized on Technology Management and Competitiveness. He founded Ultinet Systems - telephony and IT firm- and later joined Diamond Bank, Lagos where he last held the title of Banking Executive. He is the founder of the US based non-profit African Institution of Technology. Author of two books on microelectronics and electrochemistry, he co-invented a microchip used in robotics. He has organized more than thirty five seminars and workshops on technology design, innovation and diffusion across the world. Dr Ekekwe currently works in the US semiconductor industry. Well published, featured in Marquis *Who's Who in America* (2010 ed), and an invitee to major meetings like World Economic Forum and African Union congress, he has lectured (adjunct) in three African universities. He served in the United States National Science Foundation ERC/CISST E&D committee for four years. A TED fellow, he graduated top of his class with BEng in electrical and electronics engineering (Aug. 1998) from Federal University of Technology, Owerri, Nigeria.

* * *

Michael Umale Adikwu is Professor of Pharmaceutics at the University of Nigeria, Nsukka. He has researched and worked on various aspects of pharmaceutical sciences, including health sector reforms as well as raw materials production from various agricultural and forestry products. In the area of International grants, he has won a Research grant from the Royal Society of Chemistry of Great Britain (2002); a grant from the Third World Academy of Sciences (2004) and two other grants from the International Foundation for Science, Sweden (2004 and 2006). Locally, he has won several grants including one from Raw Materials Research and Development Council, Abuja, as well as from National Institute for Pharmaceutical Research and Development, Abuja. His research on the use of mucin extracted from snail for wound healing won the Nigerian Academy of Science/Nigerian Liquefied Natural Gas Limited Prize for Science in 2006. He was recently elected Fellow of the Nigerian Academy of Science.

Yves E. Amaïzo, Ph. D. and MBA, is the Director of a Think Tank *'Afrology'* , Austria and author of several papers and books. He is presently an International Consultant on Strategic Project Management and International Business Affairs, Austria. With 20 years with a United Nations specialized organization, he has advised high-level officials of governments, private sector and financial institutions. He writes in three main domains: Africa in the globalization process with special focus on the financial, economic and trade integration process; Productive capacity and capabilities' development for weak influential economies; and Mutability management and competitive advantages. His is often invited as a key speaker or a lecturer on innovative approaches to promote bottom-up approaches. He was the chairman of the group of experts for the Africa Union on the establishment of the three financial institutions as well as a member of the African union eminent experts and the *'Other Canon'*, a group of alternative economists.

Chi Anyansi-Archibong received her PhD. in Strategic Management, with minors in International Management and Entrepreneurship, an MBA and a BS degree in Accounting and Management from the University of Kansas, USA. She is currently a professor at North Carolina A & T State University in Greensboro, USA. She has been an active participant in organizing and implementing several management education programs and workshops in China, Latvia, Estonia, Russia, Nigeria, St. Kitts, Trinidad and Tobago, Barbados, The U.S Virgin Islands and the United States. She has participated in several international faculty developments in Senegal, South Africa, Hungary, Czech Republic, India, Morocco, Costa Rica, New Zealand, Ireland, etc. Dr. Anyansi -Archibong is the author of four Research and Business Case Books; Over 112 refereed journal and proceedings articles and cases; fifteen grant proposals and reports; and numerous teaching materials such as "experiential exercises" and skills development materials for Strategic, International, Ethics, and Business Environment courses (Instructional Materials). She has appeared as an invited speaker in several national and international conferences as well as held leadership roles in case research, Faculty short-term study abroad, cultural competency and case teaching workshops for both professional organizations and universities.

Fernando Audebert is a Professor in Physical Metallurgy in the Faculty of Engineering, University of Buenos Aires (UBA). His main research work is on Amorphous and Nanostructured Materials. He has worked in Belgium, Brazil, Italy, Portugal, Slovakia, Spain, and United Kingdom. Since 2001, he is Academic Visitor at the Department of Materials, University of Oxford, UK; and has been elected as Visiting Fellow 2010 at Mansfield College, University of Oxford. In 2003 he became Head of the Mechanical Engineering Department, UBA. In 2007 he created the CIDIDI (Research, Development, Innovation and Engineering Design Center) in Buenos Aires. In 2008 he has been awarded with both, the Vocational Academic Prize by the "Fundación El Libro", and with the *Pedro Vicien Medal* by the National Academy of Science of Buenos Aires. At present he is advisor of the Argentine Minister of Science, Technology and Productive Innovation.

Neslihan Aydogan-Duda is an Associate Professor of microeconomics, Department of Business Administration, Izmir University of Economics, Turkey. She served as an assistant Professor at Çankaya University, Turkey; Visiting Assistant Professor, Duquesne University, Donahue Graduate School of Business, Assistant Prof. University of Maastricht; Visiting Scholar, Carnegie Mellon University, Software Industry Center; and Visiting Assistant Professor, Department of Business Economics and Public Policy, Kelley School of Business, Indiana University. She has published on such topics as knowledge management, innovation, entrepreneurship, high-tech industries, and economic development is coauthor of Social Capital and Business Development in High-Technology Clusters (2008) and she is the editor of Innovation Policies, Business Creation and Economic Development: A Comparative Approach.

Lorena Barrientos was born in Santiago, Chile in 1980. She studied Education in Chemistry at the Universidad Metropolitana de Ciencias de la Educación, where she graduated with distinction from the first of her class. During 2004 she worked on inorganic complexes for dye-sensitized solar cells at the Universidad de Chile under the supervision of Professor Irma Crivelli. In 2009 she obtained Doctor of Chemistry degree from the Universidad de Chile, under the supervision of Professor Nicolás Yutronic working on ordered arrangement of metal nanoparticles onto supramolecular complexes and metal nano-structures (nanowires, nanofibers, thin films). At present she is Instructor Professor of the Department of Chemistry of the Universidad Metropolitana de Ciencias de la Educación. She is interested in the development of new generation solar cells combining supramolecular architectures, nanotechnologies, and nanofabrication. She has obtained the doctoral (2005) and postdoctoral (2009) fellowships from the National Commission for Scientific and Technological Research (CONICYT).

Nicola Cantore held a Doctorate in Economics at the Università Cattolica del Sacro Cuore, Milan, Italy and a Ph.D. in Environmental Economics and Management at the University of York. Now he is research fellow at the Overseas Development Institute, London. He is involved in projects and scientific papers at international level. His research field includes environmental economics, equity and agriculture issues with a particular focus on themes concerning climate change.

Carlos J. L. Constantino is Bachelor in Physics (IFQSC/USP - Sao Carlos, Brazil - 1993) and Production and Materials Engineering (UFSCar - Sao Carlos, Brazil - 1997). He took his Master in Applied Physics (IFSC/USP - Sao Carlos, Brazil - 1995) and PhD in Materials Science and Engineering (IFSC/USP - Sao Carlos, Brazil - 1999). He was Post-Doc fellow at Windsor University (Canada - 2001), EMBRAPA (Sao Carlos, Brazil – 2002) and Universidad de Valladolid (Spain – 2009). Nowadays he is Assistant Professor at Faculdade de Ciências e Tecnologia – Universidade Estadual Paulista (UNESP - Presidente Prudente, Brazil). The research fields of Dr. Constantino involve spectroscopic characterization of materials by micro-Raman and surface-enhanced Raman scattering (SERS), nanostructured thin films of organic materials for sensing and optical devices, polymeric films for engineering applications.

Ariel Dinar is a Professor of Environmental Economics and Policy and the Director of Water Science and Policy Center at the Department of Environmental Sciences, University of California, Riverside, USA. He teaches, conducts research and publishes on water economics, economics of climate change, strategic behavior and the environment, and regional cooperation over natural resources. He was the Lead Economist and Coordinator of Climate Change and Water Economics Research, Development Economics Research Group, The World Bank, Washington DC, USA, (2006-2008); Lead Economist, Agriculture and Rural Development Department, The World Bank, Washington DC, USA, (2000-2006); Principal Economist, Rural Development, The World Bank, Washington DC, USA, (1998-2000); Adjunct Professor, School of Advanced Int'l Studies, Johns Hopkins University, Washington D.C., USA, (2000-2008); Adjunct Professor, Department of Economics, George Washington University, Washington, D.C., USA, (1998-2000). He obtained PhD from The Hebrew University of Jerusalem (Agricultural and Resource Economics) in 1984.

Ahmed Driouchi holds a Ph.D in applied economics from the University of Minnesota, USA. He is currently teaching "Special Topics" to Al Akhawayn University, Morocco to those graduate students willing to discover new economic research horizons and methods. He is the Dean of the Institute of Economic Analysis and Prospective Studies (IEAPS) and the former Dean of the School of Business Administration of Al Akhawayn University (1995-2005). Professor Driouchi has published series of research papers and book chapters. These publications cover the areas of economics of knowledge with emphasis on education and technology, human development and poverty.

Augustine O. Ejiogu, Ph.D. Agricultural Economics, University of Nigeria, Nsukka; Lecturer II, Department of Agricultural Economics, Extension and Rural Development, Imo State University, Owerri, Nigeria; Formerly, Chief Agricultural Officer, Ministry of Agriculture and Natural Resources, Owerri Nigeria. Significant knowledge and experience in project/programme monitoring and evaluation, agricultural policy analysis and implementation, agricultural loan investigation and supervision, agricultural extension services, management of farms and farm settlements, formation and organization of farmers' groups and co-operative societies, data collection for social and economic research, and statistical analyses of research data.

Guillermo González is full professor at the Faculty of Sciences of the Universidad de Chile where he teaches inorganic chemistry, chemistry of materials and nanochemistry. Currently; he is also member of the Board of Directors and Titular researcher of the Center for the Development of Nanosciece and Nanotecnology. His current research is centered on the synthesis, characterization and properties of hybrid organic inorganic nanostructures particularly in layered materials and in its conversion into tubular species by hydrothermal methods. Most recent research is related to the regulation of the physical, structural, electrochemical, and optical properties of hybrids performed using semiconductors like molybdenum disulfide, vanadium pentoxide and titanium dioxide. Although work is manly directed to the creation of knowledge, the electrochemical properties of the products, especially those containing intercalated with lithium, are potentially interesting as electrode materials.

Edwin M. Igbokwe is a professor of agricultural/rural sociology and Director of the Centre for Rural Development and Cooperatives at the University of Nigeria, Nsukka, Nigeria. Born in 1950, he received a Ph.D in 1992 from the same university. He has carried out rural research in several development-related subjects in Africa including public policy, land use, indigenous knowledge and rural development and technology policies. An associate fellow of the African Institute for Applied Economics (A.I.A.E.), Enugu, Nigeria, Dr. Igbokwe has consulted widely for both local and international agencies.

Mathew O. Ilori is a professor of Technology Management at Technology Planning and Development Unit (TPDU), Faculty of Technology, Obafemi Awolowo University (OAU), Ile-Ife. He holds his B.Sc. and M.Sc. degrees from OAU in 1979 and 1982 respectively. He obtained his Ph.D. from University of Ibadan in 1990. He was two times the director of TPDU, a Post-graduate Department in the faculty of Technology of the university that runs M.Sc. and Ph.D. degrees in Technology Management. His research interest is in the areas of food processing, Technology Innovation Management and Industrial Technology Management. He has published several papers in these fields.

Annamária Inzelt is Founding Director of IKU Innovation Research Centre (1991). Member of Economic Advisory Council of Budapest Business School, prime member of doctoral school at the University of Szeged and private professor at Budapest Corvinus University. She is teaching on various national and international training seminars on STI issues. She was the first Hungarian representative in the OECD Working group of the National Experts of Science and Technology Indicators (NESTI) for 12 years. During that period she was also involved in OECD collaboration with transition economies and with emerging, promising economies from different regions (ASEAN, African, and Latin-American countries). She is an active member of EU PRIME Network of Excellence, ERAWATCH network and a scientific advisor to various EU FP projects. She has been participating and evaluating EU FP projects and coordinating several EU sub-projects. Her main research interest includes the theoretical and practical issues of the innovation systems, the innovative capabilities and performance of the different actors, business organisations and universities. She has well accumulated experiences on STI measurement issues. She is regularly publishing her scientific results.

Nazrul Islam is a Post-Doctoral Research Fellow at Cardiff University Innovative Manufacturing Research Centre (CUIMRC), UK. Recently he holds an appointment in the Welsh School of Architecture. He has received his D.Eng. in the area of innovation management focusing on nanotechnology. Dr. Islam has also received an M.Eng. and an M.Sc. in Applied Chemistry and Chemical Technology. His research interests center on the management of nanotechnological innovation, nanotechnological forecasting and life cycle assessment, and the development of new concepts and research methods for technology and innovation management. He has authored over 30 journal and conference papers including several book chapters in these areas. He serves as a peer-reviewer to the UK research council and many journals. Dr. Islam's publications have received academic awards including the 'Pratt & Whitney Canada: Best Paper Award in Innovation Management'. He is an associate member of the Institute of Nanotechnology, a member of IAMOT and ISPIM.

Olalekan A. Jesuleye heads the Service Delivery Unit at National Centre for Technology Management (NACETEM), Obafemi Awolowo University (OAU), Ile-Ife, Nigeria. He also leads NACETEM's World Bank Project on Technological Entrepreneurship being executed under the Science and Technology Post-Basic Education Project. He is about completing his doctoral degree in Technology Management at OAU, where he earlier had M. Sc. in Technology Policy and Planning in 2000. He obtained B. Sc. degree in Economics in 1988 at University of Maiduguri, Nigeria. He has considerable experience in energy modelling, analysis and policy having worked on an EEC-funded project (1989-1992) with the Institute for Energy Economics and Rational Use of Energy, University of Stuttgart, Germany, and while at the Centre for Energy Research and Development, OAU, Ile-Ife, as an Energy and Environmental Economist on a US-Country Study Project on Green-House-Gases Emissions Mitigation Strategies for Nigeria (1989-1994). He has several published papers to his credit.

Molk Kadiri holds a Bachelor's of Business Administration and a Master's of finance from Al Akhawayn University, Morocco. She is now working as Junior Research Assistant to the Institute of Economic Analysis and Prospective Studies (IEAPS), Morocco.

Johnson A. Kamalu is an Associate Professor in the Department of Family and Consumer Sciences, School of Agricultural and Environmental Sciences at Alabama A&M University, Normal, Alabama, USA. He received his B.S. and M.S. degrees from Tuskegee University, USA and a doctorate from Howard University, USA. Dr. Johnson Kamalu served as Quality Assurance Manager of Mash's Food Products, Landover Maryland, USA. He was also, Research Associate at the Human Nutrition Laboratory, Beltsville, Maryland. Prior to joining Alabama A&M University, Dr. Johnson Kamalu taught for several years at North Carolina A&T State University in Greensboro, North Carolina, USA.

Ngozi C. Kamalu is Professor of Political Science in the Department of Government and History, and the Director of the M.A. Program in Political Science at Fayetteville State University, USA. Dr. Kamalu holds B.A. and M.P.A Degrees from Texas Southern University, USA and a Ph.D. in Political Science from Howard University in Washington, D.C, USA. Before joining Fayetteville State University, Professor Ngozi Kamalu taught at Howard University and North Carolina Agricultural and Technical State University, USA.

Alfred T. Kisubi was born in Uganda and holds advanced degrees in political science, sociology and education from Makerere University, Kampala, Uganda and the University of Missouri -Kansas City, USA. He is a Distinguished Professor and Coordinator of the Human Service Leadership program and alternative chair of the Human Service and Education Leadership Department in the College of Education and Human Services at the University of Wisconsin Oshkosh, USA and on the Editorial Board of Free Inquiry, a journal of the International Council of Secular Humanism and is the international editor of Human Services Today, an online journal. He holds a Doctor of Philosophy in Administration of Higher Education with a minor in Sociology in 1993 from University of Missouri Kansas-City, USA.

Donald F. Larson is a Senior Economist working on Rural Development in the Development Research Group. Prior to his arrival at the World Bank, Mr. Larson served as an economist with the United States Department of Agriculture and with the United Nation's Food and Agriculture Organization. At the Bank, Mr. Larson served in several assignments focused on providing policy advice related to commodity and agricultural markets in Africa, Asia and Latin America. Recently, Mr. Larson also contributed to the design of the World Bank's Prototype Carbon Fund and to the International Finance Corporation's program to provide Weather Insurance in Emerging Markets. His research areas include the study of markets for tradable permits and rural development. Dr. Larson was a member of the team that launched the World Bank's first carbon fund. Mr. Larson holds a B.A. in Economics from the College of William and Mary, an M.A. in Economics from Virginia Tech, and a PhD. in Agricultural and Resource Economics from the University of Maryland.

Taiwo E. Mafimisebi is an agricultural economist and a senior lecturer in the Department of Agricultural Economics and Extension, The Federal University of Technology, Akure, (FUTA), Nigeria. He obtained a Ph.D from the University of Ibadan, Nigeria. Dr. Mafimisebi is the Associate Dean, School of Agriculture and Agricultural Technology, FUTA. He is also the Associate Director (Consulting) of the University's Centre for Research & Development. He is vastly experienced in agricultural marketing / agribusiness management and social research focusing on innovation adoption and poverty reduction among agricultural households. He has served and is still serving as consultant in many participatory community development projects for individuals, corporate bodies, governments, donor agencies and multi-nationals. His key research, professional and experiential qualifications include issues related to marketing efficiency and welfare of agricultural workers using sustainable livelihoods framework. He is widely published and has attended a good number of local and international conferences and training programmes.

Cristina Martinez-Fernandez (PhD UNSW, Doc Salamanca) is an Associate Professor and Senior Research Fellow at the Urban Research centre of the University of Western Sydney, Australia, where she leads the research program on Urban and Regional dynamics. Her research looks at the intersection of industry knowledge, innovation and urban transformation. The study of industry change, urban performance and socio-economic development is strongly anchored within the innovation imperative and the impact of global factors in cities and regions. She is also responsible for the Higher Degree research program. She has published more than 100 works and been a visitor to European and American universities and the OECD. Cristina is a foundation member of the NSW Chapter of the Australian Nano-Business Forum (ANBF) Executive Committee, the Planning Institute of Australia, NSW Policy Committee and the Australia and New Zealand Regional Sciences Association International (ANZRSAI). Cristina also participates in several industry reference panels for the Sydney Metropolitan Strategy with the NSW Planning Department.

Arijit Mukherjee is an Associate Professor and Reader in Economics at the University of Nottingham, UK, with research interests in the area of Industrial Organization and International Trade. His research works mainly focuses on foreign direct investment, innovation, patent policy, merger and technology transfer. Before joining the University of Nottingham, he was working as a Lecturer at the Keele University, UK. He has published a range of papers in internationally reputed journals including papers in European Economic Review, European Journal of Political Economy, Economics Letters, Economic Theory, International Journal of Industrial Organization, Journal of Economic Behavior and Organization, Journal of Industrial Economics, Labour Economics, Oxford Economic Papers, Review of International Economics and The Economic Journal.

Adrian Muller is a post-doctoral researcher at the Socioeconomic Institute and the Ethics Center of the University of Zurich, Switzerland. He currently works on theoretical and empirical analysis of policy instruments for environmental and resource management, with a focus on climate policy. He researches both economic and ethical aspects of policy instruments and how their analysis fruitfully combines to design optimal policies. He also works on ethical aspects of consumption reduction as a sustainability strategy in liberal societies. Earlier work dealt with environmental and resource conflicts, environmental taxation and questions of energy poverty in India. From March 1997 to April 2000, he did PhD. studies at the Institute for Theoretical Physics, University of Zurich, Switzerland.

Adolfo Nemirovsky is a technologist and a serial entrepreneur. Currently, he is board member of several technology companies. He is co-founder of nanoEDU, the Nanotechnology Education and Training Forum. Previously he launched two semiconductor companies, XStream Logic and FlowStorm, that raised over 36 million in venture and corporate funding. In both, he led technical marketing and business development efforts. He is co-founder and was the first Chair of the Benchmarking Working Group, Network Processing Forum, an industry-wide consortium. Previously he held technical positions at Analog Devices and Taligent (Apple-IBM joint venture). Mr. Nemirovsky holds a Ph.D. in Physics, and worked in Polymer and Material Sciences for over ten years. He was a Research Associate at the University of Chicago where he worked on polymeric and magnetic systems, and held a tenured faculty position at the Physics Department, Universidade de Pernambuco, Brazil. He authored over 40 scientific and technical publications, popular white papers and holds nine U.S. patents.

Alejandro Nin-Pratt is a Research Fellow at the International Food Policy Research Institute (IFPRI) since 2005. His research topics are development, economic growth and growth linkages and poverty, agricultural productivity and trade. He has also worked in international trade of livestock and livestock products and sanitary measures affecting trade and their impact on domestic markets, smallholder producers and poverty. Alejandro received his B.S. degree in Agronomy (1987) and a M.S. in International Economics from the Universidad de la República in Uruguay. He received his Ph.D. in Agricultural Economics (specialties: trade and production) from Purdue University (USA) in 2001. After obtaining his degree he worked as a Post-Doctoral Fellow in the Agricultural Economics Department at Purdue. In 2002 he moved to Africa where he worked for the International Livestock Research Institute (ILRI) in Ethiopia and Kenya. In 2005 Alejandro joined IFPRI in Washington DC, USA.

Osvaldo N. Oliveira Jr. is a Professor at the Instituto de Física de São Carlos, USP, Brazil, and has led research into the fabrication of novel materials in the form of ultra thin films obtained with the Langmuir-Blodgett and self-assembly techniques. Most of this work has been associated with fundamental properties of ultrathin films with molecular control, but technological aspects have also been addressed in specific projects. This is the case of an electronic tongue, whose response to a number of tastants is considerably more sensitive than the human gustatory system. Prof. Oliveira is a member of the editorial board of several journals and an associate editor of the J. anosci. Nanotech. He was awarded two prizes for technological innovations, in addition to the Elsevier Scopus Prize in 2006.

Evans Stephen Osabuohien is a lecturer and a Ph.D candidate in Dept. of Economics and Development Studies, Covenant University, Nigeria. He holds M.Sc and B.Sc. in the same field. His research focus include: Development Economics; Trade; and Institutions. He has participated in several conferences and workshops both within and outside Nigeria. He has over 13 scholarly publications. His recent awards include: *Guest PhD Scholar* (Swedish Institute Fellowship) 2009/2010 and *Grant for PhD Thesis Writing* (CODESRIA) 2009. His other awards/recognitions include: Selected Participant at the Africa/Asia/Latin America Collaborative Program on "Rethinking Development for the South, Vietnam, 5-6th October, 2007; Selected Participant for Workshop on Writing for Scholarly Publishing for Anglophone African Countries, CODESRIA/Makerere University, Uganda, 17th-21st September, 2007. He is a member of some learned societies/association, which include: Nigerian Economic Society (NES) and Council for the Development of Social Science Research in Africa (CODESRA). He has engaged in consultancy services for international agencies.

Louis O. Osuji is a professor of Management and Marketing at Chicago State University, Chicago, Illinois, USA where he teaches business and marketing courses. His areas of research interest include International Business & Entrepreneurship development, and Economic Development. Osuji was an assistant professor of Marketing and International Business at Tuskegee University, Alabama, USA. Prior to coming to the United States, he was an associate professor and director of Graduate School of Business, University of Nigeria, Enugu Campus (1987-89). He has contributed book chapters on management and Third World issues as well as papers on refereed journals. He has done extensive consulting and customized training work in microenterprise and entrepreneurship for delegations from different African countries.

Nicholas Ozor holds a Ph.D in Agricultural Administration from the University of Nigeria, Nsukka. He is currently a lecturer and researcher in the Department of Agricultural Extension in the same University. He has published widely and currently is working on climate change in Africa and innovation systems in agriculture.

Emilio Padilla is with the Applied Economics Department, Autonomous University of Barcelona, Spain. Research interests include environmental taxation, climate change economics, project appraisal, intergenerational equity, relationship between economic growth and environmental degradation, and distribution and environment. Dr Padilla has published several book chapters and articles in journals such as Ecological Economics, Energy Economics, Energy Policy, Environmental Values, Environmental and Resource Economics and holds the following qualifications: PhD in Applied Economics, Autonomous University of Barcelona, Spain; MA in Environmental Economics, University of East Anglia, United Kingdom; MA in Applied Economics, Autonomous University of Barcelona, Spain; BA in Economics, Autonomous University of Barcelona, Spain.

Chris Phoenix, co-founder and Director of Research of the Center or Responsible Nanotechnology, USA, has studied nanotechnology for more than 20 years. He obtained his BS in Symbolic Systems and MS in Computer Science from Stanford University in 1991. From 1991 to 1997, he worked as an embedded software engineer at Electronics for Imaging. In 1997, he left the software field to concentrate on dyslexia correction and research. Since 2000 he has focused on studying and writing about molecular manufacturing. Chris is a published author in nanotechnology and nanomedical research, and maintains close contacts with many leading researchers in the field.

Shaikh M Rahman is an Assistant Professor of Agricultural and Applied Economics at Texas Tech University, Lubbock, Texas, USA. His research focuses on various fields of applied microeconomics including economics of climate change with special emphasis on the Kyoto Protocol, industrial organization of US agriculture, and the interface of agricultural production and climate change. He holds a PhD from the University of Maryland (2007). He was formerly a consultant to the World Bank.

Elder de la Rosa is native from Chiapas, Mexico. He received the Ph.D. degree in 1998 from Centro de Investigaciones en Optica. He carried out his posdoctoral research (1999) at UNAM and joined to Centro de Investigaciones en Optica in 2000 as a fulltime professor. His research interest are the synthesis and characterization of nanostructured materials, and the fabrication and characterization of special optical fibers, both for photonics application such as solid state lighting, displays, dosimetry, sensors, biomedical applications, laser and amplifiers. He is member of OSA, SPIE, AMO and SMF. He has published more than 70 peer-reviewed scientific papers and cited more than 350 times. Since 2004 he has been organizing the Topical Meeting on Nanoscience and Nanostructured Materials in collaboration with different colleagues.

Saikou E. Sanyang obtained Ph.D and MSc at National Pingtung University of Science and Technology, Taiwan. He obtained BSc (Hons) in General Agriculture at University of Maiduguri in Nigeria, Borno State. He finished Higher Diploma in Agriculture at Gambia College. Dr. Sanyang held many senior positions at the Department of State for Agriculture in the Gambia. The positions were Subject Matter Specialist, Assistant Divisional Agricultural Coordinator, Training of Trainers, Monitoring and Evaluation officer and Principal Marketing officer at Agribusiness Department. Dr. Sanyang has published series of articles at different international journals. Dr. Saikou E. Sanyang is a Gambian by birth.

İrge Şener received her BS degree from Middle East Technical University, Faculty of Engineering in 1997. She has an MBA degree from Bilkent University. She had worked for MNCs and for both private and public companies in Turkey, where she had gained her diverse work experience. She has also worked for several research projects of the Scientific and Technological Research Council of Turkey. She is currently a PhD candidate in the Department of Management of Başkent University, Turkey. She has been an associate PhD at Çankaya University, since 2006. Her major areas of interest are strategic management, corporate governance and entrepreneurship.

Williams O. Siyanbola is the Director General/Chief Executive Officer of National Centre for Technology Management since 2005. He studied Physics at Obafemi Awolowo University (OAU), Ile-Ife, Nigeria, where he graduated as the best graduating student (Material Science Option). He bagged his M.Sc. and PhD degrees in Energy Studies and Solid State Physics at University of Sussex, Brighton, UK in 1986 and 1991 respectively. He became a Senior Research Fellow at the Centre for Energy Research and Development, OAU, Ile-Ife, 1993 and now a reader. He has published many papers and has attended workshops, and conferences in many countries of the world where he has been privileged to either chair sessions or deliver keynote addresses. Currently, he is involved in the federal government effort to reform the nation's S&T sector. He also belongs to many professional bodies like the Material Science of Nigeria, Nigerian Meteorological Society, and Nigerian Institute of Physics.

Silvanus J. Udoka is an Associate Professor and LANGURE Research Ethics Senior Fellow with joint appointment in the Department of Industrial and Systems Engineering and the Department of Business Administration at North Carolina A&T State University. His current research interests include Immersive 3-Dimensional Environments for interactive visualization, visual depiction and applications; Robotics, Automation and Integrated Manufacturing Systems Engineering; Six Sigma and Lean Enterprises; Logistics and Supply Chain Systems. He teaches in the areas of Lean Manufacturing, Robotics and Automated Production Systems, Six Sigma Quality, Project Management, Production & Operations Management, and Technological Entrepreneurship. Dr. Udoka holds a Ph. D. in Industrial Engineering and Management from Oklahoma State University.

Jarunee Wonglimpiyarat, CPA, CIA, CGAP, CFSA is a faculty member of the College of Innovation, Thammasat University, Thailand. She had working experiences at PriceWaterhouseCoopers, Standard Chartered Bank, Citibank N.A., Sussex Innovation Centre, Boston Technology Commercialization Institute and United States Securities and Exchange Commission. Dr. Wonglimpiyarat has carried out many S&T research projects at the national and international levels. She receives ProSPER.Net-Scopus Young Scientist Award for having made outstanding contributions to research in the field of sustainable development.

Index

A

adopter fatigue 129, 130, 131, 134, 135, 136, 137
adoption 129-137, 167, 168, 169, 298, 299, 300, 303, 305-313, 325, 326, 329, 332
agile and fragile economy 101
agricultural development 301, 303, 309, 313
agricultural economy 301
agricultural growth 272, 273, 278
agricultural innovations 304
agricultural production 273, 292, 293, 294, 299, 303, 311
agricultural productivity 315, 316
agricultural research 273, 276, 277, 282-286, 292, 293, 294, 296, 302, 315-320
agricultural sector 298, 301, 303, 416
agricultural technology 314, 315, 316, 317, 319, 322, 324, 326, 337
agriculture 272,-275, 280, 281, 283, 285, 287, 289-304, 309-317, 320-326, 330, 332, 333, 339
Annex B countries 193, 194, 195, 198, 199, 205, 206
assessment tool 174, 188
atomic force microscope (AFM) 175, 343, 374
atomic precision 26, 27, 31
atomic scale 167, 264

B

biased technical change 274, 275
Bioinformatics 383, 385, 396
bio-mimetic version 44
bionanotechnology 48, 49, 54

biotechnology 43, 44, 45, 48, 49, 50, 56, 59, 88, 90, 96, 209, 211, 272, 273, 281-297, 325, 326, 336, 337, 339, 382, 383, 385, 386, 390, 396
bottom-up approach 43, 44, 45, 178
brain bank 2
brain drain 414, 426
building block 212

C

capability formation 101, 104, 105, 107, 108, 109, 110, 111, 112, 113, 115, 116, 117, 118
carbon finance 226, 228, 230, 231, 234, 235, 236, 237, 238
Center for Responsible Nanotechnology (CRN) 263
Certified Emission Reductions (CERs) 193, 194, 195, 196, 197, 198, 199, 200, 201, 202, 203, 205, 206,
challenges 207, 208, 211, 219, 220, 221, 222, 223
chemical synthesis 29
Clean Development Mechanism (CDM) 193-206, 226, 231-236, 239-242
climate change 226-240, 243-250, 253-260
climate policy 226, 227, 238, 240
climatic change 302
cluster 167, 171, 172
collaboration 61-66, 69-79, 84
commercial adoption 168
commercialization 167-172
cooperative R&D 156-166